Defect Prevention

QUALITY AND RELIABILITY

A Series Edited by

Edward G. Schilling
Center for Quality and Applied Statistics
Rochester Institute of Technology
Rochester, New York

1. Designing for Minimal Maintenance Expense: The Practical Application of Reliability and Maintainability, *Marvin A. Moss*
2. Quality Control for Profit, Second Edition, Revised and Expanded, *Ronald H. Lester, Norbert L. Enrick, and Harry E. Mottley, Jr.*
3. QCPAC: Statistical Quality Control on the IBM PC, *Steven M. Zimmerman and Leo M. Conrad*
4. Quality by Experimental Design, *Thomas B. Barker*
5. Applications of Quality Control in the Service Industry, *A. C. Rosander*
6. Integrated Product Testing and Evaluating: A Systems Approach to Improve Reliability and Quality, Revised Edition, *Harold L. Gilmore and Herbert C. Schwartz*
7. Quality Management Handbook, *edited by Loren Walsh, Ralph Wurster, and Raymond J. Kimber*
8. Statistical Process Control: A Guide for Implementation, *Roger W. Berger and Thomas Hart*
9. Quality Circles: Selected Readings, *edited by Roger W. Berger and David L. Shores*
10. Quality and Productivity for Bankers and Financial Managers, *William J. Latzko*
11. Poor-Quality Cost, *H. James Harrington*
12. Human Resources Management, *edited by Jill P. Kern, John J. Riley, and Louis N. Jones*
13. The Good and the Bad News About Quality, *Edward M. Schrock and Henry L. Lefevre*
14. Engineering Design for Producibility and Reliability, *John W. Priest*
15. Statistical Process Control in Automated Manufacturing, *J. Bert Keats and Norma Faris Hubele*

16. Automated Inspection and Quality Assurance, *Stanley L. Robinson and Richard K. Miller*
17. Defect Prevention: Use of Simple Statistical Tools, *Victor E. Kane*
18. Defect Prevention: Use of Simple Statistical Tools, Solutions Manual, *Victor E. Kane*
19. Purchasing and Quality, *Max McRobb*
20. Specification Writing and Management, *Max McRobb*

Defect Prevention
Use of Simple Statistical Tools

Victor E. Kane

*Ford Motor Company
Livonia, Michigan*

Sponsored by the American Society for Quality Control

CRC Press
Taylor & Francis Group
Boca Raton London New York

CRC Press is an imprint of the
Taylor & Francis Group, an **informa** business

CRC Press
Taylor & Francis Group
6000 Broken Sound Parkway NW, Suite 300
Boca Raton, FL 33487-2742

© 1989 by Taylor & Francis Group, LLC
CRC Press is an imprint of Taylor & Francis Group, an Informa business

First issued in paperback 2019

No claim to original U.S. Government works

ISBN 13: 978-0-367-45109-7 (pbk)
ISBN 13: 978-0-8247-7887-3 (hbk)

This book contains information obtained from authentic and highly regarded sources. Reasonable efforts have been made to publish reliable data and information, but the author and publisher cannot assume responsibility for the validity of all materials or the consequences of their use. The authors and publishers have attempted to trace the copyright holders of all material reproduced in this publication and apologize to copyright holders if permission to publish in this form has not been obtained. If any copyright material has not been acknowledged please write and let us know so we may rectify in any future reprint.

Except as permitted under U.S. Copyright Law, no part of this book may be reprinted, reproduced, transmitted, or utilized in any form by any electronic, mechanical, or other means, now known or hereafter invented, including photocopying, microfilming, and recording, or in any information storage or retrieval system, without written permission from the publishers.

For permission to photocopy or use material electronically from this work, please access www.copyright.com (http://www.copyright.com/) or contact the Copyright Clearance Center, Inc. (CCC), 222 Rosewood Drive, Danvers, MA 01923, 978-750-8400. CCC is a not-for-profit organization that provides licenses and registration for a variety of users. For organizations that have been granted a photocopy license by the CCC, a separate system of payment has been arranged.

Trademark Notice: Product or corporate names may be trademarks or registered trademarks, and are used only for identification and explanation without intent to infringe.

Visit the Taylor & Francis Web site at
http://www.taylorandfrancis.com

and the CRC Press Web site at
http://www.crcpress.com

Statistics is the art and science of discovering what is at first difficult to see and later becomes obvious.

About the Series

The genesis of modern methods of quality and reliability will be found in a simple memo dated May 16, 1924, in which Walter A. Shewhart proposed the control chart for the analysis of inspection data. This led to a broadening of the concept of inspection from emphasis on detection and correction of defective material to control of quality through analysis and prevention of quality problems. Subsequent concern for product performance in the hands of the user stimulated development of the systems and techniques of reliability. Emphasis on the consumer as the ultimate judge of quality serves as the catalyst to bring about the integration of the methodology of quality with that of reliability. Thus, the innovations that came out of the control chart spawned a philosophy of control of quality and reliability that has come to include not only the methodology of the statistical sciences and engineering, but also the use of appropriate management methods together with various motivational procedures in a concerted effort dedicated to quality improvement.

This series is intended to provide a vehicle to foster interaction of the elements of the modern approach to quality, including statistical applications, quality and reliability engineering, management, and motivational aspects. It is a forum in which the subject matter of these various areas can be brought together to allow for effective integration of appropriate techniques. This will promote the true benefit of each, which can be achieved only through their interaction. In this sense, the whole of quality and reliability is greater than the sum of its parts, as each element augments the others.

The contributors to this series have been encouraged to discuss fundamental concepts as well as methodology, technology, and procedures at the leading edge of the discipline. Thus, new concepts are placed in proper perspective in these evolving disciplines. The series is intended for those in manufacturing, engineering, and marketing and management, as well as the consuming public, all of whom have an interest and stake in the improvement and maintenance of quality and reliability in the products and services that are the lifeblood of the economic system.

The modern approach to quality and reliability concerns excellence: excellence when the product is designed, excellence when the product is made, excellence as the product is used, and excellence throughout its lifetime. But excellence does not result without effort, and products and services of superior quality and reliability require an appropriate combination of statistical, engineering, management, and motivational effort. This effort can be directed for maximum benefit only in light of timely knowledge of approaches and methods that have been developed and are available in these areas of expertise. Within the volumes of this series, the reader will find the means to create, control, correct, and improve quality and reliability in ways that are cost effective, that enhance productivity, and that create a motivational atmosphere that is harmonious and constructive. It is dedicated to that end and to the readers whose study of quality and reliability will lead to greater understanding of their products, their processes, their workplaces, and themselves.

Edward G. Schilling

Preface

The intent of this book is to develop the concepts and tools necessary to establish a defect prevention system (DPS) for any type of work activity. A DPS requires effective process control and problem analysis which can be accomplished by using the simple statistical tools presented here. While this book focuses on use of these tools, the overriding importance of the organizational environment stressed by Deming (1982) in his 14 points is emphasized. The management philosophy and corresponding actions set the stage where these tools can be effectively utilized. This book does not address the quality requirements involved in the initial design of products or services. However, the importance of the initial design in an overall DPS is obvious. A superior product or service cannot result from a flawed design.

The first part of this book discusses statistical process control (SPC) concepts, emphasizing the need to establish stability of work processes. The use of control charts in a variety of industrial settings is discussed. The second part of the book presents a number of additional simple statistical tools that can be used for solving manufacturing or administrative problems. These methods have proven useful in solving a variety of problems and most importantly can be used by personnel having a wide spectrum of educational levels and backgrounds. A problem analysis system is given to provide a step-by-step approach for finding the root causes of a problem. The final part of the book gives the elements required to develop a DPS, and integrate the application of process control and problem analysis tools.

A number of themes and presentation techniques are used:

Simplicity Any individual familiar with basic math calculations should be able to utilize the tools presented here. To make this goal attainable, the presentation formats are kept simple. Essentially, all calculations are based on computing the mean, median, range, or standard deviation of a set of numbers. The statistical theory associated with most analysis procedures is not given. This is a text on statistical thinking, not statistical theory. References to appropriate texts are given. In several cases there are better statistical procedures for analyzing problems, especially when only small sample sizes are available. However, simplicity is more important here than statistical optimality. In a number of cases, minimal sample sizes are suggested as a guide for the typical user. These sample sizes are obtained from experience in attempting to balance obtaining differences that are both practically and statistically significant. An attempt is also made to keep the terminology simple. For example, a process "in statistical control" is simply referred to as being "stable".

Need for Data   The role of data in studying a process and attempting to solve problems is continually emphasized. Collecting meaningful data is a central part of working toward a DPS.

Graphical Presentations   The best way to display data is by using a simple graph. Most analyses in this book result in a graph or display. For example, classical statistical testing is performed using plotted means or standard deviations with associated 95% confidence intervals. Users are forced into graphically assessing whether differences are meaningful—not just statistically significant.

Examples   Many examples of actual manufacturing data are presented in text examples, chapter problems, and case studies. These examples illustrate the diversity of problems encountered when real applications are attempted. The data are obtained from over 20 manufacturers and represent a number of industries. In many cases, the data and specifications have been altered to eliminate any sensitivity. A few examples using artificial data are presented to illustrate techniques or concepts.

Displays   Many procedures have a self-contained boxed display of the important definitions and calculations. A step-by-step approach is used for calculations to assist individuals who have difficulty with mathematical equations. Because each display is self-contained, there is some redundancy between the text and other displays. In several cases, worksheets are provided to further assist in the computations. These displays are also useful for seminar presentations and provide a convenient reference.

One-Liners   One-line statements emphasizing major points appear in many sections. This feature reiterates important concepts and helps to ensure the material is understood.

Anecdotes   It is useful to learn from the experience of others. The anecdotes give a short presentation of actual situations that relate, often subtly, to the associated discussion in the book. In many cases, there is no clear right or wrong approach to the situation presented. Since actual situations are used, many factors must be considered to determine a reasonable course of action. Readers are encouraged to discuss the situations with others.

Training Levels   This book is organized so that it is possible to have several training levels:

Level	Chapters
Process control and analysis concepts	1, 2, 3, 8, 9
Control chart design and analysis	4, 5, 6, 7, Appendix I
Problem analysis	10, 11, 12, 13
Building a DPS	14

Each level uses material from the preceding level. Training sessions may be separated by levels or proceed sequentially through the chapters. Also, an asterisk (*) following a section heading signifies that the material either is appropriate only for special applications or is slightly more difficult than other parts of the book.

Target Audience Implementing a DPS within a company requires involvement of all organizational levels. The systems, procedures and tools presented here are intended to provide a common basis for attaining continuous improvement—the cornerstone of a prevention-oriented system. Each organization must tailor its prevention systems to its operations, but the procedures presented here can serve as a useful guide. Managers, engineers, quality specialists, and others will find this book useful in improving a company's quality and productivity performance. Students intending to work in industry will find that knowledge of this material will enable them to solve real-world problems and, more importantly, implement systems to prevent problems from occurring.

I assume full responsibility for the contents of this book. Both present and past employers assume no obligation. Also, this material does not necessarily represent standard practices for present or past employers.

Victor Kane

Contents

About the Series v
Preface vii

Part I Basic Concepts of Process Control **1**
1 Defect Prevention Concepts 3

   1.1 Introduction 3
   1.2 Some Essential Ingredients 6
   1.3 Defect Prevention System 10
   1.4 Vital Systems 10

2 Process Definition and Measurement 12

   2.1 Defining Work 12
   2.2 Defining a Process 12
   2.3 Process Flow Diagram Examples 16
   2.4 Applications of Process Flow Diagrams 23
   2.5 Measurement 26
   2.6 Gaging 28
       Problems 33

3 Control Charts 35

   3.1 Coded Measurements 36
   3.2 Location and Variation Estimates 38
   3.3 Run Chart 44
   3.4 Multiple-Point Run Chart 46
   3.5 Mean and Range Control Chart 47
   3.6 Median and Range Control Chart* 58
   3.7 Percentage Control Chart 60
   3.8 Count Control Chart* 63
   3.9 Special-Cause and Common-Cause Variation 65
   3.10 Out-of-Control Signals 67
   3.11 Control Limits Versus Specification Limits 72

	3.12	Reacting to Control Charts	73
	3.13	Case Studies	75
		Problems	83
4	Control Chart Sampling Concepts and Applications		97
	4.1	Sampling from a Process	97
	4.2	Process Control Sampling Considerations	101
	4.3	Within and Between Variation	109
	4.4	Variation and Special Causes Diagrams	114
	4.5	Troubleshooting Using Increased Sampling	120
	4.6	Some Alternatives to Isolate Process Variation	122
	4.7	Control Chart Application Problems	124
	4.8	Special Applications*	136
	4.9	Case Studies	143
		Problems	156
5	Histograms		165
	5.1	General Concepts	165
	5.2	Stem and Leaf Plots	173
	5.3	Normal Distribution and Zones A, B, and C	178
	5.4	Transformations*	180
	5.5	Histogram Applications	184
	5.6	Histograms from Control Charts	191
	5.7	Case Studies	200
		Problems	210
6	Interpretation of Control Charts		213
	6.1	Process Stability	213
	6.2	Process Instability	216
	6.3	Control Chart Patterns	220
	6.4	Interpretation of Stratified Control Charts	231
	6.5	Control Chart Simulation	238
	6.6	Summary	242
	6.7	Case Studies	244
		Problems	260
7	Process Capability		266
	7.1	General Concepts	266
	7.2	Process Potential Index C_p	268
	7.3	Process Performance Index C_{pk}	271
	7.4	Process Capability Applications	275
	7.5	Evaluation of Stability, C_p, and C_{pk}	277
	7.6	Relating Control Chart Sampling Intensity to Capability	283
	7.7	Computing Capability from Histograms	285
	7.8	New Machine Tryout*	293

7.9	Process Audits*	295
7.10	Summary	297
7.11	Case Studies	298
	Problems	313

Part II Simple Problem Analysis Tools — 317

8 Check Sheets — 319

- 8.1 The Role of Data — 319
- 8.2 Types of Check Sheets — 320
- 8.3 Checklist for Evaluating a Check Sheet — 330
- 8.4 Primary Uses of Check Sheets — 330
- 8.5 Summary — 330
- 8.6 Case Studies — 332

9 Pareto Diagrams — 344

- 9.1 General Concepts — 344
- 9.2 How to Construct a Pareto Diagram — 345
- 9.3 Directing Improvement Efforts — 349
- 9.4 Pareto Comparisons — 355
- 9.5 Pareto Diagrams for Cost — 358
- 9.6 Stratification by Two or More Factors — 360
- 9.7 Warnings About Using Pareto Analysis — 365
- 9.8 Case Studies — 365
- Problems — 377

10 Stratification and Graphs — 381

- 10.1 Two Fundamental Tools — 381
- 10.2 Bar Graphs — 387
- 10.3 Line Graphs — 395
- 10.4 Pie Chart — 398
- 10.5 Interpretation Problems with Graphs — 400
- Problems — 403

11 Comparison Methods — 405

- 11.1 Improvement and Troubleshooting Using Comparisons — 405
- 11.2 Variability Comparisons — 410
- 11.3 Location Comparisons — 419
- 11.4 Comparison of Two Groups (Different Parts) — 428
- 11.5 Case Studies — 439
- Problems — 449

12 Scatter Plots — 454

- 12.1 General Concepts — 454
- 12.2 Construction and Analysis of a Scatter Plot — 460

12.3	Scatter Plot Applications	479
12.4	Application Problems	488
12.5	Prediction Line*	490
12.6	Comparison of Two Groups (Same Parts)	495
12.7	Relationship Between Scatter Plots and Control Charts	506
12.8	Target Charts*	507
12.9	Case Studies	511
	Problems	523

13 Cause-and-Effect Diagrams and Problem Solving — 540

13.1	General Concepts	540
13.2	Concepts of the Cause-and-Effect Diagram	542
13.3	Types of Cause-and-Effect Diagrams	546
13.4	Investigation of Potential Causes	551
13.5	Benefits of a Cause-and-Effect Diagram	554
13.6	Problem Analysis System	557

Part III Defect Prevention System — 583

14 Building a Defect Prevention System — 585

14.1	Establish Open Communications	585
14.2	Change Operating Systems	588
14.3	Initiate Customer Feedback Systems	589
14.4	Develop Key Quality and Productivity Indicators	590
14.5	Utilize Problem-Solving Teams	592
14.6	Define Process Relationships	593
14.7	Develop and Implement a Process Control Plan	601
14.8	Develop an Incoming Material Defect Prevention System	609
14.9	Emphasize Management Evaluation of Systems	610
14.10	Develop a Continual Improvement Mindset	610

Selected Bibliography — 611

Appendix I Gage Evaluation — 613

I.1 Measurement Process	613
I.2 Sources of Measurement Error	614
I.3 Gage Sensitivity	633
I.4 Gage Stability	636
I.5 Gage Purchase	640

Appendix II Geometric Dimensioning and Tolerancing — 643

II.1 Introduction	643
II.2 Common Terms and Definitions	644

	II.3 Datums	644
	II.4 Datum Reference Frame	645
	II.5 Rules	648
	II.6 Main Geometric Characteristics	649
Appendix III	Tables	661
	III.1 Control Chart Constants	662
	III.2 Normal Distribution Probability Values	664
	III.3 t-Distribution Critical Values	665
	III.4 F-Distribution Critical Values	666
	III.5 Analysis of Means Constants h	668
	III.6 Variability Comparison Critical Values	670
	III.7 F_{max} Ratio Critical Values	671
	III.8 Sign Test for Correlation Critical Values	673
	III.9 Correlation Coefficient r Critical Values	674
	III.10 Random Numbers	675
	III.11 Equivalent Hardness Values for Steel	679
	III.12 Metric to U.S. Conversion Factors	680
Index		682

Defect Prevention

Part I
Basic Concepts of Process Control

1
Defect Prevention Concepts

1.1 Introduction

The basis of defect prevention is simple: a problem or defect is prevented before it occurs. The logical appeal is apparent:

1. Customers never experience a defect that is prevented.
2. A product does not need to be returned for repair by the customer if a defect is prevented.
3. A product does not need to be scrapped or repaired in-plant if a defect is prevented.
4. Productivity increases if no defects are produced.

The impact on the market share of these areas is enormous. In the longer term, the absence of defect prevention directly equates with lost sales since a customer dissatisfied with a product's quality will not repurchase from the same company. It has also been found that, on the average, one dissatisfied customer will tell of his or her bad experience to 22 people. Conversely, good experiences are related to only 8 people. Finally, defects increase the cost of a product since the costs for defective products must be covered. In short, the lack of a good system to prevent defects can adversely impact on the ability of a company to stay in business.

All these ideas are simple and appeal to common sense. However, defect prevention practices are not as widespread as this logic would dictate. Why? Many excuses may be cited, but a major contributor is emphasis on a reaction-oriented management style for solving problems that cause defects. There never seems time for prevention actions and planning. Stated simply, most management is on the defense rather than the offense in dealing with defects. The reaction-oriented approach to problem solving creates additional roadblocks for preventing defects since prevention is not part of the process, as illustrated by the problem reaction wheel in Figure 1.1. In this continually moving wheel, the emphasis is clearly on *action*, whatever the basis. Once a problem arises, containment actions must occur so that further defects are not shipped to customers. The problem must be fixed in a short period of time because of the impact on productivity. Finally, defective products must be repaired, if possible. The management reward system is often based on how fast one can circle the wheel.

The reaction wheel approach to solving problems is not bad, merely incomplete. The emphasis should be on problem *analysis* and *prevention* rather than action alone. Clearly,

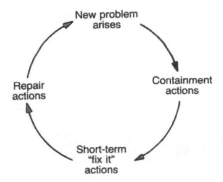

Figure 1.1 Problem reaction wheel.

short-term fixes are necessary to salvage a bad situation, but we cannot stop there. The prevention of problem causes must be a major emphasis in addressing any problem. For preventive actions to occur, effective problem analysis must occur:

Problem analysis → corrective actions → preventive actions

Why is this analysis-oriented approach not used? There are many contributing factors, but the main reason is that the problem reaction wheel is the minimum required activity. Most managers feel consumed in an endless cycle involving many problems, all managed using the reaction wheel. There seems to be too little time for activities beyond daily crises. As a result, there are few new problems; most are similar to old problems "solved" in the past. Chapter 13 discusses a more rational approach to problem solving.

Too often, circling the problem reaction wheel is thought to constitute a system that "manages" quality. The management of quality too often consists of implementing more methods to solve surface problems. These systems are generally inspection based, including such actions as extensive final product auditing. Additionally, customer surveys are utilized to identify more problem areas. In all these cases, more problems are brought to the surface, all to be addressed by the problem reaction wheel. It is not unusual that this approach dramatically improves quality, but inspection-based systems cannot progress beyond a certain quality level. Permanent solutions are not implemented so similar problems tend to recur and there is minimal control of work processes to prevent other problems from developing.

> An assembly defect kept recurring in a component that was difficult to assemble and could not readily be checked at the assembly stations. When a defective component was returned by a customer, all operators were brought together and instructed about the importance of properly assembling the component.

How can an organization make the transition from an emphasis on the problem reaction wheel to a system that stresses process control combined with problem analysis to obtain a prevention-based system? Deming (1982, 1986), Scherkenbach (1986), Juran and Gryna (1980), and others give guiding principles that help lead to this change. There are three common elements essential for the transition:

1.1 Introduction

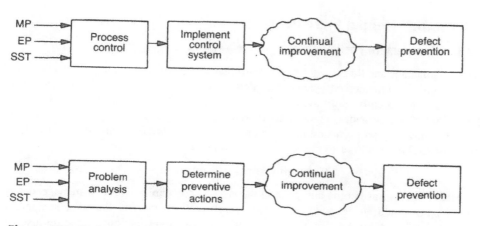

Figure 1.2 Defect prevention essentials.

1. Management participation (MP)
2. Employee participation (EP)
3. Use of simple statistical tools (SST) for process monitoring, problem solving, and communication

Each of these elements is discussed in the following sections. Many of the guiding principles for quality improvement are outgrowths of these elements, which synergistically interact to provide an atmosphere in which continual improvement is possible. Figure 1.2 shows that the three elements are inputs into the systems that generate effective process control and problem analysis, which in turn generate the continuing improvement in the organization that is essential for defect prevention. It should be emphasized that absolute defect prevention is rarely attained, but by continually improving processes through effective process control and problem analysis, we are able to come closer to a state of true defect prevention. The presence of defects provides further room for improvement and input into the problem analysis system.

As Figure 1.2 illustrates, *continual improvement* is critical to preventing defects. All too often the reaction wheel results in problems that are not really solved: short-term fixes are not replaced by a permanent problem solution. The lack of improvement perpetuates the existence of quality defects. In most cases management systems focus on improvement targets that become part of a budgeting system. This approach seems on the surface to encourage continual improvement. In fact, managing by targets can limit improvement. Managers focus on attaining the target by low-risk, easy-to-implement actions. Once an improvement target appears to be attained, attention is focused elsewhere. If continual improvement is the goal, problems must be analyzed and causes understood. Once root causes are identified, solutions are often easily implemented and will produce the permanent elimination of a problem.

> One manager was distressed that his subordinates were continually unable to meet their improvement targets. Each year the improvement actions did not attain the planned results. The manager decided to have his subordinates submit actions that totaled twice the target level of improvement.

1.2 Some Essential Ingredients

1.2.1 Management Participation

Managers manage and thus, by definition, participate in the improvement process, or so the thinking goes. All managers, when confronted by superiors concerning their commitment to quality, emphatically assert their total commitment. Unfortunately, in the early stages, management commitment is not sufficient. What is needed is leadership, which can most easily be shown by participation in the improvement process. This participation can take many forms, but the closer to the working level the better.

Commitment is not enough

Quality circles have come and gone in many U.S. companies with few managers ever having participated in an employee group meeting. Behaviorists feared a manager's presence would inhibit the free expression of ideas. However, as long as the discussion focuses on improvement, there are numerous cases in which the manager's presence is able to facilitate the improvement process. Most improvements (Deming says 85%) involve changes to the system, often small changes. The presence of a manager facilitates the changes. The absence of managers conveys a lack of commitment as well as a lack of urgency.

At a weekly group meeting, a discussion of machine downtime led to employees suggesting that the delivery of key replacement parts be expedited. The manager inquired further about why the downtime always seemed high in the department. The employees were only too happy to volunteer that a reduction in the spare parts inventory had occurred the previous year. The machine repairmen volunteered to review the spare parts list and add needed components. All components were ordered within a week and the inventory reorder points were changed. Spare parts review policies were changed.

A manager's job, even at the lowest level, is to "work on the system." This means every management system hindering improvement is considered for change. Too often, managers tend to be "problem solvers" (via the problem reaction wheel) rather than "system reformers." It is important to solve an individual problem, but it is more important to challenge the system that allowed the problem to exist in the first place. It is not possible for the manager to effectively determine which systems are in need of change unless he or she participates in improvement projects or groups. Of course, it is not practical for a manager to participate in more than a few groups. Fortunately, because system problems, by definition, influence many processes, participation in selective projects or groups is often sufficient.

A common problem with many systems is that the successful use of the system relies on individual initiative. Although this may speak well or poorly of the individual involved, it is a poor characteristic of a system. An effective system should be usable by the least qualified member who uses it. Participation by management in improvement efforts quickly points to undesirable variability in the use of a system. Often this variability is most easily corrected by changing the system since some variability in individual ability can be expected.

Management participation is also important to help identify the important problems. These problems may simply be a list of customer returns or scrap. Other concerns may involve responsiveness to a special customer request or a product design feasibility. Too often undirected group efforts digress into environmental complaints, personality issues, or

1.2 Some Essential Ingredients

other backwaters. Clear communication of problem areas facilitates the improvement process and encourages useful group interaction.

1.2.2 Employee Participation

The goal of every organization should be to have every employee contributing to the improvement of product quality. Every employee knows that increasing quality promotes customer satisfaction, which leads to increased job security. Asking employees to focus on improving quality is thus asking them to increase their job security. Many companies have attempted to encourage worker participation by using some type of employee participation group, often called quality circles. Unfortunately, many group meetings focus on productivity or cost issues from management's perspective and environmental concerns from the worker's perspective. If either focus continues for long, the results are predictably poor.

It is important to assess why many quality circles have failed after so many companies have dedicated substantial time and money to their success. Some of the major reasons for failure are:

Lack of management participation
Work not focused on quality improvement
Employees not trained in simple statistical tools

In designing quality circles, behaviorists purposely omitted these above three concerns. A certain lack of structure was thought to encourage group interaction. However, it should be noted that many Japanese circles have supervisors lead groups that attempt to improve quality by using simple statistical tools.

Most managers would agree on the desirability of employee participation, but too few make a concerted effort to foster participation. The endless cycle of the problem reaction wheel, along with other duties, often leaves little time available for interacting with employee participation groups. A manager often seems to be forced into doing the wrong job. There appears to be too little time left for encouraging and participating in employee problem-solving efforts. The system is never managed or changed. It is not possible for a few people to solve all the problems within an organization. Continual improvement demands that there are many people, at the lowest possible level in the organization, trying to solve problems and make improvements.

> One high-level production manager remarked that he was always presented with problems by his employees. Rarely were even seemingly minor problems handled by subordinates or individual workers. He said that he spent all his time addressing various crises.

One of the ironies of discussing quality in employee groups is that very often workers identify major problems in the system that have gone unnoticed by management. The workers either assume that it is part of their job to overcome the system or that management does not care. Generally, neither assumption is correct: there simply is insufficient communication on problem areas. There is, of course, another aspect to soliciting employee participation; some employees think they understand and have the solution to whatever problem is addressed. If the solution is not known, the group will

Blame a supplier
Blame another organization
Blame another shift
Become agitated

The inability to deal with these problems and to determine the root cause of a problem area discourages many managers from becoming involved in employee groups. However, we will see that data collection and simple analysis methods can avert many potentially confrontational situations.

There is a tendency to use the logic "if a little is good, more is better" for employee participation. If too many groups are attempted, it is difficult to respond to the groups' suggestions. If a manager uses a group's problem identification to help him or her focus on systems that failed, a large number of groups is not necessary. Plants in which 10% of the employees are actively involved in quality improvement groups are often more effective than plants at a 50% level. Many organizations with quality circles simply tried to progress too fast.

1.2.3 Statistical Problem Solving and Communication

Statistical methods have been used for over 50 years to help solve complex problems in industry, agriculture, and government. However, the use of statistical methods was largely restricted to specialists; even most college-trained engineers have had little statistical training. Similarly, most managers have had little statistical training. Unfortunately, the lack of use of statistical methods by managers, engineers, or workers was largely due to focus of most statistical training on statistical theory rather than on simple applications of statistical methods. It has been the successful application of simple statistical methods by the Japanese (e.g., Ishikawa, 1982) that has made many realize their importance.

Why are simple statistical methods important to the defect prevention process in Figure 1.2 for either process control or problem analysis? The logic is simple:

Defect prevention requires the elimination of defects.
   Defects are caused by variability of work processes.
      Variability is analyzed using statistical methods.

If the output of a process is the same with no variation, no defects occur. However, all processes, whether they involve machines, people, or a combination of both, exist in a state in which most process elements exhibit some degree of variability. Statistical methods enable sources and magnitudes of variability to be evaluated with the goal of reducing variability.

Variability makes defects possible

What do we mean when we say simple statistical methods? The primary focus of these methods is to arrive at a meaningful graphic display of appropriately collected data. The value of this approach cannot be overemphasized. Simple graphs not only communicate a variety of information but can be easily understood by everyone. The continual improvement required for defect prevention can most easily be attained if many people are engaged in using statistical methods. For this to occur, the approach must be kept simple.

Make a picture with data

A prominent statistician was visiting a famous Japanese engineer who had promoted the use of statistical methods in industry. The statistician wanted to see the advanced methods that were responsible for producing high-quality products. After having spent a week visiting Japanese plants, he returned to see the Japanese engineer. The statistician said, "I'm very surprised at your application of statistical methods. In most cases the

1.2 Some Essential Ingredients

methods used were quite unsophisticated—below the level of a high school graduate.'' The Japanese engineer smiled.

Some simple statistical methods are discussed by Ishikawa (1982); a list of the 10 tools discussed in this text appears in Table 1.1. An attempt is made to list the most important tools first. However, any ordering is difficult since all the methods listed have wide application. It should be noted that the list is divided into manufacturing and administrative applications.

It is unfortunate that many employees in administrative areas believe that "quality control" tools do not pertain to them. It is apparent that standard control charts do not often apply in administrative areas, but it is also apparent that variability exists and "defects" occur. Understanding control charts and the associated principles of natural variability are of equal importance in manufacturing and administrative applications.

Finally, the statistical tools in Table 1.1 are not only important to process control and problem analysis but also provide an effective communication tool. Since simple graphs can be understood by everyone, they are effective in communicating ideas. Further, since the graphs are based on data, they are believable and can form a basis for discussion—often eliminating the opportunity for confrontation. Unfortunately, most quality circle training did not provide adequate training in the use of simple statistical methods. The results of group interactions when sensitive problems were addressed are predictable.

A supervisor was having trouble with his employee meetings that were designed to address scrap performance. The defensive workers were quite hostile and disruptive. A consultant suggested that he change his approach to first having the employees decide on the data to be collected to address the scrap problem. Second, the consultant suggested that the meetings focus on the simple statistical graphs that were prepared from the data. The tone of future meetings changed.

As a final note, it is important to recognize that the widespread use of statistical methods alone within a company can serve little purpose. Hiring of statistical specialists can likewise yield poor results. Statistical methods become an important tool when manage-

Table 1.1 Simple Statistical Tools

Manufacturing applications	Administrative applications
Process flow diagram	Process flow diagram
Control chart	Cause-and-effect diagram
Comparison plot	Pareto diagram
Scatter plot	Check sheet
Histogram	Stratified graph
Cause-and-effect diagram	Scatter plot
Check sheet	Control chart
Pareto diagram	Histogram
Stratified graph	Comparison plot
ANOM plot	ANOM plot

ment and employees are working together with a continual improvement objective. The necessary environmental changes within the organization are a prerequisite for effectively using statistical tools.

>Statistical methods are not enough

1.3 Defect Prevention System

Defect prevention requires both process and problem analyses that utilize the simple statistical problem-solving tools listed in Table 1.1. How does all this fit together to form a defect prevention system (DPS)? The text is organized in three parts. Part I addresses the basic statistical tools used to attain process control. Part II presents additional tools that are used in the problem analysis system presented in Chapter 13. Part III addresses how all the elements can be combined to form a defect prevention system.

1.4 Vital Systems

When attempting to establish a defect prevention system, it is important to evaluate the total working environment to ensure that other systems contribute appropriately. It is not the intent of this text to detail all the necessary elements since many are unique to individual industries. Some vital systems for a manufacturing operation are discussed here.

1.4.1 Housekeeping

A prerequisite for producing high-quality products is maintenance of a clean, orderly workplace. If workers and management have little regard for their own environment, it is hard to believe that they are concerned about defective products received by a distant customer. Good housekeeping does not guarantee high-quality products, but poor housekeeping establishes a state of mind in which defective products can be expected.

1.4.2 Preventive Maintenance

"Prevention" is the key word in both DPS and PM. However, preventive maintenance can be accomplished more directly than prevention of defects. Yet, effective PM programs in U.S. industry are the exception. High-quality products demand that machinery be in good working order. When defects begin to be produced it is too late to start repairs. Defective products have very likely gone out the door. Few managers will argue with this logic, but the lack of PM speaks for itself.

1.4.3. Training

The training of employees has increased dramatically over the last several years. However, there is still much room for improvement. Typically, there is no evaluation of the effectiveness of the training. That is, did the employees learn the intended skills? Rarely is there any follow-up on whether the training had any impact on job-related activity. Further, much of the training is not job related or directed toward any particular goal related to an employee's job performance. Many companies do not even keep permanent records of the training received by employees. Training for training's sake is too prevalent.

High-quality products demand that employees be well trained in their jobs.

1.4 Vital Systems

Does a worker understand how to operate the machine correctly?
Does a worker understand how to evaluate the work?
Does a worker understand how to control the quality of the work?
Does a worker understand how the work influences a customer?

Many similar questions need to be addressed in an effective job-related training program.

> A plant manager was remarking to a consultant that he had trained most of his employees in statistical process control (SPC). He himself along with the entire management group had received 2 days of training more than 2 years previously. However, after funding thousands of hours of training, he had concluded that SPC was not of great value since there were few SPC charts in the plant and his plant's quality had not improved.

1.4.4 Computers

The use of computers will obviously increase in the future since computers often have a direct impact on productivity. The opportunity to use computers to enhance quality also exists, but caution must be exercised. Computers have the power to collect and quickly summarize the valuable data required to control processes. In many cases, it is not humanly possible to deal with the volume of data that can be easily summarized by computers. However, two problems must be addressed. First, workers who control an operation using computerized systems must be highly trained in how to use the computer effectively. People are naturally suspicious of the new and unknown. Many a computer sits idle in the workplace. Second, computer output needs to be tailored to individual applications. Computer output is often so inflexible that individual needs are not met. A computer program "good for all applications" may in fact be good for none.

2
Process Definition and Measurement

2.1 Defining Work

All groups or individuals within an organization perform work that consists of a series of tasks that add value to some type of input. The product or output produced by this work is then used by a customer. The customer is not only the end user of a product but also the next user of the product. The customer may be, for example, the next operation in a machining process. The work process is diagrammed in Figure 2.1.

The possible control of a process is also indicated in Figure 2.1. Inspection focuses on the output of a work task; process control focuses on the control of the tasks themselves, as well as control of the inputs. A defect prevention system (DPS) relies on process control, not inspection. It should be noted that all inspection is not necessarily bad. Deming (1982, Chap. 13) gives an equation for determining if inspection is justified. However, the inspection rule is to inspect all products or none, depending on the cost of inspection, the rate of defective parts, and the cost of a defective part during customer use. Since 100% inspection is economically warranted when the defect rate is high, the need for process control (to reduce the defect rate) is apparent.

2.2 Defining a Process

The activities associated with changing and adding value to input in the work process are called added value tasks in Figure 2.1. An important characteristic of these tasks is that they are accomplished by combining six basic elements to create an "output":

1. Personnel: People are often the basic element in accomplishing a work task. Their ability, training, and inherent differences create potential variability in process output.
2. Methods: Work procedures or methods form the template that people use to accomplish the work tasks. These methods may or may not be the best way to accomplish a task. People may interpret the methods differently, adding to the variability in process output.
3. Machines: The equipment used to accomplish work adds variability to process output. The effectiveness of preventive maintenance, setup of tools, inherent capability, and so on, all create variability.

2.2 Defining a Process

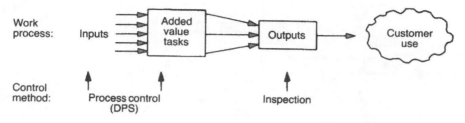

Figure 2.1 The work process with control method.

4. Materials: Raw materials or other inputs utilized in the work tasks are typically the output from other work tasks and correspondingly exhibit variability. The lack of consistency of these materials creates variability in the work tasks.
5. Measurements: At each state of a production process, the process output is evaluated by some measurement "process." The ability of the measurement process to evaluate the overall performance of a process is clearly related to process performance.
6. Environment: Changes in the work environment can alter the performance of any of the five basic elements that create variability in the work tasks.

These elements are shown in Figure 2.2. The combination of these elements is the definition of a process—The combination of personnel, methods, machines, materials, measurements, and environment that produces an output that is a given product or service.

All work is a process

The objective of a DPS is controlling and improving work output. A first step toward this end is the realization that all work is a process combining these elements. The combination of these elements can be diagrammed in a flowchart form that displays the work tasks. The importance of flowcharting the work process cannot be overemphasized. It is not unusual in a large system for many employees to have insufficient knowledge of the total process. Control actions and improvements can most easily be made if all employees

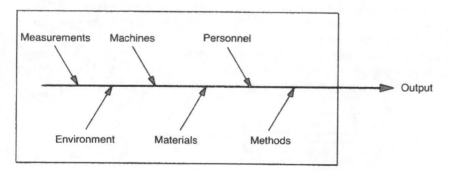

Figure 2.2 Combination of elements to form a work process.

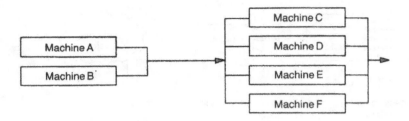

Figure 2.3 Single path flow.

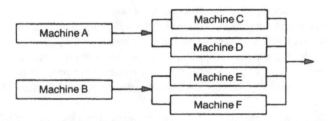

Figure 2.4 Double path flow.

understand the total work process. This enables an employee to understand the intent of his or her work, which makes self-control possible. The information on a work process can be most easily conveyed using a process flow diagram. This diagram is simply a flowchart of a work process with several key elements.

2.2.1 Inputs

All inputs to the work system should be identified. These inputs may be raw material, semifinished parts, subassemblies, or information required to perform the work tasks. In most cases the quality of the input directly influences the quality of the work output.

2.2.2 Paths

The flow of parts or information during the work process defines the "system" by which work is performed. These paths must be clearly defined. For example, the path flow in Figure 2.3 is different from the path flow in Figure 2.4.

A problem in defining path flows is the identification of alternative paths. An example of an alternative flow is a repair operation as defined below. It is not unusual for alternative paths to generate a disproportionately large number of defects.

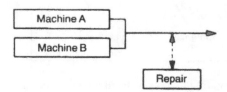

2.2 Defining a Process

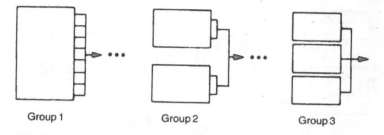

Figure 2.5 Example of different sources of variation.

2.2.3 Sources of Variation

Most flowcharts of a work process focus on the operations or tasks being performed. However, the operations in Figure 2.5 could all be producing the same part but the sources of variation are quite different. In group 1, a single machine with eight identical spindles produces a part. In group 2, two separate machines, each with two spindles, produce a part. In group 3, three separate machines produce a part. Examples are given that illustrate how multiple suppliers, machines, spindles, pallets, and other elements are sources of variability. In each case, the "common" in-process parts are derived from different processing sources.

Example 2.1 The process flow diagram shown here illustrates common in-process parts derived from different production paths.

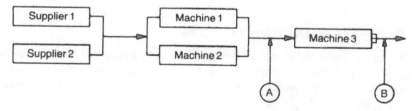

A part at location A could have derived from 4 paths (2 suppliers × 2 machines):

Supplier 1, machine 1
Supplier 1, machine 2
Supplier 2, machine 1
Supplier 2, machine 2

At location B, 12 paths exist (2 suppliers × 2 machines × 3 spindles). In the next section we see that most real manufacturing processes have many paths. Similarly, most administrative work processes have many sources of variation and alternative path flows exist.

Example 2.2 This process flow diagram shows that many characteristics contribute to alternative part paths that define some of the sources of variability in a process.

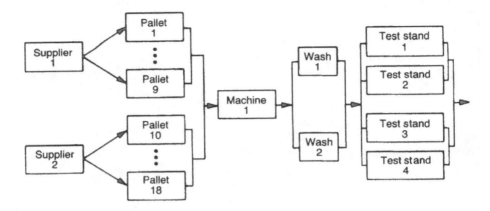

2.2.4 Outputs

In some applications, the output of a work task can be simply defined as a finished part, subassembly, or report. In other cases, multiple outputs are possible. An assembly line may produce several products, or an accounting system may have multiple reports for different customers.

2.2.5 Customer Requirements

In Figure 2.1 customer use is indicated as the purpose of producing an output. In this context "customer" refers to the next user of an output as well as the final end user. The customer requirements should be a primary focus of the control and improvements to a work process. These requirements are integrated into the DPS described in Chapter 14.

> The next user is a customer

2.3 Process Flow Diagram Examples

Since all work is a process, a process flow diagram can be prepared for any activity. Four examples are given in the following sections. It should be emphasized that the exact format of the diagram is not important. What is important is to clearly represent the inputs, paths, sources of variation, and outputs.

2.3.1 Manufacturing

An example of the machining of an engine piston is given in Figure 2.6. It is quickly apparent that the multiple paths a piston may take are large. Suppose a defective piston is found after operation 90. What is the source of the problem? A single spindle in operation 30 may be causing the problem. Notice that some operations, such as operation 60, are not manned on a full-time basis. The purpose of the DPS is to provide a system in which problems can be quickly detected (i.e., prevent a defect) without having so many charts and controls that normal operation is not possible. Most statistical process control (SPC) training addresses the control of a single source of variation, such as a single spindle on a machine, rather than control of a complete process.

2.3 Process Flow Diagram Examples

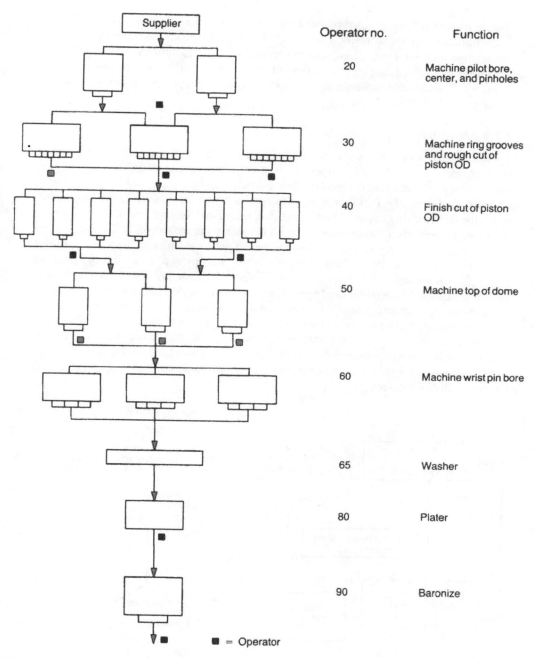

Figure 2.6 Process flow diagram for an engine piston.

A first-line machining supervisor and the department workers prepared a process flow diagram for their department. They focused on identifying all the potential sources of variation and quickly realized why the control of their quality was so difficult. Employees recognized that they were dependent to a great extent on each others' performance. Training and implementation of control procedures progressed without many of the usual difficulties.

2.3.2 Assembly

Typical assembly operations involve a combination of manual and automated assembly tasks. Aside from the normal sources of variation, a critical part of an assembly process flow diagram is identifying the product flow and actions when the assembly task cannot be accomplished correctly. Part shortages, defective components, or a defect from a previous operation are typical examples. Often a high percentage of defective assemblies result from improper actions when "exceptions" occur. These actions are generally the result of not fully understanding what actions to take (alternative paths) when exceptions occur. Figure 2.7 is an example of a simple alternative path for the attachment of a car interior head liner to the roof. Even for this simple operation, Figure 2.7 illustrates how defects adversely impact on assembly operations and how alternative paths highlight the problems encountered in dealing with exceptions. A fundamental consideration is whether the worker has sufficient time to deal with the exceptions or whether the defect is allowed to pass by so that the worker is not pressured by the next unit up the line. It should be noted that it is not the intent of these alternative paths to accept or condone defects. The objective should always be elimination of the defects, but our planning using process flow diagrams must address the possibility of a defect.

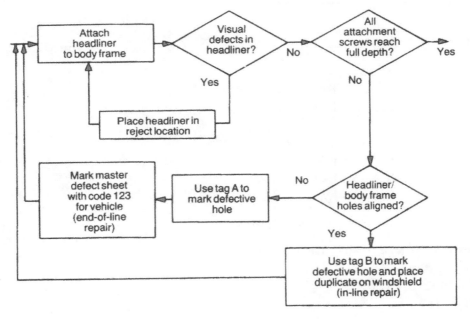

Figure 2.7 Alternative paths for attachment of a car headliner.

2.3 Process Flow Diagram Examples

2.3.3 Batch Process

Another major type of manufacturing involves a batch processing industry, such as a chemical or steel process. In these cases, "batches" of material are combined and processed to form an end product. Other examples that are not purely a batch process include foundry and forging operations. In these cases the casting is derived from a process that produces castings in groups from a batchlike process. Figure 2.8 shows that a foundry can have many sources of variation. Note that the furnace variability can involve zones or conveyor tracks within a furnace. Figure 2.9 shows a forging operation. The team preparing the process flow diagram indicated the major sources of variation for each operation. Chapter 13 shows a process cause-and-effect diagram that assists in identifying sources of variability. Also, note that the dashed line indicates a possible recycling of material. It is important to have appropriate controls for this practice.

2.3.4 Administrative

Most administrative systems are run by a large volume of detailed procedures that few people have read, fewer understand, and even fewer use. In practice, most "systems" are run at each stage of processing based on what is needed by the next user or level of authority. Thus, most systems are "run" by the individual initiative of many employees. It is not unusual to find that large amounts of time are spent correcting errors, reprocessing forms, or placing telephone calls to obtain required information.

A local manager decided to evaluate how his clerks were spending their time. He found that over half of a clerk's time was spent on the phone. What he first thought was a poor work habit was really a sincere effort to complete the information on standard forms. For some types of forms, every form submitted required a telephone call since users did not understand what information was being requested. In other cases, clerks were receiving calls from employees asking how particular forms were to be completed.

A process flow diagram for an administrative system should not focus on what is supposed to happen but on what actually happens. Constructing this diagram can almost always lead to improvements. For many company systems there is a "black-market" system of how to get things done. This activity is generally an attempt to get things done faster and not have to deal with a bureaucracy. For example, Figure 2.10 gives a process flow diagram of how an employee purchases an item. In most cases, the employee's objective is to minimize the time delay for the item being purchased. The purchasing department's objective is to minimize cost. Most organizations believe that this adversarial relationship minimizes cost. In practice, these systems foster the attitude that the company is not concerned about whether an individual has the items needed to perform the job.

The purchasing example in Figure 2.10 illustrates how customer requirements are not the driving force behind many administrative systems. Does the purchasing department realize that a purchase request may be stalled for 2–3 weeks if the assigned buyer is on vacation? Do they know that many "emergency" requests are submitted to circumvent the system (creating less control)? Do they realize that no communication with the employee submitting the request occurs until the purchased item arrives? The lack of customer focus is clear to all who must use the system.

Constructing a process flow diagram also enables a focus on quantifying the performance of an administrative system. Typically, few meaningful data focusing on customer requirements are available. For example, the purchasing department undoubtedly has

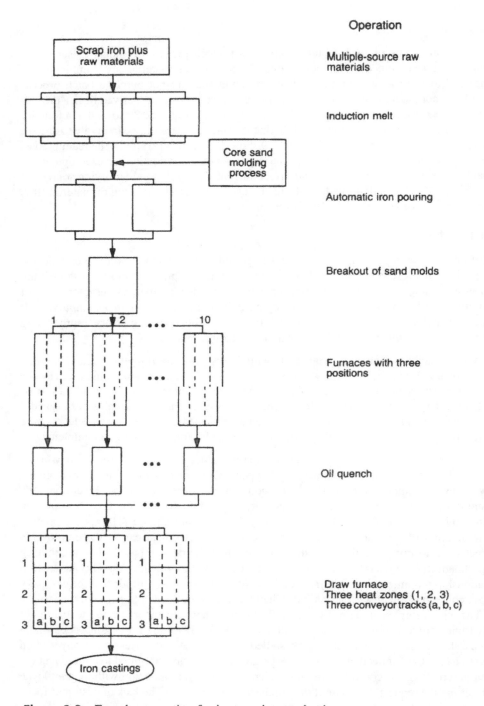

Figure 2.8 Foundry operation for iron casting production.

2.3 Process Flow Diagram Examples

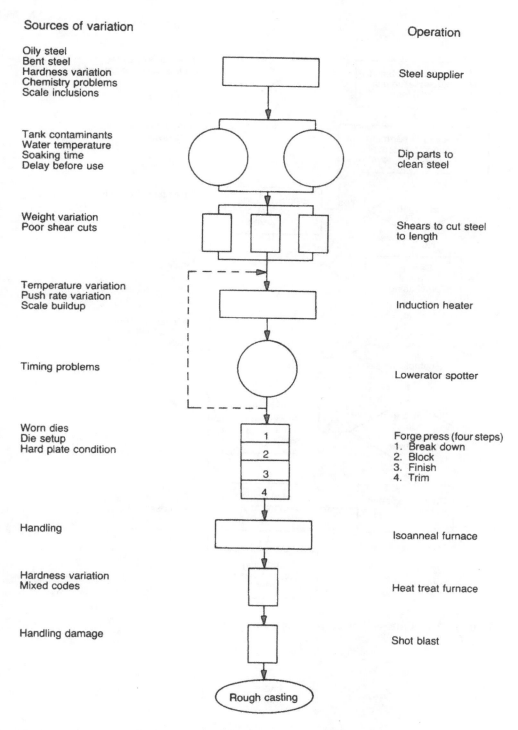

Figure 2.9 Forging example for a transmission parking gear.

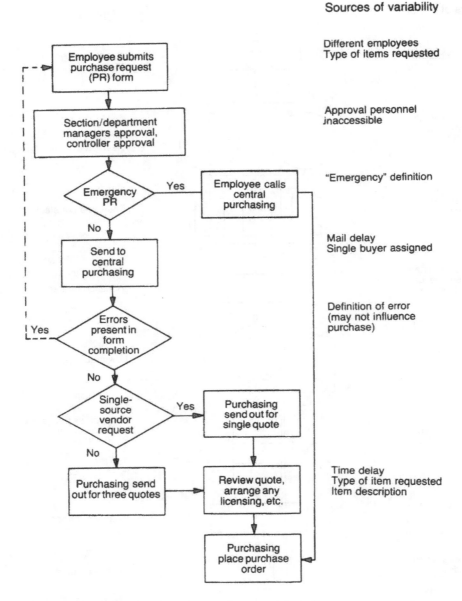

Figure 2.10 Process flow diagram for material purchase.

2.4 Applications of Process Flow Diagrams

detailed cost-related data. However, what is the average time for an employee to receive an order? What type of items are most delayed? What types of errors cause delays in placing an order?

2.4 Applications of Process Flow Diagrams

The discussion of Figure 2.1 emphasized that to implement the concepts of defect prevention it is necessary to focus on controlling the work process rather than the work output. The process flow diagram is the first step in this direction:

The work process is clearly defined.
Major sources of variability are identified.

The examples in the previous section indicated that in most real applications the work process is complex, with many sources of variability. The process flow diagram has other benefits as discussed here.

2.4.1 Target Flows

An engineer was confronted with a problem of dealing with a process with very old machinery that constantly produced rejects. Part of the process is diagrammed in Figure 2.11. The engineer's solution to the problem was to target the flow of material. The targeting enabled machine operators to "fine tune" a machine to the output of a minimal number of machines performing previous operations. The targeting principle involves reducing the number of sources of variability to reduce the variability of the output. The major disadvantage is that most systems are purposely designed with no targeting since flexibility is reduced. If machine 7 in Figure 2.11a becomes inoperable, the impact on the system output is minimal. However, in the target flow case in Figure 2.11b, machine 8 would need to double its production rate (i.e., machines 9 and 10 are not connected to machine 5). In practice, there are often ways to circumvent any lack of flow-through flexibility.

In some operations it is not immediately obvious how to target the flow of material. If an odd number of machines faces an even number of machines, the situation must dictate whether targeting is possible.

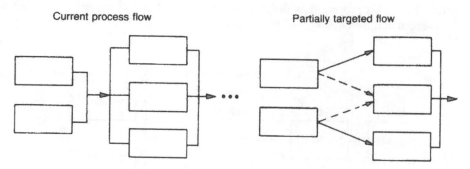

In some cases a partial targeting is feasible.

An example of targeting that at first seems difficult is a conveyor system in which long distances may separate sequential operations. A team effort solved this problem by placing colored tags on the parts when they were placed on the conveyor.

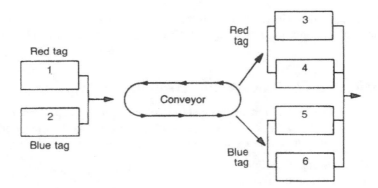

In this example, machine 1 output was processed by machines 3 and 4 (red tag) and machine 2 output by machines 5 and 6 (blue tags).

A significant side benefit of targeted process flows is that troubleshooting a problem becomes easier. Suppose a problem developed on the output of machine 7 in the targeted flow case in Figure 2.11. It would be necessary only to check machines 7, 5, 1, and 2. Without targeting, three additional machines (6, 3, and 4) could be causing the problem. Minimizing the response time for solving problems is an important part of developing a DPS.

2.4.2 Upstream Control

In most manufacturing situations, the output of an operation influences the output of several subsequent downstream operations. A quantitative method to assess the degree to which this is true is presented in Chapter 12. A process flow diagram can be used to indicate potential influences and thus assist in troubleshooting. In Example 2.3, two characteristics of a part are considered. If a statistical chart at position I showed a problem, machines 5, 2, 3, and 4, along with the supplier's material, would need to be checked. A and B in the example indicate operations in which upstream operations influence the final A or B characteristics. For example, machine 1 might drill a pilot hole and machines 2, 3, and 4 drill the final finished hole size.

Example 2.3 This process flow diagram shows that final part characteristics (A and B) are influenced at different stages of the process.

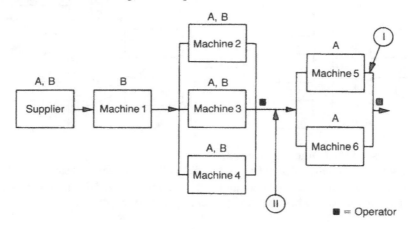

2.4 Applications of Process Flow Diagrams

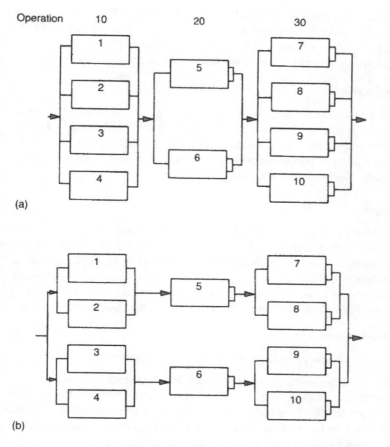

Figure 2.11 Examples of process flows: (a) typical process flow; (b) targeted process flow.

Because of the importance of upstream control, employees sometimes "spot" parts. This practice is simply color coding upstream operations. In this process flow diagram, the output from machine 5 and 6 are color coded with a small paint dot to help detect problems in the output of machines 10, 11, and 12.

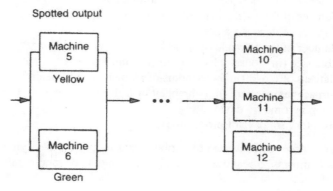

Employees believed that the output of machines 5 and 6 significantly influenced the output of machines 10, 11, and 12, so a spotting system was implemented. It is common practice to use spotting to target the flow of parts. In this example, machine 10 could run yellow parts, machine 12 green parts, and machine 11 a combination of parts.

2.4.3 Simulate Defects

Since a process flow diagram highlights sources of variability, it is possible to conduct "what if" exercises. In Figure 2.11, what if one of the spindles on machine 5 started producing poor-quality output? How long would it take to notice that a problem existed? Generally, employees realistically evaluate whether the current system is adequate. It is convenient that no defects had to occur to conduct this exercise.

2.5 Measurement

Upon first being exposed to some of the simple statistical techniques described in later chapters, many employees return to their work areas to "collect some data." Although this is often helpful for practicing use of the techniques, it can lead to confusion. What is thought to be information is really a list of "figures." We have just seen that it is important to first understand the process flow. Second, the measurement method must be evaluated. This section gives some of the considerations needed to evaluate the measurement system.

2.5.1 Types of Measurement Data

There are two basic types of data considered when evaluating quality:

1. Attribute data: Measurement consists of evaluating whether an individual part or item has a particular feature (e.g., is a part defective or nondefective?). The measurements are usually summarized by noting the number of parts with the feature versus the number without the feature. Measurement may also consist of counting the number of occurrences of the feature on a part.
2. Variable data: Measurement is a physical feature, such as a dimension or temperature recorded using some measurement scale, such as inches, pounds, or degrees.

Note that the term "defective" simply means a part does not conform to stated requirements. The part may not be defective from the functional standpoint. Some examples of attribute and variable data follow:

Attribute   Number of items within specification limits versus number outside specification limits
Number of items rejected versus number passed
Number of returned parts versus number usable parts
Number of dents on a body panel
Number of errors made by a clerk
Variable   Hole size (in thousandths of an inch or millimeters)
Flatness of a part (in thousandths of an inch or millimeters)
Furnace temperature (in Fahrenheit or Celsius degrees)
Shipping time (in days)
Inventory (in days of production)

Variable data can often be converted to attribute data simply by specifying what range of measurements conforms to requirements and what range is beyond requirements. Thus,

2.5 Measurement

variable measurements are more flexible than attribute measurements. Also, we see in later chapters that variable measurements convey more information about the process.

Why should we ever use attribute data? In some cases it is not possible or meaningful to provide variable data, for example, the number of visual defects from a machine, dents on a body panel, scratches in paint, or forms with errors. In some cases, attribute data are collected when variable data might be more meaningful. For example, for windshield cracks, it would be possible to measure the crack location and size instead of simply noting the presence or absence of a crack. Also, it has been manufacturing practice to use attribute data because of the widespread use of go/no-go gaging. This gaging provides minimal information (with minimal cost). It is difficult, if not impossible, to adequately adjust or control most machines using attribute gaging. Modern quality standards and DPS concepts necessitate the use of a reasonable amount of variable gaging (see Chap. 14).

2.5.2 Operational Definitions

Deming (1982, Chap. 15) discusses the importance of operationally defining performance characteristics. To conduct any type of business, it is necessary for the producer of a product or service to meet the intent of a customer's purchase. It is thus necessary to use specifications to define this intent. These specifications make it possible to evaluate the product or service.

Operational definitions are equally important when considering measurements to be made on a process. The question is simple: Are all measurements operationally defined? Some typical examples of inadequate definitions appear below:

Machining: Diameter of a hole, burr-free part
Process: Temperature of a furnace
Assembly: Number of defects
Administrative: Number of errors on a form

In each case, it is possible for two people to perform the required measurement differently. Whenever misunderstanding is possible, it will occur.

Consider the first example. What are the diameters of these holes?

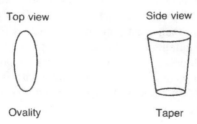

In the first case, the ovality of the hole makes a single diameter measurement meaningless. In the second case, the taper along the depth of the hole makes a single diameter measurement meaningless. A correct specification uses a circularity or cylindricity measurement (see App. II). However, these measurements require more elaborate gaging. In practice an operational definition is to measure the minimum diameter at a specified depth. Other issues that may need to be addressed concern temperature and cleanliness of the part. Standard manufacturing tolerances in the automobile industry are often in the range of .001–.0005 inch, where .001 inch is about the width of a human hair. Proper operational definitions are essential for controlling a process since they lead to the necessary process information:

Operational definition → measurement → process information → process control

Consider the remaining examples. Does a "burr-free part" mean that the edge of a part is not sharp? Does it mean no metal slivers can be seen under a certain magnification? Is the temperature of an oven uniform throughout the furnace? Does an error have the same meaning for all people?

In many of these examples, the best way to define the intent of a measurement is to give actual examples. Actual parts or photographs of what is (and what is not) a burr are useful. Parts with a certain defect may be needed for proper assessment. Examples of what is and what is not an error on a form could be listed.

An operational definition communicates the same measurement intent to all people

The need for operational definitions in administrative applications is as great as the need in manufacturing applications. Probably the simplest example comes from service industries or service organizations within larger companies. In most cases, management often reminds employees of the importance of the customer and that their organization is in the business of providing a service. Such slogans as "Serve the Customer" are used to remind employees of the organization's purpose. However, how does the employee operationally define the intended behavior? In most cases, customers place a premium on how problems are handled. The automobile dealer is often evaluated not on the politeness during the sale (it is expected), but on the attention given when any subsequent problem occurs. The opinion of an entire organization is frequently based on employees lowest on the organizational ladder. The salesclerk, serviceman, waitress, teller, or receptionist form the basis of the interface with the public. Yet these individuals typically receive minimal training and have the lowest company seniority. They are often ill prepared to handle the customer with a problem: no operational definition of "Serve the Customer" exists. Management's intent is therefore lost.

Another type of situation is in making measurements when they are easy to make rather than when they are useful (to a customer). For example, the controller records total department tool costs rather than use by machine, which is required for the production manager to manage the operation. Also, the number of products shipped is often reported, but the stability of the shipping schedule or changes in product complexity are not measured. The impact of schedule stability and complexity is great on most production operations, yet there is often no attempt to operationally define what is important to the customer (i.e., the production manager).

A material control manager remarked that he had just "cleaned house" and shipped all subassemblies to an assembly plant. Also, boxcars of material would not be unloaded until the next day. It was the last of the month, and his performance was measured by the inventory on hand on the last day of the month.

2.6 Gaging

A gage is required to produce measurements to obtain a set of variable data values. These values form the basis for evaluating the performance of the process. It is important that the

2.6 Gaging

measurement process be addressed prior to any data collection. Operational definitions should be established and the gaging methods understood and analyzed. This section only introduces some basic concepts in gaging. A complete analysis of the gaging process requires the use of many of the 10 tools and is deferred to Appendix I.

2.6.1 Gaging Concepts

There are three basic concepts that form a basis for understanding any gaging process:

1. Accuracy: the degree of agreement between a gage's measurements and a reference standard measurement device
2. Repeatability: the degree of variability between repeated measurements of a single part using one gage and one operator
3. Reproducibility: the degree of variability between different operators measuring a single part using one gage

It should be noted that the reference standard referred to here is typically a measurement procedure that is significantly better than the gage being evaluated but is not practical to use routinely in a production setting (e.g., a coordinate measuring machine). A precise definition of the three concepts appears in Appendix I.

A large number of manufacturing problems arise from inadequate gage accuracy, repeatability, or reproducibility. An important aspect of a DPS is to quantify the three gage parameters *prior* to any problems having occurred. An understanding of the three concepts can be gained by considering the following examples.

Example 2.4 Suppose five repeated measurements of a single part using gage 1 are

5.0 3.5 6.3 5.4 4.8

and a reference standard gage measures the part to be 4.4. A convenient way to plot the measurements is

```
                  reference standard
                  x      ↓ x x    x           x
Gage 1:  ─────────────────────────────────────
            2      3      4      5      6      7
```

Suppose five repeated measurements of the part using a second gage are

3.1 4.1 3.6 4.8 5.7

```
                  reference standard
               x    x    x ↓ x          x
Gage 2:  ─────────────────────────────────────
            2      3      4      5      6      7
```

We would say that gage 2 is more accurate than gage 1 since its measurements are more centered around the reference standard.

Example 2.5 Suppose five repeated measurements of a single part using gage 1 are

10.0 7.1 6.5 11.1 9.2

and five measurements of the part using gage 2 are

7.9 8.7 9.3 8.5 9.6

Again, plotting the gage measurements makes comparison easy:

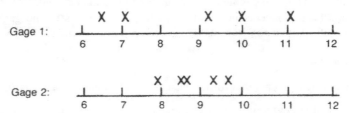

It is apparent that the spread of the repeated measurements for gage 1 is greater than for gage 2. Gage 2 is said to have better repeatability than gage 1: measurements of the same part *repeat* more closely. Note that the value of the reference standard is not important for assessing repeatability. In this example the reference standard could have a value of 6, 10, or 12 and the assessment of repeatability would be the same.

Example 2.6 Suppose we again measure the same part five times using two gages, but we now have two operators perform the measurements:

	Operator 1	Operator 2
Gage 1	2.3, 1.5, 6.1, 4.2, 2.5	2.7, 1.9, 5.9, 6.5, 3.2
Gage 2	2.7, 4.1, 3.4, 2.2, 3.1	6.1, 2.5, 5.2, 4.7, 5.9

To plot these measurements, it is useful to plot operator 1 (x) and operator 2 (o) measurements using different codes:

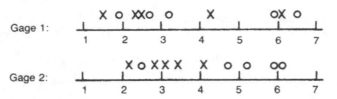

The overlap of the measurements from operator 1 and operator 2 is greater for gage 1 than for gage 2. Thus gage 1 is said to be more reproducible than gage 2.

The following plots represent various combinations of gage accuracy and repeatability. Assume that for each case five measurements of a single part have been made.

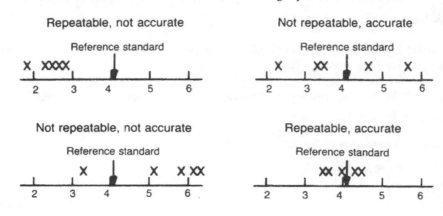

2.6 Gaging

A spring supplier was continually receiving returned rejected shipments from an assembly plant. The sample checks made by the plant frequently detected springs outside the specification limits. The spring supplier thought that their parts were acceptable but the internal control system did not provide adequate verification. Generally, the supplier would 100% sort the rejected shipment and send back the material to the assembly plant. Often even these shipments were returned. It was later learned that neither the supplier nor the assembly plant had adequate gaging to measure springs with tight specifications. The inadequate repeatability of the gages made good parts appear to be beyond the specification limits.

2.6.2 Manufacturing Considerations

There are several important considerations in evaluating a gaging system. A standard ruler is adequate to perform most carpentry work although the same ruler would be inadequate for machining an engine piston. A general rule of thumb is to require 10–20 gage scale divisions to be within the tolerance limits.

A minimum of 10 scale divisions should be within specification limits

Thus, a .001 inch tolerance should have scale divisions of at least .0001 inch. Failure to meet this criterion often results in difficulty in evaluating the performance of a process.

Relate gaging to specification limits

The measures of accuracy, repeatability, and reproducibility must also be related to the spread of the specification limits for the characteristic being measured. For example, a gage whose accuracy is .01 inch is useless for measuring a characteristic on an engine piston but may be acceptable for furniture manufacturing. If the three sources of gage variability are considered jointly, the criteria in Figure 2.12 can be used as a rule of thumb in evaluating the gage.

The method to evaluate the percentage of gage variability is given in Appendix I. The gaging system should also consider the ability of a process to consistently produce parts within specification limits (process capability, Chap. 7). Intuitively, a process that easily meets specification requirements may not require sophisticated gaging.

Relate gaging to process performance and requirements

The purpose of any gage is to enable operators to control a process. Only go/no-go gaging is often present for a characteristic that is directly controlled by a machine adjustment. This practice often leads to problems. Typically, several parts are checked after an

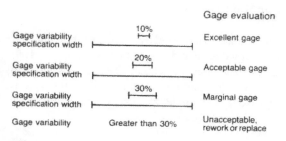

Figure 2.12 Gage evaluation criteria.

adjustment. Assuming these parts pass the go/no-go gage, the process is allowed to run. Small changes in the process or normal process variability can lead to unnoticed rejects. Again, building a DPS requires adequate gaging to collect variable data.

Go/no-go gaging should not be used
where machine adjustments are necessary

Another aspect of relating gaging to process requirements concerns how a part is located in a gage. An operator must use a gage to control a process so the gage measurement system should simulate how the characteristic is produced. For example, suppose a machine cuts the top portion of the block below and uses the slots to hold the part in the machine:

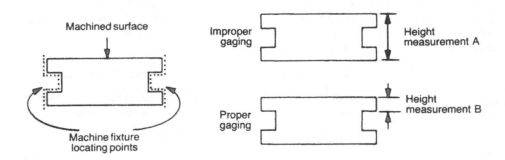

This operation can only control height measurement B, so design of the gage should use the machine locating points and measure height B. Typically, another dimension (e.g., height A) is important to product function, so to reduce gaging cost a gage to measure only height A is purchased. This practice can lead to two problems. First, the operator who generates height B (using a gage to measure height A) cannot control the process since he or she has no way to measure height B. Also, if the operator does attempt to control height A, he or she must try to compensate for all the sources of machining variability prior to the operation. This is impossible and leads to the operator making too many adjustments to the process, leading to a lack of consistency in height A (along with height B and other dimensions). Quality must be "built up" in each operation that leads to the control of height A. From the product engineering viewpoint, it may also be advisable to have selected gages that simulate product function. Ideally, product and manufacturing engineers should work closely together to develop machining processes that correspond to controlling characteristics important to product function.

Locate a part in a gage using the same part-locating positions
as used in the process

Many gage designers consider a gage to be adequate if repeated gaging of the same part (often a specially produced master part) yields values that do not differ by more than 10% of the specification limit range of the characteristic being measured. The problem with this approach is that repeated gaging of a single part does not introduce representative within-part variability to simulate production gage use. For example, the outer diameter (OD) of a gear blank is to be measured.

Problems

One method of measurement would be to place the gear over a post that centralizes the part and uses an electronic sensor located on an arm that moves back to allow loading of the part.

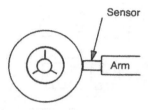

A gage designer's test would typically be to move the arm back and forth several times, attempting to repeat the same measurement. This approach does not consider the potential noncircularity of the OD and how a machine operator would use the gage. In order to make a machine adjustment, an operator would rotate the part and obtain a maximum and minimum OD. The average of the maximum and minimum would be used as the OD of the part. Evaluation of the gage should then repeat finding the maximum and minimum OD of a number of production parts. This is a much more severe evaluation of the gage. For example, the ability of the gage to consistently centralize the part becomes important. Thus, gage design should consider variability within the part characteristic being gaged and how the gage output is used. Gage evaluation should repeatedly measure a variety of different parts, removing the parts each time from the gage. Appendix I gives an evaluation procedure.

> Gage design must address variability of a characteristic
> within a single part

Problems

2.1 Consider two different arrangements of eight machining spindles that all produce the same part:

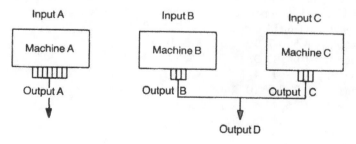

Suppose a problem is indicated at the following output positions, list the potential sources of variability: (a) output A; (b) output D; (c) output B. How would knowledge of the spindle that generated the problem part be useful in each case?

2.2 How many paths can a part have followed in this process flow diagram if the output is checked at I?

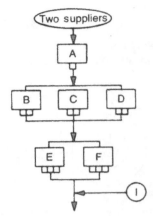

2.3 How many paths could a part have followed if a part is checked at the end of the line in Figure 2.6?

2.4 Explain how the potential paths for a part at I can be reduced in this example. Give the reduction in paths.

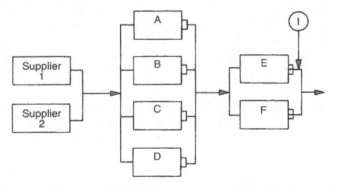

2.5 Compare gage 1 and gage 2 for accuracy and repeatability. Assume a reference standard of 12.2 and six repeated measurements of the same part.
   (a) Gage 1:   11.8, 14.6, 9.2, 13.7, 10.2, 10.9
         Gage 2:   8.5, 9.7, 10.7, 8.7, 9.3, 10.3
   (b) Gage 1:   10.7, 13.3, 11.5, 11.8, 12.7, 11.2
         Gage 2:   13.3, 14.4, 8.5, 10.7, 14.7, 9.6
   (c) Gage 1:   9.2, 9.9, 10.2, 8.5, 9.5, 8.2
         Gage 2:   13.5, 14.4, 13.8, 12.5, 13.2, 14.5

3
Control Charts

Measurements made on the output of a process are for the purpose of collecting information about the process. This process "information" exhibits variability for two reasons. First, the process parameters (men, methods, machines, materials, measurements, and environment) are always changing. These changes cause process output to vary. Second, any observed collection of measured units is a "sample" or subset of possible units from a process. Different measurements are observed if the units are collected at different times. For example, suppose plan 1 measured every fifth unit and plan 2, every fourth unit. The measurements might appear as follows:

Plan 1:   4, 7, 3, 1, 9, 5, . . .
Plan 2:   1, 5, 8, 11, 2, 7, . . .

Alternatively, plan 1 might be to collect the first 25 units and plan 2, the next 25 units. Even if a process remains unchanged, two samples (e.g., plan 1 versus plan 2) are not the same, except occasionally when a small number of units is considered.

A process always shows variation

One purpose of making measurements is to assess the degree of variation present. Also, we wish to determine whether the variation is predictable (controlled) or unpredictable (uncontrolled). Obviously, a defect prevention system (DPS) requires controlled variation to reduce the possibility of defects that can arise in the two ways shown in Figure 3.1. Uncontrolled variability always has the potential for producing parts outside the specification limits (i.e., defects). Controlled variability can also produce defects if the process variability is too large. One of the main purposes of the control charts discussed in this chapter is to establish consistent, measurable criteria to assess whether a process exhibits predictable, controlled variation. Chapter 7 discusses methods to assess process capability, an evaluation of the extent to which controlled variation is within specification limits.

The discussion of Figure 2.1 addressed the importance of process control, which utilizes control charts and other tools throughout the process. High-quality output from the process requires that controlled variation must exist at each stage of the process. Many applications of control charts fail when only the output of a process is addressed. Controlled variation of the output of a process cannot exist on an ongoing basis when intermediate process steps exhibit uncontrolled variation. Preparing a process flow diagram is useful in identifying

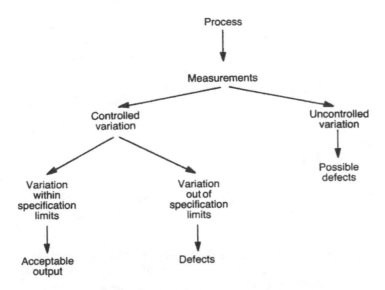

Figure 3.1 Relating defects to process variation.

where in the process control charts would be beneficial. A comprehensive plan is given in Chapter 14.

3.1 Coded Measurements

A coded measurement arises when the value read from a gage is related to the value of a part characteristic by a change of scale or offset. There are many situations in which it is convenient to use coded data when collecting and analyzing measurements. No information is lost when using coded data, and it is often much easier to perform calculations using coded data. However, it is critical that the user be able to translate the results back into noncoded form for interpretation.

Coded data are useful, provided the code is understood

The simplest type of coding is a simple scaling, where

Coded value	Actual value
1	.001
2	.002
3	.003
.	.
.	.
.	.

which can be represented as

3.1 Coded Measurements

Actual value = .001 × coded value

In this case the coded values simply change the units of measurement. For example, a gage scale may have a set of numbers 1–10 that represent .001 through .010. It is of course also important to note the measurement scale being used (e.g., inches, millimeters, or degrees Fahrenheit or Celsius).

The simplest example of scaling is a revolutions per minute (RPM) gage on a car (tachometer) that reads 1–6 but has a note on the gage that reads ×1000 RPM. Thus, the coded gage and actual RPM data are

Gage RPM	Actual RPM
1.0	1000
1.5	1500
2.0	2000
2.5	2500
.	.
.	.
.	.
6.0	6000

The coding system can be represented as

Actual RPM = 1000 × gage RPM

Similar examples occur when measuring force, temperature, pressure, and speed, for example.

Dial indicator gages are often used to measure the distance a part dimension is from the desired target position. Gages are designed to give a positive value called a total indicator reading (TIR), that is, distance between the actual and desired positions. For a thousandths indicator gage, values are typically recorded as 1.0, 2.5, 3.2, and so on, which represent .001, .0025, and .0032. For a one ten-thousandths indicator gage, the same measurements would be .0001, .00025, and .00032. The following scales are commonly used:

Code value	Tenth actual value (scale = .1)	Hundredth actual value (scale = .01)	Thousandth actual value (scale = .001)	Ten-thousandths actual value (scale = .0001)
.5	.05	.005	.0005	.00005
1.0	.10	.010	.0010	.00010
1.5	.15	.015	.0015	.00015
2.0	.20	.020	.0020	.00020
2.5	.25	.025	.0025	.00025
.	.	.	.	.
.	.	.	.	.
.	.	.	.	.

Another type of coded data arises from measuring the deviation a part is from a target value, which is often the midpoint between the specification limits. This target or nominal value is often represented as 0 on the gage. Values greater than the target are positive (+) measurements, and values less than the target are negative (−). A gage to measure a desired size of 5.24 ± .10 inches typically marks the upper specification limit (USL), nominal specification limit, and lower specification limit (LSL) as follows:

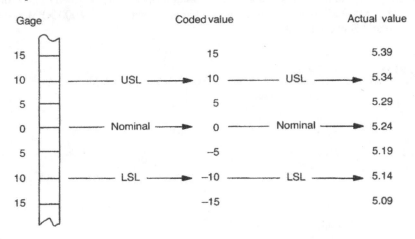

The coding system can be represented as

Actual value = 5.24 + .01 × gage value

The general gage coding system can be stated as

Actual value = offset + (scale × gage value)

This equation is sometimes abbreviated by using the convention 0 = offset and 1 = scale. Thus, the coding system in this example may be stated as 0 = 5.24 and 1 = .01.

3.2 Location and Variation Estimates

There are two important properties that must be evaluated for any process: process location and variability. These two properties quantify the essential information for the output of most processes:

Location: The central-most part of the process output
Variability: The spread or range of the process where most values occur

There are two common measures for quantifying both location and variability:

Location	Variability
Mean $\bar{X}$	Standard deviation s
Median $\tilde{X}$	Range R

3.2 Location and Variation Estimates

The symbols are used to denote each measure. The definition of each measure appears below, where it is assumed that we have measured three values denoted by

X_1 = measurement 1
X_2 = measurement 2
X_3 = measurement 3

Mean = average value of the measurements

$$= \frac{X_1 + X_2 + X_3}{3}$$

$$= \bar{X}$$

Median = middle value when all measurements are placed in order

$$= \tilde{X}$$

Standard deviation = s

Standard deviation is computed in six steps:

1. Compute mean $\bar{X}$.

2. Subtract all values from $\bar{X}$:

    $X_1 - \bar{X}$
    $X_2 - \bar{X}$
    $X_3 - \bar{X}$

3. Square each value computed in step 2:

    $(X_1 - \bar{X})^2$
    $(X_2 - \bar{X})^2$
    $(X_3 - \bar{X})^2$

4. Add the values computed in step 3:

    $(X_1 - \bar{X})^2 + (X_2 - \bar{X})^2 + (X_3 - \bar{X})^2$

5. Divide the result of step 4 by 1 less than the number of measurements:

    $$\frac{(X_1 - \bar{X})^2 + (X_2 - \bar{X})^2 + (X_3 - \bar{X})^2}{2}$$

6. Take the square root of the result of step 5:

    $$\sqrt{\frac{(X_1 - \bar{X})^2 + (X_2 - \bar{X})^2 + (X_3 - \bar{X})^2}{2}}$$

Range = highest value minus lowest value

$$= X_{highest} - X_{lowest}$$

$$= R$$

The three measurements 5, 7, and 2 are used to illustrate these calculations.

$$\text{Mean} = \frac{5 + 7 + 2}{3} = 4.67 = \bar{X}$$

Median = middle value of 5, 7, 2

Ordered values are 2, 5, 7: middle value is 5.

$\tilde{X} = 5$

Standard deviation:

1. $\bar{X} = 4.67$
2. $5 - 4.67 = 0.33$
   $7 - 4.67 = 2.33$
   $2 - 4.67 = -2.67$
3. $(0.33)^2 = 0.11$
   $(2.33)^2 = 5.43$
   $(-2.67)^2 = 7.13$
4. $0.11 + 5.43 + 7.13 = 12.67$
5. $\dfrac{12.67}{2} = 6.34$
6. $\sqrt{6.34} = 2.52 = s$
   Range $= 7 - 2 = 5 = R$

Most of the procedures in this text utilize one or more of these four simple quantities for appropriately collected data.

Procedure 3.1 is a general procedure for computing the four values assuming there are N measurements:

$X_1 =$ measurement 1
$X_2 =$ measurement 2
.
.
.
$X_N = N^{\text{th}}$ measurement

The Nth measurement is the last measurement; N can be any integer (1, 2, 3, . . .). It should be noted that many calculators can easily compute the mean $\bar{X}$ and standard deviation s, so the calculations in Procedure 3.1 are rarely performed by hand. Also, the median of an even number of values is the average of the two values closest to the middle. Thus, the median of 4, 7, 2, and 5 is computed by ordering the values

2, 4, 5, 7

$$\text{Median} = \frac{4 + 5}{2} = 4.5$$

and computing the average.

Procedure 3.1 Computing Measures of Location and Variability

Let N measurements from a process be denoted by $X_1, X_2, \ldots, X_N$. The following are methods of estimating location and variability measures.

Location

$$\text{Mean } \bar{X} = \frac{X_1 + X_2 + \cdots + X_N}{N}$$

Median $\tilde{X}$ = middle value when all measurements are placed in order

If N is even, the median is the average of the two middle values.

Variability

$$\text{Standard deviation } s = \sqrt{\frac{(X_1 - \bar{X})^2 + (X_2 - \bar{X})^2 + \cdots + (X_N - \bar{X})^2}{N - 1}}$$

1. Compute mean $\bar{X}$.
2. Subtract all values from $\bar{X}$:

    $(X_1 - \bar{X})$
    $(X_2 - \bar{X})$
    .
    .
    .
    $(X_N - \bar{X})$

3. Square each value computed in step 2:

    $(X_1 - \bar{X})^2$
    $(X_2 - \bar{X})^2$
    .
    .
    .
    $(X_N - \bar{X})^2$

4. Add the values computed in step 3:

    $(X_1 - \bar{X})^2 + (X_2 - \bar{X})^2 + \cdots + (X_N - \bar{X})^2$

5. Divide the result of step 4 by $N - 1$:

    $$\frac{(X_1 - \bar{X})^2 + (X_2 - \bar{X})^2 + \cdots + (X_N - \bar{X})^2}{N - 1}$$

6. Take the square root of the result in step 5.

    Range $R = X \text{ (highest)} - X \text{ (lowest)}$

Suppose we have the following $N = 30$ measurements that were collected in production order every 15 minutes from a process:

Time	Value	Time	Value	Time	Value
1	13	11	11	21	9
2	11	12	8	22	7
3	10	13	10	23	10
4	8	14	10	24	11
5	11	15	11	25	10
6	10	16	11	26	7
7	9	17	10	27	10
8	11	18	7	28	8
9	9	19	10	29	9
10	8	20	10	30	10

The values of the four quantities are

$$\bar{X} = \frac{13 + 11 + 10 + \cdots + 8 + 9 + 10}{30} = 9.63$$

Median = middle value of the 30 ordered values: 7, 7, 7, 8, 8, 8, 8, 9, 9, 9, 9, 10, 10, 10, 10, 10, 10, 10, 10, 10, 10, 11, 11, 11, 11, 11, 11, 11, 11, 13

$$\tilde{X} = \frac{10 + 10}{2} = 10$$

$s = 1.43$ use a calculator!

Range = X (highest) $-$ X (lowest)

$R = 13 - 7$
$= 6$

As previously noted, s and R were related to the "spread" of a process. A more precise way of stating the relationship between s and the process spread is that most (99.73%) of the measurements from a process are within a $6s$ interval. The upper and lower limits of the process are

Upper process limit (UPL) = $\bar{X} + 3s$
Lower process limit (LPL) = $\bar{X} - 3s$

and the spread is

Process spread = upper process limit $-$ lower process limit
$=$ UPL $-$ LPL
$= (\bar{X} + 3s) - (\bar{X} - 3s)$
$= 6s$

3.2 Location and Variation Estimates

This equation assumes that the process does not exhibit uncontrolled variation, as discussed later in this chapter, and the distribution of the measurements is the familiar normal, bell-shaped histogram (Chap. 5).

Using the $N = 30$ data example,

$$\text{UPL} = 9.63 + (3 \times 1.43) = 13.92$$
$$\text{LPL} = 9.63 - (3 \times 1.43) = 5.34$$

Process spread $= 6 \times 1.43 = 8.58$

These results can be represented in a plot:

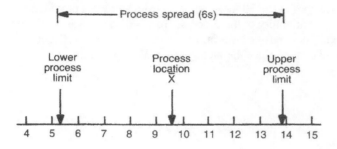

The range can also be related to the process spread using the d constant in Table 3.1:

$$\text{Process spread} = \frac{6R}{d}$$

Table 3.1 Constant d for Estimating Process Variability using the Range of N Measurements[a]

N	d
10	3.18
15	3.55
20	3.81
25	3.99
30	4.14
35	4.27
40	4.37
45	4.46
50	4.55
60	4.68
70	4.80
80	4.89
90	4.98
100	5.05

[a]The table assumes a single group of N measurements is used to compute the range R.

Note that d depends on N. Generally, the spread using s or R agrees reasonably well, but for N greater than about 10, s is the preferred method to estimate process variability. For the $N = 30$ data example, using the range method,

$$\text{Process spread} = \frac{6 \times 6}{4.14} = 8.70$$

which is close to the value obtained (8.58) using s.

3.3 Run Chart

The previous analysis of the 30 measurements from a process provided a method to assess both location and variability of the process. However, there was no evaluation of whether the process exhibits controlled variation through time. Generally, a process can be expected to vary over time since the factors that make up a process:

Men
Methods
Machines
Materials
Measurements
Environment

can be expected to change over time. Operator assignments, machine setup methods, machine performance, incoming material, gage condition, and seasonal changes are all factors that change over time. Our previous analysis did not assess this very important aspect of any process.

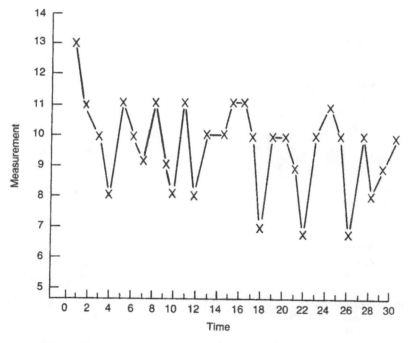

Figure 3.2 Run chart for $N = 30$ measurements.

3.3 Run Chart

All process factors vary over time

The simplest method of evaluating a process over time is to plot the measurements in the sequence in which the parts were produced. The plot of the 30 measurements appears in Figure 3.2, where change over time can be easily examined. This sample chart is called a run chart since the measurements are plotted in the run order in which they were produced in the process.

Figure 3.3 shows three examples of various patterns that can be detected using a run chart. In the top chart, an upward trend is apparent as might be expected for part size when a tool is wearing over time. The second chart shows cycles that might occur over an 8-hour

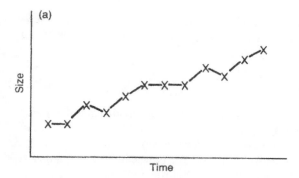

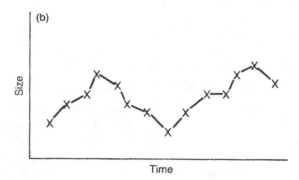

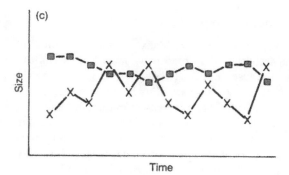

Figure 3.3 Sample run chart patterns (× = machine 1; ■ = machine 2): (a) trends; (b) cycles; (c) stratification.

shift due to machine or environmental temperature changes. The bottom chart shows how the output from several machines (or fixtures, spindles, pallets, and so on) could be represented on a run chart.

Some of the applications of a run chart are as follows:

A run chart is the best method for process analysis when changes are expected to occur over an interval associated with producing a part. For example, consecutive parts are typically measured when studying tool wear or tool compensation problems.

In short-term studies in which every part produced is measured, a run chart is useful. For example, in initial trial runs of machinery every part is often measured.

Run charts are a useful introductory tool to individuals just learning how to study a process. Later we see how that run charts can be expanded to more informative analyses.

An important practice in collecting data for a run chart is that approximately the same time interval should be used between measurements. This practice makes interpretation of patterns much easier.

3.4 Multiple-Point Run Chart

Suppose that we are interested in monitoring a dimensional characteristic (in millimeters) of an engine crankshaft. During monitoring a part is measured every 2 hours. The run chart in Figure 3.4 is obtained. This figure is somewhat informative since it shows the process variation over time. However, what we really want to know is whether the variability is sufficiently large to cause us concern. That is, is the process sufficiently stable to ensure the predictability that future output is acceptable? Confronted with this question, most people ask that the specification limits be defined. This is entirely the wrong approach. For the data in Figure 3.4, the specification limits could be (5.00, 5.50), (5.10, 5.40), or (5.20, 5.30) and nothing would be known about whether the process was predictably stable. The specification limits are set by people and have nothing to do with process stability.

Specification limits have nothing to do with process output

Suppose we wish to ask whether the process meets the specification limits of (5.10, 5.40)? All the points plotted in Figure 3.4 are within the stated specification limits, so can we confidently assume that our process is meeting the specification limits? No! No real criteria have been used to evaluate whether the process is stable. If it is not stable, then out-of-specification parts may occur between the parts that were measured, or they may occur in the future.

Without stability there is no predictability

What is a reasonable criterion to judge process stability? Many ideas are possible, but most people would first try to assess the definition of a "reasonable level" of variability. We know that the process output varies because of normal changes over time in the six process elements, but what is the expected level of variability? The opportunity for variability increases as the time between measured parts increases; minimum variability typically occurs when consecutive parts are measured. Thus, process variability over the long run can be expected to be at least as large as the variability quantified using parts produced over a short period of time (e.g., consecutive parts). Our definition of stability requires that the process does not change over the long run in location or variability more than what would be expected from short-term variability fluctuations. Stability implies that

3.5 Mean and Range Control Chart

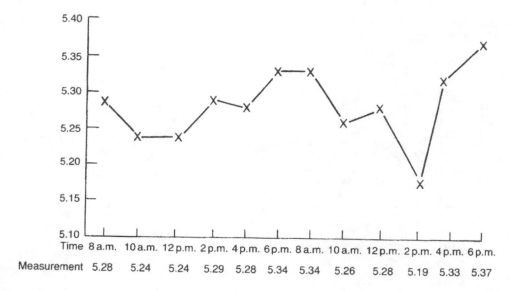

Figure 3.4 Run chart for an engine crankshaft dimensional characteristic.

the process elements do not change the process output over time more than expected from short-term fluctuations.

> A process is stable if its output does not change location or variability over the long run more than that expected from short-term variability

Suppose we want to use our run chart in Figure 3.4 but with measurements made on five consecutive pieces at the 2 hour intervals. The plot in Figure 3.5 for this situation is called a multiple-point run chart because multiple measurements are plotted at each time point. Generally, this type of data is represented in coded form as appears in the bottom of Figure 3.5. Both actual and coded measurements appear in Figure 3.5 to allow comparison to the run chart in Figure 3.4. Notice the median of the five measurements is circled and that the medians are connected to show the central location of the process. The plot in Figure 3.5 provides more information than Figure 3.4. It appears that the minimum range between consecutive parts is about .05, with a number of time points having ranges between .05 and .10. An indication of a process problem appears in two time periods during which the very unusual measurements of 5.55 and 5.44 were observed (these are called outliers). However, there are as yet no clear criteria to judge process stability. A dot control chart form for collecting and plotting multiple measurements is given in Chapter 7.

3.5 Mean and Range Control Chart

It is possible to eliminate much of the subjectivity in evaluating the stability of a process using control charts, the most common being the $\bar{X}$ and R chart. These two charts are very simple alterations of the multiple-point run chart discussed in the previous section. Consider a group of five measurements collected over a short interval of time. This group of measurements is called a subgroup; in Figure 3.5 there are 12 subgroups. The subgroups

48 3 Control Charts

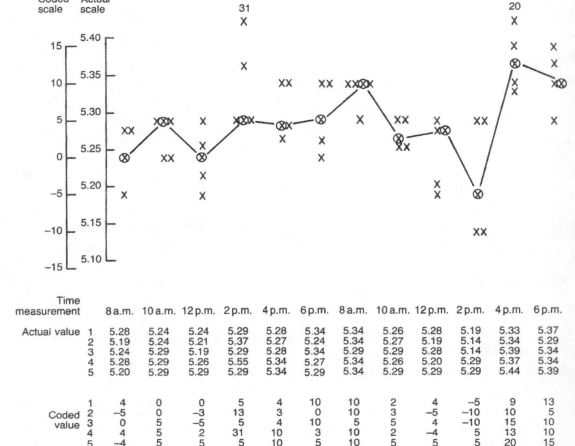

Figure 3.5 Multiple-point run chart for an engine crankshaft characteristic.

provide two types of information about the process over the short period of time in which measurements were made. The five measurements can be used to evaluate both location and variability. Consider the eighth subgroup in Figure 3.5; the two measures of location $\hat{X}$ and $\bar{X}$ and the two measures of variability s and R discussed earlier in this chapter can easily be computed.

Subgroup measurements	Subgroup 8	Location	Variability
1	2	$\hat{X} = 3.4$	$s = 1.52$
2	3	$\bar{X} = 3.0$	$R = 3.0$
3	5		
4	2		
5	5		

3.5 Mean and Range Control Chart

It is common practice to use $\bar{X}$ and R to evaluate the location and variability of a process. In Figure 3.6, the $\bar{X}$ and R values are computed for each of the 12 subgroups and plotted on the two graphs. To assist in interpreting these graphs, it is useful to plot the average of the 12 subgroup $\bar{X}$ values (denoted by $\bar{\bar{X}}$) and the average of the 12 subgroup R values (denoted by $\bar{R}$) on the plot. These averages are

$$\bar{\bar{X}} = \frac{-0.2 + 3.0 + -0.2 + 11.8 + \cdots + 13.4 + 10.6}{12} = 5.03$$

$$\bar{R} = \frac{9 + 5 + 10 + 26 + \cdots + 11 + 10}{12} = 10.1$$

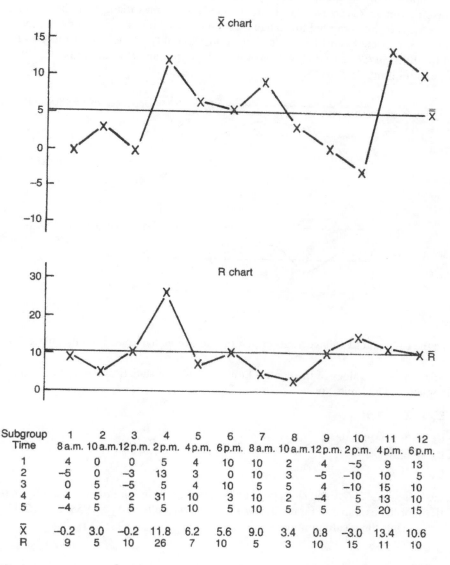

Subgroup	1	2	3	4	5	6	7	8	9	10	11	12
Time	8 a.m.	10 a.m.	12 p.m.	2 p.m.	4 p.m.	6 p.m.	8 a.m.	10 a.m.	12 p.m.	2 p.m.	4 p.m.	6 p.m.
1	4	0	0	5	4	10	10	2	4	-5	9	13
2	-5	0	-3	13	3	0	10	3	-5	-10	10	5
3	0	5	-5	5	4	10	5	5	4	-10	15	10
4	4	5	2	31	10	3	10	2	-4	5	13	10
5	-4	5	5	5	10	5	10	5	5	5	20	15
$\bar{X}$	-0.2	3.0	-0.2	11.8	6.2	5.6	9.0	3.4	0.8	-3.0	13.4	10.6
R	9	5	10	26	7	10	5	3	10	15	11	10

Figure 3.6 Plot of $\bar{X}$ and R values for an engine crankshaft dimensional characteristic.

The graphs in Figure 3.6 provide information similar to that in the multiple-point run chart in Figure 3.4, but the information on location (measured by a plot of the $\bar{X}$ values) and variability (measured by a plot of the R values) are separated. This is a very important feature of $\bar{X}$ and R charts since problem causes for changes in location can be different from problem causes for changes in variability. It is also important to use both charts together. For example, subgroup 4 has $\bar{X} = 11.8$ and $R = 26$ and subgroup 12 has $\bar{X} = 10.6$ and $R = 10$. Thus, the high mean for subgroup 4 is largely due to the high range (due to the measurement of 31), where subgroup 12 has a number of high values since the range of 10 is about equal to the average range of all subgroups ($\bar{R} = 10.1$) yet the mean is quite high.

Our goal for a criterion to evaluate the stability of a process has not yet been met. However, the chart in Figure 3.6 can be used as a basis to develop a criterion. Recall that our five samples in the subgroup were collected over a reasonably short period of time so that the average range $\bar{R}$ is an average of the expected short-term variability of the process. A stable process is not permitted to change location or variability over the long run more than would be expected considering its short-term variability. Thus, for stability to be attained for the $\bar{X}$ and R values, all values of the subgroup $\bar{X}$ and R values must be within certain "control limits":

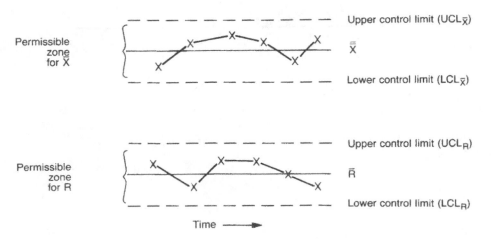

Note that the UCL and LCL have a subscript $\bar{X}$ or R to denote the limits for either the $\bar{X}$ or R plot.

Calculation of the values of the control limits is based on statistical theory but is quite easy in practice. The limits are

$$\text{UCL}_{\bar{X}} = \bar{\bar{X}} + A_2 \bar{R}$$
$$\text{LCL}_{\bar{X}} = \bar{\bar{X}} + A_2 \bar{R}$$
$$\text{UCL}_R = D_4 \bar{R}$$
$$\text{LCL}_R = D_3 \bar{R}$$

where the values of A_2, D_4, and D_3 depend on the number of samples n used in a subgroup (App. III). For the example in Figure 3.6, $n = 5$, which results in $A_2 = 0.58$, $D_4 = 2.11$, and $D_3 = 0$. Thus the resulting limits are

3.5 Mean and Range Control Chart

$$UCL_{\bar{X}} = 5.03 + (0.58 \times 10.1)$$
$$= 5.03 + 5.86$$
$$= 10.89$$
$$LCL_{\bar{X}} = 5.03 - (0.58 \times 10.1)$$
$$= 5.03 - 5.86$$
$$= -0.83$$
$$UCL_R = 2.11 \times 10.1$$
$$= 21.31$$
$$LCL_R = 0 \times 10.1$$
$$= 0$$

The easiest method of evaluating whether the subgroup $\bar{X}$ and R values are within the calculated control limits is to draw dashed lines on the chart. Using the dashed lines, it is easy to assess whether the process is stable by determining whether any values are beyond the control limits. Other indications of instability are discussed in a later section of this chapter. Limits for the crankshaft example are drawn on the chart in Figure 3.7. Note that when LCL_R is 0, no line is drawn since an R value below 0 is not possible. The figure shows

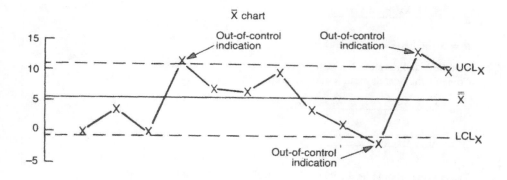

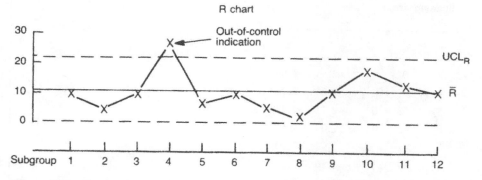

Figure 3.7 Control limits for a camshaft dimensional characteristic.

Procedure 3.2 Mean $\bar{X}$ and Range R Control Chart

Let a constant group of measurements n be made on a process for at least 20–25 time periods k, where the values are arranged as follows:

Time	1	2	...	k	
Measurement 1	X_1	X_1	...	X_1	n measurements form each subgroup
2	X_2	X_2	...	X_2	
.	.	.		.	
.	.	.		.	
.	.	.		.	
n	X_n	X_n	...	X_n	
Mean $\bar{X}$	$\bar{X}_1$	$\bar{X}_2$	...	$\bar{X}_k$	A subgroup mean and range are computed for each time period
Range R	R_1	R_2	...	R_k	

The subgroup size n is generally 3, 4, or 5 but may be greater. The amount of time represented by the measurements forming a subgroup defines "short-term" variability for the process. The following steps are used to evaluate whether the process is stable.

1. Compute the mean and range for each subgroup:

$$\bar{X} = \frac{X_1 + X_2 + \cdots + X_n}{n}$$

$$R = X\text{ (highest)} - X\text{ (lowest)}$$

2. Plot the $\bar{X}$ and R values for each subgroup on control chart paper.
3. Compute the overall mean $\bar{\bar{X}}$ and the average range $\bar{R}$ for the k time periods:

$$\bar{\bar{X}} = \frac{\bar{X}_1 + \bar{X}_2 + \cdots + \bar{X}_k}{k}$$

$$\bar{R} = \frac{R_1 + R_2 + \cdots + R_k}{k}$$

These values should be drawn on the control chart with a solid line.

4. Compute the control limits using the values of A_2, D_3, and D_4 that correspond to the subgroup size n.

n	A_2	D_3	D_4	d_2
3	1.02	0	2.58	1.69
4	.73	0	2.28	2.06
5	.58	0	2.12	2.33
6	.48	0	2.00	2.53
7	.42	.08	1.92	2.70
8	.37	.14	1.86	2.85
9	.34	.18	1.82	2.97
10	.31	.22	1.78	3.08

3.5 Mean and Range Control Chart

Upper control limit for $\bar{X}$:

$$\text{UCL}_{\bar{X}} = \bar{\bar{X}} + A_2 \bar{R}$$

Lower control limit for $\bar{X}$:

$$\text{LCL}_{\bar{X}} = \bar{\bar{X}} - A_2 \bar{R}$$

Upper control limit for R:

$$\text{UCL}_R = D_4 \bar{R}$$

Lower control limit for R:

$$\text{LCL}_R = D_3 \bar{R}$$

Appendix III contains a more complete list of tabled values.
5. Using a dashed line, draw $\text{UCL}_{\bar{X}}$, $\text{LCL}_{\bar{X}}$, UCL_R, and LCL_R.
6. Assess whether the range chart and then the $\bar{X}$ chart are stable by determining if a plotted $\bar{X}$ or R value is beyond the control limits computed in step 4. Note that additional criteria for evaluating stability are given in Procedure 3.7. On an ongoing basis, all process changes should be indicated on the chart.
7. If the process is judged stable, compute an estimate of the process standard deviation and process spread. Process standard deviation estimate:

$$\hat{\sigma} = \frac{\bar{R}}{d_2}$$

Process spread $= 6\hat{\sigma}$

Chapter 5 discusses the interpretation of the process standard deviation estimate $\hat{\sigma}$.

---

that both the range chart (subgroup 4) and $\bar{X}$ chart (subgroups 4, 10, and 11) have points beyond the control limits. Thus, the crankshaft process cannot be judged stable.

Stability requires all $\bar{X}$ and R values to be within control limits

Why is a control chart a logical method to assess process stability? Notice that all the control limits depend on $\bar{R}$. The value of $\bar{R}$ is simply the average of our subgroup ranges that quantify short-term variability. In a stable process, the subgroup means $\bar{X}$ measure process location, which cannot vary more than $A_2\bar{R}$ around $\bar{\bar{X}}$. Similarly, the subgroup ranges R measure process variability, which cannot vary more than $D_4\bar{R}$ up or $D_3\bar{R}$ down. The output of a stable process varies predictably between these limits. A stable process is also said to be in a state of "statistical control."

It is apparent that the value of $\bar{R}$ is critical in determining the control limits so that the range stability should be evaluated first. If the range is not stable, the value of $\bar{R}$ is incorrect, so all control limits are incorrect and stability cannot be meaningfully assessed. Another general rule is that at least 20–25 subgroups should be collected before stability is assessed so that $\bar{\bar{X}}$ and $\bar{R}$ (and the resulting control limits) are based on a meaningful amount of data. The crankshaft example with only 12 subgroups was used only for illustration.

First evaluate range stability

Procedure 3.3 Mean $\bar{X}$ and Standard Deviation s Control Chart

Let a constant group of measurements n be made on a process for at least 20–25 time periods k, where the values are arranged as follows:

Time	1	2	...	k	
Measurement 1	X_1	X_1	...	X_1	n measurements form each subgroup
2	X_2	X_2	...	X_2	
.	.	.		.	
.	.	.		.	
.	.	.		.	
n	X_n	X_n	...	X_n	
Mean $\bar{X}$	$\bar{X}_1$	$\bar{X}_2$	...	$\bar{X}_k$	A subgroup mean and range are computed for each time period
SD s	s_1	s_2	...	s_k	

The subgroup size n is generally at least 3, 4, or 5. Typically, the $\bar{X}$ and R chart is used for these small subgroup sizes, with the $\bar{X}$ and s chart used for subgroup sizes of 10 or more. The amount of time represented by the measurements forming a subgroup defines "short-term" variability for the process. The following steps are used to evaluate whether the process is stable.

1. Compute the mean and standard deviation for each subgroup:

$$\bar{X} = \frac{X_1 + X_2 + \cdots + X_n}{n}$$

$$s = \sqrt{\frac{(X_1 - \bar{X})^2 + (X_2 - \bar{X})^2 + \cdots + (X_n - \bar{X})^2}{n - 1}}$$

Procedure 3.1 gives the steps to use for computing s.

2. Plot the $\bar{X}$ and s values for each subgroup on control chart paper.
3. Compute the overall mean $\bar{\bar{X}}$ and the average range $\bar{R}$ for the k time periods:

$$\bar{\bar{X}} = \frac{\bar{X}_1 + \bar{X}_2 + \cdots + \bar{X}_k}{k}$$

$$\bar{s} = \frac{s_1 + s_2 + \cdots + s_k}{k}$$

These values should be drawn on the control chart with a solid line.

4. Compute the control limits using the values of A_3, B_3, and B_4 that correspond to the subgroup size n.

n	A_3	B_3	B_4	c_4
3	1.95	0	2.57	.89
4	1.63	0	2.27	.92
5	1.43	0	2.09	.94
6	1.29	.03	1.97	.95
7	1.18	.12	1.88	.96
8	1.10	.19	1.82	.97
9	1.03	.24	1.76	.97
10	.98	.28	1.72	.97

3.5 Mean and Range Control Chart

Upper control limit for $\bar{X}$:

$$\text{UCL}_{\bar{X}} = \bar{\bar{X}} + A_3 \bar{s}$$

Lower control limit for $\bar{X}$:

$$\text{LCL}_{\bar{X}} = \bar{\bar{X}} - A_3 \bar{s}$$

Upper control limit for s:

$$\text{UCL}_s = B_4 \bar{s}$$

Lower control limit for s:

$$\text{LCL}_s = B_3 \bar{s}$$

Appendix III contains a more complete list of tabled values.
5. Using a dashed line, draw $\text{UCL}_{\bar{X}}$, $\text{LCL}_{\bar{X}}$, UCL_s, and LCL_s.
6. Assess whether the s chart and then the $\bar{X}$ chart are stable by determining if a plotted $\bar{X}$ or s value is beyond the control limits computed in step 4. Note that additional criteria for evaluating stability are given in Procedure 3.7. On an ongoing basis, all process changes should be indicated on the chart.
7. If the process is judged stable, compute an estimate of the process standard deviation and process spread. Process standard deviation estimate:

$$\hat{\sigma} = \frac{\bar{s}}{c_4}$$

Process spread $= 6\hat{\sigma}$

Chapter 5 discusses the interpretation of the process standard deviation estimate $\hat{\sigma}$.

In the crankshaft example, the subgroup size was $n = 5$. Although a subgroup size of 5 is often a good choice, some process considerations make other values more reasonable, as discussed in Chapter 4. If there is no apparent reason to select a particular n, values of $n = 3, 4,$ or 5 are typically used. When measurements are difficult to make or destructive tests are involved, $n = 3$ can be used. Otherwise $n = 4$ or $n = 5$ is preferable. Generally, $n = 2$ is not recommended. Larger subgroup sizes (say, $n = 10$ or more) arise only in special applications.

A time line can be used to represent how we plan to collect measurements to evaluate process stability. Figure 3.8 shows a control chart time line. The time interval between subgroups should be small enough to force at least three, four, or five subgroups to be collected for each major source of variability. For example, if a tool change is important to maintaining process stability, at least three subgroups should be collected per tool. Other examples are operator changes or shift changes. Another important factor is the capability of the process, which is discussed in Chapter 7.

Procedure 3.2 provides a systematic method for calculating and plotting data on an $\bar{X}$ and R chart. Also, recall that either R or s can be used to quantify the spread or variability of the process. Procedure 3.3 is a method for an $\bar{X}$ and s chart. Generally, this chart is not used unless the subgroup size is 10 or more. The R chart is much easier to use than the s chart for most applications.

Procedure 3.4 Median $\tilde{X}$ and Range R Control Chart

Let a constant group of measurements n be made on a process for at least 20–25 time periods k, where the values are arranged as follows:

Time		1	2	$\cdots$	k	
Measurement	1	X_1	X_1	$\cdots$	X_1	n measurements form each subgroup
	2	X_2	X_2	$\cdots$	X_2	
	.	.	.		.	
	.	.	.		.	
	.	.	.		.	
	n	X_n	X_n	$\cdots$	X_n	
Mean $\tilde{X}$		$\tilde{X}_1$	$\tilde{X}_2$	$\cdots$	$\tilde{X}_k$	A subgroup median and range are computed
Range R		R_1	R_2	$\cdots$	R_k	for each time period

The subgroup size n is generally an odd number ($n = 3, 5, 7, 9,$ or 11) to make calculation of the median easy. The amount of time represented by the measurements forming a subgroup defines "short-term" variability for the process. The following steps are used to evaluate whether the process is stable.

1. Compute the median and range for each subgroup: $\tilde{X}$ = middle value of $X_1, X_2, \ldots, X_n$ when all measurements are placed in order; $R = X$ (highest) $- X$ (lowest). Note that the median $\tilde{X}$ for a subgroup is often obtained by plotting individual subgroup measurements and circling the center point, which is the median.
2. Plot the $\tilde{X}$ and R values for each subgroup on control chart paper.
3. Compute the overall mean $\tilde{\tilde{X}}$ of all measurements and the average range $\bar{R}$ for the k time periods:

$$\tilde{\tilde{X}} = \frac{\text{sum of all measurements}}{nk}$$

$$\bar{R} = \frac{R_1 + R_2 + \cdots + R_k}{k}$$

These values should be drawn on the control chart with a solid line.

4. Compute the control limits using the values of A_6, D_3, and D_4 that correspond to the subgroup size n.

n	A_6	D_3	D_4	d_2
3	1.19	0	2.58	1.69
5	.69	0	2.12	2.33
7	.51	.08	1.92	2.70
9	.41	.18	1.82	2.97
11	.35	.26	1.74	3.17

3.5 Mean and Range Control Chart

Upper control limit for $\bar{X}$:

$$\text{UCL}_{\bar{X}} = \bar{\bar{X}} + A_6\bar{R}$$

Lower control limit for $\bar{X}$:

$$\text{LCL}_{\bar{X}} = \bar{\bar{X}} - A_6\bar{R}$$

Upper control limit for R:

$$\text{UCL}_R = D_4\bar{R}$$

Lower control limit for R:

$$\text{LCL}_R = D_3\bar{R}$$

5. Using a dashed line, draw $\text{UCL}_{\bar{X}}$, $\text{LCL}_{\bar{X}}$, UCL_R, and LCL_R.
6. Assess whether the R chart and then the $\bar{X}$ chart are stable by determining if a plotted $\bar{X}$ or s value is beyond the control limits computed in step 4. Note that additional criteria for evaluating stability are given in Procedure 3.7. On an ongoing basis, all process changes should be indicated on the chart.
7. If the process is judged stable, compute an estimate of the process standard deviation and process spread. Process standard deviation estimate:

$$\hat{\sigma} = \frac{\bar{R}}{d_2}$$

Process spread = $6\hat{\sigma}$

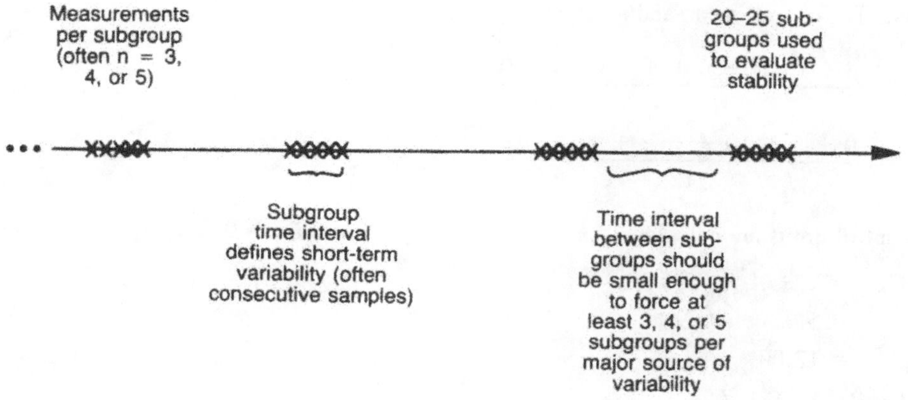

Figure 3.8 Control chart time line.

3.6 Median and Range Control Chart

We have seen from the discussion of process flow diagrams in Chapter 2 that many processes can be complex and thus difficult to control. Development of a DPS requires use of various types of control charts, which can be time consuming for machine operators. A method of reducing the effort involved in an $\bar{X}$ and R chart is to use medians of the subgroup measurements. Typically, each individual measurement is plotted on the control chart when a median chart is used so that it is quite easy to pick the middle (i.e., median) value visually. The following illustrates the ease of the median and range calculation.

```
Subgroup      Control
 values      chart plot        Control chart values

   4            5
  -5            4   XX
   0            3
   4            2
  -4            1
                0    X         Middle (median)       Range
               -1              value                 R = 9
               -2              X̃ = 0
               -3
               -4    X
               -5    X
```

The median chart was originally developed by Ferrell (1953) and modified by Clifford (1959). A number of variations for using medians are available (e.g., Nelson, 1982 or Wadsworth, Stephens and Godfrey, 1986 Chap. 7).

The crankshaft example in Figure 3.6 appears in Figure 3.9 using a median control chart. Following the steps in Procedure 3.4, all medians and ranges were computed and plotted. The overall mean and average range can be computed and plotted:

$$\bar{\bar{X}} = \frac{4 + (-5) + 0 + 4 + (-4) + 0 + \cdots + 10 + 15}{5 \times 12} = 5.03$$

$$\bar{R} = \frac{9 + 5 + 10 + \cdots + 11 + 10}{12} = 10.1$$

The control limits are calculated for $n = 5$ using $A_6 = 0.69$, $D_3 = 0$, and $D_4 = 2.12$.

$$\text{UCL}_{\tilde{X}} = \bar{\bar{X}} + A_6 \bar{R}$$
$$= 5.03 + (0.69 \times 10.1)$$
$$= 12.00$$

$$\text{LCL}_{\tilde{X}} = \bar{\bar{X}} - A_6 \bar{R}$$
$$= 5.03 - (0.69 \times 10.1)$$
$$= -1.94$$

$$\text{UCL}_R = D_4 \bar{R}$$
$$= 2.12 \times 10.1$$
$$= 21.41$$

$$\text{LCL}_R = D_3 \bar{R}$$
$$= 0$$

3.6 Median and Range Control Chart

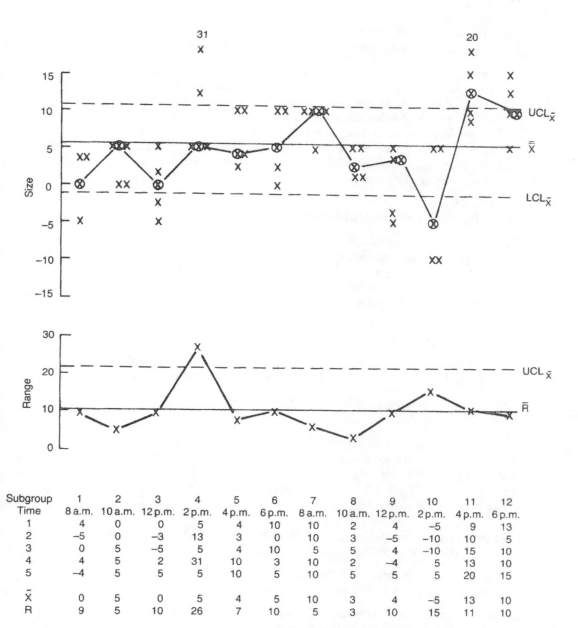

Figure 3.9 Plot of $\tilde{X}$ and R values for an engine crankshaft dimensional characteristic.

These control limits are plotted in Figure 3.9. It must be emphasized that the limits should be used only for the median values (i.e., the circled points). Individual measurements can be expected to fall outside the control limits. A comparison of Figure 3.9 with Figure 3.7 shows that the median chart is similar to the mean chart. An advantage of the median chart is that the individual measurements appear on the chart. However, it is also possible to plot individual values on the $\bar{X}$ chart. Unfortunately, there is a tendency to try to use control limits for the individual measurements. A disadvantage of the median chart is its difficulty in application for an even number of measurements n in a subgroup.

After receiving training on how to use control charts, a machine operator said he would try to chart a characteristic on his machine that had been a problem. However, he stressed he knew the problem was that his machine needed to be overhauled. His charts showed excess variability and a process that was out of control. After showing the chart to a maintenance supervisor, approval was received for overhauling the machine. However, a day prior to the overhaul the variability on the R chart was significantly lower. After inquiry, the operator determined that the central machine coolant fluid had just been replaced by new fluid.

3.7 Percentage Control Chart

The previous chapter discussed two classifications of data—variable and attribute data. The analysis of location and variability in the previous sections was for variable data only. The next two sections give control charts for attribute data. The most common type of attribute data is observing whether a part (or some other sampling unit) does or does not meet some criteria. Often this can be stated in terms of a "good" or a "defective" part:

Good: Part meets stated criteria
Defective: Part does not meet stated criteria

As noted in Chapter 2, it is important to have a good operational definition of the criteria being studied. The percentage P of defective parts of n parts observed is typically studied:

$$P = 100 \times \frac{\text{number of defective parts}}{n}$$

The percentage of defective parts at a given point in time may be of interest, but it is often more important to observe a process over time and monitor the percentage of defective parts. This practice gives an indication of whether a process is improving, stable, or getting worse. Table 3.2 contains the number and percentage of defective snap rings found over a 12 day period.

The percentage of defective parts could be plotted on a run chart, but important questions would remain. Defect prevention thinking suggests that we address two questions:

Is the process that generates defective snap rings stable?
Can the percentage of defects be reduced?

Why do we care about these defects since an operator uses a snap ring to assemble a component: effectively we have 100% inspection!

Table 3.2 Defective Snap Ring Data

Day	7/29	7/30	8/2	8/4	8/5	8/6	8/7	8/9	8/10	8/11	8/12	8/16
Inspected	1400	1400	1400	1400	1400	1400	1400	1400	1400	1400	1400	1400
Defects	35	51	33	17	26	26	22	31	18	17	14	19
%	2.5	3.6	2.4	1.2	1.9	1.9	1.6	2.2	1.3	1.2	1.0	1.4

3.7 Percentage Control Chart

100% inspection is less than 100% effective

It is apparent to anyone who has tried 100% inspection that customers receive defective products at an unacceptably high rate. We see here that a control chart for the percentage of rejects allows evaluation of whether the process generating defects is stable. This approach can lead to a reduction in the defect rate and elimination of inspection, as discussed in the snap ring case study in Section 3.13.2. Using the control chart approach, an increase in the percentage of defects (out-of-control signal on the high side) indicates the process is becoming worse and additional containment actions may be necessary. Also, a decrease in the percentage of defects (out-of-control signal on the low side) is evidence that improvement actions were effective.

It is possible to compute control limits for the percentage of defects using the method in Procedure 3.5. These charts are referred to as P charts. Using the snap ring data from Table 3.2, control limits can be computed as follows:

$$\tilde{P} = 100 \times \frac{35 + 51 + \cdots + 19}{1400 + 1400 + \cdots + 1400}$$

$$= 100 \times \frac{309}{16,800}$$

$$= 1.84$$

$$A_0 = 3\sqrt{\frac{1.84 \times (100 - 1.84)}{1400}}$$

$$= 1.08$$

$$UCL_p = 1.84 + 1.08 = 2.92$$

$$LCL_p = 1.84 - 1.08 = 0.76$$

The plotted data and control limits appear in Figure 3.10. On 7/30 a value above the control limits was observed so the process cannot be judged stable.

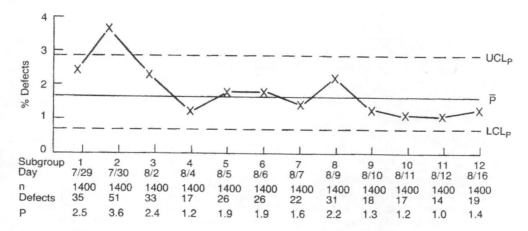

Figure 3.10 Percentage control chart (P chart) for defective snap rings.

Procedure 3.5 Percentage P Control Chart

Decide upon an operational definition of what is to be called defective, and instruct all personnel on the evaluation criteria. Decide upon a minimum sample size n to be evaluated for every time period. The process should be observed for 20–25 time periods k, where the values are arranged as follows:

Time	1	2	...	k
Sample size n	n_1	n_2	...	n_k
Defects	D_1	D_2	...	D_k
%	P_1	P_2	...	P_k

It is convenient to have n constant for each time period. Also, n should be large enough so that five or more rejects per time period can be expected. Typical sample sizes are $n = 50$ to $n = 200$.

The following steps are used to evaluate whether the process is stable.

1. Compute the percentage of defects for each subgroup:

    $$P = 100 \times \frac{\text{number of defects}}{\text{number inspected}}$$

    $$P_1 = 100 \times \frac{D_1}{n}, \quad P_2 = 100 \times \frac{D_2}{n}, \quad \ldots$$

2. Compute the overall average percentage of defects:

    $$\bar{P} = 100 \times \frac{\text{total number of defects}}{\text{total number inspected}}$$

    This value should be drawn on the control chart with a solid line.

3. Compute the control limit factor A_0 in the following sequence.

    a. $\bar{P}(100 - \bar{P})$

    b. $\dfrac{\bar{P}(100 - \bar{P})}{n}$

    c. $\sqrt{\dfrac{\bar{P}(100 - \bar{P})}{n}}$

    d. $A_0 = 3 \times$ value obtained in step c.

    In summary,

    $$A_0 = 3\sqrt{\frac{\bar{P}(100 - \bar{P})}{n}}$$

4. Compute the control limits using the control limit factor A_0.

    Upper control limit for P:

    $$\text{UCL}_P = \bar{P} + A_0$$

    Lower control limit for P:

    $$\text{LCL}_P = \bar{P} - A_0$$

    Note that if A_0 is larger than $\bar{P}$, set $\text{LCL}_P = 0$.

5. Using a dashed line, draw UCL_P and LCL_P.
6. Assess whether the P chart is stable by determining if a plotted P value is beyond the control limits computed in step 5. Note that additional criteria for evaluating stability are given in Procedure 3.7. On an ongoing basis, all process changes should be indicated on the chart.

3.8 Count Control Chart

In practice it is sometimes difficult that the number of units inspected be the same for all time periods. There are two approaches typically used to deal with varying n:

1. Average subgroup size. Compute the average subgroup size for the k time periods

$$\bar{n} = \frac{n_1 + n_2 + \cdots + n_k}{k}$$

where n_1 is the subgroup size for time 1, n_2 the size at time 2, and so on. If all n are between $0.75\bar{n}$ and $1.25\bar{n}$, the value of $\bar{n}$ can be used in computing A_0 in Procedure 3.5.

2. Individual control limits. A separate value of A_0 can be computed for each subgroup so that

$$A_0 = 3\sqrt{\frac{P(100-P)}{n_i}}$$

and a different dashed line is drawn for each subgroup. In practice, this method is difficult to use and to explain to others.

The best practice is to keep the subgroup sizes about the same so that the first method can be used.

A major application of P charts is in management systems in which various types of rejects are tracked on a routine basis. The control limits for the P chart allow a realistic assessment of whether a process is stable, improving, or deteriorating. Overreaction to changes in various percentage indicators is not uncommon.

As part of a management information system, the percentage of units failing a testing procedure was routinely discussed at weekly management meetings. Corrective action plans were required if any percentage was above the prior week's level.

3.8 Count Control Chart*

A second type of attribute data that arises in both manufacturing and administrative areas is count data in which the counts typically represent the number of defects on a unit. Several examples follow:

Manufacturing: number of defects per transmission;
number of flaws on a bolt;
number of scratches or dents, for example, on a car
Administrative: number of keypunch errors per hour;
number of recording errors on a form;
number of accounting errors in a department

As with the application of P charts, we are interested in determining whether the process generating the defects is stable. Ultimately, the objective is to reduce the number of defects.

Consider the 12 time periods listed in Table 3.3, which represent the number of dents, scratches, or other defects on a car body.

Table 3.3 Number of Defects on the Exterior of a Car Body

Time	1	2	3	4	5	6	7	8	9	10	11	12
Defects	8	5	6	5	9	6	11	6	3	4	3	3

Procedure 3.6 Count C Control Chart

Decide upon an operational definition of what is to be called defective, and instruct all personnel on the evaluation criteria. The size of the unit area being evaluated must be the same for all time periods. The process should be observed for 20–25 time periods k, where the values are arranged as follows:

Time	1	2	$\cdots$	k
Number of defects per unit	C_1	C_2	$\cdots$	C_k

The C values are the count of the number of defects observed on a sampling unit. The following steps are used to evaluate whether the process is stable.

1. Compute the average number of defects:

$$\bar{C} = \frac{C_1 + C_2 + \cdots + C_k}{k}$$

   This value should be drawn on the control chart with a solid line.

2. Compute the control limit factor A_0 in the following sequence:

   a. $\sqrt{\bar{C}}$

   b. $A_0 = 3 \times$ value obtained in step a

   In summary,

   $A_0 = 3\sqrt{\bar{C}}$

3. Compute the control limits using the control limit factor A_0.

   Upper control limit for C:

   $$\text{UCL}_c = \bar{C} + A_0$$

   Lower control limit for C:

   $$\text{LCL}_c = \bar{C} - A_0$$

   Note that if A_0 is larger than $\bar{C}$, set $\text{LCL}_c = 0$.

4. Using a dashed line, draw UCL_c and LCL_c.
5. Assess whether the C chart is stable by determining if a plotted C value is beyond the control limits computed in step 3. Note that additional criteria for evaluating stability are given in Procedure 3.7. On an ongoing basis, all process changes should be indicated on the chart.

3.9 Special-Cause and Common-Cause Variation

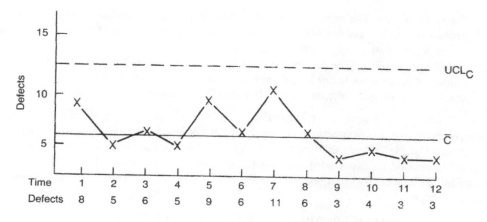

Figure 3.11 Count control chart (C chart) for exterior car body defects.

The method for computing control limits is given in Procedure 3.6. Using the data in Table 3.3, the control limit computations are as follows:

$$\bar{C} = \frac{8 + 5 + \cdots + 3}{12}$$

$$= \frac{69}{12}$$

$$= 5.75$$

$$A_0 = 3\sqrt{5.75}$$

$$= 7.19$$

$$\text{UCL}_c = 5.75 + 7.19$$

$$= 12.94$$

$$\text{LCL}_c = 5.75 - 7.19$$

$$= -1.44 \quad \text{set LCL}_c = 0$$

The control chart for these data is shown in Figure 3.11. The process appears to be stable. However, the control limits are based on only 12 subgroups rather than the 25 or more that should be used in real applications.

Note that if the sampling unit varies on which the number of defects is counted, a C chart is not appropriate. A U chart (Grant and Levenworth, 1980, p. 262) should be used if the size of the sampling unit is not constant over time.

3.9 Special-Cause and Common-Cause Variation

The process by which work is accomplished can be quite complex, as seen in Chapter 2, where the process flow diagram identified some major sources of variability. Two alternatives are available for sources of variability to influence the output of a process:

1. **Stable variability:** The process sources of variability are controlled and influence the process output in a predictable, consistent manner. The process (or system) is said to be stable, exhibiting only controlled variation.
2. **Unstable variability:** The process sources of variability are not controlled, and the output of the process is not completely predictable. However, the process may, for short periods of time, exhibit seemingly controlled variation.

Establishing a DPS obviously requires that a process exhibit only stable variability. This term seems, at first, a contradiction in terms, but both words have an explicit interpretation. Recall that "stable" means that the long-term variability does not exceed the short-term variability beyond an expected magnitude (as quantified on a control chart). The term "variability" emphasizes that process output will vary: a stable process does not imply constant output.

Defect prevention requires stable variability

The stable variability exhibited on a control chart is called common-cause variation. This term simply means that the common, expected system sources of variability are responsible for the variation exhibited by the process output. Conversely, special causes of variation are responsible for variation beyond common-cause variation. These special causes of variation (also called assignable causes) result in out-of-control points on a control chart. Special causes of variation may result in high or low out-of-control points on either a location or variability chart. Figure 3.12 shows examples of each case. Also, we

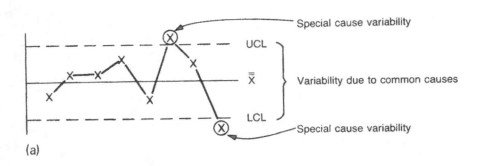

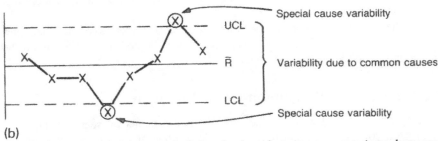

Figure 3.12 Special-cause (circled points) and common-cause (area between dashed lines) variability; (a) $\bar{X}$ chart; (b) R chart.

see in the next section that it is possible to obtain a process out-of-control signal from certain patterns of points within the control limits. These out-of-control signals are similarly due to special causes.

It is necessary to eliminate the special causes of variation to obtain a stable process. If these causes of excessive process variation are not eliminated, they continue to make the process output unstable. It is sometimes difficult to determine the sources for the special-cause variability, but elimination of these problems is necessary to obtain a stable process. A common mistake is to conduct an intensive training and implementation program to use control charts and then provide no follow-up system to address special causes of variation. Also, as discussed in Chapter 14, some control charts can be eliminated when the special causes of variation have been removed and a stable, capable process attained.

<blockquote>Defect prevention requires ongoing efforts
to eliminate special causes of variation</blockquote>

A first-line supervisor was pleased with his accomplishment in maintaining a large number of control charts. Upon looking at one chart, an observer asked the supervisor about several out-of-control points on the chart. The supervisor replied that those values were unusual and that the chart always seemed to return to the normal range so that he believed he was controlling his process. The observer remarked, "Control charts don't *control* anything!"

3.10 Out-of-Control Signals

In the previous sections we saw how a single point beyond control limits indicated a special cause of variation was acting on the process, causing a lack of process stability. This out-of-control signal is very useful in a defect prevention system since process instability can result in defective output. There are other indications of special causes based on runs of points or the distribution of points in control chart zones. It is advisable to use these additional indicators since they collectively provide an early-warning system by indicating process instability before a point beyond the control limits is obtained.

A run of points on a control chart is simply a group of consecutive points on a control chart. For example,

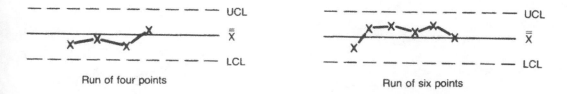

Run of four points   Run of six points

Consecutive points on a control chart correspond to consecutive points in time. Thus, runs provide information on how the process is changing over time. Since the control limits (LCL and UCL) and $\bar{\bar{X}}$ were calculated from previous data obtained from the process, runs allow comparison of the current process output to a historical benchmark. A stable process having only common-cause sources of variation varies randomly and typically does not exhibit runs having certain patterns. Three types of unusual patterns indicating a special-cause problem are discussed here:

1. Runs above or below the centerline
2. Runs up or down
3. Distribution of points in control chart zones

A run above or below the centerline is simply a series of consecutive points all either above or below $\bar{\bar{X}}$; a point exactly on the centerline is ignored in counting the length of the run:

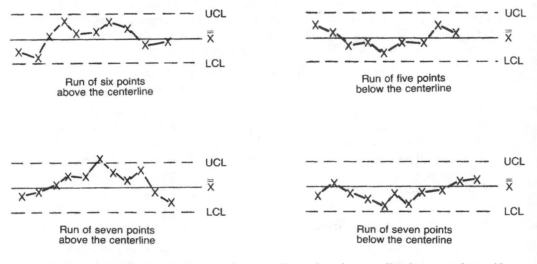

Another type of run pattern is a group of consecutive points that are all going up or down; if two consecutive points are the same, only one point is used to count the length of the run:

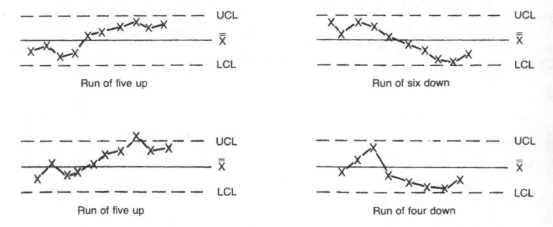

Procedure 3.7 indicates which run patterns are unusual. A run of eight or more consecutive points on one side of the centerline or a run of six or more consecutive points up or down is considered unusual. The justification of these rules is simple. Consider having eight consecutive points on one side of the centerline. Since one-half of the points should be above the centerline, the chance that eight consecutive points would be above the centerline is

3.10 Out-of-Control Signals

$$\tfrac{1}{2} \times \tfrac{1}{2} \times \tfrac{1}{2} \times \tfrac{1}{2} \times \tfrac{1}{2} \times \tfrac{1}{2} \times \tfrac{1}{2} \times \tfrac{1}{2} = \frac{1}{256} = 0.0039$$

This is the same as getting eight heads in a row on a coin toss! The chance of eight consecutive points below the centerline is likewise .0039. Thus, the chance of a run of eight above *or* below the centerline is $0.0039 + 0.0039 = 0.0078$, assuming no process change has occurred. Clearly an unusual event, possibly due to a special cause, has occurred. A similar argument applies to runs up or down.

Another type of control chart pattern involves zones between UCL and LCL. There are six equal zones on an $\bar{\bar{X}}$ chart, which appear as follows:

```
— — — — — — — — — — — UCL
       . . . Zone A . . .
       . . . Zone B . . .
       . . . Zone C . . .
_____   ̿X
       . . . Zone C . . .
       . . . Zone B . . .
       . . . Zone A . . .
— — — — — — — — — — — LCL
```

After studying several stable control charts, it is apparent that there are few points in zone A and most points are in zone C. The expected percentage of points in either the upper or lower zones is

Zone	Expected % of points
A	2
B	14
C	34

In Procedure 3.7 we see that too many or too few points in any of the zones indicates a special cause of variation is probably acting on the process.

It should be noted many criteria are used to indicate an out-of-control condition. Nelson (1984, 1985) and Grant and Levenworth (1980, Chap. 3) list a number of additional criteria. Procedure 3.7 includes common indicators that are easy to apply and provide a reasonable warning for the existence of special causes. However, the user must always realize that mechanically searching for out-of-control signals is not the main intent of implementing control charts. A control chart is an ongoing record of the past and present process performance. The insight gained from this information, as well as the state of stability, must be assessed in the context of the overall process. Only then can the process be understood and future performance controlled. This is one of the main objectives in using control charts.

When using control charts to monitor a process, the question frequently arises of when control limits should be recalculated. Technically, there is never a need to recalculate control limits of a stable process that exhibits only common-cause variation. However,

Procedure 3.7 Out-of-Control Signals

Consider a control chart with control limits calculated from the process while it was stable exhibiting only common-cause variability. The out-of-control signals listed are indicators that a special cause of variation is acting on the process. Only certain signals apply to each type of control chart.

Chart type	Signals
Mean $\bar{X}$, median $\tilde{X}$	1–5
Range R, standard deviation s	1–5 (if $n \geq 4$)
Percentage P, count C	1, 3, 4

Note that only signals 1-3 above the centerline should be used for the R or s chart if the subgroup size n is 2 or 3. The "center" value in the figures refers to $\bar{X}$, $\tilde{X}$, $\bar{R}$, $\bar{s}$, $\bar{P}$, or $\bar{C}$.

Signal 1   One or more points above UCL or below LCL:

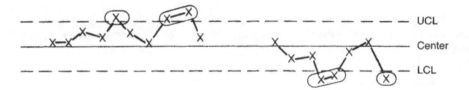

Signal 2   Run of two of three consecutive points in the same zone A or beyond; the third point may be anywhere:

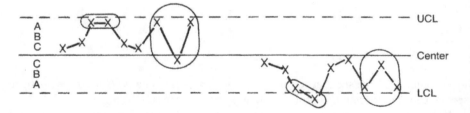

Signal 3   Run of eight or more points above or below centerline:

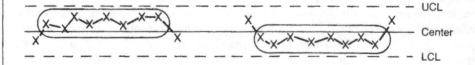

Signal 4   Run of six or more points up or down:

3.10 Out-of-Control Signals

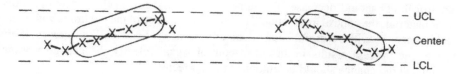

Signal 5 Too many or too few points in zones A, B, or C. Note that the percentage of points in each zone should be calculated over at least 25 subgroups.

```
- - - - - - - - - UCL
   Zone A    2%
   .................
   Zone B    14%
   .................
   Zone C    34%
                                Center
   Zone C    34%
   .................
   Zone B    14%
   .................
   Zone A    2%
- - - - - - - - - LCL
```

Conventions for out-of-control signals:

1. Out-of-control points are circled on the control chart.
2. For signal 1, a point on the UCL or LCL line is not considered out of control.
3. For signal 3, a point on the centerline is not counted in the run.
4. For signal 4, two or more equal consecutive points are counted as one point in a run.
5. For signals 2 and 5, a point on a zone line is considered in the zone closer to the centerline.

Note that out-of-control signals may be due to a lack of measurement sensitivity. Wheeler and Chambers (1986) suggest a criterion that sensitivity is inadequate if four or fewer values are possible within the range control limits.

practically it is a good practice to recalculate control limits on a periodic basis. Typical periods in industry are every 8 to 16 weeks. This interval keeps personnel involved in the charting process and familiar with the procedures. Also, when limits are recalculated, the width of the control limits should never be increased. The width of the control limits for an $\bar{X}$ chart is

$$\text{Width of control limits} = \text{UCL} - \text{LCL}$$
$$= (\bar{\bar{X}} + A_2\bar{R}) - (\bar{\bar{X}} - A_2\bar{R})$$
$$= 2A_2\bar{R}$$

Thus, an increased width of the control limit means that the $\bar{R}$ has increased, which implies the process has increased variability. Occasionally, the width of the control limits increases and the cause is not known. The old limits should be retained, and an effort should be made to identify the special cause that introduced increased variability in the process. In these cases, it is often useful to increase the sampling from the process to identify the special causes (see Chap. 4 on troubleshooting).

Do not increase the width of control limits

Control limits should be recalculated when major process changes have occurred, such as machine overhauls or new procedures. Since the process is being improved, the width of the control limits should decrease. For minor process changes, such as new operators, tools, or material, the control limits should not be changed. Use of the old limits with minor process changes provides a useful comparison of the "new" and "old" processes. If any change to the process results in an out-of-control signal, clearly the change was not beneficial and the change should be eliminated or corrected.

When a study is performed and control limits are calculated, the question arises of what to do with an out-of-control subgroup. Including an out-of-control subgroup that is beyond the control limits increases the width of the newly recalculated control limits. This is undesirable since wider control limits make it more difficult to identify future special causes of variation. An accepted practice is to omit out-of-control subgroups from control limit calculations if the special cause of variation that resulted in the out-of-control signal is identified and eliminated. The logic is simply that a special cause that has been eliminated will likely not be present in the future and will not be part of the common-cause variation in the process.

> Omit only those subgroups corresponding to known special causes of variation

3.11 Control Limits Versus Specification Limits

When some individuals are introduced to control limits, there is a tendency to confuse control limits with specification limits. This confusion stems from the historical practice of trying to run processes so the output was within the specification limits. Further confusion arises in an attempt to relate control limits to specification limits in some way. Stated simply, there is no relationship between the two sets of limits. A simple argument can be used to see why this is true. Control limits are derived from the values of $\bar{\bar{X}}$ and $\bar{R}$, which are obtained from the process output. Thus, the process performance determines the position of the control limits (LCL and UCL). Conversely, the specification limits are obtained by human judgment based on past practices and final product requirements. Specification limits are easily changed with no change to the manufacturing process!

> Control limits are not in any way related to specification limits

Determining the stability of a process has nothing to do with engineering specification limits. Stability depends only on the control chart limits, which are determined from the process output regardless of the specification limits. This point is largely misunderstood since specification limits have traditionally been used to evaluate processes. The importance of specification limits is not for evaluating process stability but for assessing process capability. A stable process may or may not produce most parts within the specification limits. A process is said to be "capable" if it has two features:

1. The process is stable (thus predictable).
2. Essentially all parts are within specification limits.

A complete discussion of capability is delayed until Chapter 7. The important point here is there are two important process characteristics that utilize different limits.

Stability → control limits
Capability → specification limits   stability required

> Process stability does not depend on specification limits

3.12 Reacting to Control Charts

A bad practice that stems from the confusion between specification and control limits is to draw specification limits on $\bar{X}$ charts. Aside from leading to confusion, this practice does not make technical sense. We see in Chapter 5 that the subgroup means on the $\bar{X}$ chart show less variability than the individual measurements. Thus, an attempt to visually relate the subgroup $\bar{X}$ values to the specification limits is not meaningful. However, the process spread of individual measurements can be related to the width of the control limits by

Process spread = $\sqrt{n}$ (UCL − LCL)

This relationship for varying n becomes

Subgroup size (n)	Process spread of individual measurements (UCL − LCL)
3	1.7
4	2.0
5	2.2
6	2.4
7	2.6
8	2.8

Thus, for a subgroup size of $n = 4$ we simply double the width of the control limits to get the spread of the individual measurements.

3.12 Reacting to Control Charts

Many employees have received training on the techniques described in this chapter, often referred to as statistical process control (SPC) methods. Yet in many cases the overall improvement in final product quality has been slower than most would have liked. Explanations for the lack of progress commonly cite two causes:

1. The usefulness of SPC methods was largely oversold.
2. Employees who were responsible for preparing the charts did not do their job.

These explanations are really consequences of the real causes:

There was a lack of management follow-up on implementation of SPC.
Employees do not connect use of SPC with improved quality or believe that even when problems are identified, nothing will be done to correct problems.
There was a lack of a comprehensive system to address controlling the whole process.

The first two causes can be addressed by building an appropriate reaction system discussed here for control charts. The third cause is addressed by attempting to build a comprehensive DPS (Chap. 14).

Control charts provide a basis for action. It is the collective responsibility of employees and management to participate in a process that corrects problems generating special causes of variation. Unfortunately, an experience that is all too common is to review a control chart on the manufacturing floor and see an out-of-control situation in which no action has been taken. Ideally, the control chart review process should funnel down from employees to top management.

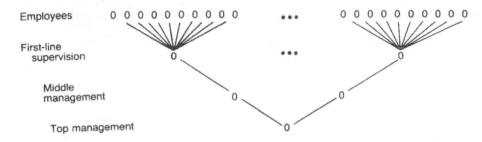

The responsibility of each group is given in Table 3.4. Notice that work must involve eliminating special-cause problems that cause out-of-control signals. The elimination of these problems ultimately results in a stable process. However, the effort must also involve reducing common-cause variability. The goal here is to reduce the width of the control limits by eliminating common-cause sources of variability. This practice results in minimizing process variability.

A plant manager held two 1 hour meetings per week to review SPC charts. At these meetings production employees and first-line supervisors reviewed their progress on

Table 3.4 Responsibilities for Reacting to Control Charts and Eliminating Causes of Variation

Level	Responsibility	% of Variation addressed	
		Special cause	Common cause
Employees	Collect data and maintain charts; mark all process changes on the chart; attempts should be made to identify causes of out-of-control signals; suggestions for improvement should be made	15	0
First-line supervisor	Train employees on chart use; review charts each day; identify special-cause problems; request special assistance to identify and eliminate special causes	50	15
Middle management	Selectively review charts; organize problem-solving teams to address problems; work on common-cause problems by changing the system	35	35
Top management	Selectively review charts; review team problem-solving efforts; change the system	0	50

3.13 Case Studies

reducing variability in their processes. The entire management team participated in these meetings. Often employees would seek help in making changes that required management's assistance—managers volunteered to investigate needed system changes. The quality performance of the plant improved steadily.

Control charts do not fix problems—people do

Unfortunately, after receiving initial SPC training some employees believe that out-of-control signals are in some way bad and indicate poor performance on their part. Using the follow-up system described in Table 3.4, the presence of a special-cause problem becomes a useful indicator that can result in process improvement.

Special causes indicate opportunities for improvement

For improvements to occur, it is then critical that action take place when an out-of-control signal is detected. Employees and management should seek to identify "what changed." A good practice is to indicate all process changes on the chart. It then becomes much easier to identify causes of problems.

Mark all process changes on control charts

Eliminating special causes and attaining process stability is not easy.

Stability, or the existence of a system, is seldom a natural state. It is an achievement, the result of eliminating special causes, one by one on statistical signals, leaving only the random variation of a stable process. (Deming, 1982, p. 119)

A significant effort is often required to make improvements. Typically, the most difficult task is convincing people to change past practices that have worked "well enough." Establishing a mindset in which current practices are questioned, change for improvement sake is desired, and defects are not tolerated is the essence of the continual improvement process that is the cornerstone of defect prevention.

3.13 Case Studies

3.13.1 Engine Crankshaft

The reject rate for an engine crankshaft was unacceptably high for a dimensional characteristic, so a team was formed to address the problem. The team consisted of process engineers and maintenance and production personnel. The team decided to use an $\bar{X}$ and R chart to analyze the process. Only one of three machines responsible for machining the characteristic was studied since all machines were thought to have similar problems. Five samples were collected over a shift for 25 shifts. The data appear in Chart 3.1. The process exhibited an out-of-control range and $\bar{X}$ chart that confirmed the high reject rate.

The out-of-control points did not correspond to changes in material, tools, or any other apparent cause. Since the range chart showed excess variability and out-of-control points, the team decided to examine the machining process for the causes of variability. They found that the part-locating system prior to machining was causing excess variability. (Inconsistent locating and clamping of parts is often a cause of machining variability.) The existing process had an initial "soft" prepositioning of the part in a machining fixture prior to a "hard" clamping for machining. The variability in hydraulic pressure used in the clamping frequently caused the hard clamp to engage before the soft positioning was

76　　　　　　　　　　　　　　　　　　　　　　　　　　　　3　Control Charts

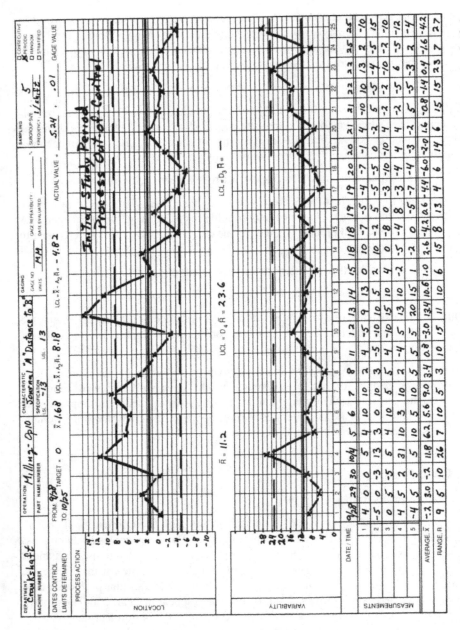

Chart 3.1　Control chart for variability data: data collected to study the engine crankshaft problem in Section 3.13.1.

3.13 Case Studies

completed. Additional investigation indicated that each of the three machines had different controls to activate the clamping sequence.

Once the problem was identified, the solution was easily implemented. A limit switch was installed to ensure that soft clamping was completed prior to hard clamping. The three machines were reworked to make the control systems identical. The same sampling plan was used to verify that the true cause of the problem had been identified. The control chart in Chart 3.2 was dropped after about a month's evaluation period since the process was stable and capable and the dimensional characteristic was related to another characteristic that was being monitored. The reject rate was reduced by a factor of 4.

A better approach to studying this problem would have been to use consecutive piece subgroups over a shorter period of time. A sampling plan of $n = 3, 4,$ or 5 could be used collecting a subgroup every 2 hours. After 7–10 days the lack of control would have been obvious. Also, the behavior of consecutive pieces would have varied appreciably since the hydraulic pressure changes were erratic. Thus, the problem would have been identified more quickly. Also, the same charting procedure should have been used on all three machines. This would probably have indicated machine differences in the clamping controls, further assisting in troubleshooting.

3.13.2 Clutch Snap Ring

As part of a clutch assembly operation, a snap ring was placed in a groove of a clutch body. An assembler was rejecting about 1–8% of clutches during assembly because the snap ring was not seated properly in the groove. The employee team decided that the problem should be addressed. Their first step was to institute a P chart at the assembly operation. Chart 3.3 shows that the early process was not stable and the percentage of rings not seating was unacceptably high. The team invited both the machining department for the clutch body and the snap ring supplier to their meetings. Both groups initiated $\bar{X}$ and R control charts on ring groove location and snap ring flatness. The combined group of charts showed that the snap ring's not seating was caused by a machining problem with the groove location as well as snap ring flatness. Toward the end of Chart 3.3 the reject rate was reduced to below 1%. The group continued to work on improvements, and after 9 months the reject rate was less than 0.1%.

3.13.3 Customer Service Time

In all service industries, an important part of customer satisfaction is the amount of time a customer waits for service. This and the next case study are from completely different services, but both involve a service waiting time. It is important to realize the quality of the service produced is usually more important than the service time. We often hear, "I would rather wait a little longer and get what I want." However, the stability of the service system is important. A system that has unstable waiting times with many out-of-control points undoubtedly produces dissatisfied customers.

Report Requests A department provided a service that involved obtaining reports and various types of other information for "customers" from other departments within a company. A team analyzing this system prepared the process flow diagram in Figure 3.13. Unfortunately, the direct sequence on the process flow diagram rarely occurs in any practical situation. Figure 3.13 is the way the manager thinks the process operates. Many questions are left unanswered:

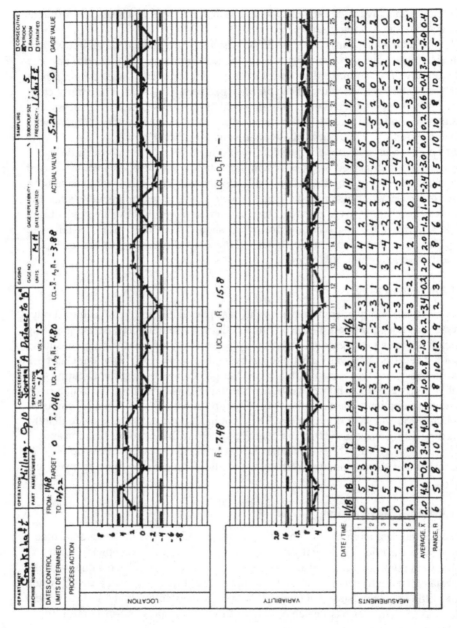

Chart 3.2 Control chart for variable data: the control chart was dropped after a month's evaluation period because the process was stable and capable. See Section 3.13.1.

3.13 Case Studies

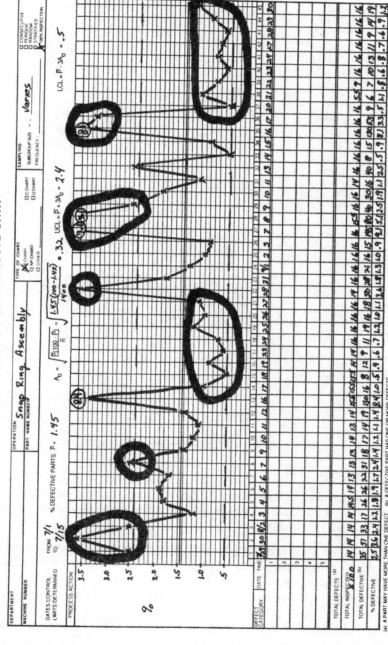

Chart 3.3 Control chart for attribute data: the P chart was instituted at the assembly operation discussed in Section 3.13.2.

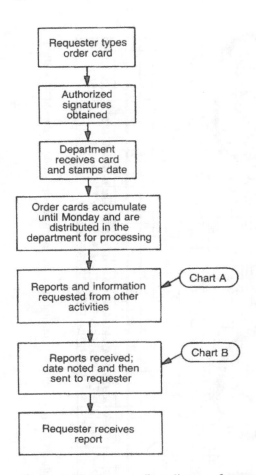

Figure 3.13 Process flow diagram for report request processing.

What happens when an order form is incomplete, unsigned, or not readable, for example? What happens when a report title is incorrect or the author's name is missing or incorrect? What if reports cannot be found?

In most administrative systems it is not the usual flow-through process that consumes most employees' efforts. The Pareto principle (Chap. 9) applies: 80% of the work effort comes from 20% of the requests. Clearly, the process flow diagram in Figure 3.13 does not address potential problems. Unfortunately, many managers view their systems in a flow-through manner.

However, even the poorly prepared diagram in Figure 3.13 can be useful. For example, why are the request cards saved and distributed only on Monday? Clearly this practice is not for the department's customers. Figure 3.13 can also be used to suggest various useful types of data:

How many order cards have errors, and which errors occur most frequently?
How much time do employees spend correcting errors via phone calls or returned forms, for example?

3.13 Case Studies

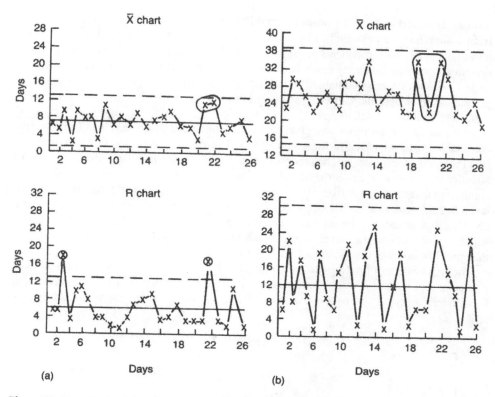

Figure 3.14 Report request order processing control charts: (a) days to order report; (b) days to receive report.

How many incorrect reports are received, and what was the cause?

Check sheets and Pareto analysis, discussed in later chapters, can be used to obtain and analyze answers to these questions.

Data were collected on the time required for a report to be requested by the service department (Figure 3.14a) and the time required for the ordering department to receive a requested report (Figure 3.14b). Note that from the customer's viewpoint only the time from completing the order card to receipt of the report is of interest. The sampling plan consisted of randomly selecting $n = 3$ order cards per day and noting the days required for processing. The charts in Figure 3.14 show that both the $\bar{X}$ and R charts for the days required to order a report are out of control. For two subgroups, over 3 work weeks were required to place an individual order. It is not difficult to believe that errors play a major role in this process. Although the average number of days required to order is stable, the mean is about 7 days, which seems quite high from the customer's viewpoint. In Figure 3.14b there are no points beyond the control limits so on the surface the process may appear stable. However, the up-and-down pattern on the range chart indicates that some orders are received quickly but others require up to 1 month for processing. The $\bar{X}$ chart also exhibits a lack of stability. The cause of the variability should be identified so that the system can be improved.

Railcar Transit Times and Computerized Systems The transit time of a railroad car from a supplier to an assembly plant is a service time for the railrod. This time is very important since all components need to be sent for a complete assembly. An unstable system could cause assembly plant downtime if a component is not available. In the past, large inventories were maintained to protect against any delivery delays and upstream production interruptions. Current just-in-time inventory control systems demand a stable system of component deliveries.

Railroad car transit times were reported by a computerized system. The time for five consecutive trips was used to form a subgroup. The control chart appears in Figure 3.15 (see Prob. 3.12 for the data). The process does not have a stable mean trip time. This lack of stability will probably cause the assembly plant to maintain more inventory. Also, it is significant that the mean delivery time was $\bar{X} = 82$ hours for an earlier period (not shown), so an overall deterioration has occurred. Currently, there appears to be a shift around trip 25. This type of data can be used as a basis for improvement by involving assembly plant, supplier, and railroad personnel. A cause-and-effect diagram (Chap. 13) for some factors that could be addressed appears in Figure 3.16.

The trip times in this example were collected by an automated system that generated computer reports for the over 750 routes to various assembly plants. The computer output for a month's period was over a foot thick! Personnel assigned to "monitor" the system

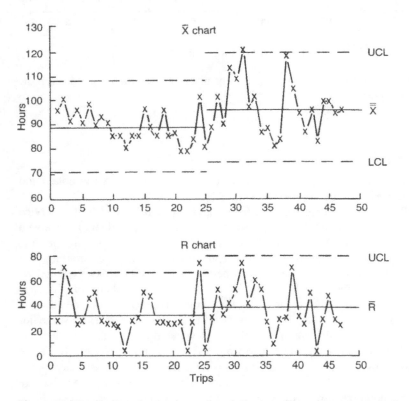

Figure 3.15 Railroad trip times for delivery of supplier components to an assembly plant.

Problems

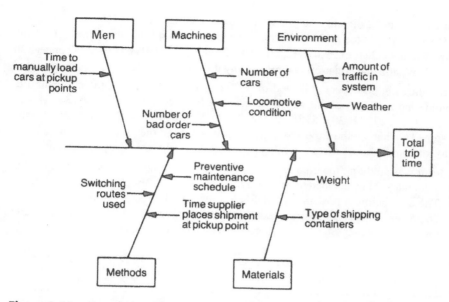

Figure 3.16 Cause-and-effect diagram for factors influencing rail transit.

had no clear criteria to define good and bad performance. The system was modified to use control charts and print only those cases in which out-of-control signals indicated a deteriorating performance. There were several benefits. Only about 20 routes in any one month needed to be addressed so the work load decreased drastically (as did the volume of computer output). Employees had an operational definition of unacceptable performance (i.e., out-of-control signals). It was possible to correct problems before they affected assembly schedules (defect prevention) since out-of-control signals indicated any system change. Meetings between the assembly plant, supplier, and railroad companies could focus on data (i.e., the out-of-control charts) and seek to identify and eliminate special causes. Emotional discussions were thus minimized.

Problems

3.1 For the following lists of numbers, calculate the mean $\bar{X}$, median $\tilde{X}$, standard deviation s, and range R (use a calculator if needed): (a) 7, 2, 4; (b) 1.2, 9.7, 3.4, 1.6; (c) 7.1, 10.2, 4.1, 5.2, 7.7; (d) 3, −1, 5; (e) 0, 4, −7, −3, −10; (f) 4.61, 7.24, −1.12; (g) 9, 7, 6, 10, 11; (h) −3.1, 6.2, 0.

3.2 Suppose a furnace thermocouple reads in deviations from 450°F. (a) What are the actual temperatures for gage readings of −25, 3, 102, −17, and −0.5? (b) How would the gage read temperatures of 420, 502, 463, 413.4, and 473?

3.3 A gage reading in thousandths of an inch (.001 inch) produces readings of 2.4, 17.8, 1.5, and 0.7. What are the actual dimensional values?

3.4 The midpoint of a dimensional specification (nominal) is 29.42 inches, with a specification width of ±.001. A gage at the operation is calibrated to read in ten-thousandths of an inch (1.0 on the gage equals .0001 inch), with 0 equal to 29.42 inches.

(a) What are the actual dimensional values for gage readings of 1.0, 15.5, 0, −3.5, and −16.0? (b) What would the gage read at the upper and lower specification limits? (c) How would dimensional values of 29.42062, 29.4191, 29.423, 29.419, and 29.42057 appear on the gage? (d) What is the equation that relates any dimensional value to the gage reading?

3.5 The following data were collected over 20 consecutive hours on the compatibility (%) of mold sand used for pouring engine blocks in a foundry: 39, 40, 37, 36, 34, 34, 43, 43, 41, 38, 38, 36, 27, 42, 28, 35, 38, 40, 28, 43. (a) Compute $\bar{X}$, $\bar{\bar{X}}$, s, and R. What is the process spread? What are the expected high and low values of the process? What do these calculations assume about process stability? (b) Plot a run chart. Is the process stable?

3.6 A team was studying the compatibility (%) of mold sand used for pouring engine blocks in a foundry. They decided to collect five samples randomly per hour to monitor the process. The data below represent a period before and after several improvements were made to the process. It was considered desirable to have values below 38%. Prepare $\bar{X}$ and R charts for the before and after periods separately, and compare the charts.

		Before						After
Day	Hour	Data	Day	Hour	Data	Day	Hour	Data
1	1	39, 39, 40, 40, 38	5	1	40, 41, 41, 40, 43	1	1	35, 40, 36, 36, 42
	2	40, 28, 31, 32, 37		2	41, 41, 41, 40, 38		2	35, 35, 32, 36, 45
	3	37, 38, 35, 32, 41		3	42, 43, 41, 35, 36		3	32, 32, 32, 30, 32
	4	36, 37, 35, 37, 33		4	40, 36, 36, 33, 37		4	33, 30, 28, 38, 39
	5	34, 33, 35, 30, 27		5	32, 31, 42, 41, 41		5	40, 35, 35, 46, 43
	6	34, 29, 36, 39, 38		6	37, 41, 43, 41, 41		6	38, 36, 39, 35, 37
	7	43, 42, 37, 39, 33		7	36, 36, 34, 35, 36		7	35, 39, 30, 39, 31
	8	43, 44, 43, 41, 43		8	41, 34, 31, 32, 28		8	38, 35, 38, 37, 37
2	1	41, 38, 38, 39, 38	6	1	41, 29, 38, 39, 38	2	1	42, 40, 37, 41, 35
	2	38, 39, 38, 40, 40		2	43, 36, 44, 40, 43		2	41, 36, 40, 36, 40
	3	38, 41, 36, 33, 36		3	46, 34, 29, 38, 38		3	36, 34, 30, 36, 37
	4	36, 37, 34, 37, 34		4	45, 40, 43, 43, 40		4	42, 42, 29, 35, 33
	5	27, 34, 36, 34, 42		5	34, 36, 44, 42, 40		5	37, 36, 40, 34, 31
	6	42, 37, 36, 32, 32		6	31, 40, 44, 44, 37		6	39, 36, 35, 36, 30
	7	28, 35, 32, 34, 35		7	41, 42, 44, 42, 44		7	31, 34, 33, 39, 44
	8	35, 31, 32, 32, 35		8	35, 30, 40, 41, 42		8	38, 38, 40, 40, 40
3	5	38, 36, 38, 33, 25	7	5	38, 41, 40, 38, 38	3	1	35, 34, 39, 41, 39
	6	40, 40, 39, 38, 37		6	43, 44, 42, 36, 42		2	34, 30, 37, 35, 37
	7	28, 32, 32, 33, 33		7	45, 46, 46, 46, 42		3	37, 44, 39, 41, 37
	8	43, 34, 34, 40, 37		8	48, 44, 42, 42, 41		4	31, 37, 39, 40, 36
4	1	38, 37, 35, 38, 29					5	32, 36, 35, 30, 35
	2	34, 27, 30, 35, 31					6	34, 37, 39, 40, 36
	3	34, 41, 40, 37, 31					7	30, 36, 40, 38, 39
	4	27, 29, 25, 26, 26					8	38, 41, 34, 29, 34
	5	34, 35, 33, 31, 34						
	6	45, 26, 26, 45, 36						
	7	31, 32, 32, 31, 31						
	8	42, 37, 35, 31, 35						

Problems

3.7 A bearing diameter was machined on a lathe that had two identical stations on the right and left sides of the machine. A monitoring sampling plan was being used where $n = 4$ parts were measured each shift. The data appearing below are as coded so that $1 = .001$ inch. (a) Prepare an $\bar{X}$ and R chart for both sides. Is the process stable? (b) Calculate $\hat{\sigma}$ and the process spread for the left side. How useful is the estimate of variability? (c) The specification is ± 4. Is this an acceptable process? (d) Is the gaging acceptable?

No.	Left	Right	No.	Left	Right
1	0, 1, 2, 2	−2, −2, −1, −1	14	0, 0, −2, −2	1, 1, −2, −2
2	−2, 2, −1, 2	2, 1, 0, 1	15	0, −1, 1, 0	1, 0, 2, 1
3	−3, −2, 1, 1	−3, −2, 1, 1	16	0, 0, −1, −1	2, 1, −1, −1
4	−3, −2, 1, 1	−3, −1, 0, 0	17	0, 1, 0, 1	0, 0, 2, 1
5	−1, −2, −2, 2	1, −2, 1, 1	18	−1, −1, −1, −1	1, −1, −1, −1
6	−2, −2, −1, −1	0, −1, −1, −1	19	−1, −1, −1, −1	1, −1, −1, −1
7	−3, −1, 0, 0	−2, 0, 1, 1	20	−1, 1, 1, 1	−1, 1, 1, 1
8	−1, 0, 1, 2	−1, 0, −1, 1	21	0, 0, 0, 0	0, 0, 0, 0
9	−3, −1, 0, 2	−3, 0, −3, 0	22	1, 2, 1, 1	1, −2, −1, −1
10	−2, 1, −1, −1	−2, −1, −2, 0	23	0, 1, 2, 0	−1, −1, 1, 0
11	0, −1, 1, 0	0, −1, 0, 1	24	0, −1, 0, 0	−2, 0, −2, −2
12	−2, −2, −2, −2	−1, −1, −1, −1	25	1, 1, 0, 0	0, 1, 1, 0
13	1, 1, 1, 1	0, 0, 0, 0			

3.8 The surface finish on one face of an aluminum transmission component is important since it mates to another component. Too rough a finish would lead to potential leakage in the hydraulic system of the transmission. A mill is used to hold the surface finish below 40 microinches. A control chart is kept to monitor the process by sampling five parts randomly during a production shift. The data appear below. (a) Prepare an $\bar{X}$ and R control chart. Is the process stable? (b) Does the sampling plan seem appropriate? (c) Is it meaningful to use a lower control limit for the $\bar{X}$ chart since only an upper specification limit is given?

No.	Microinches	No.	Microinches
1	23, 27, 29, 20, 31	21	27, 24, 32, 20, 30
2	30, 28, 21, 24, 29	22	26, 18, 29, 34, 31
3	27, 24, 20, 32, 30	23	28, 23, 29, 30, 34
4	26, 28, 24, 30, 32	24	31, 27, 21, 33, 30
5	30, 21, 28, 32, 21	25	26, 29, 26, 28, 31
6	28, 33, 26, 20, 31	26	25, 28, 34, 27, 31
7	25, 29, 33, 20, 31	27	28, 23, 29, 34, 36
8	31, 30, 28, 32, 26	28	24, 27, 33, 20, 31
9	23, 32, 34, 28, 21	29	21, 30, 31, 18, 38

No.	Microinches	No.	Microinches
10	26, 29, 20, 36, 32	30	35, 27, 25, 33, 24
11	31, 27, 34, 28, 20	31	32, 24, 28, 31, 30
12	23, 32, 35, 20, 29	32	26, 33, 31, 23, 25
13	27, 25, 32, 34, 21	33	32, 40, 29, 26, 31
14	24, 31, 27, 33, 22	34[a]	46, 40, 48, 48, 44
15	31, 26, 27, 20, 32	35	37, 29, 34, 41, 25
16	19, 23, 32, 18, 25	36	29, 33, 22, 39, 36
17	28, 24, 31, 18, 22	37	31, 40, 40, 33, 34
18	23, 26, 29, 20, 19	38	27, 36, 41, 33, 40
19	31, 22, 41, 38, 26	39	19, 16, 18, 20, 19
20	30, 31, 29, 21, 30	40	44, 32, 33, 37, 31

[a]Tool changed.

3.9 Aluminum transmission valves are anodized so that the valves have a hard surface coating. The thickness and consistency of the coating is important for proper transmission shifting. The specification for the coating is 9–21 μm. Control of the thickness depends on a number of process parameters, such as temperature of the anodizing bath, concentration of the bath, anodizing time, amperage, and bath agitation. (a) These control chart measurements were collected to evaluate the anodizing process. Five parts were measured from each hanger used in the anodizing bath. Prepare a "before" control chart for this data, and interpret the chart. Circle the out-of-control points.

No.	Thickness					$\bar{X}$	R
1	15	14	15	14	15	14.6	1
2	15	15	19	15	15	15.8	4
3	12	14	12	14	15	13.4	3
4	15	16	15	16	17	15.8	2
5	12	13	12	12	12	12.2	1
6	14	15	14	15	16	14.8	2
7	12	14	14	16	15	14.2	4
8	13	18	13	15	14	14.6	5
9	15	14	14	14	14	14.2	1
10	12	13	12	13	15	13.0	3
11	16	16	15	16	17	16.0	2
12	13	14	14	12	13	13.2	2
13	14	12	14	12	13	13.0	2
14	15	14	13	13	15	14.0	2

No.	Thickness					$\bar{X}$	R
15	17	17	16	16	15	16.2	2
16	17	18	14	14	15	15.6	4
17	13	16	13	12	13	13.4	4
18	13	15	17	20	13	15.6	7
19	15	15	16	14	15	15.0	2
20	17	17	17	17	13	16.2	4

(b) Although no parts were observed beyond the specification limits, it was decided that the overall transmission shift quality would improve if the process variability were reduced. An overhaul of the process was conducted and a computerized control system installed. Using the same sampling plan, these data were obtained. Prepare an "after" control chart for the data, and interpret the chart.

No.	Thickness					X	R
1	17	15	17	15	15	15.8	2
2	15	14	15	14	15	14.6	1
3	16	15	14	15	15	15.0	2
4	16	16	16	16	15	15.8	1
5	14	14	14	13	13	13.6[a]	1
6	16	15	16	15	16	15.6	1
7	16	16	15	15	14	15.2	2
8	14	14	15	14	15	14.4	1
9	16	16	15	16	16	15.8	1
10	16	15	15	14	15	15.0	2
11	15	17	15	17	15	15.8	2
12	13	14	15	15	14	14.2	2
13	15	15	17	16	15	15.6	2
14	15	16	15	14	15	15.0	2
15	16	15	15	15	15	15.2	1
16	16	14	15	15	15	15.0	2
17	16	15	15	15	17	15.6	2
18	15	16	15	16	15	15.4	1
19	15	16	15	16	15	15.4	1
20	14	14	15	16	15	14.8	2

[a]Special cause due to anodizing time.

3.10 A forging plant produced a differential gear for a car rear axle. One of the operations involved piercing a hole that had an upper specification of .080 inch TIR. Another plant that machined the gear complained that gears with high TIR, although not out of specification, were difficult to machine. (a) Prepare a control chart from the "before" data shown here; subgroups were formed by $n = 5$ part per hour.

No.	Day			TIR			$\bar{X}$	R	Action
1	1	60	62	50	50	40	52.4	22	
2		13	32	18	50	30	28.6	37	
3		55	42	28	66	46	47.4	38	
4		13	15	15	14	10	13.4	5	
5		30	40	50	40	37	39.4	20	
6	2	25	40	50	45	55	43.0	30	
7		47	54	56	50	47	50.8	9	
8	3	42	45	40	42	20	37.8	25	
9		60	72	70	80	68	70.0	20	
10		48	45	62	51	67	54.6	22	
11		56	73	68	55	52	60.8	21	
12		80	47	71	77	72	69.4	33	Changed #3 die
13		52	50	58	50	65	55.0	15	Changed #3 punch
14		50	38	46	37	42	42.6	13	
15	4	34	13	42	20	38	29.4	29	
16		68	10	66	67	55	53.2	58	
17		64	68	62	40	60	58.8	28	Changed #3 punch
18		58	55	42	55	47	51.4	16	
19		30	37	35	15	20	27.4	22	Found broken tool holder

(b) During the before period, operators made numerous process changes, such as new tools, transfer mechanism modifications, and cleaning, but no positive results were obtained and operators began to question the usefulness of control charts. One day while changing a tool an operator noticed a small crack in a tool holder. After replacing the holder the "after" data were obtained. Prepare a control chart, and comment on the results.

No.	TIR					$\bar{X}$	R
1	30	25	40	35	25	31.0	15
2	40	30	20	35	25	30.0	20
3	28	33	43	32	38	34.8	15
4	15	25	35	30	35	28.0	20
5	25	19	27	27	29	25.4	10

No.	TIR					$\bar{X}$	R
6	19	22	26	10	18	19.0	16
7	10	14	9	12	8	10.6	6
8	25	19	28	30	23	25.0	11
9	30	28	18	25	20	24.2	12
10	15	20	15	10	20	16.0	10
11	18	33	18	23	22	22.8	15
12	20	20	25	15	25	21.0	10
13	15	10	15	10	5	11.0	10
14	15	12	16	15	18	15.2	6
15	25	15	13	19	21	18.6	12
16	13	25	11	20	21	18.0	14
17	15	12	12	20	15	14.8	8
18	30	10	20	15	20	19.0	20

3.11 The data are an artificial set that will be used to illustrate several control chart sampling concepts in later chapters. (a) Using a subgroup size of $n = 5$, prepare a control chart using the subgroups starting with numbers marked with an asterisk. (b) In part a, only $k = 25$ subgroups were plotted. Suppose all $k = 45$ subgroups of size $n = 5$ were used, would the results change appreciably? (c) Suppose you found out that there were three different spindles on the machine producing the 225 parts that were measured. What type of chart would you use to study the process? Compare the out-of-control signals obtained from this chart with the chart obtained in part a.

No.	Value	No.	Value	No.	Value	No.	Value	No.	Value
1*	11	46*	10	91*	12	136*	10	181*	10
2	14	47	12	92	16	137	15	182	14
3	12	48	9	93	7	138	12	183	12
4	11	49	13	94	11	139	9	184	10
5	16	50	13	95	12	140	15	185	13
6	11	51	9	96	10	141	11	186	9
7	12	52	7	97	11	142	8	187	11
8	12	53	16	98	14	143	15	188	15
9	9	54	8	99	11	144	9	189	8
10	9	55	11	100	8	145	10	190	10
11*	12	56*	14	101*	13	146*	14	191*	16
12	9	57	10	102	10	147	9	192	9
13	10	58	14	103	12	148	11	193	10
14	11	59	15	104	13	149	14	194	13

No.	Value	No.	Value	No.	Value	No.	Value	No.	Value
15	10	60	10	105	9	150	10	195	10
16	10	61	7	106	10	151	10	196	6
17	11	62	14	107	13	152	13	197	14
18	8	63	8	108	12	153	8	198	9
19	11	64	10	109	10	154	9	199	15
20	13	65	13	110	15	155	14	200	11
21*	8	66*	14	111*	10	156*	11	201*	9
22	10	67	12	112	8	157	9	202	11
23	14	68	13	113	13	158	15	203	13
24	14	69	11	114	11	159	13	204	8
25	10	70	13	115	11	160	10	205	9
26	12	71	13	116	12	161	13	206	13
27	11	72	9	117	10	162	8	207	12
28	11	73	12	118	10	163	9	208	9
29	13	74	14	119	15	164	12	209	16
30	11	75	9	120	10	165	11	210	11
31*	11	76*	10	121*	10	166*	8	211*	11
32	12	77	14	122	15	167	13	212	15
33	10	78	8	123	15	168	9	213	10
34	10	79	11	124	10	169	15	214	12
35	12	80	14	125	15	170	14	215	14
36	12	81	12	126	11	171	11	216	8
37	11	82	9	127	10	172	11	217	11
38	15	83	14	128	14	173	15	218	13
39	8	84	9	129	11	174	14	219	11
40	10	85	12	130	8	175	11	220	10
41*	14	86*	12	131*	13	176*	15	221*	12
42	11	87	9	132	12	177	11	222	12
43	11	88	11	133	9	178	9	223	7
44	15	89	15	134	16	179	14	224	17
45	9	90	8	135	11	180	9	225	9

3.12 The transit times (hours) between a supplier and an assembly plant discussed in the case study appear below. (a) Calculate the control limits for the 47 subgroups. Is the process stable? (b) Calculate the control limits for the first group of 25 and last group of 22 separately. Does the process appear stable?

No.	Transit times	No.	Transit times
1	108, 105, 81, 104, 80	26	85, 78, 85, 108, 84
2	152, 80, 84, 81, 107	27	83, 105, 131, 105, 78
3	83, 81, 129, 78, 84	28	77, 100, 83, 82, 109
4	107, 83, 106, 82, 103	29	126, 127, 104, 86, 122
5	83, 103, 79, 107, 79	30	128, 106, 103, 77, 129
6	82, 125, 100, 80, 104	31	148, 126, 106, 74, 149
7	81, 126, 84, 79, 76	32	121, 102, 103, 81, 80
8	78, 105, 101, 79, 105	33	103, 102, 87, 138, 78
9	104, 105, 81, 80, 83	34	83, 80, 108, 55, 106
10	79, 103, 83, 78, 83	35	79, 103, 77, 77, 103
11	102, 78, 85, 82, 79	36	86, 82, 84, 80, 77
12	80, 78, 82, 81, 80	37	78, 77, 80, 105, 81
13	105, 83, 84, 77, 81	38	103, 112, 131, 133, 111
14	79, 84, 77, 107, 82	39	80, 127, 151, 82, 82
15	107, 83, 131, 81, 84	40	77, 100, 102, 109, 82
16	81, 81, 77, 124, 82	41	82, 81, 106, 85, 83
17	79, 83, 79, 80, 105	42	83, 81, 104, 131, 82
18	107, 103, 79, 83, 107	43	82, 83, 82, 83, 83
19	80, 80, 80, 105, 81	44	78, 104, 107, 102, 103
20	82, 106, 83, 81, 83	45	80, 103, 128, 82, 106
21	59, 83, 83, 83, 85	46	103, 77, 82, 106, 106
22	77, 81, 81, 81, 79	47	82, 103, 106, 106, 82
23	80, 82, 78, 79, 103		
24	82, 83, 80, 155, 106		
25	80, 80, 82, 80, 86		

3.13 A lathe was used to machine the inner diameter (ID) of the bore of a transmission support component that had a specification of 49.933 ± .025 mm. Subgroups of five consecutive pieces were collected four times a day. The bore ID was measured at each end about 10 mm apart so that two measurements were made on each part using the coding

Actual dimension = 49.933 + .0025 × gage reading

(a) Prepare a median and range chart for each bore position. Do the measurements from the two positions appear to be related? (b) A problem developed when a sleeve that was pressed into the bore provided too loose a fit. Both parts measured by this system were within specification limits. How can this result be explained if the specification limits are assumed correct?

No.	Day	Time	Position 1	Position 2
1	1	7:35	−3, −3, 0, −2, −1	−3, −2, −2, −2, 0
2		10:00	−2, −4, −6, −4, −4	−2, −1, −4, −3, −3
3		12:30	3, 2, 1, 2, 4	3, 3, 5, 4, 5
4		3:00	−4, −2, −3, −5, −4	−3, −4, −4, −4, −5
5	2	8:00	−4, −4, −3, −4, −3	−4, −4, −6, −2, −2
6		10:30	3, 6, 6, 2, 5	4, 5, 6, 3, 6
7		12:00	5, 4, 2, 3, 3	6, 5, 5, 4, 6
8		2:15	2, 3, 1, 2, 2	4, 3, 1, 2, 4
9	3	7:30	−3, −4, −4, −3, −4	−3, −4, −4, −2, −3
10		10:00	−3, −2, −4, −3, −1	−4, −2, −5, −4, −4
11		12:30	−3, −4, −3, −5, −2	−4, −5, −4, −6, −3
12		2:30	−4, −5, −1, −1, −1	−5, −4, −2, −2, −3
13	4	7:30	−4, −4, −3, −5, −3	−4, −2, −2, −5, −5
14		10:30	−4, −3, −6, −5, −4	−3, −2, −4, −5, −3
15		1:00	−3, −4, −3, −1, −2	−2, −4, −4, −4, −4
16		3:15	−5, −3, −1, −2, −3	−4, −6, −2, −2, −5
17	5	7:40	−1, −2, −3, −4, −3	−4, −4, −3, −3, −4
18		9:20	−4, −4, −4, −5, −5	−4, −5, −1, −5, −4
19		11:00	−2, −4, −2, −4, −4	−1, −4, −3, −4, −3
20		2:00	−3, −4, −5, −5, −4	−2, −2, −5, −3, −4

3.14 In an engine component housing, the heater hole (diameter specifications, 15.67–15.72 mm) and mating tube (diameter specifications, 15.77–15.88 mm) were machined. To target the best fit conditions, both characteristics were charted together. A subgroup size of $n = 3$ was collected randomly over an 8 hour period. The data are coded so that

Actual value (mm) = 15.64 + .01 × recorded value

(a) Calculate $\bar{X}$ and R chart values for both characteristics. Is the process stable? (b) The optimal fit from the blueprint specification was a hole nominal specification of 15.695 mm and a tube nominal specification of 15.825 mm. Was the process targeted correctly? Why or why not? How does process stability influence the targeted fit?

Day	Hole	Tube	Day	Hole	Tube
1	7.1, 7.9, 7.6	20.0, 20.2, 19.7	12		20.0, 20.5, 19.2
	7.1, 7.3, 7.4	20.0, 20.5, 20.5			21.0, 20.1, 19.1
2	6.5, 7.1, 7.1	20.0, 20.5, 20.5	16	7.8, 7.5, 7.5	
	7.1, 7.9, 7.8	20.0, 19.5, 20.0		7.4, 7.4, 7.4	
3	6.0, 7.0, 7.6	19.5, 20.0, 20.0	17	7.2, 7.8, 7.4	
	6.8, 7.8, 7.6	20.0, 19.6, 19.5		7.1, 7.9, 7.8	

Problems

Day	Hole	Tube	Day	Hole	Tube
4	7.8, 7.6, 7.0 5.8, 5.6, 6.2	19.9, 20.0, 19.7 19.9, 20.5, 20.7	18	7.6, 7.5, 7.7 7.9, 7.2, 7.3	20.5, 20.3, 20.0 20.1, 19.6, 19.8
5	7.0, 6.6, 5.8 6.8, 6.6, 7.4	19.9, 19.8, 20.1 19.5, 20.0, 19.6	19		20.0, 19.7, 19.5 19.3, 20.1, 20.0
8	6.8, 7.2, 7.4 7.3, 7.5, 7.3	20.3, 20.0, 20.6 20.2, 20.8, 20.5	29	7.2, 7.8, 7.7 7.5, 7.5, 7.6	18.5, 19.7, 19.7 19.8, 19.8, 19.5
9	7.1, 7.2, 7.2	19.0, 19.5, 19.0 20.0, 20.0, 20.5	30 —[a]	7.9, 7.5, 7.7 5.8, 5.4, 5.6	19.8, 19.8, 19.5 19.5, 19.5, 19.5
10	8.3, 8.5, 8.6	20.0, 19.0, 19.5 20.0, 19.0, 20.2	31	5.4, 5.4, 5.6 5.8, 5.5, 5.4	21.6, 20.7, 20.7
11	7.8, 7.6, 7.9	20.0, 18.5, 18.7 19.7, 20.5, 20.1	32		19.5, 19.3, 20.0
			33	5.6, 5.2, 5.9	19.5, 19.3, 20.0 20.3, 19.9, 20.3

[a]Changed reamer.

3.15 Suppose that 100 items are examined in 25 time periods and the following number of defective items were found:

10, 11, 15, 8, 2, 7, 13, 4, 12, 11, 18, 7, 3, 7, 11, 9, 14, 16, 7, 5, 8, 12, 10, 9, 5

Prepare a control chart with control limits. Is the process stable?

3.16 A simple classroom exercise was used to illustrate how a P chart works. Fifteen students tossed a coin. Those students having a head raised their hands. This exercise was repeated for $k = 25$ trials. The numbers from two different runs of the exercise appear below.

Run 1:

9, 10, 7, 9, 8, 6, 8, 9, 9, 6, 4, 9, 9, 6, 9, 8, 10, 5, 8, 8, 8, 6, 6, 8, 9

Run 2:

8, 7, 8, 6, 9, 9, 8, 10, 6, 8, 9, 12, 5, 8, 4, 7, 9, 9, 8, 7, 9, 7, 8, 8, 7

(a) Compute the P chart values for both runs. Does the $k = 25$ subgroups produce similar control chart values? Are the processes stable? (b) What is the highest and lowest number of heads that would not be considered unusual?

3.17 A stamping operation was experiencing an unacceptably high scrap rate for a car hinge reinforcing plate. A departmental team was formed to address the problem. They studied the process and concluded from the "before" control chart data that a process change was needed rather than addressing individual defects as had been past practice. A draw die was reworked and an intensive incoming stock control method was implemented. A control chart for the "after" period indicated a significant improvement. This improvement process continued, and after 4 months the average scrap rate was 0.1%. (a) Calculate the control chart limits for the before and after periods. The sampling plan consisted of evaluating 500 pieces several times a day. (b) How did the team interpret the control chart to conclude a system change was needed?

Before		After	
Day	Rejects	Day	Rejects
1	3	10	2
	5	11	1
	5	12	0
	6		1
2	0		0
	5	13	0
	5		0
	3		0
3	5	14	0
	5		2
4	5		0
	5	15	3
	1		1
5	9		0
	4	16	0
	5		2
6	2		0
7	16	17	0
	15		0
	4		1
8	4	18	0
	9		0
	4		6[a]
9	5	19	0
	8		0

[a]Old stock used.

3.18 Porosity in an aluminum transmission housing was a chronic problem. The porosity problem is difficult to address since it cannot be detected effectively in a raw casting. Only after machining is it possible to detect the porosity defect in the metal. A P chart was kept by department personnel to provide effective feedback to the casting vendor. The number of rejects and the daily production are given here. Prepare a P chart. Is the process in control?

No.	Production	Rejects	No.	Production	Rejects
1	3194	6	26	2550	6
2	2959	5	27[a]	1610	39
3	2606	2	28	2717	9
4	3136	1	29	3019	10

No.	Production	Rejects	No.	Production	Rejects
5	1069	0	30	3218	3
6	3294	4	31	3228	6
7	3206	4	32	2622	4
8	3145	1	33[a]	2519	42
9	2826	1	34[a]	2610	23
10	3129	7	35	3724	20
11	1743	2	36	3212	12
12	2503	5	37	3054	6
13	2472	4	38	1458	10
14	2704	10	39	3131	6
15	2080	5	40	2930	8
16	3058	4	41	3037	6
17	3649	2	42	2729	8
18	3158	7	43	2701	7
19	2138	8	44	2356	12
20	2545	0	45	2840	9
21	1061	3	46	2997	6
22	3191	3	47	2735	5
23	3341	2	48	2979	6
24	3414	5	49	3240	6
25	3306	9	50	1302	3

[a]Different model of housing produced.

3.19 The hydraulic circuit in a transmission was tested by an air test prior to final assembly in a transmission. The initial rejection limit was increased on day 9 owing to a vehicle correlation study that showed the initial limit was unnecessarily high. (a) Is it meaningful to use all subgroups to calculate control limits? Why or why not? (b) Calculate control limits for the percentage of rejects. Is the process stable?

Day	No. inspected	No. rejected	Day	No. inspected	No. rejected
1	242	99	13	460	42
2	310	98	14	637	92
3	480	134	15	520	67
4	539	151	16	440	23
5	590	268	17	400	50
6	558	311	18	912	117

Day	No. inspected	No. rejected	Day	No. inspected	No. rejected
7	630	227	19	930	105
8	399	174	20	564	29
9[a]	800	78	21[b]	798	318
10	700	58	22[b]	345	124
11	650	45	23	750	102
12	530	64	24	1141	292

[a] New rejection limit instituted.
[b] Test equipment malfunction.

3.20 Continuing the example from Table 3.2 on the number of car body defects, the following data were collected:

8, 5, 6, 5, 9, 6, 11, 6, 3, 4, 3, 3, 8, 6, 4, 3, 5, 6, 6, 2, 5, 7, 2, 4, 3

What are the new control limits? Plot the control chart. Is the process stable?

4
Control Chart Sampling Concepts and Applications

The previous chapter introduced control charts by discussing process stability (also referred to as process control) and giving the basic rules for constructing and using a control chart. Since control charts are our primary method of assessing whether a process is stable, they are a major element in constructing a defect prevention system (DPS) based on overall process stability. This chapter addresses basic sampling concepts that help explain how a control chart works. These concepts provide a guide for adapting control chart methods to a variety of industrial applications. We see throughout this chapter that there are many different ways to use control charts to adapt to different process factors. These factors include

Manufacturing conditions
Gaging capability
Personnel preferences

Practical experience has shown that many applications of control charts fail because rigid implementation rules are used for the sake of uniformity. These rules often make the required flexibility and experimentation with different approaches impossible. The objective of continuing improvement of process quality is replaced by satisfying a management requirement for ''charts.'' Control charts then cease to be a tool to help employees do their work better and are discontinued at the earliest opportunity.

> A team in a forging operation had received (SPC) training and wanted to address their high scrap rate. However, the environment in their work area was quite warm and the protective gear they wore made it difficult to write or converse, so a standard control chart was not practical. The team's solution was to construct a large (4 × 8 foot) board that had a control chart grid painted on it. Large colored pins were used to mark measured values on a median range chart. The chart was mounted on wheels and moved to a rest area to allow discussion of the charted results. Predictable results followed.

A control chart is a tool for doing a better job—flexibility is required

4.1 Sampling from a Process

Before collecting data we have seen that it is important to first define the process that generates the data using a process flow diagram. It is then necessary to define operationally

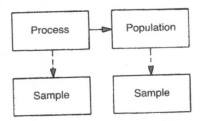

Figure 4.1 Sampling from a process or population.

what is being measured and to assess the gaging. We may now collect measurements at some point in the process or from an end-of-line population of items produced by the process. To evaluate different sampling alternatives, the relationship between process, population, and sample must be understood. A simple example of a population is a lot of material. For example, a screw machine supplier might produce from the manufacturing *process* a tub of 10,000 bolts that could be said to define a *population* of bolts. Generally, we would not consider measuring all bolts to assess whether the lot of 10,000 is acceptable. A *sample* of bolts could be collected, measured, and evaluated to determine whether the lot is acceptable. Thus, the terms can be defined:

Process: a collection of resources that act on various inputs to produce some output
Population: a collection of units or lot of material produced from a process
Sample: a collection of units from a process or population

It should be noted that a population can be defined more generally as any collection of items or units, but we are concerned primarily with the relationship of a process and a population of units produced from the process. Figure 4.1 illustrates how it is possible to obtain a sample of items from either a process or a population. The critical difference is that it is possible to retain the production order of parts when we sample from the process.

The question then arises of whether it is preferable to collect a sample from a process or a population. Considering the lot of bolts, would it be better to sample from the bolt production process or from the final lot of bolts? The answer is apparent when process stability is considered. Sampling from the process using a control chart approach is the only method to assess stability and is thus the preferred approach. If the process is stable, the population- or lot-sampling method is acceptable. However, if we can assure process stability using control charts, there is no need for lot sampling.

Sampling from production lots cannot assure process stability

There are several methods by which samples can be collected from a process. There is no best sampling method; selection of the method depends on the manufacturing conditions.

4.1.1 Consecutive Process Sampling

Items are selected for sampling that are in production order:

Subgroup 1   Subgroup 2   Subgroup 3

X X X ⊗⊗⊗⊗⊗ X X X X X X X ⊗⊗⊗⊗⊗ X X X X X X ⊗⊗⊗⊗⊗ X X X · · ·

Production order ⟶

4.1 Sampling from a Process

Normally, consecutive samples are used to form subgroups for control charts. This sampling plan provides the minimum variability within a subgroup since samples within a subgroup are separated by a minimal amount of time.

4.1.2 Periodic Process Sampling

Items are selected for sampling that are separated by a constant unit of time:

x x x x ⊗ x x x x ⊗ x x x x ⊗ x x x x ⊗ x x x x ⊗ x x x x ⊗ x x ...

Production order ⟶

Periodic sampling can be used with control charts when samples are collected periodically with a certain number forming a subgroup. For example, "collect a sample every 15 minutes and let the four samples within an hour form a subgroup." Subgroups formed using periodic samples span more time than consecutive sampling so the variability within a subgroup is generally greater. This type of sampling also can be useful for short-term studies of a process (Chap. 11) to identify sources of a problem.

4.1.3 Random Process Sampling

Each item from the process has an equally likely chance of being sampled:

x ⓧ x x x x x x ⓧ x ⓧ x x x x x ⓧ x x x x ⓧ x x ⓧ x x x x ...

Production order ⟶

Random sampling is not generally used with control charts since it is more difficult to implement and most people prefer a fixed, constant time interval for part collection and measurement. This sampling can be useful in short-term studies in which some type of "representative" sample is desirable (see Deming, 1986 p. 295). The use of a random-number table used to collect random samples is discussed in Appendix III.

4.1.4 Stratified Process Sampling

Items are selected at a point in time from a particular group of machines, fixtures, or pallets, for example (strata), to form a subgroup:

It is apparent that stratified sampling is a very economical approach to sampling since separate control charts can be combined into a single chart. Although this sampling plan can be quite useful, it should be implemented by first using a "baseline study," discussed in a later section.

Every user of a control chart would like to know the best sampling approach for collecting control chart subgroups. Unfortunately, control chart sampling must be adapted to the process so it is not possible to suggest one approach for all situations. However, it is possible to recommend consecutive piece sampling for troubleshooting. Figure 4.2 shows

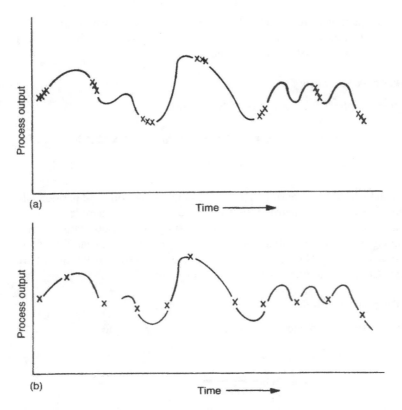

Figure 4.2 Comparison of (a) consecutive and (b) periodic sampling.

an out-of-control process in which the mean is constantly changing. A consecutive and a periodic sampling plan are compared in the figure.

Consecutive sampling is best for troubleshooting

Because consecutive samples span a minimal amount of time, the range and resulting $\bar{R}$ are small compared with the $\bar{R}$ obtained from periodic sampling. A smaller $\bar{R}$ results in tighter control limits and more out-of-control signals due to special causes:

Sampling plan	Range ($\bar{R}$)	Control limits	Out-of-control signals	Special causes identified
Consecutive →	Smaller $\bar{R}$ →	Narrower →	More →	More
Periodic →	Larger $\bar{R}$ →	Wider →	Fewer →	Fewer

There is thus more opportunity to identify and eliminate special causes using consecutive piece sampling.

The question then arises, if consecutive piece sampling helps to obtain special causes,

4.2 Sampling Considerations

why not use it all the time? The difficulty with using consecutive sampling is that it requires the greatest sampling intensity. With complex manufacturing processes, such as that shown in Figure 2.6, the consecutive sampling of all sources of variability is often not practical. Also, consecutive samples are not spread out over the time period of a production run so that the total process output may not be represented. Fortunately, once a process is in control, a variety of sampling methods can be used to monitor process output.

4.2 Process Control Sampling Considerations

The procedure given in Chapter 3 for preparing a control chart is not difficult. However, the usefulness of an individual chart depends on where in the process the measurements are to be collected and how we decide to sample from the process. Developing an overall process sampling plan (Chap. 14) prior to data collection can significantly improve the usefulness of an individual chart.

4.2.1 Rational Subgroups

It is desirable to use rational subgroups when establishing a control chart. In a rational subgroup, the parts used to form a subgroup are collected from the process such that the chance for variation within a subgroup is minimized and the chance for variation between subgroups is maximized. Thus, rational subgroups result in comparatively small R values:

```
Subgroup no.    1    2    3   . . .
               10    3    8         ←  Want minimum variation
               15   12    9            within a subgroup
                7    2   12            (i.e., small R)
                4    7   10         ←

               ↑____↑
               Want maximum variation
               between subgroups
```

Rational subgroups have several advantages:

Small R values quantify the smallest possible variability that can be expected from the process. This variability can be used to assess process capability.

A small R value results in narrow control limits, which causes more out-of-control signals. These signals assist in identifying special causes of variation more quickly.

Reducing the time interval over which samples within a subgroup are collected makes interpretation of out-of-control signals easier for both the $\bar{X}$ and R charts. Rational subgroups seek to have a minimal number of sources at variation within a subgroup, so causes of R chart out-of-control signals can more easily be determined.

and few disadvantages:

To obtain a minimal amount of variability within a subgroup often requires selecting the smallest possible sampling unit within a process. For example, selecting every machine spindle in the Figure 2.6 example would result in 66 charts if only one characteristic per spindle is considered.

Minimizing variability within a subgroup often requires consecutive piece sampling to minimize the time interval between parts within a subgroup. To get a representative sample over a day's production would require too many subgroups for a particular chart.

It is desirable to develop a procedure for starting control charts that has the advantages just listed during the initial period of start-up and troubleshooting but does not require too many charts or subgroups on a particular chart for the purpose of monitoring a process.

4.2.2 Stratified Control Charts

Procedure 4.1 suggests a method that can be used as a guide for starting control charts in a complex manufacturing operation in which multiple streams of output, also called strata (e.g., fixtures, pallets, or spindles) may exist. Central to this procedure is the initial use of rational subgroups on the smallest sampling units. Once stability is established, it is possible to combine various control charts into a stratified control chart and to adopt a monitoring sampling plan. Other methods for addressing multiple streams, such as group control charts, are discussed by Burr (1976, Sec. 7.5.3) and Nelson (1986).

The procedure for combining several baseline control charts into a single stratified control chart appears here. Consider the case in which there are separate stable baseline control charts on machines, spindles, or fixtures, for example, that all produce the same characteristic.

1. Compute the grand averages $\bar{\bar{X}}$ and $\bar{R}$ using values from the separate control charts: $\bar{\bar{X}}_{base}$ = average of $\bar{\bar{X}}$ values for baseline charts; $\bar{R}_{base}$ = average of $\bar{R}$ values for baseline charts.
2. Compute the estimated standard deviation. Let d_{2base} denote the d_2 constant for the baseline unit control charts:

$$\hat{\sigma} = \frac{\bar{R}_{base}}{d_{2base}}$$

3. Compute the new $\bar{R}_{strata}$ for the stratified control chart. Let $d_{2strata}$ denote the d_2 constant for the stratified control chart. The subgroup size corresponds to the number of baseline units:

$$\bar{R}_{strata} = d_{2strata} \hat{\sigma}$$

4. Compute the new control limits using $\bar{R}_{strata}$ with A_2, D_3, and D_4 corresponding to the subgroup size equal to the number of baseline units. Note that

$$\bar{\bar{X}}_{strata} = \bar{\bar{X}}_{base}$$

Example 4.1 Consider a three-spindle machine with a stable control chart on each spindle:

The following values were obtained using $n = 5$ consecutive piece sampling:

Procedure 4.1 Process Control Chart Start-Up Procedure

The following list of items should be addressed when initially starting a control chart:

1. Process flow diagram. A flowchart of the entire process should be prepared. All major sources of variation should be identified along with any alternative path flows.
2. Measurement analysis. The characteristic being investigated should have a clear operational definition. Customer requirements should be assessed for both in-process and end-user satisfaction. The gaging method should be evaluated (App. I).
3. Upstream control. Stability must be assessed for upstream operations influencing a characteristic being considered. Incoming material and all prior processing must be evaluated.
4. Select smallest sampling unit. Using the process flow diagram, the smallest sampling unit (e.g., machine, spindle, or fixture) should be selected for sampling.
5. Intensive sampling. Sample intensively (e.g., every ½–2 hours) using rational subgroups so that changes with time can be evaluated. Do not study too many operations. It is better to study a few operations thoroughly.
6. Establish a baseline. Obtain stability for each unit being studied. All $\bar{X}$ and $\bar{R}$ values should be about the same when stability is attained. An average $\bar{\bar{X}}$ and $\bar{\bar{R}}$ should be computed for the baseline study:

    $\bar{\bar{X}}_{base}$ = average of $\bar{\bar{X}}$ values for baseline units

    $\bar{\bar{R}}_{base}$ = average of $\bar{R}$ values for baseline units

    Attempt to eliminate all special causes of variation.

7. Increase sampling interval. If after a period of time during which no unusual problems are present, increase the interval of time between subgroups (say every 4 hours). Maintain same subgroup size n.
8. Combine charts. After a reasonable period of time during which stability has been maintained, combine the separate control charts into a single stratified control chart. Control limits should be calculated from

    $$\bar{R}_{strata} = \frac{d_{2strata}}{d_{2base}} \bar{R}_{base}$$

    where $d_{2strata}$ and d_{2base} are the d_2 constants (App. III) for the subgroup size n used in the stratified and baseline control charts, respectively.

9. Evaluate capability. The ability of the process to meet engineering specifications ultimately influences the frequency at which subgroups should be collected (Chap. 14). The process should be monitored using the same frequency as in step 7 to assess process capability (Chap. 7).
10. Integrate into DPS. The purpose of the defect prevention system is to integrate the control of all characteristics in the process. The Operation Control Plan in Chapter 14 should be used to evaluate the need for control charts on an ongoing basis.

	Spindle 1	Spindle 2	Spindle 3
$\bar{X}$	10.2	10.3	9.8
$\bar{R}$	3.1	2.7	3.2

The following computational procedure uses the steps just described.

1. Compute $\bar{\bar{X}}_{base}$ and $\bar{R}_{base}$:

$$\bar{\bar{X}}_{base} = \frac{10.2 + 10.3 + 9.8}{3} = 10.1$$

$$\bar{R}_{base} = \frac{3.1 + 2.7 + 3.2}{3} = 3.0$$

2. Since a subgroup size of $n = 5$ was used to evaluate the baseline units, $d_{2base} = 2.33$:

$$\hat{\sigma} = \frac{3.0}{2.33} = 1.29$$

3. Since there are three baseline units (spindles), $n = 3$ for the stratified control chart and $d_{2strata} = 1.69$. Thus,

$$\bar{R}_{strata} = 1.69 \times 1.29 = 2.18$$

4. Using $\bar{R}_{strata}$, A_2, D_3, and D_4 for $n = 3$ baseline units gives the following control limits:

$$UCL_{\bar{X}} = \bar{\bar{X}}_{strata} + A_2 \bar{R}_{strata} = 10.1 + (1.02 \times 2.2) = 12.3$$
$$LCL_{\bar{X}} = \bar{\bar{X}}_{strata} - A_2 \bar{R}_{strata} = 10.1 - (1.02 \times 2.2) = 7.9$$
$$UCL_R = D_4 \bar{R}_{strata} = 2.58 \times 2.2 = 5.7$$

The new stratified control chart is

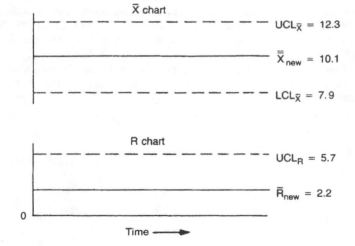

4.2 Sampling Considerations

Subgroup	1	2	3	4	5	6	...
Spindle 1							
Spindle 2							
Spindle 3							

4.2.3 Failure to Adapt Control Charts to the Process

The method described in Procedure 4.1 is more time consuming than the usual applications of control charts in which a chart is simply placed at an operation with little consideration of the overall process. What are the risks of the usual shortcut method? The risk of not adapting control charts to the process is that a process may be more likely to appear stable when, in fact, special causes of variation are present. Two common mistakes are (1) not matching the subgroup size with the major sources of variation, and (2) not establishing stability of the individual baseline units that represent the source of variation. The data for Problem 3.11 can be used to illustrate both points.

The data set in Problem 3.11 is a typical example in which measurements are taken in their order of production:

11, 14, 12, 11, 16, 11, 12, 12, 9, 9, 12, 9, 10, 11, 10, ... , 7, 17, 9

On the surface there seems nothing wrong with using a subgroup size of $n = 5$ and preparing an $\bar{X}$ and R chart as was done in Chart 4.1. Although there are no points beyond the control limits, there are possibly too many points in zone C. This pattern is usually referred to as "hugging the centerline." From Chapter 3 we would expect the points to be distributed as 4% zone A, 28% zone B, and 68% zone C. Dividing the width between each control limit and $\bar{\bar{X}}$ into thirds gives

Zone	Computed limit	No. of points in interval	Expected % of points
Zone A (UCL)	14.8	0/25 = 0%	2
Zone B	13.8	2/25 = 8%	14
Zone C	12.7	11/25 = 44%	34
$\bar{\bar{X}}$	11.6	10/25 = 40%	34
Zone C	10.5	2/25 = 8%	14
Zone B	9.4	0/25 = 0%	2
Zone A (LCL)	8.4		

Combining the results for each zone gives

Zone	Actual % of points	Expected % of points
A	0	4
B	16	28
C	84	68

106   4 Control Chart Sampling

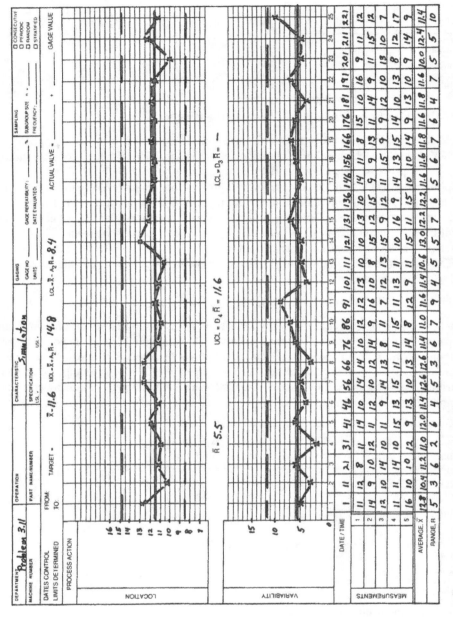

Chart 4.1 Control chart for variable data using a subgroup size of $n = 5$.

4.2 Sampling Considerations

There appears to be some tendency to have too many points in zone C, but the indication is not totally clear.

Suppose that you now learned that there were three main sources of variability and that the single stream of data was really three individual streams. If three machine spindles were involved, the data would appear as follows:

```
        Machine
    |     |     |
   11    14    12
   11    16    11
   12    12     9
    9    12     9
   10    11    10
    .     .     .
    .     .     .
    .     .     .
    7    17     9
```

To use a subgroup size of $n = 5$ would not seem logical and is, in fact, a bad practice. Adapting the control chart to the process would suggest using a subgroup size of $n = 3$, which gives Chart 4.2. Hugging the centerline is more apparent as the table indicates:

Zone	Computed limit	No. of points in interval	Expected % of points
Zone A (UCL)	16.1	0/25 = 0%	2
Zone B	14.6	0/25 = 0%	14
Zone C	13.0	13/25 = 52%	34
$\bar{\bar{X}}$	11.4	10/25 = 40%	34
Zone C	9.8	2/25 = 8%	14
Zone B	8.2	0/25 = 0%	2
Zone A (LCL)	6.7		

Combining the results for each zone gives

Zone	Actual % of points	Expected % of points
A	0	4
B	8	28
C	92	68

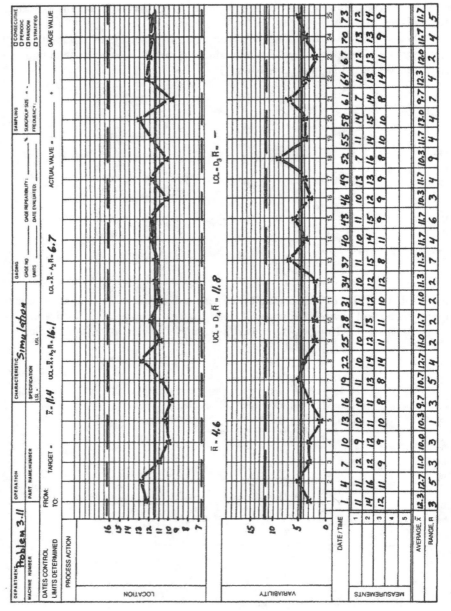

Chart 4.2 Control chart for variable data using a subgroup size of $n = 3$.

The out-of-control pattern for hugging the centerline seems clearer. As expected, the control chart gives a clearer indication of a special-cause problem when the subgroup sampling plan is adapted to the process. It may seem illogical that an $n = 5$ sample size would be used for this problem in which there are three major sources of variability. However, typically the major sources of variability (e.g., vendors, machines, or operators) are not identified so it is not possible to monitor stratification factors. As discussed in Chapter 2, it is often possible to develop an appropriate sampling method to indicate the major sources of variability (e.g., part spotting).

Adapt the subgroup sampling plan to the process

The second common problem in the start-up of a control chart is in not establishing a baseline control chart for the major sources of variability. Combining major sources of variability on a control chart makes identification of seemingly obvious problems difficult. Returning to the Problem 3.11 data, a baseline chart was prepared for each spindle separately with the following values obtained:

Spindle	$\bar{X}$ chart			R chart		
	LCL	$\bar{\bar{X}}$	UCL	LCL	$\bar{R}$	UCL
1	7.6	10.3	13.0	0	2.7	6.9
2	11.8	13.7	15.6	0	1.8	4.7
3	7.1	10.1	13.1	0	3.0	7.7

The individual plots appear in Chart 4.3, where it is apparent that spindle 2 has a mean significantly above that of the other spindles. Combining major sources of variability that have different location ($\bar{\bar{X}}$) or variability ($\bar{R}$) values causes control limits to be too wide, which makes discovery of the problem more difficult. Use of separate baseline charts for a short period of time, as discussed in Procedure 4.1, enables the assessment of common location and variability values. Combining control charts then enables a reduced sampling effort using control limits that are appropriate to indicate any future process changes.

It should be noted the $n = 3$ sampling plan did indicate an out-of-control situation. Hugging the centerline is an indication of a difference between the means of different sources of variability being sampled. This situation is called a stratification problem since the out-of-control signal is due to the different output of the process strata (spindles). It should be emphasized that all out-of-control signals in Procedure 3.7 should be used to detect special causes of variation.

That all points are within the control limits does not assure the absence of special causes of variation

4.3 Within and Between Variation

We have already seen that a way to classify major sources of variability is to use the categories

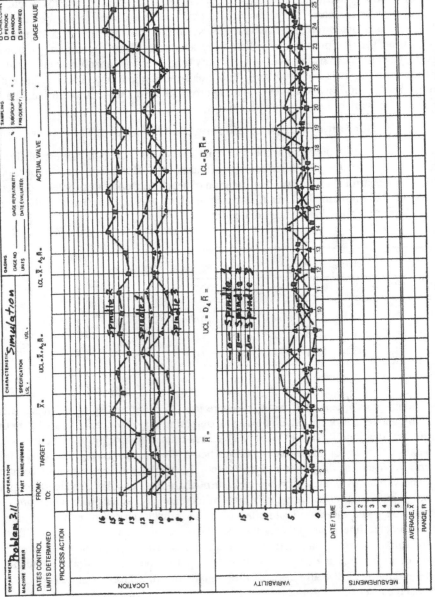

Chart 4.3 Control chart for variable data with a baseline chart prepared separately for each spindle.

4.3 Within and Between Variation

Men
Methods
Machines
Materials
Measurements
Environment

To further break down the sources of variability, the terms "within" and "between" are used. For example, if there are two similar machines, the output of each machine may be slightly different—there exists "between-machine" variability. Also, all parts produced from a single machine are not exactly the same—there exists "within-machine" variability. The figure illustrates these concepts:

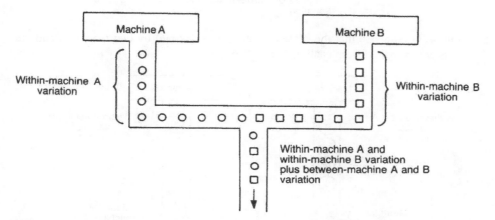

Notice that when the output of both machines is combined, both within and between variation comprises the total variation.

The terms "within" and "between" variation can be defined as follows:

Within variation
  Variability in the units produced by a particular sources of variability is usually measured by s or R.
  Variability of these units (using s or R) measures "within-machine" variation.

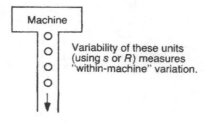

Between variation
  Variability between the mean output from different sources of variability is usually measured by s or R of the means.

112   4   Control Chart Sampling

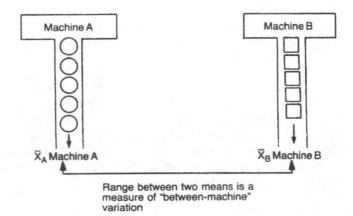

Example 4.2 Suppose that the same operation is performed by two machines and that five parts are measured from each machine:

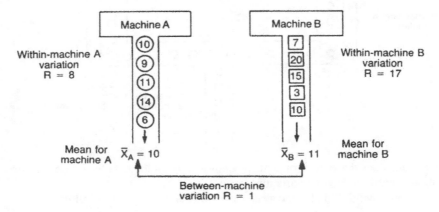

Combined process:

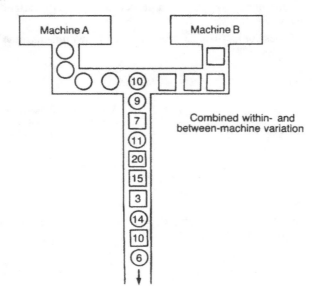

4.3 Within and Between Variation

One of the difficulties in controlling a manufacturing operation is that the sources of variability are seemingly endless. Some examples appear here.

Example 4.3 Consider the same process repeated on different days (days = source of variation):

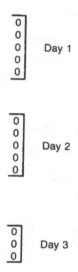

In *any* process there exists variation within a day's production and between days.

Example 4.4 Consider one machine with three spindles performing the same operation (spindles = source of variation):

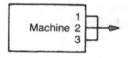

No two spindles can produce exactly the same output (between-spindle variation). The output from an individual spindle is not always constant (within-spindle variation).

Example 4.5 Consider different operators of the same process (operators = source of variation):

No two people are exactly alike (between-operator variation), and no person can perform in exactly the same manner (within-operator variation).

One of the objectives of developing a sampling plan for a control chart is to highlight major potential sources of variability.

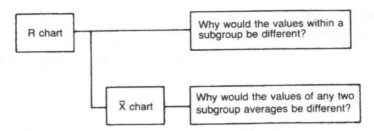

Figure 4.3 Construction of a variation diagram. Note that the R chart is completed first and sources of variation for the R chart are not repeated on the $\bar{X}$ chart.

4.4 Variation and Special Causes Diagrams

In previous sections, we have seen that the location of sampling for a control chart and the method of forming a subgroup are important factors in starting a control chart. Once the sampling plan is established, most people attempt to make measurements and start the control chart. When out-of-control conditions are found, the source of the special causes of variation are then identified, it is hoped. However, it is possible to diagram how a control chart will respond to process changes prior to data collection by using a variation diagram. In many cases the construction of these diagrams can contribute significantly to understanding the control chart.

A variation diagram simply attempts to identify the sources of variability that would cause any two subgroups on an $\bar{X}$ or R chart to have different values. The within and between sources of variability are listed in a diagram that is constructed in the format shown in Figure 4.3.

Example 4.6 A single machine is sampled by collecting five consecutive parts every hour. A single supplier of raw material is used. The process flow and variation diagrams are given here. The control chart is located at position A.

Process flow diagram:

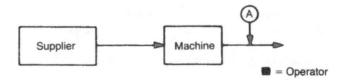

Variation diagram:

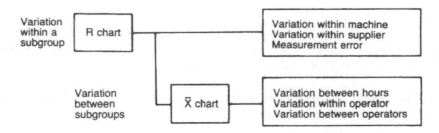

4.4 Variation and Special Causes

Example 4.7 A single machine with three spindles is sampled by collecting three parts once every 2 hours from each spindle; four subgroups are collected per day. Two suppliers of raw material are used. A stratified control chart is located at position A.

Process flow diagram:

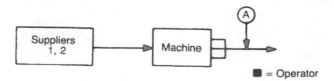

Variation diagram:

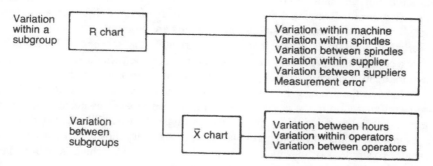

These examples illustrate several features of a variation diagram. Recall that the ideal design of a control chart utilizes rational subgroups in which there are a minimal number of sources of variation associated with the R chart. The more sources of variability associated with the R chart, the wider are the control limits and the more difficult it becomes to identify special causes of variation. The stratified control chart in Example 4.7 is a compromise since one chart is used rather than three (one for each spindle), but a baseline study is required.

If a particular chart has too many sources of variation, additional charts can be added, as Example 4.8 illustrates.

Example 4.8 This is the same situation as in Example 4.7, but four charts are used.

Process flow diagram:

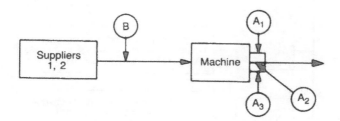

Variation diagram for A charts:

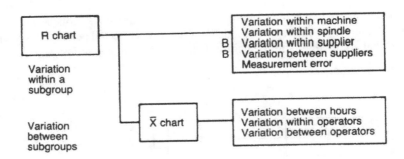

Note from Example 4.8 the variation within and between suppliers would still be present on the *A* charts, but chart *B* is used to monitor supplier variation. The *B* on the left side of the variation diagram indicates that chart *B* monitors supplier variability. Notice that the variation diagram for Example 4.8 shows fewer sources of variability than are present in Example 4.7. Although this is desirable, more charts are required.

A variation diagram is a tool that helps to identify the sources of variability that influence a control chart. A special causes diagram can be used after a variation diagram is prepared to identify specific special causes that may result in an out-of-control condition. Figure 4.4 shows the arrangement of special causes in the special causes diagram. Using the sources of variation in the variation diagram, it is possible to evaluate what could go wrong (i.e., special causes) in the process. This is a type of "what-if" exercise that greatly increases understanding of how a particular control chart responds to process changes.

Consider Example 4.6, in which the output from a single machine is being charted. Examples of individual sources of variation might be:

R chart
    Variation within machine
        Motor bearings worn
        Loose or worn part fixture
        Improper clamping force for part fixturing
        Low oil

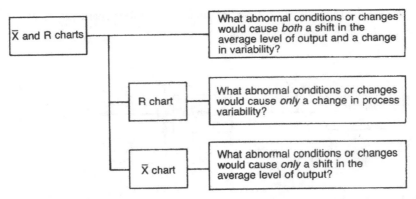

Figure 4.4 Construction of a special causes diagram.

4.4 Variation and Special Causes

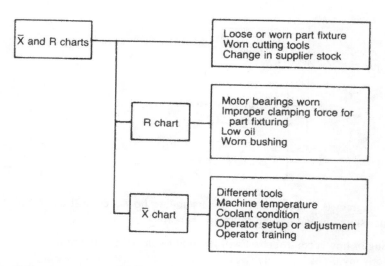

Figure 4.5 Special causes diagram for Example 4.6.

    Worn cutting tools
    Worn bushing
Variation within supplier
    Dimensional variation
    Improper metal hardness
    Part dirty
Measurement error
    Gage not repeatable
    Gage biased
$\bar{X}$ chart
    Variation between hours
    Different tools
    Change in supplier stock
    Machine temperature
    Coolant condition
Variation within operator: operator setup or adjustment
Variation between operator: operator training

Using knowledge of the application, it is possible to determine whether a particular special cause will influence the $\bar{X}$ or R chart alone or both charts. A special causes diagram for Example 4.6 appears in Figure 4.5. It should be noted that a clear grouping is not always possible: judgment is necessary.

A helpful aid in constructing a special causes diagram is to ask, if x changes, how would the control chart be affected? An alternative method is to use a cause-and-effect diagram (Chap. 13) to list special causes for process changes and then assign the causes in a special causes diagram. Note that it may be desirable to separately list special causes that result in high and low $\bar{X}$ chart out-of-control signals.

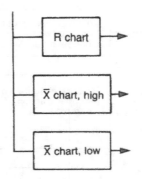

A special causes diagram is a useful tool in understanding how a control chart will function in a particular situation and has the following benefits:

Increase understanding of how a control chart will respond to changes in the process
Can be prepared before any problems actually exist
Allows evaluation of when sampling is inadequate or additional charts are needed
Provides a convenient method of identifying causes for out-of-control conditions
Can be prepared most effectively by a group, which increases the process knowledge of individual members
Can be used on the manufacturing floor (e.g., on the back of a chart)
Can be used to collect experience from problem-solving efforts

One of the advantages of a variation or special causes diagram is its use in studying how control charts respond to process changes using different sampling plans. A typical example is to study the consequences of spreading out samples within a subgroup over varying amounts of time. Consider the following sampling plans:

Plan 1: Five consecutive pieces once per hour
Plan 2: Five pieces periodically collected over a 1 hour interval
Plan 3: Five pieces periodically collected every hour over a 4 hour interval
Plan 4: Five pieces periodically collected over an 8 hour interval

Each plan has some advantages; plan 4 is commonly used in manufacturing operations since it requires the fewest measurements. However, this is the wrong justification for any plan. An alternative justification is that the wider control limits require fewer process out-of-control signals to be addressed. Again the reasoning is incorrect. However, plan 4 can be acceptable if the process capability is sufficiently high, as discussed in Chapter 14.

Evaluation of the sampling plans is possible using the variation diagram. Consider plan 1:

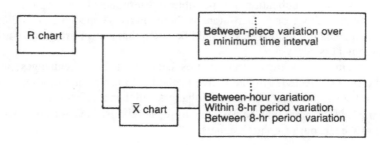

4.4 Variation and Special Causes

and plan 4:

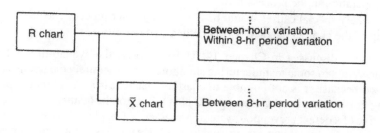

These diagrams illustrate how the variability of a tool changed every 2 hours would be evaluated. For plan 1, significant differences between tools would cause out-of-control signals on the $\bar{X}$ chart. For plan 4, normal tool change variability would increase the range and widen the control limits. Only an extreme condition would cause the range and/or the mean to be out of control. Which sampling plan is best? Table 4.1 compares the four plans.

> The wider the sampling interval for a subgroup,
> the more difficult it becomes to identify special causes

Spreading out subgroup measurements over time has a risk other than increasing the width of the control limits. Chapter 6 discusses the often unrecognized fact that only very large changes in the process mean or standard deviation are likely to be detected immediately. Smaller, but often important, changes in the process (e.g., a 1 standard deviation shift) are likely to go undetected for several subgroups. Thus, if subgroup samples are spread out over an 8 hour interval as in plan 4, several days may be required for an out-of-control signal to be obtained. During this period the process is operating in an unstable state and possibly producing defective parts.

Some of the considerations that help determine what sampling plan to use are as follows:

Importance of the characteristic
Capability of the operation

Table 4.1 Comparison of Four Sampling Plans

	Plan 1	Plan 2	Plan 3	Plan 4
Sampling method	Consecutive, once per hour	Periodic, over 1 hour interval	Periodic, over 4 hour interval	Periodic, over 8 hour interval
Number of measurements in 8 hour period ($n = 5$)	40	40	10	5
Minimum time to obtain a control chart signal when process is out of control	Minutes	1 hour	4 hours	8 hours
Width of control limits	Narrowest ←			→ Widest
Chance of detecting a special cause	Greatest ──────────────────────────────→ Least			

Consequence of a special cause going undetected for some time period
Position of operation in the process flow
Chance that likely special causes would be detected by other process checks
Relationship of the charted characteristic to other characteristics

These factors are discussed in Chapter 14. Unfortunately, the availability of labor is a prime consideration in some manufacturing situations. This is entirely the wrong approach, which is used because it is not possible to chart all characteristics in complex processes. The missing link is the existence of an overall plan to control the quality of a process—this is the objective of a defect prevention system, presented in Chapter 14. Also, control charts are only one of several tools that can be used to monitor a process. Other tools are discussed in later chapters.

> A young engineer was very pleased with achieving stability of a process. Upon explaining his accomplishment to a famous consultant, the consultant replied "So what? I can make any process appear to be stable by merely spreading out the samples over time!" The consultant went on to explain to the engineer that he must be sure he is considering which characteristics are important to the overall process and that he must utilize a meaningful sampling plan.

4.5 Troubleshooting Using Increased Sampling

When a control chart indicates an out-of-control condition, it is important to identify the special cause that resulted in a problem, as discussed in Chapter 3. However, the identification task may be difficult. One action that often proves useful is to take additional subgroups. This additional information can lead to identifying a source of variability or in some way generate an idea of the cause of the problem. For example, suppose a monitoring control chart sampling $n = 5$ periodically every 2 hours identifies the existence of a special cause problem. If the cause cannot be determined, it may prove useful to start sampling $n = 5$ parts consecutively as often as is practical. At the very least, the additional data will help quantify the extent of the problem by providing better estimates of the change in location or variability. Ideally, the pattern of the data when compared to past process performance will suggest a possible source of the problem.

Unfortunately, the control chart is often not used to help identify a problem. Employees sometimes are not aware that the control chart can suggest problem causes. For example, each pattern in Figure 4.6 would suggest potential problems in an operation. The employees in the operation in many cases could use the patterns to help suggest possible problem areas. If a special cause diagram has been prepared, this task has already been completed and referral to the diagram will provide a source of potential problems to investigate.

The only difficulty in changing the sampling plan to assist in troubleshooting is that the control limits are not necessarily correct. As discussed in the previous section, the consecutive piece sampling plan used for troubleshooting should have narrower control limits since the variability within a subgroup involves a minimal amount of time. However, the control limits are not as important when using additional sampling since we already suspect a special cause of variation is present.

4.5 Troubleshooting with Increased Sampling

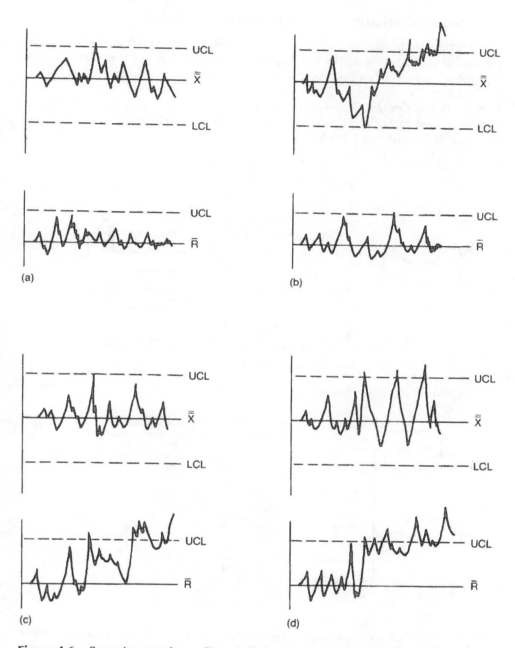

Figure 4.6 Some increased sampling problem patterns: (a) freak problem process location and range unchanged; (b) process location increasing with no change in variability; (c) process location unchanged with variability increasing; (d) increased variability causes changing mean.

4.6 Some Alternatives to Isolate Process Variation

It is not possible to eliminate all sources of variability from a process. However, it is possible to control variability by isolating sources of variability and controlling them at acceptable levels. Four methods are discussed here.

Method 1: Add Control Charts to the Operation. Consider the process flow diagram with a stratified control chart at A, where four samples are collected every 2 hours from each spindle on machines 5 and 6.

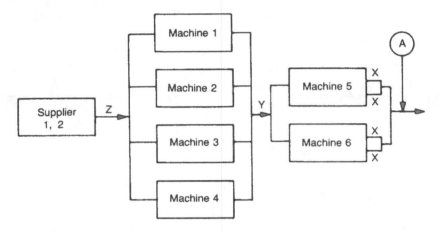

The variation diagram for monitoring machines 5 and 6 is

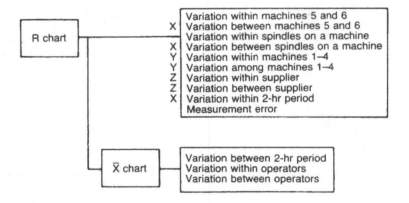

Placing four control charts, using five parts sampled periodically over 2 hours to form a subgroup, at the X location on the process flow diagram separately monitors the X sources of variation marked on the variation diagram. Thus, adding control charts to an operation separates the sources of variability monitored by individual charts.

Method 2: Add Control Charts in the Process. In the process flow diagram from Method 1, if charts were added at the Y and Z positions, upstream sources of variability would be controlled separately. Adding control charts to the process provides additional

4.6 Alternatives to Isolate Process Variation

checkpoints so that comparison of the X, Y, and Z charts can help to isolate the sources of variability. For example, if the Y and Z charts are stable and the X charts have out-of-control signals, the special causes are most likely due to within-machine or within-spindle variability for machine 5 or 6.

Method 3: Change the Process Flow. A direct method of reducing the sources of variability is to alter the process flow. The alteration can be a physical change of the process by which part flows are permanently altered to isolate process variability. In our example, the new targeted flow might be altered as follows:

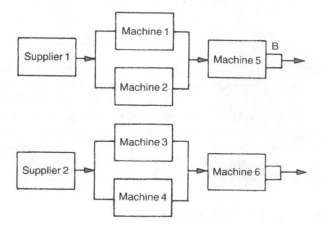

Alternatively, the same effect can be obtained by spotting parts using color codes so that no physical alteration is required. For example, supplier 1 parts might be coded blue and supplier 2, red. Only blue parts would be run on machines 1, 2, and 5 and only red parts on machines 3, 4, and 6. The variation diagram for a chart at position B using five consecutive pieces per hour would be

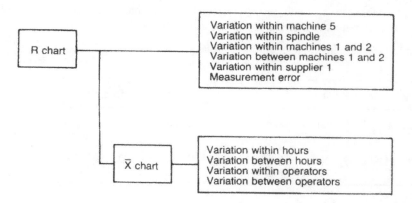

Additional charts could be added to monitor supplier and machine 1 and 2 variability.

Method 4: Alter the Sampling Plan. It is possible to alter the sampling plan to more clearly isolate variability by changing the interval between samples within a subgroup or by changing the time interval between subgroups. Consider the sampling plan in which five samples are collected periodically over a 4 hour period. Part of the variation diagram appears as follows:

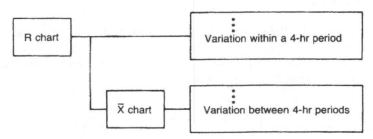

This sampling plan could be altered to collect five pieces periodically every 1 hour period. The variability within a 1 hour period is less than (or possibly equal to) that within a 4 hour period. Thus, fewer special causes would be monitored by the R chart. The narrower control limits would help to isolate the sources of variability. For example, if we assume that the initial machine start-up temperature contributes to excess variability, a periodic sample over a 4 hour period would not detect a problem. The 1 hour period would show the first hour's output to be different from that of other time periods. Thus, the temperature effect is isolated by the change in the sampling plan.

4.7 Control Chart Application Problems

4.7.1 Adjustable Versus Nonadjustable Characteristics

It is important to make a distinction between adjustable and nonadjustable characteristics in a process:

An adjustable characteristic is any characteristic that can have its mean level altered by a simple process adjustment.
A nonadjustable characteristic is any characteristic that has its mean level controlled by process parameters that are not easily adjusted.

The table gives some simple examples that help clarify the definitions:

Characteristic	Kitchen	Car	Machining
Adjustable	Oven temperature	Radio volume	Size from grinding operation
Nonadjustable	Stove temperature at "high" setting	Engine idle speed	Surface finish

Unfortunately, the distinction is not always clear. For example, the hole size from a drill is not considered adjustable by a machine operator. However, the drill size can be altered by a special sharpening operation. Basically, an adjustable characteristic can be controlled directly but a nonadjustable characteristic must be controlled indirectly through other process parameters.

4.7 Application Problems

4.7.2 Operator Overcontrol

In an effort to precisely "control" a process, operators sometimes repeatedly "fine-tune" a process by making many process adjustments for an adjustable characteristic. Unfortunately, the well-meaning operator overcontrols the process, which results in increased rather than reduced process variability. To do nothing in the way of process adjustments is the best method to minimize variability in many situations, but this is difficult for human beings to accept. We see here that the best method for making process adjustments to a stable process is to base an action on an out-of-control signal from a control chart. The only exception to this rule is when the mean level varies, as in tool wear applications (discussed later).

> Make process adjustments to a stable process based only on out-of-control signals

To illustrate the overcontrol situation, consider the $N = 30$ run chart from Chapter 3. These data were obtained from a stable process with mean $\bar{X} = 10$ and $s = 1.715$. Suppose the operator attempts to "improve" the performance of the process by adjusting the process to have a mean of 10 based on a measurement made on the process. The first measurement is 13, so the process adjustment is -3. Since the process is stable, the next measured part would have been 11, which is now 8 owing to the -3 adjustment. Now since an 8 was obtained the new operator adjustment is $+2$, and so on. The sequence of adjustment for the $N = 30$ measurement is as follows:

No.	1	2	3	4	5	6	7	8	9	10	11	12	13	14	15
Stable process	13	11	10	8	11	10	9	11	9	8	11	8	10	10	11
Adjusted process	13	8	9	8	13	9	9	12	8	9	13	7	12	10	11
Adjustment	-3	$+2$	$+1$	$+2$	-3	$+1$	$+1$	-2	$+2$	$+1$	-3	$+3$	-2	0	-1
Accumulated adjustment	-3	-1	0	$+2$	-1	0	$+1$	-1	$+1$	$+2$	-1	$+2$	0	0	-1

No.	16	17	18	19	20	21	22	23	24	25	26	27	28	29	30
Stable process	11	10	7	10	10	9	7	10	11	10	7	10	8	9	10
Adjusted process	10	9	7	13	10	9	8	13	11	9	7	13	8	11	11
Adjustment	0	$+1$	$+3$	-3	0	$+1$	$+2$	-3	-1	$+1$	$+3$	-3	$+2$	-1	-1
Accumulated adjustment	-1	0	$+3$	0	0	$+1$	$+3$	0	-1	0	$+3$	0	$+2$	$+1$	0

Stable process: $\bar{X} = 9.63$, $s = 1.43$.
Adjusted process: $\bar{X} = 9.90$, $s = 1.99$.

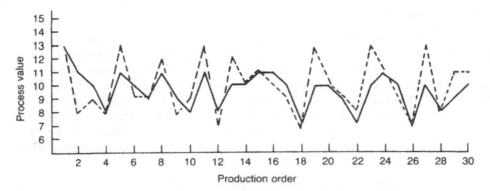

Figure 4.7 Operator overcontrol by adjusting every measurement: (—) stable process; (---) adjusted process.

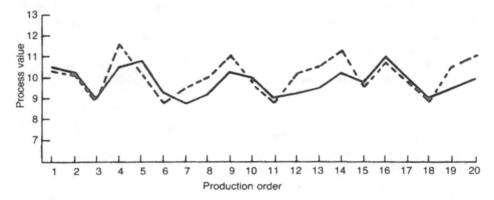

Figure 4.8 Stable process altered by a mean adjustment procedure: (—) stable process; (---) adjusted process.

The means for the stable and adjusted process are about the same, but the variability of the stable process is 39% lower. A plot of both processes appears in Figure 4.7, where the increased variability in the adjusted process can be seen.

This example is extreme in that no operator would adjust a process on every measurement. (Although some automatic machine-compensating systems implement this adjustment system!) A more typical system for an operator is to adjust the process when an out-of-specification condition arises. Problem 4.4 implements a system in which compensation occurs if a value is at or beyond the specification limits of 8 and 12. The standard deviation of the adjusted process is $s = 2.08$, which is 45% greater than the stable process.

Another misconception is that by basing process adjustments on the means of several measurements we are assured a better process. A common adjustment method is to make a process adjustment if the mean is above or below some arbitrary limits. Consider an adjustment system in which the process is adjusted if the mean of $n = 4$ measurements is 9 or below, or 11 or above. Again the data are generated from a process (Chap. 6) with $\bar{X} = 10$ and $s = 1.715$. The individual subgroup values and adjustments are as follows:

4.7 Application Problems

Subgroup	1	2	3	4	5	6	7	8	9	10
Stable process	13	11	9	10	10	9	10	9	10	10
	11	10	8	10	13	7	7	10	11	10
	10	9	11	11	10	10	10	8	11	12
	8	11	8	11	10	11	8	10	9	8
R	5	2	3	1	3	4	3	2	2	4
$\bar{X}$	10.50	10.25	9.00	10.50	10.75	9.25	8.75	9.25	10.25	10.00
Adjusted process	13	11	9	11	9.5	8.5	10.75	9.75	10.75	9.75
	11	10	8	11	12.5	6.5	7.75	10.75	11.75	9.75
	10	9	11	12	9.5	9.5	10.75	8.75	11.75	11.75
	8	11	8	12	9.5	10.5	8.75	10.75	9.75	7.75
R	5	2	3	1	3	4	3	2	2	4
$\bar{X}$	10.50	10.25	9.00	11.50	10.25	8.75	9.50	10.00	11.00	9.75
Adjustment	—	—	+1	−1.5	—	+1.25	—	—	−1	—
Accumulated adjustment	—	—	+1	−.5	−.5	+.75	+.75	+.75	−.25	−.25

Subgroup	11	12	13	14	15	16	17	18	19	20
Stable process	5	7	11	9	10	10	9	8	9	11
	8	9	6	11	10	11	9	11	12	9
	12	13	9	10	10	10	13	7	10	10
	11	8	12	11	9	13	9	10	7	10
R	7	6	6	2	1	3	4	4	5	2
$\bar{X}$	9.00	9.25	9.50	10.25	9.75	11.00	10.00	9.00	9.50	10.00
Adjusted process	4.75	8	12	10	9.75	9.75	8.75	7.75	10	12
	7.75	10	7	12	9.75	10.75	8.75	10.75	13	10
	11.75	14	10	11	9.75	9.75	12.75	6.75	11	11
	10.75	9	13	12	8.75	12.75	8.75	9.75	8	11
R	7	6	6	2	1	3	4	4	5	2
$\bar{X}$	8.75	10.25	10.50	11.25	9.50	10.75	9.75	8.75	10.50	11.00
Adjustment	+1.25	—	—	−1.25	—	—	—	+1.25	—	−1
Accumulated adjustment	+1	+1	+1	−.25	−.25	−.25	−.25	+1	+1	0

Stable process: $\bar{X} = 9.79, s = 1.63$.
Adjusted process: $\bar{X} = 10.08, s = 1.71$.

These results show that using the mean results in only a slight increase in variability for this example. A plot of both processes appears in Figure 4.8. In each of these examples, doing nothing in terms of process adjustments produced superior results to adjusting the process. When the process is stable, any type of adjustment procedure based on past measurements increases variability.

When confronted with a process that is adjusted often, the best method to study the process is to discontinue any adjustments to assess process stability. If the process is stable, then future adjustments should be made only when an out-of-control signal is given on a control chart.

4.7.3 Targeting Process Location

For most adjustable characteristics, it is desirable to target the process mean at a value within the specification limits. When the process has difficulty producing parts having measurements within the specification limits, a target of the midpoint of the specification produces the fewest rejects. When the process has sufficiently reduced variability (see Chap. 7), targeting the process location at a level that optimizes the function of the product is desirable. Also, it is often possible to target mean levels for assembly ease. Targeting process levels encourages employees to optimize some aspect of the process. Making the best possible product is clearly preferable to merely meeting specification limits.

> A visitor asked an employee why he thought the values on his control chart varied so drastically. The employee replied that all his parts were within specification limits and the control chart proved it! The visitor inquired further as to what might be done to reduce the variability. The employee had a number of suggestions of what could be done, but he was confused as to why merely meeting specifications wasn't adequate.

The danger of attempting to target a process mean without using control charts is that there is a natural human tendency to overcontrol the process. An employee obtains a slightly high measurement and in an attempt to do a good job adjusts down the process. What follows is a constant series of adjustments, increasing the overall variability of the product. However, it is possible to use control charts to target the process level of an adjustable characteristic. Using a target mean level with a control chart is not difficult. After determining the target level T, simply adjust the process mean to T. Standard control chart methods can then be used.

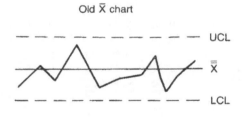
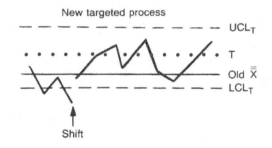

The new control limits are

$$T = \text{centerline}$$
$$\text{UCL}_T = T + A_2 \bar{R}$$
$$\text{LCL}_T = T - A_2 \bar{R}$$

Often a new symbol (for example, a dotted line) is used to denote T on the $\bar{X}$ chart.

The range chart does not change. Using the new $\bar{X}$ chart, a process adjustment should not be made unless the process exhibits an out-of-control signal. Care should be taken not to adjust the process mean arbitrarily when an out-of-control signal is obtained: the cause of the problem should be assessed. Using this system makes operator overcontrol unlikely and targets the process at a desired level.

Targeting without control charts can lead to increased process variability

4.7 Application Problems

Grubbs (1983) presents an optimal adjustment procedure for moving the process to the targeted level. The first move adjusts the process the full distance required (i.e., $T - $ old $\bar{\bar{X}}$). The new process mean at stage 2 is new $\bar{\bar{X}}_2$. The second move is $\frac{1}{2}(T - $ new $\bar{\bar{X}}_2)$, the third $\frac{1}{3}(T - $ new $\bar{\bar{X}}_3)$, and so on. This procedure can be used for means or individual measurements.

4.7.4 Within-Part Variability

It is apparent that the diameter of a hole is not constant. For a sufficiently sensitive measurement device, any hole could be shown to vary between some minimum and maximum diameter. Similarly, the case hardness of a metal surface varies across a section: it is not constant. Most physical characteristics vary provided a sufficiently sensitive measurement instrument is used. Fortunately, in most applications, it is not necessary to address the within-part differences of a characteristic. The range between the minimum and maximum values is small compared with the tolerance. However, in other cases it is necessary to operationally define how within-part differences are to be addressed (see Problem 3.13). For example, in some machining applications the clearance between two assembled components may be only .001 inch. A difference between the minimum and maximum within a part of .0005 inch would surely be important.

The preferred method of operationally defining a part feature is to use two blueprint specifications, one for absolute location and one for variability. For example, the size of a hole could be specified as $1.500 \pm .005$ inch. Also, a circularity (roundness) specification of .002 inch could be used, which means that for any cross section of the hole the radius must not vary more than 0.002 inch. However, it is still necessary to define operationally how the hole diameter should be recorded. Most operators use an average of the minimum and maximum. What is the diameter of the hole on the right?

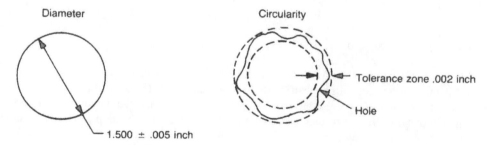

When a within-part variability specification is not given, it is usually assumed that the minimum measurement must be above the lower specification limit and the maximum measurement below the upper specification limit. Unfortunately, a common mistake is made when parts are actually measured. The average of the minimum and maximum is recorded. If the within-part variability is sufficiently large, this practice can cause problems. Clearly, the average could be well within the specification limits and the minimum and/or maximum beyond the limits. Problem 7.8 gives an example of computing process capability considering within-part variability.

The function of the part characteristic in the final product must be considered to select the best measurement approach. Several alternatives are possible:

Average of minimum and maximum
Minimum, maximum, or both
Difference between maximum and minimum

Combinations of these possibilities are sometimes used. Frequently it is necessary to conduct a study to evaluate which approach best monitors a process. It may be necessary to use two charts to monitor a single characteristic. A case study illustrates this situation.

Appendix III presents a Geometric Dimensioning and Tolerancing system which can address within part variation using more complete part specifications.

4.7.5 Inadequate Measurement Sensitivity

Recall from Chapter 2 that it was recommended that at least 10 units of measurement should be within the specification limits. The lack of measurement precision influences a control chart by making the range have many 0 values since the measurement system cannot differentiate between similar parts. This problem is sometimes associated with consecutive piece sampling since parts have a minimum opportunity to exhibit variability. The R chart may appear as follows:

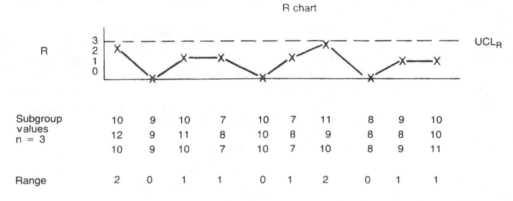

Subgroup values $n = 3$	10	9	10	7	10	7	11	8	9	10
	12	9	11	8	10	8	9	8	8	10
	10	9	10	7	10	7	10	8	9	11
Range	2	0	1	1	0	1	2	0	1	1

The main difficulty when a reasonably large number of R values are 0 (say, 20% or more) is that the $\bar{R}$ value will be artificially low, which causes the control limits to be too narrow. This results in incorrect out-of-control signals and a chart that is not useful (App. I). The simplest solution is to increase the time interval over which samples are collected for a subgroup. As discussed in Section 4.4, this increases the sources of variability represented on the R chart and often reduces the influence of any lack of measurement sensitivity. It may also be useful to increase the subgroup sample size n. Wheeler and Chambers (1986) say that inadequate measurement sensitivity exists if the range chart has four or fewer possible values within the range control limits.

An alternative solution to the problem is to obtain a more sensitive measurement system. However, this approach is not only costly but is often unnecessary. If the capability of the process is sufficiently high, why purchase expensive gaging to precisely quantify an already acceptable situation? Alternative control methods, such as a variable check sheet (Chap. 8) or a dot control chart (Chap. 7), may be better tools for monitoring the operation.

4.7.6 Modified Control Limits for a Shifting Mean*

A problem that arises in some manufacturing operations is a shifting process mean. For example, suppose a reamer is used to drill a hole in a repetitive operation. An individual

4.7 Application Problems

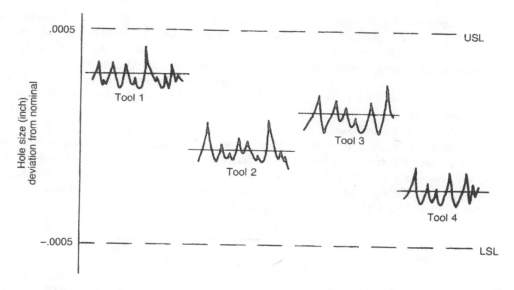

Figure 4.9 Shifting mean run chart for a reamer hole size problem.

reamer may be left in a machining station for a day or a week, depending on the production rate and the quality of the hole produced. A typical tolerance for a reamed hole is ±.0005 inch or a total specification width of .001 inch. An individual reamer generally cuts a hole to within about .0002 inch of the nominal size of the reamer. Reamer manufacturers produce tools that typically use about one-half the .001 inch allowable spread. The system discussed here would produce the run chart that appears in Figure 4.9. The shifting mean with each change of tool is apparent. If a standard $\bar{X}$ and R chart is used, the range for an individual tool is about .0002 inch, so the $\bar{X}$ chart would appear totally unstable with many out-of-control points. It is not technically feasible to improve the process by producing all tools at exactly the same value. Even if this were possible, the tool sharpening operation would change the mean.

A common strategy is not to use control charts, but simply to check a few setup pieces. However, reaming operations can have many sources of variability:

Source of size variability	Chart
Operator setup of tool	$\bar{X}$
Dull tool	R
Worn spindle bearings	R
Coolant distribution or condition	R
Ovality of hole	R

Abandoning the use of control charts is often not desirable, but the standard approach is not meaningful, either.

One method of addressing the shifting mean problem is to use "movable control limits." In a number of situations, an overlay is used on the control chart so that the control limits for the $\bar{X}$ values can be moved with each new tool. This approach can often be easily implemented in a shifting mean situation. The advantage of this approach is that a standard control chart format is maintained and process stability can be evaluated. A disadvantage is that a reasonable number of subgroups must be collected before it is apparent how to position the overlay.

Another method to address the shifting mean problem is to use modified control limits. The method of computing the rejection lines is given in Procedure 4.2. The rejection lines are simply lines that reduce the width of the tolerance so that the chance of a value beyond the specification limits is small. The theoretical basis is given in Burr (1976, Sec. 7.2, Modified limit B). The figure below shows the basic concept of rejection lines (URL = upper rejection line; LRL = lower rejection line).

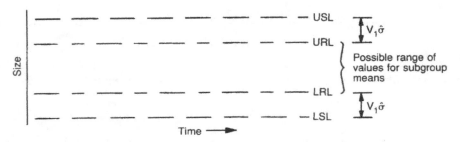

The following examples illustrate the rejection limit calculations.

Example 4.9 (a) Suppose that the range of a process was judged stable with $\bar{R} = 5$ using a subgroup size of $n = 5$. From step 6 of Procedure 4.2,

$$\hat{\sigma} = \frac{\bar{R}}{d_2} = \frac{5}{2.33} = 2.15$$

Now from step 7, using a specification limit of USL = 10 and LSL = -10 gives

URL = $10 - .91 \times 5 = 5.45$

LRL = $-10 + .91 \times 5 = -5.45$

Note that the control limits for this process would be, for $\bar{\bar{X}} = 0$,

UCL = $0 + .58 \times 5 = 2.9$

LCL = $0 - .58 \times 5 = -2.9$

(b) Suppose that the specification limits were changed to USL = 6 and LSL = -6 in part a; the new rejection limits would be

URL = $6 - .91 \times 5 = 1.45$

LRL = $-6 + .91 \times 5 = -1.45$

A plot of the specification, rejection, and control limits appears in Figure 4.10 for Example 4.9. Notice that the relationship between the rejection and control limits depends

4.7 Application Problems

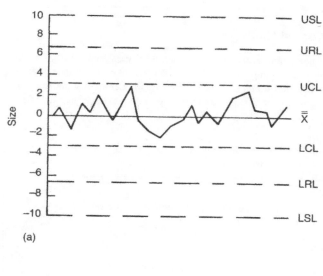

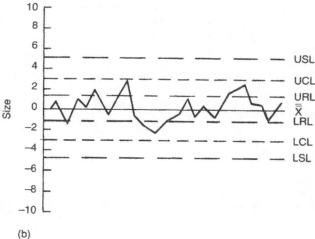

Figure 4.10 Rejection and control limits for Example 4.9: (a) specification limits of ±10; (b) specification limits of ±6.

on the specification limits. This emphasizes that there is no relationship between process stability and rejection lines. Note in Figure 4.10b that if operators were to make process adjustments based on the rejection limits, an overcontrol situation would exist and variability would increase. In this case rejection limits would not be appropriate. Unfortunately, modified control limits have been largely misused in manufacturing by using procedures for which standard $\bar{X}$ and R charts are appropriate. In these cases, not only is process stability not established, but continual improvement (see Chap. 1) is abandoned. No longer is the emphasis on narrowing the control limits and continually improving the process. The mind-set of "what we're doing is good enough" is established rather than "we must continually improve."

Procedure 4.2 Modified Control Limits

Consider a process for which it is not meaningful to establish stability (i.e., statistical control) for the mean level. However, it is desirable to implement a procedure that makes the chance of parts beyond the upper and lower specification limits (LSL and USL) small. Let a constant group of measurements n be made on the process for at least 20–25 time periods k, where the values are arranged as follows:

Time	1	2	$\cdots$	k	Comment
Measurement 1	X_1	X_1	$\cdots$	X_1	
2	X_2	X_2	$\cdots$	X_2	n measurements from each subgroup
.	.	.			
.	.	.			
.	.	.			
n	X_n	X_n	$\cdots$	X_n	
Mean $\bar{X}$	$\bar{X}_1$	$\bar{X}_2$	$\cdots$	$\bar{X}_k$	A subgroup mean and range are computed
Range R	R_1	R_2	$\cdots$	R_k	for each time period

The subgroup size n is generally 3, 4, or 5, but it may be greater. The following steps are used to compute the modified control limits.

1. Compute the mean and range for each subgroup:

$$\bar{X} = \frac{X_1 + X_2 + \cdots + X_n}{n}$$

$$R = X_{\text{highest}} - X_{\text{lowest}}$$

2. Plot the $\bar{X}$ and R values for each subgroup on control chart paper.
3. Compute the overall mean $\bar{\bar{X}}$ and average range $\bar{R}$ for the k time periods:

$$\bar{\bar{X}} = \frac{\bar{X}_1 + \bar{X}_2 + \cdots + \bar{X}_k}{k}$$

$$\bar{R} = \frac{R_1 + R_2 + \cdots + R_k}{k}$$

These values should be plotted on the control chart with a solid line.

4. Compute the upper and lower control limits for the range.

Upper control limit for R:

$$\text{UCL}_R = D_4 \bar{R}$$

Lower control limit for R:

$$\text{LCL}_R = D_3 \bar{R}$$

4.7 Application Problems

n	D_3	D_4	d_2	V_1
3	0	2.58	1.69	1.09
4	0	2.28	2.06	.97
5	0	2.12	2.33	.91
6	0	2.00	2.53	.86
7	.08	1.92	2.70	.83
8	.14	1.86	2.85	.81
9	.18	1.82	2.97	.79
10	.22	1.78	3.08	.77

Appendix III contains a more complete table of values.

5. Using dashed lines, plot UCL_R and LCL_R and assess whether the range is stable using Procedure 3.7.
6. If the process range is judged stable, compute an estimate of the process standard deviation:

$$\hat{\sigma} = \frac{\bar{R}}{d_2}$$

7. Compute the upper and lower rejection limits.

Upper rejection limit:

$$URL = USL - V_1 \bar{R}$$

Lower rejection limit:

$$LRL = LSL + V_1 \bar{R}$$

Draw these dashed lines on a $\bar{X}$ chart control chart. Lines should be labeled URL and LRL to emphasize that process stability is not being established for the $\bar{X}$ chart. Note that if URL − LRL is less than UCL − LCL from Procedure 3.2, standard control limits should be used, not rejection limits.

Process stability is not established using modified control limits

In a few situations modified control limits do provide a useful alternative when standard charting methods are not meaningful. The shifting mean is a common example, as illustrated by the reamer operation discussed earlier. Other examples include batch mixture problems, in which the variability between batches is much larger than the variability within a batch. In many of these examples, it is very important to use the R chart since it is an indicator of many problems. Recall in the reamer example that the R chart was the most useful indicator of process problems.

4.8 Special Applications*

4.8.1 Tool Wear

In some machining processes the cutting tool is designed to wear slightly as a part is machined. As the tool wears, the size characteristic being machined gradually becomes either larger or smaller. The following two figures are examples of each case:

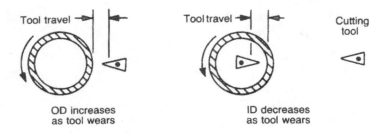

To compensate for the tool wear, various machine adjustments are made. In this example, the part is rotating in a stationary position so that the tool travel is increased as the tool wears. The rate of wear depends on the type of cutting operation. Some tools need adjustment after 10 or (fewer) parts, and some last 1000 or more parts.

Typical run chart patterns for a tool wear situation in which size increases appear in Figure 4.11. In many manufacturing situations, the general type of pattern is predictable. If a linear wear trend occurs for an individual tool, then other similar tools exhibit a linear trend. However, the type of linear trend often varies. For example, slight changes in the hardness of the metal being cut cause a variable rate of tool wear. Hardness is generally consistent within one melt of metal, but different melts have slightly different hardnesses. The following figure shows that, as expected, similar tools wear faster with harder materials:

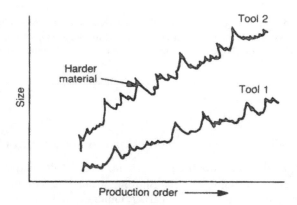

There is no "best" method of controlling a tool wear situation. Each case must consider

Predictability of tool wear pattern
Frequency of adjustments
Frequency of tool change
Availability of personnel

4.8 Special Applications

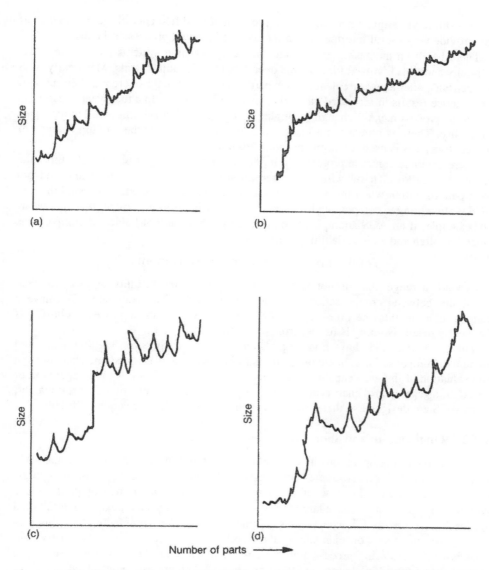

Figure 4.11 Tool wear examples in which size is increasing: (a) linear trend; (b) break-in; (c) chipped tool; (d) variable.

and adapt the method of control to the situation. There are two common approaches. First, if there is a constant linear trend across tools and material, for example, the regression control limits given in Grant and Leavenworth (1980) are appropriate. Unfortunately, as noted earlier many machining examples cannot easily be adapted to this approach. Second, modified control limits can be used since the process mean is always changing. In this case, the process stability for the mean is not established as is the case with the regression approach. In either case, it is still meaningful to use the R chart to establish the stability of piece-to-piece variation. As noted in the discussion of modified control limits, the R chart can provide very useful information for monitoring potential process problems.

There has been increasing use of automatic compensating systems (ACS) due to the difficulty and time involved for machine operators to make adjustments. Many early ACS compensated after a measurement of each part. As was seen in the overcontrol discussion, this practice results in increased variability in a stable process. In a tool wear situation in which the process mean is changing, single-part compensating systems result in increased variability. There is now a trend to use ACS when the mean or the moving average of consecutive parts is used for compensating (Prob. 4.5).

There are three common problems in implementation and use of many ACS. First, the process is not allowed to run without compensation to establish the baseline R chart and tool wear pattern. Often when the range of consecutive pieces is studied, it is found that the ACS is attempting to compensate for excess variability in the system rather than tool wear. For example, if an ACS attempts to compensate for a loose tool holder, it compensates much too often and adds variability to the system.

Let the process run without adjustments

Second, a range chart is not kept on the process so the stability of piece-to-piece variability between compensation times is not evaluated. The comment "We have a compensator on this machine so we don't need to worry about process control" is frequently heard. Nothing is further from the truth.

Third, an ACS must have a gaging system to monitor part size. The frequency and amount of compensation is based on the measurements made by this system. There is often no evaluation of the gage capability. Because this gaging is so critical to the operation of the ACS, a gage control chart is appropriate (App. I) for many operations. Unfortunately, many ACS are designed so that evaluation of the gaging capability is quite difficult.

4.8.2 Rapid Output and Short Run

Some manufacturing operations involve short production runs of parts owing to the high-volume output of a process. Some common examples are springs, bolts, small stampings, forging, and castings. Because of the high-volume output, thousands of parts can be produced in a few hours. Since large tubs of material are produced quickly, the traditional approach has been to use acceptance sampling methods. As discussed earlier, this procedure does not focus on maintaining the stability of the process. Two alternative approaches for monitoring process stability are discussed here. The first uses a common control chart for multiple parts, and the second uses a Gillette parts control system.

In many forging or casting operations, batches of parts are run together in short production runs. Only a few hours may be needed to produce the production requirements for an individual part. Many other manufacturing processes result in short production runs of many different parts. In these cases it is often not possible or even meaningful to use a control chart for every part. A solution that can be used in many cases is to combine control charts for multiple parts into a single chart. This approach then focuses attention on control of the process rather than on monitoring an individual part as is usually the case.

For example, the Brinell hardness of forgings is a primary concern of machining operations since the machinability of a forging is related to hardness. The parameters controlling the hardness of a part include a variety of metal temperatures and composition, as well as many other forging process parameters unrelated to the part being produced. It is

4.8 Special Applications

thus meaningful to combine the hardness testing results from multiple parts produced during a production run.

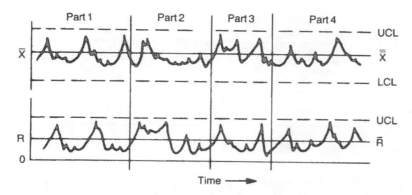

This approach assumes that the target level of hardness is the same for all parts. If the target levels are different for every part (as with a dimensional characteristic), a difference control chart discussed in the next section could be used. In this case, the plotted value would be

Difference = actual value − target value

In many rapid-output applications, machine setup is a critical factor. For example, if a stamping die is set up properly, the piece-to-piece variability is small unless the die breaks at some point during the production run. If die breakage occurs, most of the parts will be defective. Segregation of parts into small lots makes practical sense and can be used as an effective control procedure. The Gillette parts control system is often used to segregate parts into separate small lots:

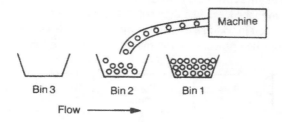

For small parts, the bins or lots can be arranged on a carousel. Alternatively, artificial "bins" are sometimes created by segregating output into layers using cardboard or canvas, for example, in a large tub:

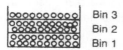

It is useful to use the Gillette system with control charts by forming a subgroup using the last n pieces going into a bin (in some cases the first n pieces are also used). For a production run, 20–25 bins can be created and the stability of the entire run assessed. Thus, the stability of the control chart provides a natural criterion for acceptance of the production run.

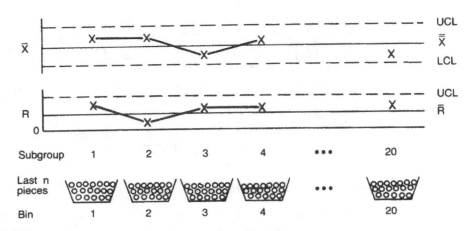

4.8.3 Job Shop

A job shop typically has many short production runs involving only a few identical parts. For example, a shop producing prototype engine or transmission parts may produce only a single part. Most people believe that this type of operation does not lend itself to using control charts. However, the question should not be whether to use a control chart, but whether a stable, predictable process exists by which work is accomplished. The charts are only a tool to evaluate this criterion. If we focus on assessing the quality of the process by which work is accomplished, we can then select the right tool.

Consider the situation in which a machining operator builds various prototype parts for a transmission. A simplified process for the machine operator might appear as follows:

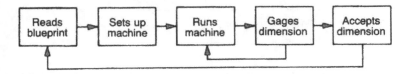

The process is repeated for each dimension on the part or for other different parts. The relevant question becomes, "How does the operator know if he's done a good job?" A reasonable criterion is to evaluate whether each dimension is close to the targeted setup, usually the midpoint of the specification limits (nominal). Thus,

Difference = actual value − target value

is a measure of the "quality" of the output. The dimensions with a similar specification range can be combined to form a subgroup. Suppose we decide to select three dimensions (with similar specification ranges) from every part produced. Subgroups of size $n = 3$ could be formed as follows:

4.8 Special Applications

	Subgroups		
Dimension	1	2	3
1	Diff$_1$ (dimension A, part 1)	Diff$_1$ (dimension D, part 2)	...
2	Diff$_2$ (dimension B, part 1)	Diff$_2$ (dimension E, part 2)	
3	Diff$_3$ (dimension C, part 1)	Diff$_3$ (dimension F, part 2)	

Consider the following example.

Example 4.10 Suppose we have four parts and decide to measure three dimensions. The target and actual values are as follows:

Dimension	Actual	Target	Difference	Part
A	9.1	10.5	−1.4	1
B	51.7	52.1	−.4	1
C	8.2	7.6	.6	1
E	18.5	17.3	1.2	2
F	9.2	11.0	−1.8	2
G	94.3	92.4	1.9	2
H	63.4	61.2	2.2	3
I	59.1	57.0	2.1	3
J	4.3	5.6	−1.3	3
K	101.2	102.5	−1.3	4
L	41.6	42.1	−.5	4
M	178.5	176.5	2.0	4

The subgroup used to evaluate the operator's performance would be these differences, which would be arranged as follows:

Subgroup	1	2	3	4	...
1	−1.4	1.2	2.2	−1.3	
2	−.4	−1.8	2.1	−.5	
3	.6	1.9	−1.3	2.0	
$\bar{X}$	−.40	.43	1.00	.07	...
R	2.0	3.7	3.5	3.3	...

The important point of the prototype example is that the process flow helps to identify the repetitive activity in the process. The difference control chart then focuses on evaluating the quality of the process. Establishing defect prevention in a job shop thus focus on

Identification of the repetitive part of the work process
Identification of the quality criteria for the process

In this example, only a single part was manufactured. However, in most cases a job shop runs very short production runs of, say, 10 to several hundred parts. In these cases, the machine setup is often the most time consuming and critical part of the operation. During a day, a job shop operation would have a number of different jobs, each requiring various machine setups. If the setup operation is most critical, the first five production parts could be measured for a characteristic. On the next job, a similar characteristic would be measured, again using the first five parts. The sampling plan would appear as follows:

```
   Job 1         Job 2      Job 3          Job 4             Job 5        Job 6
•XXXXX───────►•XXXXX•XXXXX────────►•XXXXX─────────────►•XXXXX─•XXXXX──────────►•
```

Alternatively, one may wish to sample parts throughout the run or the last parts as in the Gillette system.

The selected characteristics need not, of course, have the same target value. Just as in the prototype example the target value is subtracted from an actual value, so only differences from an actual value are considered:

Difference = actual value − target value

The following example shows how subgroups are formed.

Example 4.11 Suppose the first two runs of a machine in a job shop have target values of 96.3 for job 1 and 72.1 for job 2 for an important characteristic being machined. The first five parts from each run are

Job 1   94.2, 97.8, 98.1, 93.7, 96.4
Job 2   74.2, 74.1, 70.9, 72.7, 71.5

The first two subgroups are formed by subtracting the actual value from the target value. The $\bar{X}$ and R are then computed in the usual manner.

Subgroup	Differences					$\bar{X}$	R
1	−2.1,	1.5,	1.8,	−2.6,	.1	−.26	4.4
2	2.1,	2.0,	−1.2,	.6,	−.6	.58	3.3

An out-of-control signal indicates a job that has not been setup correctly. A mean significantly different from zero indicates bias in the setup procedure. Of major interest is the R chart since it indicates the variability in the process setup. The only difficult part of this procedure is to select characteristics that can be meaningfully compared.

Another example of a job shop activity is a testing laboratory that analyzes a wide variety of chemicals in submitted samples. Each sample is analyzed by a rapid method, and selected samples are divided after thorough mixing and analyzed by a very precise but slow measuring procedure. An abbreviated process flow is as follows:

4.9 Case Studies

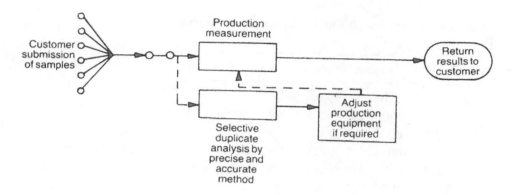

The question is how to assess whether the production equipment is operating properly. The magnitude of the measurements of a customer's samples may vary widely, but the relative percentage of variation between the two measurement methods should remain stable. Thus, the quantity

$$\text{Relative \% difference} = 100 \; \frac{\text{production method} - \text{accurate method}}{\text{accurate method}}$$

is a reasonable measure of how the system is performing. Selecting subgroups of, say, $n = 3$ randomly selected over a 2 hour period provides a good method of assessing process stability. Generally, this approach using production samples is a good method to evaluate the quality of the process. When only laboratory standards are used, sample preparation, handling systems, and other parts of the production process are not effectively evaluated. This is another example of how identifying the repetitive operation (chemical measurement) and quality characteristic (relative percentage difference) enables an effective evaluation of process stability.

4.9 Case Studies

4.9.1 Comparison of Two Sampling Plans

A special study was conducted in a foundry to determine what type of control chart sampling plan provided the best approach for monitoring the process. The characteristic of interest was the Brinell hardness of an iron casting, which had specification limits of 255 to 187 BHN (Brinell hardness number). The hardness of the casting related to the material composition and a number of process parameters, such as cooling rate of the heated part. Two sampling plans were considered:

Plan 1 (periodic): one part was collected every 30 minutes.
Plan 2 (consecutive): five consecutive parts were collected every hour.

In both cases a subgroup size of $n = 5$ was used to form subgroups for an $\bar{X}$ and R chart. The data were collected over the same 3 day time period. For plan 1, 20 subgroups were collected, and plan 2 had 49 subgroups.

If the foundry process was perfectly stable with completely controlled variation, there would be no difference between the results of the two sampling plans. The data for the two

Table 4.2 Measurements Obtained from Two Sampling Plans

No.	Subgroup values					$\bar{X}$	R
\multicolumn{8}{c}{Periodic sampling plan}							
1	229	229	235	223	229	229.0	12
2	223	223	217	223	229	223.0	12
3	223	201	229	223	229	221.0	28
4	223	217	229	223	223	223.0	12
5	229	217	223	223	229	224.2	12
6	229	223	217	223	235	225.4	18
7	241	229	241	223	212	229.2	29
8	217	223	229	223	223	223.0	12
9	229	217	212	223	235	223.2	23
10	241	235	229	235	235	235.0	12
11	235	235	217	197	212	219.2	38
12	217	223	217	207	235	219.8	28
13	223	241	235	223	229	230.2	18
14	229	235	229	229	223	229.0	12
15	235	229	223	212	223	224.4	23
16	212	223	212	217	217	216.2	11
17	223	223	223	223	229	224.2	6
18	229	223	235	255	241	236.6	32
19	248	248	229	235	217	235.4	31
20	229	217	223	229	223	224.2	12
\multicolumn{8}{c}{Consecutive sampling plan}							
1	229	229	229	229	229	229.0	0
2	235	229	229	223	229	229.0	12
3	229	229	229	235	229	230.2	6
4	223	223	235	223	229	226.6	12
5	223	223	229	229	229	226.6	6
6	229	229	229	229	229	229.0	0
7	223	229	229	229	229	227.8	6
8	235	235	229	223	223	229.0	12
9	217	217	223	207	207	214.2	16
10	229	229	223	217	207	221.0	22
11	229	223	223	223	217	223.0	12
12	217	212	212	212	207	212.0	10
13	229	223	223	217	217	221.8	12
14	229	229	229	217	217	224.2	12

4.9 Case Studies

No.	Subgroup values					$\bar{X}$	R
	Consecutive sampling plan						
15	229	229	223	217	217	223.0	12
16	235	235	235	229	229	232.6	6
17	235	235	229	229	229	231.4	6
18	223	223	217	217	207	217.4	16
19	235	229	229	223	223	227.8	12
20	229	223	223	223	217	223.0	12
21	229	223	217	217	207	218.6	22
22	229	229	223	223	223	225.4	6
23	241	235	229	229	223	231.4	18
24	235	235	235	223	229	231.4	12
25	235	235	229	229	229	231.4	6
26	235	229	223	223	217	225.4	18
27	229	229	223	223	217	224.2	12
28	285	229	229	217	207	233.4	78
29	235	229	223	223	223	226.6	12
30	229	229	223	217	217	223.0	12
31	229	223	223	217	217	221.8	12
32	223	223	223	217	217	220.6	6
33	229	223	223	223	217	223.0	12
34	235	229	229	223	223	227.8	12
35	212	229	229	229	229	225.6	17
36	223	223	229	229	229	226.6	6
37	207	217	241	229	229	224.6	34
38	223	223	229	229	229	226.6	6
39	235	248	229	229	229	234.0	19
40	229	229	229	229	235	230.2	6
41	235	235	241	241	229	236.2	12
42	229	229	229	223	223	226.6	6
43	235	235	235	229	229	232.6	6
44	229	229	223	223	223	225.4	6
45	241	229	229	223	197	223.8	44
46	223	223	223	217	217	220.6	6
47	229	223	223	217	212	220.8	17
48	229	229	223	223	217	224.2	12
49	229	223	223	223	217	223.0	12

sampling plans are given in Table 4.2 with the corresponding control charts in Figures 4.12 and 4.13. The control chart values are as follows:

	Plan 1, periodic	Plan 2, consecutive
$\bar{\bar{X}}$	225.8	225.6
UCL	237	233
LCL	215	219
$\bar{R}$	19.1	11.9
UCL	40.5	25.2

Subgroup 28 was deleted from the calculations owing to a special cause.

A comparison of the two control charts provides insight into how the sampling plan influences the interpretation of the behavior of the process. The periodic sampling plan in Figure 4.12 is not totally stable since subgroups 18 and 19 are both in zone A, but the process does not indicate erratic behavior. The process standard deviation is estimated by

$$\hat{\sigma} = \frac{\bar{R}}{d_2} = \frac{19.1}{2.33} = 8.2$$

and the upper and lower process limits are

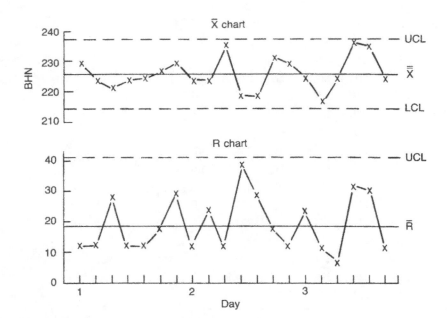

Figure 4.12 Control chart for periodic sampling plan.

4.9 Case Studies

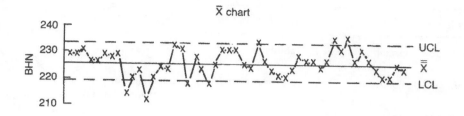

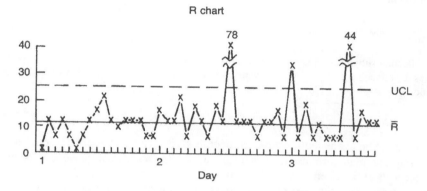

Figure 4.13 Control chart for consecutive sampling plan.

$$\text{UPL} = \bar{\bar{X}} + 3\hat{\sigma} = 225.8 + (3 \times 8.2) = 250.4$$
$$\text{LPL} = \bar{\bar{X}} - 3\hat{\sigma} = 225.8 - (3 \times 8.2) = 201.2$$

The process limits are within the specification limits so we would probably feel reasonably confident about the process. (This argument assumes that the process is stable and that any special cause associated with subgroups 18 and 19 could be eliminated.) The conclusion derived from the consecutive sampling plan in Figure 4.13 is quite different from that derived from the periodic sampling plan in Figure 4.12. The process appears quite unstable, with many out-of-control points. Since the process was sampled over the same period of time, we must conclude that Figure 4.13 gives the clearer picture of the process behavior.

Which sampling plan is best? There is no clear answer to this question. The sampling plan must be adapted to the process. In the foundry example the results of the two sampling plans are not unexpected. A foundry runs as a batch process. Adjacent parts collected in consecutive sampling are quite similar and generally show low ranges for a subgroup. Variability between different batches is expected to be much larger than within a batch of material. This variability is often considered a natural part of the process since it may be difficult, if not impossible, to eliminate. Consecutive piece sampling, in which between-subgroup variability is much greater than within-subgroup variability, can lead to too many process adjustments (operator overcontrol) and increased process variability. Conversely, the example clearly illustrates that several severe out-of-control conditions were completely missed by the periodic sampling plan. A summary of some of the advantages and disadvantages of the two plans follows:

Comparison issue	Consecutive	Periodic
Ability to detect special causes	Greatest ability	Lesser ability
Possibility of overcontrol	Greatest if between-batch variability is large	Less likely depending on time between samples
Reaction time to out-of-control condition	Smallest	Can be large if time between samples within a subgroup is large
Sampling effort and cost	Greatest	Can be adjusted to any intensity
Ease of collecting subgroup samples for unmanned operation	Easy since operator returns periodically to collect an entire subgroup	Can be difficult since operator must return to operation at regular intervals

These issues must be evaluated for different operations and the best approach selected. If periodic sampling plans are used, it is often useful to occasionally use consecutive sampling for a short time period to evaluate the process in more detail.

Comparison between the two control charts also illustrates why consecutive piece sampling is useful for troubleshooting. The possibility of observing a special cause is greatest with a consecutive sampling plan. Elimination of the special cause is the intent of the troubleshooting effort.

4.9.2 Controlling Eight Fixtures

A transmission valve body grinding machine removes about 0.060 inch from one side of an aluminum valve body casting. An eight fixture machine (see Chap. 7, Sec. 7.11 for a description) was used to grind the parts. This grinding operation had an in-process specification of .003 inch flatness of the ground surface. A downstream finish grinding operation had a .001 inch flatness specification. It was not practical to maintain a control chart for each fixture, yet the in-process flatness stability was important since many downstream finish characteristics depended on the in-process flatness.

It was decided that a stratified control chart would be used with two subgroups of $n = 8$ collected per shift. This sampling plan was operationally convenient for the machine operator since eight consecutive parts, one from each fixture, could easily be collected. Prior to using the stratified control chart, a baseline study was performed to ensure that the means and ranges of the individual fixtures were about the same. Using subgroups of size $n = 3$ consecutive samples, the fixture means and ranges are listed as follows:

Fixture	1	2	3	4	5	6	7	8	Baseline average
Mean $\bar{\bar{X}}$	21.6	22.7	19.5	22.1	22.1	23.1	19.4	22.3	21.6
Range $\bar{R}$ ($n = 3$)	7.5	9.4	8.0	10.1	9.8	6.3	7.2	8.1	8.3

4.9 Case Studies

A comparison of the fixture means was performed using the procedure discussed in Chapter 11 and it was concluded that the fixtures were as close as could be reasonably expected. To compute the control limits for the $n = 8$ stratified control chart, we first compute the estimated standard deviation from the baseline study:

$$\hat{\sigma} = \frac{\bar{R}_{base}}{d_2} = \frac{8.3}{1.69} = 4.9$$

Since $n = 8$ for the stratified chart, we now compute the expected $\bar{R}$ for the stratified chart:

$$\bar{R}_{strata} = d_2\hat{\sigma} = 2.85 \times 4.9 = 14.0$$

The control limits for the new stratified $\bar{X}$ and R charts are as follows:

$$UCL_{\bar{X}} = \bar{\bar{X}} + A_2\bar{R}_{strata} = 21.6 + (.37 \times 14) = 26.8$$
$$LCL_{\bar{X}} = \bar{\bar{X}} - A_2\bar{R}_{strata} = 21.6 - (.37 \times 14) = 16.4$$
$$UCL_R = D_4\bar{R} = 1.86 \times 14 = 26.0$$
$$LCL_R = D_3\bar{R} = .14 \times 14 = 2.0$$

Chart 7.4, using a dot control chart format given in Chapter 7, is the resulting stratified control chart for 25 subgroups. The process appears reasonably stable. The control limits for the 25 subgroups are as follows:

	$\bar{X}$ chart	R chart
UCL	27.4	26.8
Mean	22.1	14.4
LCL	16.8	2.0

These values agree well with the original control limits.

4.9.3 A Good and a Bad Sampling Plan

The process flow diagram for a car spindle-machining department appears in Figure 4.14. The second machining operation used two double-index chucker machines to cut several dimensions on the spindle. The distance from the backface of the spindle to a washer seat was difficult to cut since eight spindles on each machine needed to be monitored. A single operator ran both machines. The specification limit spanned .014 inch, but downstream processing was improved if only the upper half of the specification range was used. The data are coded so that 0 is the target dimension. The control chart for this characteristic located at A in Figure 4.14 appears in Chart 4.4. It is apparent that the process is unstable. Unfortunately, the sampling plan of $n = 5$ samples per day provided little information beyond indicating the lack of control. The multiple sources of variability indicated on the process flow diagram made it difficult to determine the special causes of variation.

A more intensive sampling plan was used to assist in troubleshooting. The eight spindles could conveniently be divided into two groups of four spindles (1, 7, 5, 3 and 8, 6, 4, 2) based on common machine parameters. Subgroups of $n = 4$ part were collected four times a

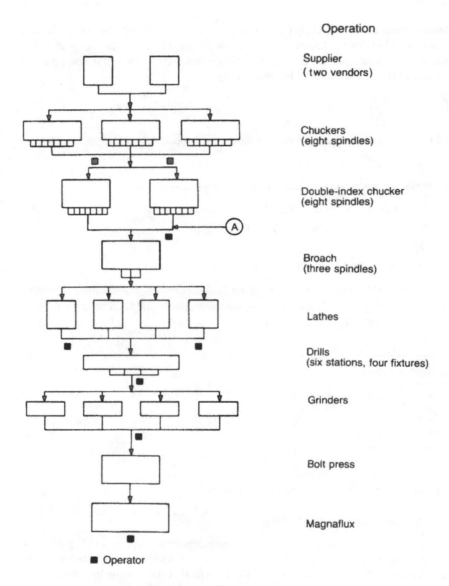

Figure 4.14 Process flow diagram for a car spindle-machining department.

day for 5 days. The control charts appear in Charts 4.5 and 4.6. The lack of control in both charts is apparent. However, the stratified control chart arrangement makes evaluation of the spindles possible. In Chart 4.5, spindle 3 has a higher mean than the other spindles and spindle 5 may have a lower mean. In Chart 4.6 spindle 8 has a high mean and spindle 2 a lower mean.

Clearly, this lack of a common mean for the spindles contributes to process variability. However, there are very likely other sources of variability that need to be removed, but the different spindle means makes their detection much more difficult. For example, on Chart

4.9 Case Studies

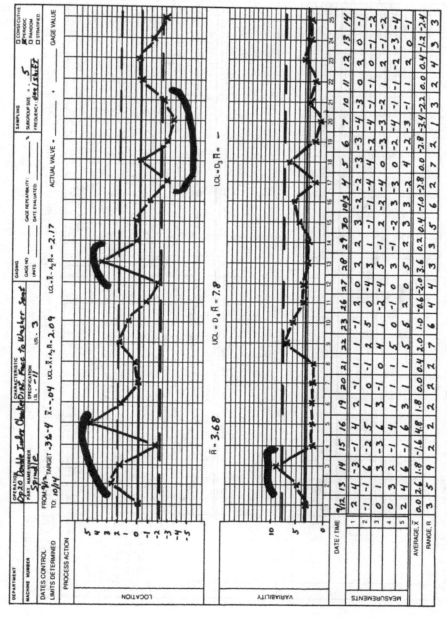

Chart 4.4 Control chart for variable data. The process is unstable, but the sampling plan of $n = 5$ provides little information beyond indicating the lack of control.

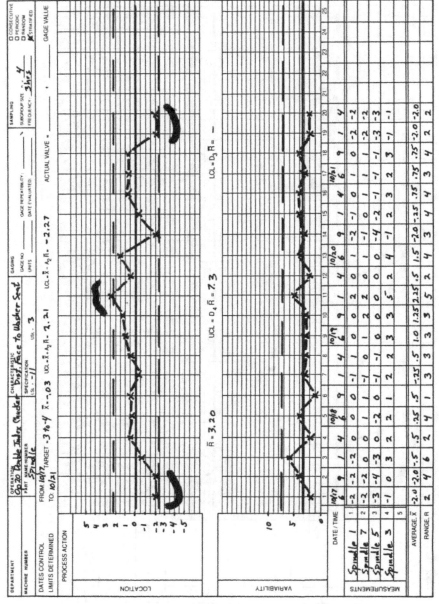

Chart 4.5 Control chart for variable data after dividing the eight spindles into two groups of four each. Here spindle 3 has a higher mean than the other spindles, and spindle 5 may have a lower mean.

4.9 Case Studies

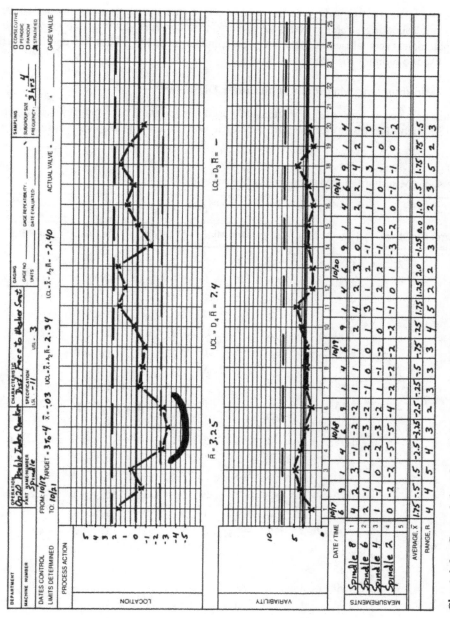

Chart 4.6 Control chart for variable data after dividing the eight spindles into two groups of four each. Here spindle 8 has a high mean and spindle 2 a lower mean.

4.4 we would want to know the process variables associated with the high mean on 9/16. A good troubleshooting strategy would be to perform a baseline study on the spindles to attempt to detect special causes. After the eight spindles are stable, a stratified control chart can be used to monitor the process.

4.9.4 Minimum and Maximum of an Outer Diameter

An aluminum spacer ring outer diameter (OD) used in equipment cabinets had a tolerance of ±.008 inch. The quality of the part was difficult to control, and operators frequently

Table 4.3 Minimum and Maximum OD Measurements

| Subgroup | Day | Time | OD measurements ||||||||||
			Min/Max		Min/Max		Min/Max		Min/Max		Min/Max	
1	2/21	8:00	−2	−1	−3	−2	−2	−1	−1	0	−4	0
2		9:00	0	1	0	1	0	2	0	1	−1	1
3		10:00	1	3	1	2	1	3	0	3	0	3
4		10:45	−1	2	−1	0	−2	0	−2	0	−2	0
5		11:30	1	2	−1	0	−1	2	−2	2	−2	0
6		12:30	−2	2	−2	1	−3	2	−2	2	−2	1
7		3:00	−2	1	0	3	−3	2	0	2	−2	2
8		3:30	−1	2	0	2	−2	1	0	3	−2	2
9	2/24	8:15	−1	1	−3	−1	−2	1	−2	0	−2	0
10		1:00	−1	0	−3	1	−2	2	−2	1	−2	1
11		1:45	−1	2	−3	3	−2	0	−2	1	−2	0
12		3:00	−1	1	−2	1	−1	2	−3	2	−2	0
13	2/25	8:30	−5	0	−7	−1	−5	−2	−5	−3	−5	−2
14		10:20	−5	−2	−7	−2	−6	−1	−5	−1	−5	−3
15[a]		11:45	−2	0	−3	1	−3	0	−5	1	−4	1
16		12:20	−4	2	−3	0	−3	−1	−5	3	−4	0
17		2:15	−5	2	−4	0	−2	0	−3	1	−4	1
18		3:35	−2	−1	−3	−1	−4	0	−2	−1	−4	0
19	2/26	8:00	−3	4	−1	4	0	2	−2	4	−4	4
20		9:00	−1	2	0	3	−2	3	−1	2	−1	2
21		11:30	−2	2	−1	1	−1	2	−2	−2	−2	2
22	2/27	8:30	1	6	2	6	2	6	1	7	1	6
23[b]		8:45	−1	4	0	5	−1	4	0	3	2	5
24		9:50	−1	3	−2	4	−2	3	−3	5	−3	4
25		11:00	0	4	−1	4	1	3	1	2	0	2

[a]Process adjusted up by 2 units.
[b]Process adjusted down by 2 units.

4.9 Case Studies

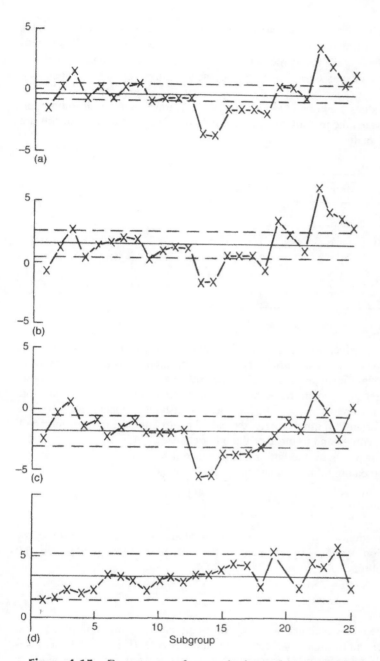

Figure 4.15 Four types of control charts for a minimum/maximum measurement: (a) Average of maximum and minimum OD; (b) maximum OD; (c) minimum OD; (d) difference (max OD − min OD).

adjusted the process. The average of the maximum and minimum OD was used to monitor the process. However, it was realized that significant within-part OD variability existed. The data in Table 4.3 were collected by measuring the minimum and maximum OD of five consecutive parts approximately every hour the machine was running. The data are coded so 1 = .001 inch.

The control chart values are computed for four cases: average, maximum, minimum, and the difference between the maximum and the minimum. The process adjustments are ignored for this initial analysis.

Measurement	Mean			Range	
	LCL	$\bar{\bar{X}}$	UCL	$\bar{R}$	UCL
Average	−1.0	−.2	0.6	1.3	2.8
Maximum	.3	1.5	2.7	2.0	4.2
Minimum	−3.1	−1.9	−.7	2.1	4.5
Difference	1.5	3.3	5.1	3.1	6.6

The $\bar{X}$ control charts are given in Figure 4.15. The range control charts were generally stable indicating the consecutive part variability was not a problem. However, the charts in Figure 4.15 indicate a significant process problem exists. Clearly, there is a lack of location control, but the difference between the maximum and minimum is comparatively more stable than the other measures. Thus, the process location instability appears not to be due to its within-part OD variability. Interestingly, the out-of-control points for the difference $\bar{X}$ chart do not generally correspond to those of the other charts.

The difference chart points to another problem. Since the range of the differences is generally stable, it is meaningful to evaluate the process spread:

$$\text{Process spread of difference} = 6\hat{\sigma} = 6\frac{\bar{R}}{d_2} = 6\frac{3.1}{2.33} = 8.0$$

Since the total specification width is 16, one-half the total allowable variability is used by the within-part variation. Process improvement efforts must address this problem, as well as the location instability.

Problems

4.1 An eight spindle screw machine was used to rough cut a transmission pinion gear blank. The concentricity of the inner diameter to the outer diameter (USL = .006 inch) was a critical characteristic for subsequent processing. The data are coded as actual value = .001 × coded value. (a) The following data were collected from a machine twice per 8 hour shift. Prepare an $\bar{X}$ and R control chart and discuss the results.

Problems

	Spindle								$\bar{X}$	R
No.	1	2	3	4	5	6	7	8		
1	3.5	3.2	2.2	2.0	0.3	1.7	0.7	3.5	2.1	3.2
2	1.2	0.5	0.7	4.2	2.1	1.5	1.9	1.7	1.7	3.7
3	1.5	1.2	1.5	3.0	2.5	3.2	2.0	1.5	2.1	2.0
4	1.0	1.8	1.7	1.0	1.8	0.4	2.5	0.8	1.4	2.1
5	1.1	2.5	1.4	1.1	0.6	1.0	1.0	1.4	1.3	1.9
6	1.8	1.0	2.3	2.4	0.3	1.5	0.8	2.0	1.5	2.1
7	3.8	1.6	2.3	0.5	0.5	1.2	1.3	0.2	1.4	3.6
8	3.2	2.0	1.7	5.5	2.5	3.5	1.5	5.5	3.2	4.0
9	0.8	1.3	1.5	1.5	0.7	1.0	0.8	2.0	1.2	1.3
10	2.5	2.7	2.3	2.2	2.7	0.8	1.0	0.9	1.9	1.9
11	1.6	3.8	1.5	3.7	1.8	1.9	0.5	1.0	2.0	3.3
12	0.5	0.5	2.5	3.0	1.0	0.4	0.6	0.4	1.1	2.6
13	2.0	2.0	0.5	4.0	2.5	4.0	1.8	2.0	2.4	3.5
14	3.0	3.5	1.5	1.2	1.6	1.0	1.0	0.7	1.7	2.8
15	1.1	2.5	1.0	2.8	0.3	0.2	1.5	0.7	1.3	2.6
16	2.5	4.0	2.5	2.5	0.5	2.0	1.0	1.0	2.0	3.5
17	0.5	3.5	4.0	0.5	1.0	3.0	0.2	4.5	2.2	4.3
18	0.8	2.0	3.5	3.5	2.5	2.5	0.8	4.0	2.5	3.2
19	3.0	3.7	4.0	1.3	2.0	2.5	1.0	2.8	2.5	3.0
20	4.0	2.0	4.6	2.4	1.0	2.6	2.0	1.5	2.5	3.6
21	0.6	5.5	1.0	1.5	0.9	0.5	1.5	2.0	1.7	5.0
22	3.5	3.5	3.1	1.6	0.9	4.0	3.0	2.0	2.7	3.1
23	1.0	4.0	1.5	3.0	1.0	2.5	1.3	3.0	2.2	3.0
24	1.1	2.8	3.2	3.5	3.4	2.0	1.4	4.0	2.7	2.9

(b) Calculate the means and standard deviations for each spindle. Comment on the results.
(c) Is this a good application for a stratified control chart? Was the chart implemented properly?

4.2 Surface finish micromeasurements (microinches) of the inside of a reamed bore are important to the ease with which a valve moves in the bore. The sampling plan consisted of measuring five parts randomly over a 4 hour period. The specification limit is 100 μinches.
(a) Prepare a $\bar{X}$ and R chart using the data through day 6. Identify some possible causes for the lack of stability. (b) Does the sampling plan seem appropriate for this process?

Day/shift	Microinches	Day/shift	Microinches
1/1	50, 54, 47, 63, 49, 52, 56, 46, 48, 72	6/2	70, 76, 67, 91, 64, 73, 77, 88, 93, 65
2/1	54, 58, 46, 62, 66, 60, 52, 45, 68, 77	7/2	49, 45, 46, 44, 57, 48, 50, 45, 39, 42
2/2	68, 77, 77, 71, 70, 70, 71, 80, 73, 68	8/1	54, 49, 86, 75, 77, 51, 82, 73, 76, 78
3/1	46, 52, 58, 45, 41, 49, 56, 60, 43, 40	8/2	69, 85, 76, 52, 64, 72, 78, 60, 68, 84
3/2	65, 77, 74, 58, 68, 71, 65, 72, 78, 81	9/1	52, 45, 40, 37, 70, 46, 44, 65, 58, 38
4/1	71, 74, 61, 53, 65, 73, 69, 63, 59, 57	9/2	78, 82, 72, 79, 72, 81, 74, 71, 75, 68
4/2	87, 85, 88, 72, 84, 78, 66, 87, 76, 55	10/1	63, 73, 71, 78, 51, 72, 70, 79, 66, 50
5/1	47, 52, 60, 57, 64, 53, 56, 59, 62, 65	10/2	44, 53, 64, 66, 67, 40, 51, 66, 44, 44
5/2	54, 48, 60, 66, 65, 50, 52, 70, 63, 57	11/1	64, 50, 54, 81, 73, 53, 67, 69, 77, 57
6/1	50, 77, 74, 79, 43, 52, 49, 80, 78, 49	11/2	73, 58, 61, 51, 55, 68, 93, 65, 69, 94

4.3 In a casting operation the mold density (%) is a critical characteristic in determining the quality of a final casting. An abbreviated process flow for a casting operation follows:

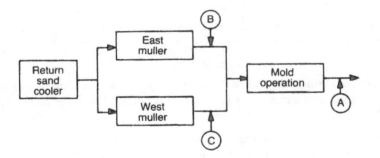

The original control chart was placed at A but did not prove useful in solving production problems. Chart A was replaced by charts B and C, which identified some operating problems. For chart A, $n = 3$ samples were collected once per day; for charts B and C, $n = 3$ samples were collected twice a day. The process specification was 37–47%. (a) Prepare an $\bar{X}$ and R chart for the position A data. Is the process stable? (b) Prepare an $\bar{X}$ and R chart for the position B and C data, separately. Is the process stable? (c) What can be said about the current performance of the process? (d) Would the sampling plan at B and C or at A be

Problems

more likely to demonstrate an out-of-control signal? Would that sampling plan therefore be undesirable?

Subgroup	Position A	Position B	Position C
1	36, 48, 31	48, 36, 31	33, 33, 45
2	29, 30, 29	42, 36, 38	34, 30, 40
3	33, 33, 45	40, 42, 34	34, 36, 42
4	37, 42, 40	42, 40, 37	38, 39, 32
5	42, 36, 38	40, 45, 41	39, 32, 33
6	40, 39, 38	35, 42, 37	34, 41, 37
7	39, 30, 40	44, 43, 34	32, 31, 39
8	41, 32, 31	38, 35, 41	37, 38, 35
9	40, 42, 34	40, 44, 46	37, 34, 35
10	31, 28, 33	40, 42, 39	39, 38, 29
11	38, 38, 38	41, 37, 34	34, 39, 36
12	38, 39, 32	35, 38, 38	38, 38, 38
13	39, 32, 33	37, 42, 40	37, 36, 37
14	40, 45, 41	40, 39, 38	34, 36, 35
15	34, 41, 37	41, 32, 31	29, 30, 29
16	32, 31, 39	35, 30, 46	31, 28, 33
17	37, 38, 35	40, 33, 33	38, 38, 38
18	35, 42, 37	36, 35, 41	37, 37, 38
19	44, 43, 34	41, 45, 32	31, 34, 38
20	34, 39, 34	40, 40, 41	31, 36, 33
21	40, 33, 33	36, 38, 42	37, 36, 31
22	36, 35, 41	38, 40, 41	34, 39, 34
23	31, 30, 35	38, 30, 37	31, 30, 35
24	41, 45, 32	38, 36, 37	40, 35, 38
25	34, 35, 38	41, 36, 37	34, 41, 38

4.4 For the $n = 30$ data used in the operator overcontrol discussion, assume that the operator adjusts the process only if a value occurs at or beyond the 8–12 specification limit. For example, a value of 7 causes a $+3$ to be added; 8 results in $+2$ to be added; with 9, 10, or 11 nothing is done; and 12 results in -2 to be added. (a) What are the mean and standard deviation of the adjusted process? Is the adjusted process better than the stable process? (b) Based on the number of values at or beyond the specification limits for the stable process, eight adjustments would be made. How many adjustments are made, and why is this number greater than 8? (c) Plot the stable and adjusted process on the same plot.

4.5 An automatic compensation system was purchased for a finish sizing operation of an engine connecting rod. Parts are 100% checked, and a compensation of $+3$ units (A) to the machine tool occurs if the average of the last five pieces is -3 units or less. Similarly, a compensation of -3 units (S) occurs if the average is $+3$ units or greater. (a) Plot a run chart of the following results noting the points where compensation occurred. Comment on the process. (b) Calculate the average range from successive points where no adjustment occurred. Does the range appear stable?

1.	−1	26.	0	51.	0	76.	6
2.	0	27.	−1	52.	−2	77.	1
3.	−3	28.	−1	53.	1	78.	−6
4.	−1	29.	−2	54.	0	79.	−3
5.	0	30.	−4	55.	−3	80.	1
6.	−1	31.	−4	56.	0	81.	−1
7.	−2	32.	−3	57.	−1	82.	−2
8.	−3	33.	−1	58.	1	83.	−7
9.	−5	34.	−4A	59.	2	84.	−4
10.	−2	35.	−4	60.	0	85.	−5A
11.	−1	36.	−8	61.	−2	86.	−2
12.	0	37.	−5A	62.	−2	87.	−4
13.	0	38.	−4	63.	−4	88.	−3
14.	0	39.	−3	64.	−2	89.	1
15.	−7A	40.	1	65.	−6A	90.	−1
16.	−2	41.	−1	66.	−5	91.	−5
17.	−3	42.	3	67.	−4	92.	−5
18.	−7A	43.	−4	68.	−4A	93.	−3
19.	−5	44.	2	69.	−6A	94.	−3A
20.	−2	45.	4	70.	−6A	95.	−9A
21.	0	46.	0	71.	3	96.	2
22.	2	47.	2	72.	6	97.	−1
23.	0	48.	−2	73.	8	98.	0
24.	1	49.	3	74.	5S	99.	−4
25.	−1	50.	−1	75.	7S	100.	3

Note that after a compensation, the compensation system adjusted the average of the last four pieces to reflect the adjustment.

4.6 A piston bore diameter of a small aluminum part had a specification of 20.125 ± .010 mm. A tool wear study was conducted by measuring five consecutive parts about once every hour. The data are coded so that

Actual value = 20.125 + .001 × coded value

(a) Prepare a standard $\bar{X}$ and R chart. Are the limits appropriate for the charts? (b) What are the modified control limits? Is it appropriate to use these limits? What is a reasonable tool change frequency?

Problems

Day	Time	Measurement	Day	Time	Measurement
1	7:30	−4, −6, −5, −3, −4	3	7:40	3, 3, 2, 2, 2
	8:30	−5, −7, −6, −4, −4		8:45	2, 1, 3, 3, 2
	9:30	−4, −6, −5, −6, −4		9:40	3, 4, 3, 2, 2
	10:30	−7, −3, −4, −5, −5		10:40	4, 3, 3, 4, 2
	11:30	−4, −5, −3, −3, −4		11:45	5, 4, 3, 3, 2
	1:00	−3, −2, −2, −4, −5	4	7:15	5, 4, 4, 3, 4
	2:00	−3, −3, −1, −4, −2		8:10	4, 3, 2, 4, 5
	3:00	−2, −2, −4, −3, −2		9:15	5, 5, 4, 6, 3
2	7:10	−1, −2, −3, −2, −1		10:20	6, 4, 4, 5, 6
	8:15	−2, 0, −3, −1, 0			
	9:15	−1, 0, −2, −2, −1			
	10:20	0, 1, −1, −2, 1			
	11:15	2, −1, 0, 1, −1			
	12:45	0, 2, 2, 1, −1			
	1:45	2, 1, 1, 2, 0			
	2:45	3, 2, 1, 0, 2			

4.7 A lapping operation is used to finish grind the outer diameter (OD) of a pin with a specification of .5657–.5660 inch. The pin is used to position a transmission gear where a roller bearing rides on the inner diameter (ID) of the gear and OD of the pin. The sampling plan consists of measuring five pieces randomly per hour. The data are coded so that

Actual value = .56585 inch + .0001 × coded value

(a) Prepare an $\bar{X}$ and R chart. Is the process stable? (b) The parts are measured after machining and are still hot. At normal room temperature the OD shrinks about .0001 inch. Is the process centered correctly? (c) Is the gaging adequate? What should be done to the gaging?

Day	Size	Day	Size	Day	Size
1	3, 3, 2, 2, 1	4	3, 3, 4, 4, 2	5	1, 1, 1, 1, 2
	1, 1, 1, 1, 1		2, 2, 2, 3, 2		1, 0, 0, 0, 2
	0, 0, 0, 0, 0		2, 2, 1, 0, 3		2, 2, 1, 0, 1
	3, 1, 1, 4, 1		1, 1, 2, 1, 1		1, 2, 1, 0, 0
2	3, 3, 2, 2, 3		1, 1, 1, 1, 1		1, 2, 1, 2, 1
	3, 3, 3, 3, 3		1, 3, 3, 3, 3		1, 2, 2, 1, 1
	3, 2, 1, 1, 1		3, 3, 2, 3, 2		2, 2, 2, 1, 3
	1, 1, 1, 1, 1		0, 0, 0, 0, 0		2, 1, 3, 2, 2
3	1, 2, 2, 3, 1			6	2, 1, 1, 2, 2
	2, 1, 2, 1, 2				1, 1, 2, 3, 2
	2, 3, 0, 1, 1				2, 0, 0, 1, 2
	1, 1, 0, 3, 2				2, 1, 1, 2, 1
	3, 3, 1, 0, 1				3, 0, 3, 1, 1
					2, 1, 2, 0, 2

4.8 Consider a foundry attempting to control the Brinell hardness of a malleable iron casting to a specification range of 3.8–4.1 mm (diameter of the ball indentation used to check hardness). Diameters of 3.8, 3.9, 4.0, and 4.1 can be read within the range of the specification. The gage repeatability of this process was about 40% (App. I). The sampling plan consisted of an end-of-the-line inspector collecting $n = 10$ castings (all at the same time) prior to the parts falling into shipping containers. A varying number of subgroups were collected each day. (a) Prepare an $\bar{X}$ and R control chart, and interpret the results. (b) Suggest an alternative sampling procedure that might provide a more useful chart.

Day	Measurements										$\bar{X}$	R
1	3.9	3.9	3.9	3.9	3.9	4.0	4.0	4.0	4.0	4.0	3.95	0.1
2	3.9	3.9	3.9	4.0	4.0	4.0	4.0	4.0	4.1	4.1	3.99	0.2
	3.9	3.9	3.9	4.0	4.0	4.0	4.0	4.0	4.0	4.1	3.98	0.2
3	3.8	3.9	3.9	3.9	3.9	3.9	3.9	4.0	4.0	4.0	3.92	0.2
	3.9	3.9	3.9	3.9	3.9	3.9	3.9	3.9	3.9	4.0	3.91	0.1
	3.9	3.9	3.9	3.9	3.9	4.0	4.0	4.0	4.0	4.0	3.95	0.1
	3.9	3.9	3.9	3.9	3.9	3.9	3.9	4.0	4.0	4.0	3.93	0.1
4	3.8	3.8	3.9	3.9	3.9	3.9	3.9	4.0	4.0	4.0	3.91	0.2
	4.0	4.0	4.0	4.0	4.0	4.1	4.1	4.1	4.1	4.1	4.05	0.1
5	3.8	3.8	3.8	3.8	3.8	3.9	3.9	4.0	4.1	4.1	3.90	0.3
	3.9	3.9	3.9	3.9	3.9	3.9	3.9	3.9	3.9	4.0	3.91	0.1
	3.9	3.9	3.9	3.9	3.9	3.9	3.9	4.0	4.0	4.0	3.93	0.1
	3.8	3.8	3.9	4.0	4.0	4.0	4.0	4.0	4.0	4.0	3.95	0.2
	3.8	3.9	3.9	3.9	3.9	4.0	4.0	4.0	4.0	4.0	3.94	0.2
	3.8	3.8	3.8	3.9	3.9	3.9	3.9	4.0	4.0	4.0	3.90	0.2
	3.9	3.9	3.9	3.9	3.9	3.9	3.9	4.0	4.0	4.0	3.93	0.1
	3.8	3.9	3.9	3.9	3.9	3.9	3.9	4.0	4.0	4.0	3.92	0.2
	3.9	3.9	3.9	3.9	3.9	3.9	4.0	4.0	4.0	4.0	3.94	0.1
6	3.9	3.9	3.9	3.9	4.0	4.0	4.0	4.0	4.0	4.0	3.96	0.1
	3.9	4.0	4.0	4.0	4.0	4.0	4.0	4.0	4.0	4.0	3.99	0.1

4.9 A testing laboratory provided analyses of emissions by a production method that gave the analysis quickly (method A). Results of the analyses were selectively verified by an accurate but time-consuming procedure (method B). For the quick method to operate properly, it needs to be set up properly at the beginning of the day. To examine the setup process, four samples were run by both methods. Since the magnitude of the measurements differed greatly between any two samples, it was decided to use the percentage difference calculated as follows:

$$\% \text{ Difference} = 100 \times \frac{\text{quick method result} - \text{standard method result}}{\text{standard method result}}$$

The data follow. Prepare an $\bar{X}$ and R chart. Is the process stable? If you were a user of the lab, would you be satisfied with the results?

No.	% Difference	No.	% Difference
1	44, 2, −4, 24	14	−26, −11, −1, 7
2	−1, 7, 14, 10	15	3, 5, 59, −4
3	10, 4, 10, 10	16	14, 5, −5, 3
4	2, 11, 25, −18	17	3, −6, −19, 16
5	14, 31, 0, −14	18	3, −7, 2, −11
6	1, 35, −17, 18	19	14, −7, 33, 22
7	3, −7, 23, 36	20	30, 2, −13, 8
8	0, 4, 1, 71	21	−2, −7, 5, 9
9	45, −11, 12, −3	22	30, 2, −22, 1
10	23, −7, −10, −11	23	7, 5, 5, −3
11	10, −6, 37, −7	24	−21, 15, 32, 71
12	13, 5, −9, −5	25	7, 16, 47, −4
13	20, 23, 3, 1		

4.10 A metal stamping operation could produce thousands of washers over a production shift. Most production runs ranged from 1 to 4 hours. The finished washers fell into a large tub after the stamping operation. Rather than use a lot-sampling method, the company used a Gillette lot control system along with a standard control chart. A canvas divider was placed in the tub after every several hundred washers. A subgroup of size $n = 3$ was collected from each division. The samples in a subgroup were collected at the beginning, middle, and end of the parts making up a division. After a subgroup was plotted and evaluated, the divider was removed if the process was judged stable. If process adjustments were necessary, only the washers in the last division of parts (i.e., those on top of the canvas) needed to be inspected. The washer OD (specification 0.62–0.64 inch) for a production run follow, where

Actual value = .6 + .001 × recorded value

(a) Prepare an $\bar{X}$ and R chart. Is the process stable? (b) Why do you think the sampling plan called for sampling at the beginning, middle, and end of the production run for each division? (c) If this was your process would you spend money for a more sensitive gage?

No.	Measurement	No.	Measurement
1	24, 24, 25	11	23, 24, 24
2	25, 24, 24	12	24, 23, 23
3	23, 24, 23	13	24, 23, 24
4	24, 23, 24	14	24, 25, 24
5	23, 24, 24	15	23, 24, 23
6	23, 25, 25	16	24, 23, 22
7	24, 23, 22	17	23, 24, 23
8	23, 22, 25	18	24, 23, 24
9	24, 23, 25	19	22, 23, 22
10	23, 25, 25		

4.11 Consider a job shop machining situation in which a lathe produces one-of-a-kind parts. The capability of the lathe is thought to be ±.001 inch. To evaluate the performance of the machine and operator, three unrelated dimensions are checked on each part. The performance is evaluated by determining how close the actual part measurement is to the machining target using the following formula:

Difference = actual value − target value

(a) Prepare a control chart for the differences. Is the machining process stable? (b) Is the stated capability of the lathe reasonable?

Part	Actual	Target	Difference	Actual	Target	Difference	Actual	Target	Difference
1	9.9984	10	−.0016	6.9984	7	−.0016	6.0020	6	.0020
2	11.9991	12	−.0009	0.0010	0	.0010	18.9998	19	−.0002
3	6.9990	7	−.0010	56.9972	57	−.0028	58.9996	59	−.0004
4	8.0021	8	.0021	75.0020	75	.0020	6.9986	7	−.0014
5	4.9987	5	−.0013	12.9987	13	−.0013	61.0009	61	.0009
6	1.9978	2	−.0022	17.0011	17	.0011	40.9993	41	−.0007
7	14.9976	15	−.0024	21.9992	22	−.0008	15.9986	16	−.0014
8	76.9991	77	−.0009	56.9979	57	−.0021	39.0043	39	.0043
9	22.9977	23	−.0023	2.9991	3	−.0009	42.0000	42	.0000
10	4.9992	5	−.0008	12.9968	13	−.0032	64.0002	64	.0002
11	1.9991	2	−.0009	0.9996	1	−.0004	51.0005	51	.0005
12	7.0000	7	.0000	66.9986	67	−.0014	26.9981	27	−.0019
13	8.9990	9	−.0010	80.9985	81	−.0015	50.0013	50	.0013
14	16.9996	17	−.0004	21.9997	22	−.0003	1.0005	1	.0005
15	21.0001	21	.0001	51.0006	51	.0006	5.0010	5	.0010
16	54.9987	55	−.0013	7.0000	7	.0000	2.0020	2	.0020
17	17.0000	17	.0000	22.0010	22	.0010	40.9992	41	−.0008
18	35.9985	36	−.0015	16.9987	17	−.0013	65.0000	65	.0000
19	4.9988	5	−.0012	8.9999	9	−.0001	86.9996	87	−.0004
20	42.0006	42	.0006	10.9998	11	−.0002	16.9993	17	−.0007

5
Histograms

5.1 General Concepts

The control charts discussed in previous chapters provide a basis for evaluating the stability of a process over time. Other simple statistical techniques, such as histograms, provide additional process information. Histograms are a graphic display of the number of times a measurement has occurred at a particular value or within an interval of the measurement scale. The following examples illustrate a procedure for preparing a histogram.

Example 5.1 Suppose we obtain $N = 15$ measurements from a process:

1, 5, 7, 9, 7, 8, 5, 9, 7, 1, 7, 5, 9, 9, 3

A histogram is started by placing an "x" over the value on the measurement scale each time the measurement occurs. Indicating the first 2 measurements gives the results in Figure 5.1a; all 15 measurements would appear as shown in Figure 5.1b. Computing the standard measures of location and variability for the 15 measurements gives

Location	Variability
Mean $\bar{X} = 6.13$	Standard deviation $s = 2.75$
Median $\tilde{X} = 7$	Range $R = 8$

Example 5.2 Suppose the following $N = 15$ measurements had been obtained from this process:

8, 1, 7, 9, 6, 5, 4, 6, 7, 5, 7, 6, 7, 6, 8

The histogram appears as shown in Figure 5.2. The measures of location and variability are

Location	Variability
Mean $\bar{X} = 6.13$	Standard deviation $s = 1.92$
Median $\tilde{X} = 6$	Range $R = 8$

166   5   Histograms

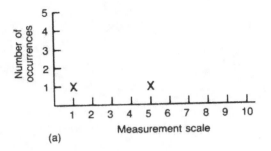

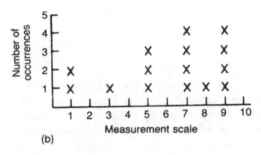

Figure 5.1 Constructing a histogram for Example 5.1: (a) first 2 measurements; (b) all 15 measurements.

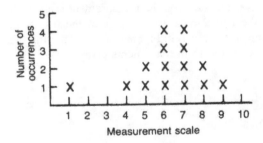

Figure 5.2 Histogram for Example 5.2.

The coded data frequently used in manufacturing make it easy to prepare a histogram. The format of a histogram may vary. Sometimes the number of occurrences is boxed in to produce a more visually appealing graph. The two displays in Figure 5.3 are equivalent histograms for Example 5.1. Figure 5.3a is a common format with the shaded "bars." Notice that "frequency" is often used rather than "number of occurrences" on the left-hand scale. When preparing a histogram of a large set of data, it is convenient to use slashes for measurements one through four, and large slash for the fifth occurrence of a measurement. Thus, the frequency scale is created automatically.

5.1 General Concepts

Frequency	Symbol
1	/
2	//
3	///
4	////
5	̶/̶/̶/̶/̶
6	̶/̶/̶/̶/̶ /

Examples 5.1 and 5.2 illustrate the importance of a histogram. Suppose these examples represented two processes—which would be preferable? Both have the same mean $\bar{X}$ and range R. However, the histograms clearly show the two processes are different. In Example 5.1 there is no clear pattern, but Example 5.2 shows a smooth "bell-shaped" pattern with a single extreme point at 1. The reduced variability in Example 5.2 is also indicated by a smaller standard deviation. A histogram showing some smooth patterns with a single "peak" is more desirable than a multi-peaked process. Thus, it is the pattern of the histogram that is important in studying a process. We see here that many patterns are possible, each indicating a different process situation. For the histogram to be representative of the process pattern, $N = 50$ or more values is preferable. The $N = 15$ values used in the examples are for illustration only.

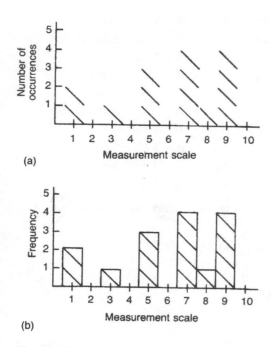

Figure 5.3

Histogram patterns indicate process problems

In some cases the measurements must be combined so that one bar on the histogram represents an interval on the measurement scale. The number of bars or subdivisions on the measurement scale should be related to N, the number of measurements. The following guidelines have proven useful:

Number of measurements N	Number of scale intervals
<50	5–7
50–100	6–10
100–250	7–12
>250	10–20

Thus, the larger the number of measurements, the more intervals it is possible to use to form the histogram. These guidelines are used to ensure that the number of intervals is not too large. Using too many intervals makes it difficult to evaluate histogram patterns. A procedure to create intervals of the measurement scale appears in Procedure 5.1.

Example 5.3 Consider the following set of $N = 100$ measurements given in Ishikawa (1982, p. 8)

3.56	3.46	3.48	3.50	3.42	3.43	3.52	3.49	3.44	3.50
3.48	3.56	3.50	3.52	3.47	3.48	3.46	3.50	3.56	3.38
3.41	3.37	3.47	3.49	3.45	3.44	3.50	3.49	3.46	3.46
3.55	3.52	3.44	3.50	3.45	3.44	3.48	3.46	3.52	3.46
3.48	3.48	3.32	3.40	3.52	3.34	3.46	3.43	3.30	3.46
3.59	3.63	3.59	3.47	3.38	3.52	3.45	3.48	3.31	3.46
3.40	3.54	3.46	3.51	3.48	3.50	3.68	3.60	3.46	3.52
3.48	3.50	3.56	3.50	3.52	3.46	3.48	3.46	3.52	3.56
3.52	3.48	3.46	3.45	3.46	3.54	3.54	3.48	3.49	3.41
3.41	3.45	3.34	3.44	3.47	3.47	3.41	3.48	3.54	3.47

Since the data span a wide range of the measurement scale, the interval method of producing a histogram is appropriate.

1. The largest value is 3.68 and the smallest 3.30.
2. For $N = 100$ we can have 6–10 intervals; we select 10.
3. The initial interval size is then

$$I = \frac{3.68 - 3.30}{10} = .038$$

4. A convenient number close to .038 is .05, so set $I = .05$.
5. Since the smallest number, 3.30, is a convenient number, we use it to form the intervals so $X_{low} = 3.30$. The intervals become

5.1 General Concepts

$3.30 \qquad\qquad\quad 3.3$
$3.30 + 1 \times .05 = 3.35$
$3.30 + 2 \times .05 = 3.40$
$3.30 + 3 \times .05 = 3.45$

interval 1
interval 2
interval 3

.
.
.

$3.30 + 7 \times .05 = 3.65$
$3.30 + 8 \times .05 = 3.70$ } last interval

6. The plotting scale is then

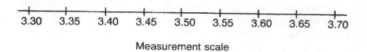

7. The histogram appears in Figure 5.4, which shows a reasonably bell-shaped curve typical of the output of many manufacturing processes.

An advantage of arranging measurements in histograms is the ease with which the mean and standard deviation can be computed. This approach is used primarily when there are

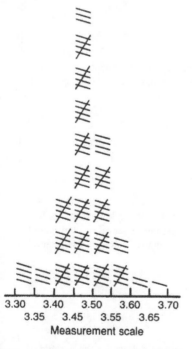

Figure 5.4 Histogram for Example 5.3.

Procedure 5.1 Histograms

A histogram is a display of the number of times a measurement occurred at a particular value or within an interval of the measurement scale. As a general rule, the number of intervals of the measurement scale should not be too large. The guidelines are as follows:

Number of measurements N	Number of scale intervals
<50	5–7
50–100	6–10
100–250	7–12
>250	10–20

There are two methods for constructing histograms. The quick method is adequate for many sets of measurements but may produce a histogram with too many intervals. In this case, the interval method should be used. An alternative approach is to construct a histogram using a stem and leaf plot, described in Procedure 5.3.

Quick Method

A histogram can be constructed using the following steps:

1. Find the largest and smallest measurement.
2. Construct the measurement scale between the largest and smallest measurement.

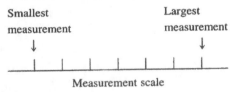

3. Place a slash (/) on the scale in step 2 for each occurrence of a measurement. Every fifth occurrence should use a longer slash (/ / / /). A space should be skipped between groups of five.

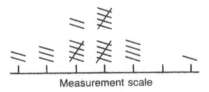

4. Evaluate whether there are too many intervals in the main body of the histogram. A few extremely large or small measurements should not be counted in the interval evaluation but should be noted in some manner on the plot. If there are too many intervals, the interval method should be used.

Interval Method

A histogram can be constructed using the following steps:

1. Find the largest and smallest measurement.
2. Choose a value for the number of intervals using the preceding guidelines for the number of measurements to be plotted in the histogram.

3. Compute the interval size:

$$I = \text{interval size} = \frac{\text{largest measurement} - \text{smallest measurement}}{\text{number of scale intervals}}$$

4. Round I to a convenient number. For example, .512 becomes .5, 2.89 could be 2 or 3, and 7.32 could be 5, 10, or 7.

5. Choose a convenient number slightly smaller than or equal to the smallest measurement; call it X_{low}. Compute the following intervals:

$$\left.\begin{array}{l} X_{low} \\ X_{low} + I \\ X_{low} + 2I \\ X_{low} + 3I \\ \quad \vdots \end{array}\right\} \begin{array}{l}\text{interval 1} \\ \text{interval 2} \\ \text{interval 3}\end{array}$$

until an interval covers the largest measurement.

6. Construct a scale to be used in plotting that corresponds to the intervals computed in step 5:

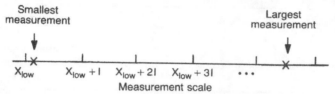

7. Place a slash (/) in the interval in step 6 for each occurrence of a measurement. A measurement equal to an interval end point should be placed in the highest interval. Every fifth occurrence of a measurement should use a large slash (⧅). A space should be skipped between groups of five:

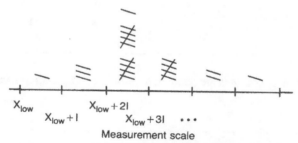

For either the quick or the interval method, shaded bars are often drawn around the slashes. A frequency (number of occurrences) scale is typically used.

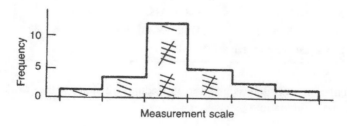

A few extremely large or small measurements can be noted on the plot but not used in steps 1–7.

not too many subdivisions of the measurement scale and the quick method in Procedure 5.1 is used to construct the histogram. Consider an example in which the histogram appears as shown in Figure 5.5.

The number of measurements (frequency) at each interval of the measurement scale is

Interval	1	2	3	4	5
Measurement	9	10	11	12	13
Frequency	3	7	15	4	6

These 35 values could be entered into a calculator to obtain $\bar{X} = 11.1$ and $s = 1.17$, but there is a simpler approach.

The procedure to calculate the $\bar{X}$ and s directly is first to complete five columns of numbers:

Column interval	1 Measurement x	2 Frequency f	$3 = 1 \times 2$ Measurement × frequency xf	$4 = 1 \times 1$ Measurement × measurement xx	$5 = 2 \times 4$ Frequency × measurement × measurement fxx
1	9	3	27	81	243
2	10	7	70	100	700
3	11	15	165	121	1815
4	12	4	48	144	576
5	13	6	78	169	1014
Total		$N = 35$	$A = 388$		$B = 4348$

The totals of three columns are denoted N, A, and B. These totals can be used to compute $\bar{X}$ and s.

Step	Formula	Example calculation
2	$\bar{X} = A/N$	$\bar{X} = 388/35 = 11.1$
3	$E = AA/N$	$(388 \times 388)/35 = 4301.26$
4	$(B - E)/(N - 1)$	$(4348 - 4301.26)/(35 - 1) = 1.37$
5	$s = $ square root of step 4	$s = 1.17$

The calculation procedure is given in Procedure 5.2 along with a work sheet that can be used for the computations.

Example 5.4 The mean and standard deviation are calculated for the histogram in Figure 5.6 using a work sheet format described in Procedure 5.2.

5.2 Stem and Leaf Plots

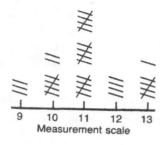

Figure 5.5 Histogram to illustrate $\bar{X}$ and s calculations: $N = 35$; $\bar{X} = 11.1$; $s = 1.17$.

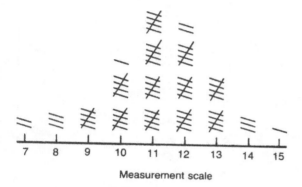

Figure 5.6 Histogram for Example 5.4.

5.2 Stem and Leaf Plots

A stem and leaf plot is a type of histogram in which the values of the numbers are retained in the plot. The procedure for constructing a plot is given in Procedure 5.3. In many cases it is quicker to construct a histogram using a stem and leaf plot rather than using the interval method described in Procedure 5.1. The histograms may not appear exactly alike since the intervals used to plot the measurements may be different. However, with a sample size of $n = 50$ or more, the general shapes of the plots are about the same.

Aside from providing a recording of the data and simplifying the construction of some histograms, a stem and leaf plot enables a detailed analysis of the actual measurements. Consider the data from Example 5.3 plotted in Figure 5.7. The intervals are the same as those in Example 5.3. Examination of Figure 5.7 reveals there are few odd numbers (23 of 100) and the interval 3.5a has almost all even numbers. Perhaps the measurement scale had even-number scale divisions:

Inconsistent rounding would lead to few odd numbers. The operational definition of how to use the measurement scale may have been unclear.

Work Sheet to Compute Mean and Standard Deviation from Histogram Frequencies

Interval	1	2	3	4	5	6	7	8	9	10	11	12	13	14	15
Measurement x	7	8	9	10	11	12	13	14	15						
Frequency f	2	3	5	11	20	17	10	3	1						

Step-by-step computational approach:

1. Complete the following table:

Interval	x	f	xf	xx	fxx
1	7	2	14	49	98
2	8	3	24	64	192
3	9	5	45	81	405
4	10	11	110	100	1100
5	11	20	220	121	2420
6	12	17	204	144	2448
7	13	10	130	169	1690
8	14	3	42	196	588
9	15	1	15	225	225
10					
11					
12					
13					
14					
15					
Sum		$N = 72$	$A = 804$		$B = 9166$

2. Mean computation $\bar{X}$:

$$\bar{X} = \frac{A}{N} = \frac{804}{72} = 11.17$$

3. Then perform the standard deviation computations. Compute the first intermediate quantity:

$$E = \frac{AA}{N} = \frac{804 \times 804}{72} = 8978$$

4. Compute the second intermediate quantity:

$$\frac{B - E}{N - 1} = \frac{9166 - 8978}{71} = 2.6479$$

5. Compute the standard deviation:

s = square root of step 4 = 1.63

Procedure 5.2 Computing the Mean and Standard Deviation from Histogram Frequencies

It is convenient to use this approach for calculating the mean $\bar{X}$ and standard deviation s for data appearing in histogram format. A standard work sheet can be used for the calculations. The data from the histogram can be arranged as follows:

Interval	1	2	3	...	k
Measurement x	x_1	x_2	x_3	...	x_k
Frequency f	f_1	f_2	f_3	...	f_k

If the internal method is used to prepare the histogram, replace the measurement row by the midpoint of the interval. The following steps are used to calculate $\bar{X}$ and s:

1. Complete the five columns in the following order:

Interval	x	f	xf	xx	fxx
1					
2					
.					
.					
.					
k					
Totals		N =	A =		B =

   The totals for the three columns should be computed.

2. Compute the mean:

$$\bar{X} = \frac{A}{N}$$

3. Compute the first intermediate quantity:

$$E = \frac{AA}{N}$$

4. Compute the second intermediate quantity:

$$\frac{B - E}{N - 1}$$

5. Compute the standard deviation by taking the square root of the number in step 4:

   s = the square root of step 4

Procedure 5.3 Stem and Leaf Plots

A stem and leaf plot is a type of histogram in which the actual data values appear in the plot. The procedure for constructing the plot consists of first forming the "stems" and then plotting the "leaves."

Forming the Stems

The stems are the plotting intervals on the measurement scale. Guidelines for the number of stems are related to the number of measurements:

Number of measurements N	Number of stems
<50	5–7
50–100	6–10
100–250	7–12
>250	10–20

The following steps can be used to determine the stems of the plot.

1. Starting from the left side of the measurement digits, determine the first digit that changes in the set of measurements to be plotted. For example, consider the two sets of $N = 10$ measurements:

   Set 1   5.29, 5.37, 5.29, 5.55, 5.29, 5.28, 5.27, 5.28, 5.34, 5.34

   Set 2   13, 11, 10, 08, 11, 10, 09, 11, 09, 08

   Notice a 0 was added in front of the single-digit numbers so all numbers have the same number of digits. The second digit in set 1 and first digit in set 2 are the first changing digits.

2. Write the numbers from the smallest to the largest digit determined (including only the first changing digit) in Step 1.

   Set 1   5.2   5.3   5.4   5.5

   Set 2   0   1

   These are trial values for the stems.

3. If the number of stems in step 2 is too small using these guidelines, divide the stems into two or five subdivisions for each stem. Each divisor of the stem represents different numbers in the original measurement scale.

		Set 1		Set 2
	Stem	Numbers represented	Stem	Numbers represented
One stem	5.2	5.20, 5.21, 5.22, 5.23, 5.24, 5.25, 5.26, 5.27, 5.28, 5.29	0	00, 01, 02, 03, 04, 05, 06, 07, 08, 09
Two stems	5.2a 5.2b	5.20, 5.21, 5.22, 5.23, 5.24 5.25, 5.26, 5.27, 5.28, 5.29	0a 0b	00, 01, 02, 03, 04 05, 06, 07, 08, 09
Five stems	5.2a 5.2b 5.2c 5.2d 5.2e	5.20, 5.21 5.22, 5.23 5.24, 5.25 5.26, 5.27 5.28, 5.29	0a 0b 0c 0d 0e	00, 01 02, 03 04, 05 06, 07 08, 09

5.2 Stem and Leaf Plots

4. Arrange the stems determined in step 3 to form a measurement scale. The scale can be drawn either across or down the page:

   Set 1
   5.2
   5.3
   5.4
   5.5

   Set 2

   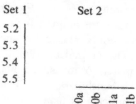

5. The leaves of the plot are the remaining digits plotted next to the correct stem:

   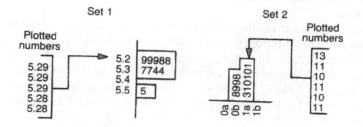

   Notice that bars are sometimes placed around the leaves to form a traditional histogram.

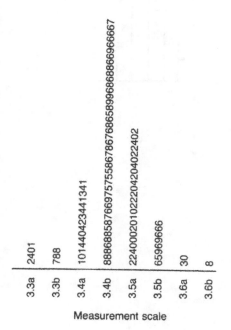

Figure 5.7 Stem and leaf plot for Example 5.4.

5.3 Normal Distribution and Zones A, B, and C

Many manufacturing situations produce a histogram in the familiar bell shape. This shape may be flat or peaked, but the bell shape is nevertheless apparent, as seen in Figure 5.8. (Note that the slashes used to form the bars of the histogram are omitted for ease of presentation.) The bell-shaped pattern of the histogram can often be reasonably approximated by a normal distribution, as shown in Figure 5.9. The normal distribution is a statistical curve that provides a theoretical basis to study a pattern of measurements. The smooth pattern of the normal distribution is often used as a benchmark to evaluate the performance of a manufacturing process. Two common applications using the normal distribution are comparison of different sources of process variation and evaluation of the number of measurements in intervals of the measurement scale. These applications are discussed here.

We wish to evaluate the output of a machine. How do the measures of location ($\bar{X}$) and variation (s) relate to the histogram and normal distribution? As shown in Figure 5.10, the center of the normal distribution is the process mean $\bar{X}$. Changing $\bar{X}$ shifts the center of the normal distribution. The spread of the process is related to the measure of variability s. Recall from Chapter 3 that the spread of the process is (process spread =) $6s$, so the spread of Figures 5.10a and b is 12 and the spread of Figures 5.10c and d is 18. Notice that the end points of the smooth normal curve are at the process limits:

Upper process limit (UPS) = $\bar{X} + 3s$
Lower process limit (LPS) = $\bar{X} - 3s$

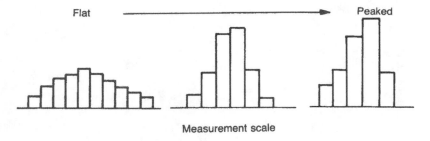

Figure 5.8 Histograms with a bell-shaped pattern.

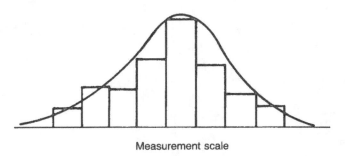

Figure 5.9 Normal distribution curve approximating a bell-shaped histogram.

5.3 Normal Distribution

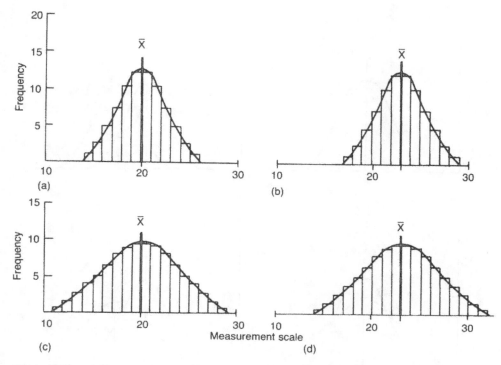

Figure 5.10 Relating changing location and variability to histograms and the normal distribution curve: (a) base process; $\bar{X} = 20$, $s = 2$; (b) increasing location; $\bar{X} = 23$, $s = 2$; (c) increasing variability; $\bar{X} = 20$, $s = 3$; (d) increasing location and variability; $\bar{X} = 23$, $s = 3$.

Suppose we wish to compare the output of two machines that produce the same part. Figure 5.11 gives four cases in which the machines are compared using the normal distribution curve. Various changes in location and/or variability can be easily evaluated.

A standard convention is to denote the process standard deviation by the Greek letter sigma (σ). We never know the exact value of σ (unless we simulate the process as in Chap. 6); however, it is possible to estimate the process standard deviation from the measurements obtained from a process. The estimate of the process standard deviation is denoted by $\hat{\sigma}$. Two methods have been discussed for obtaining $\hat{\sigma}$. The standard deviation of a number of measurements (Procedure 3.1 or 5.2) or the range from a stable control chart (Procedure 3.2):

$$\hat{\sigma} = s \quad \text{or} \quad \hat{\sigma} = \frac{\bar{R}}{d_2}$$

(a third method is discussed in Procedure 3.3). Each method gives similar results for a process with a stable normal distribution.

Interpretation of the standard deviation is possible by noting

Process spread = $6s$

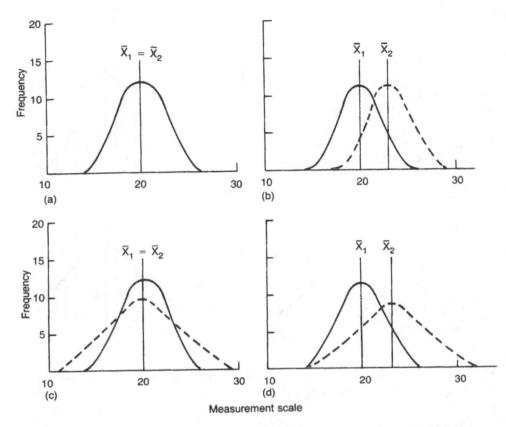

Figure 5.11 Comparing the location and variability of two machines: (a) identical machine output; $\bar{X}_1 = \bar{X}_2 = 20$, $s_1 = s_2 = 2$; (b) machine 2 shifted higher, common variability; $\bar{X}_1 = 20$, $\bar{X}_2 = 23$, $s_1 = s_2 = 2$; (c) machine 2 more variable than machine 1, common location; $\bar{X}_1 = \bar{X}_2 = 20$, $s_1 = 2$, $s_2 = 3$; (d) machine 2 shifted higher and more variable than machine 1; $\bar{X}_1 = 20$, $\bar{X}_2 = 23$, $s_1 = 2$, $s_2 = 3$; (———) machine 1; (---) machine 2.

For a process having a normal distribution, 99.73% of the process measurements are within the 6σ spread. A related approach for interpreting σ and the normal distribution is to divide the process spread into six equal zones around the process mean $\bar{X}$. These zones are shown in Figure 5.12, with the expected percentage of measurements within each zone. These percentages provide a benchmark for evaluating how well a histogram is approximated by a normal distribution. Of course, most manufacturing processes do not have exactly a normal distribution, but this distribution often serves as a reasonable approximation.

5.4 Transformations*

Some characteristics exhibit histograms that are not bell shaped but can be made approximately bell shaped by changing the measurement scale. A typical example of different

5.4 Transformations

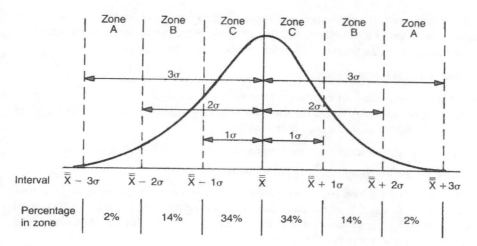

Figure 5.12 Zones of a normal distribution.

scales of measurement involves determining the hardness of a metal. Appendix III, Table K, lists two scales of hardness, the Brinell and the Rockwell. However, even using one measurement scale, different expressions of the data are possible. For example, Brinell hardness uses a ball and measures the indentation diameter of the ball under load on the tested metal surface (typically, a 10 mm ball and a 3000 kg load). It is possible to equivalently express measurements using the indentation diameter or the Brinell hardness number. In the ideal case, one measurement scale produces an approximately bell-shaped histogram, as illustrated in Figure 5.13.

When the transformed data are related to the original data by the linear form

Transformed measurement = constant + factor × original measurement

it makes no difference which scale of measurement is used. The coded data examples in Chapter 3 were of the linear form

Actual value = offset + scale × coded value

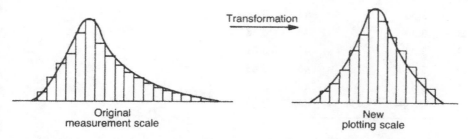

Figure 5.13 Transforming the original measurement to obtain a bell-shaped histogram: (left) skewed, non–bell-shaped histogram; (right) bell-shaped histogram.

which is equivalent to the previous form. Thus, using coded data or the actual values do not influence the shape of a histogram. Another example is the use of metric versus standard English units. For example, the conversion of millimeters to inches uses

Inches = 0.0394 × millimeters

which is the same as the previous linear form with a zero offset.

Another approach is re-expressing (transforming) the original measurement by using the square root or logarithm of the measurements:

Square root measurement = $\sqrt{\text{measurement}}$

Log normal measurement = ln (measurement)

(It is possible to use log base 10, but log base e (ln) is more common.) If the histogram has a longer tail on the right side of the plot, these transformations tend to make the histogram more bell shaped. The square root transformation is used for moderate skewness to the right, and the ln transformation for more extreme cases.

Example 5.5 Consider the micromeasurements in Problem 3.8. The microvalues range between 16 and 48. The values in the original microscale and the ln scale follow:

Original scale	ln scale	Histogram interval	Original scale	ln scale	Histogram interval
		← 2.75			
16	2.77		32	3.47	
17	2.83	← 2.85	33	3.50	
18	2.89		34	3.53	← 3.55
19	2.94	← 2.95	35	3.56	
20	3.00		36	3.58	
21	3.04	← 3.05	37	3.61	
22	3.09		38	3.64	← 3.65
23	3.14	← 3.15	39	3.66	
24	3.18		40	3.69	
25	3.22	← 3.25	41	3.71	
26	3.26		42	3.74	← 3.75
27	3.30		43	3.76	
28	3.33	← 3.35	44	3.78	
29	3.37		45	3.81	
30	3.40		46	3.83	
31	3.43		47	3.85	← 3.85
	← 3.45		48	3.87	

Note that as the micro measurements increase, more measurements on the original scale are combined in the histogram interval for the ln transformation. The histograms for both groups of measurements appear in Figure 5.14. The histogram using the ln transformation

5.4 Transformations

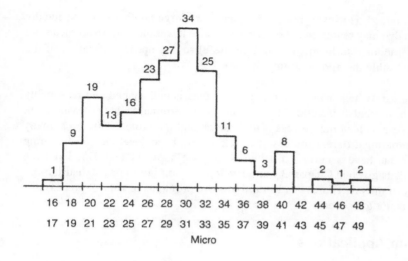

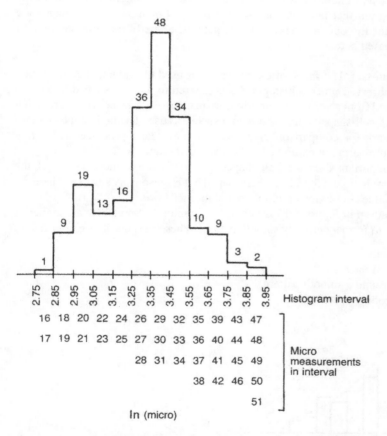

Figure 5.14 Histograms of micromeasurements in original and natural log (ln) scale.

appears more bell shaped. However, notice that there are a large number of unexpectedly low measurements that may represent a special cause. Since lower micro measurements are desirable (i.e., a smoother surface), it would be useful to attempt identification of the special cause for possible incorporation into the process.

Why should we care whether the histogram of our data is bell shaped and reasonably approximated by the normal distribution? In many cases, the original scale of measurement is best since it represents how the process is understood and monitored. However, many procedures for comparing different sources of variation have been developed by assuming that the measurements have a normal distribution (e.g., Chaps. 11 and 12). Even the control chart constants given in Chapter 3 were developed assuming a normal distribution. Fortunately, many of the procedures can be used to solve problems or control processes if the nonnormality is not too severe.

5.5 Histogram Applications

Histograms can be used in a number of ways. In all cases, it is the pattern of measurements formed by the histogram that is important, so that $N = 50$ or more measurements are generally recommended to obtain a representative pattern. Some of the common uses of histograms are discussed here.

Shape and Smoothness. The shape of the histogram is used to evaluate the output of a process. Often the observed process histogram is compared to the bell-shaped histogram that would be expected from a normal distribution. In many cases, a process histogram that differs greatly from a bell-shaped curve is due to a special cause. The bell-shaped pattern serves as a benchmark for comparison of process output. Many processes exhibiting random common cause variation exhibit a bell-shaped histogram.

Probably more important than the bell-shaped pattern is the "smoothness" of the histogram. The pattern in Figure 5.15 is not bell shaped but is reasonably smooth. There are no extreme outlier values or obvious multiple populations.

Notice that the pattern in Figure 5.15 is not perfectly smooth. Some bars are higher or lower than expected to form a perfectly smooth pattern. These irregularities can be due to

Too few measurements
Too narrow histogram intervals
Random variation around a smooth pattern
Special causes indicating a problem

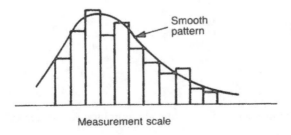

Figure 5.15 Smooth histogram pattern.

5.5 Applications

Selecting $N = 50$ or more measurements is generally adequate to evaluate the histogram pattern. The guidelines in Procedure 5.1 provide an appropriate number of intervals. The real challenge is to determine if the lack of a smooth pattern is due to a process problem (special cause) or to random variation. Two options are available to determine if a problem exists. First, more measurements can be collected and new histograms made. Second, a control chart can be used. This option is preferred since histograms do not assess any changes over time and thus do not evaluate process stability. When a problem is suspected, a control chart is a very useful problem-solving tool, as we have seen. The special cause that may produce the lack of smoothness in the histogram is indicated on the control chart by an out-of-control signal.

Histograms do not assess process stability

Comparison to Specification Limits. In manufacturing, one of the common applications of histograms is to compare the process output to the specification limits. Some examples are given in Figure 5.16. This practice can be informative, particularly as a screening method to determine which characteristics in the process may need further study. A manufacturing process may have many characteristics that exhibit good process performance, as shown in Figure 5.16a. Although these characteristics may not be stable, characteristics that have a location or variability problem, as shown in Figure 5.16c and d, have a higher priority for receiving immediate attention.

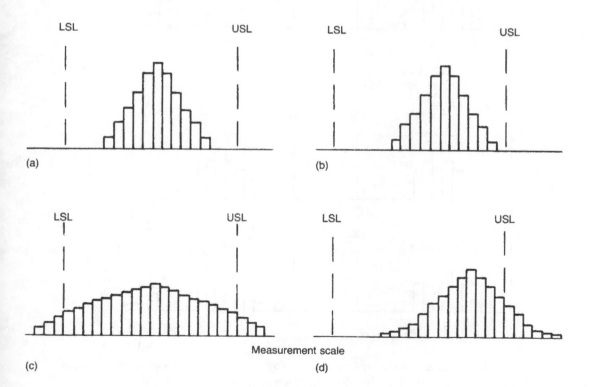

Figure 5.16 Using histograms with specification limits.

Comparison of Sources of Variability. As previous process flow diagrams indicated, manufacturing processes have many sources of variability. Often different machines, spindles, fixtures, or pallets, for example, produce the same part. A major improvement strategy is to compare like sources of variation—the comparison principle:

> All common units (e.g., machines) producing a characteristic can be improved to at least the level of the best-performing unit. Making the location and variability of all units the same reduces overall process variability.

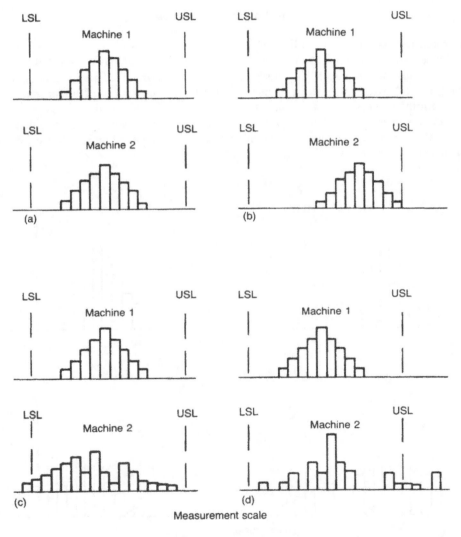

Figure 5.17 Using histograms to make comparisons: (a) both machines performing similarly; (b) machine 2 off location; (c) machine 2 excessively variable; (d) machine 2 probably unstable.

5.5 Applications

We have already seen one application of the comparison principle in Procedure 4.1, when baseline control charts were used to obtain approximately the same $\bar{\bar{X}}$ and $\bar{R}$ for various units prior to forming a stratified control chart. Chapter 11 discusses the comparison principle in more detail. An approach not focusing on process stability, as with the baseline control charts, is to obtain histograms for like units. For example, suppose two machines produce the same part; Figure 5.17 illustrates some comparisons. It is not necessary to show the specification limits as in Figure 5.17, but the limits can be used to provide a reference. As Figure 5.17 indicates, histograms can be used to assess both process location and variability. Chapter 11 provides a more formal approach to this problem, but a simple histogram can be used to make an initial evaluation.

Multiple Populations. To implement the comparison principle, we need to identify all common producing units and compare individual histograms. In some cases different units are not identified so only a single combined set of measurements is present. If the histograms are combined for the two machines in each Figure 5.17 plot, it is difficult to detect the type of process problem that is present.

The ease with which multiple populations location differences can be identified depends on the separation of the populations. Figure 5.18 shows the appearance of varying degrees of overlap. In general, it is difficult to identify the following three cases:

1. Similar location (major overlap) of common units
2. Different variability of common units
3. Sampling variability from a process having a single non–bell-shaped histogram

Often knowledge of the characteristic being measured can be used to select the correct

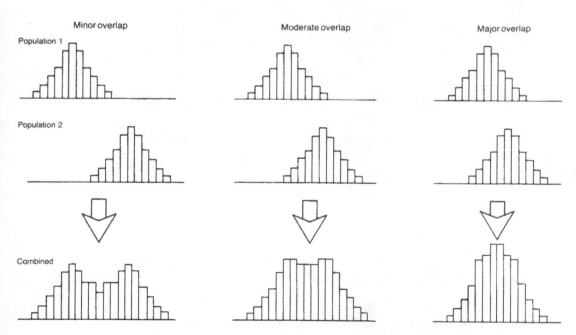

Figure 5.18 Detecting multiple populations with varying overlap.

alternative. For example, a size characteristic may be assumed to have a bell-shaped histogram, so a non–bell-shaped histogram could be the result of multiple populations. Conversely, a non–bell-shaped histogram may be typical of a surface finish characteristic. Fortunately, when severe differences in location exist, the combined histogram shows different peaks. However, moderate differences in location appear as increased variability, as Figure 5.18 indicates. This effect is even more pronounced when more than two units are considered. Thus, what may appear as excess process variability may be due to differences in location (or variability) of several units producing a common characteristic.

Differences in location can be difficult to detect, but differences in variability between common units can be even more difficult. Even with only two units, large differences cannot be easily discovered unless the different units are plotted separately. In Figure 5.17c, the combined machine 1 and 2 histogram does not indicate the existence of variability differences.

Why are multiple populations important? The comparison principle for improvement indicates that we can improve a process by making the means of like units the same and making the variability of all units similar to the unit with the lowest variability. Thus, any indication of multiple populations must be pursued to eliminate differences in location or variability. If differences between units are suspected, the measurements should be separated and compared using separate histograms or the methods discussed in Chapter 11.

Differences in common units indicate an opportunity for improvement

Outlier Detection. An outlier is a value beyond the normal smooth pattern that appears in a histogram. It does not relate to measurements beyond the specification limits. If measurements exist that do not conform to the histogram pattern, they are called outliers. These measurements are generally the result of a special cause of variation and are often caused by an unstable process, as shown in Figure 5.19a. Occasionally, a small group of outlier measurements is caused by a repeating special cause. Stratification may exist in which one unit (e.g., a spindle or fixture) is producing output at a different mean level. In

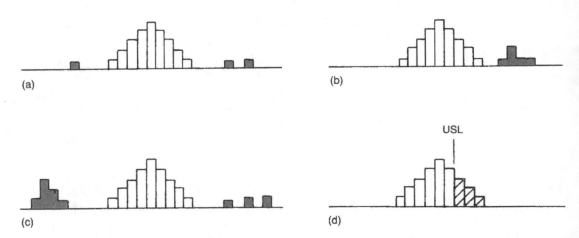

Figure 5.19 Using histograms to identify process problems indicated by outliers: (a) outlier measurements, special causes present; (b) outlier population, possible stratification; (c) combined outlier problems; (d) incapable process not due to outliers.

5.5 Applications

this case, it is apparent the process is not stable. Notice in Figure 5.19d that the hatched area is not an outlier since it is part of the smooth pattern of the histogram. Since a control chart is not used, we cannot assess process stability even though the histogram has a smooth pattern.

Before and After Comparison. Teams attempting to improve a process can use histograms to

Assess the current process state (before)
Evaluate the effectiveness of an improvement (after)

In a simple histogram, the effectiveness of an improvement action can be determined, as shown in Figure 5.20. Using this histogram format as a communication tool forces the improvement team into good problem-solving discipline. The problem is initially defined by the poor "before" histogram. The verification of any improvement is quantified by an "after" histogram.

Assembled Components. Many manufacturing processes result in the assembly of components. Histograms can be used to evaluate the relationship between components that are to be assembled. For example, suppose a pin is to be placed in a drilled hole with a target clearance of .0015 inch (.0005 to .0025). Histograms of the results of both processes can be plotted together, as shown in Figure 5.21. The minimum and maximum clearance

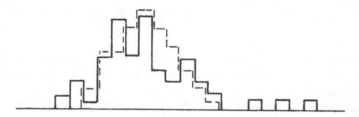

Figure 5.20 Using histograms to make a before (———) and after (---) comparison.

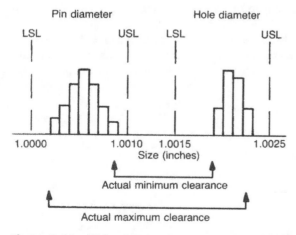

Figure 5.21 Using histograms to evaluate mating components.

between the pin (specification 1.000–1.001 inches) and the hole (specification 1.0015–1.0025 inches) can be determined and related to the target clearance of .0015 inch. Too much clearance results in a loose pin, and too little clearance results in possible interference during assembly. The histogram can be used to help target the pin and hole processes at the desired value. Often the process can be targeted to improve the ease of assembly without impacting on the performance of the product. These histograms can be obtained from control chart data (Problem 3.14).

An important consideration for clearance evaluation is to properly evaluate within-part variability. For example, if a two-flute reamer is used to drill a hole, three lobes can be expected. An exaggerated view of the hole appears as follows:

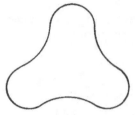

It is not uncommon for the degree of out of roundness to be .0002–.0004 inch. Clearly, the measurement procedure must address this within-part variability to meaningfully evaluate clearances. In this case, the minimum diameter is typically measured, which may not correspond to the standard average diameter used to monitor the process. This problem was discussed in Chapter 4.

Measurement and Recording Process. Histograms can be used to detect problems with a measurement process. Recall from Figure 5.7 how a stem and leaf plot was used to detect a lack of interpolation between tick marks on a measurement scale. For example, the gage readings in Figure 5.22 are subject to interpretation of the measurement procedure. The choice of the best method depends on the application, but the operational definition for the measurement process should address how to read the measurement scale. This lack of consistency results in the comblike histogram shown in Figure 5.23a. Again, we wish to keep our measurement process consistent so that we can effectively evaluate the performance of the process.

Another histogram pattern results from not reporting all measurements. Occasionally, employees do not report values beyond the specification limits so that a histogram appears to have a cliff, as shown in Figure 5.23b. This case occurs when an out-of-specification

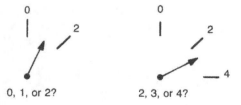

Figure 5.22 Ambiguous gage readings.

5.6 Histograms from Control Charts

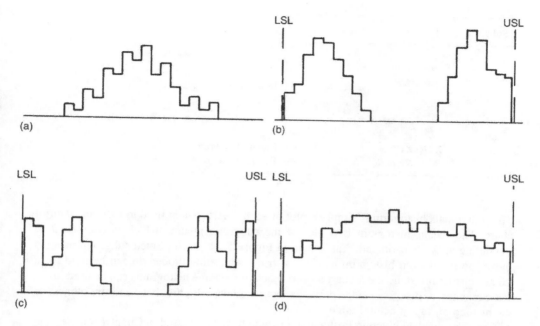

Figure 5.23 Histogram patterns: (a) comblike histogram measurement; (b) clifflike histograms, nonrecording, sorting; (c) clifflike histogram, adjustment; (d) clifflike histogram, sorting.

part is simply not recorded. Sometimes an out-of-specification part is "adjusted" to be recorded just inside the limits. These histograms appear as shown in Figure 5.23c. In either case, the reason employees believe that out-of-specification parts should not be reported needs to be addressed.

A single or double clifflike histogram in Figure 5.23d is also obtained when parts are measured from a production lot that has been sorted. The histogram tails have been eliminated by the sorting.

5.6 Histograms from Control Charts

We have discussed the separate use of control charts and histograms. Although it is possible to use the techniques independently, it is good practice to use the tools together. The independent use of control charts or histograms can lead to overlooking important problems.

There are three types of histograms that can be obtained from typical $\bar{X}$ and R chart data:

Histogram of individual measurements (individuals histogram)
Histogram of subgroup means (means histogram)
Histogram of subgroup ranges (ranges histogram)

Each histogram can be obtained by selecting the values to be plotted from the subgroup data blocks on a control chart.

Subgroup	1	2	3	$\cdots$	k	
1						
2						
3					$\cdots$	Individuals histogram
4						
5						
Sum						
$\bar{X}$	$\bar{X}_1$	$\bar{X}_2$	$\bar{X}_3$		$\bar{X}_k$	$\leftarrow$ Means histogram
R	R_1	R_2	R_3		R_k	$\leftarrow$ Ranges histogram

The individuals histogram is simply a plot of each measurement used to form the control chart. The means histogram is a plot of the subgroup means, which is equivalent to "standing an $\bar{X}$ chart on end," as shown in Figure 5.24. Each plotted subgroup mean is moved to the bottom histogram as if on a string. The same procedure can be used for a ranges histogram: stand the R chart on end. Each of the three histograms can be used in the same manner, as discussed in the previous section. Generally, the shape and smoothness of the histograms are of central importance.

Consider the data given in Problem 3.11, which was discussed in Chapter 4 in Section 4.2.3. The individuals, means, and ranges histograms appear in Figure 5.25. The individuals histogram in Figure 5.25a has all $N = 225$ measurements and shows a non–bell-shaped pattern with a peak at 11 and another at 14. For these data there were three spindles, with spindles 1 and 3 having means of $\bar{\bar{X}}_1 = 10.3$ and $\bar{\bar{X}}_3 = 10.1$ and spindle 2 with $\bar{\bar{X}} = 13.7$. The multiple populations (i.e., spindles) are indicated on the individuals histogram. The means histogram in Figure 5.25b uses a subgroup size of $n = 3$ and plots $N = 75$ subgroup means. The resulting histogram is bell shaped. The ranges histogram in Figure 5.25c uses the subgroup ranges, which have a non–bell-shaped pattern.

Suppose the $N = 225$ measurements used in the individuals histogram in Figure 5.15a were plotted by individual spindles. Figure 5.26 shows the histogram for each spindle. It is apparent that spindle 2 has a higher mean. The variability does not appear to differ between the spindles. This simple method of comparing sources of variability is an effective method of identifying problems and is much easier to interpret than the multiple populations in Figure 5.25a. Unfortunately, too many process studies do not record possibly important stratification factors (e.g., spindle) to enable an effective comparison. It is interesting to note that the data for each of the spindles were generated by simulating an exact normal distribution using the method described in Chapter 6. Notice that these histograms do not have a perfect bell-shaped histogram but rather show the influence of random variation.

Another practice of using histograms with control charts is to divide the control chart into meaningful time segments. It is then possible to make a histogram for each time period and make a before and after comparison. Typically, this procedure is used to evaluate the impact of a process change. In Figure 5.27, the mean shifted lower and the variability was reduced.

5.6.1 Properties of Control Chart Histograms*

Three important properties relate control charts and various histograms obtained from control charts. Understanding these properties increases understanding of the two tools.

5.6 Histograms from Control Charts

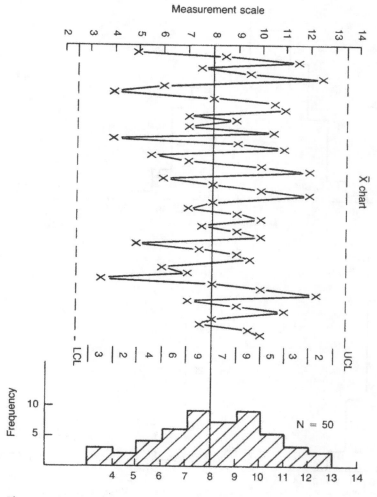

Figure 5.24 Standing a control chart on its end.

Property 1. The histogram obtained from subgroup means is bell shaped and follows a normal distribution if the subgroup size n is sufficiently large. Generally, a subgroup size of $n = 3$, 4, or 5 is considered large enough. The amazing part of this property is that no matter how the histogram of individual measurements is shaped, the subgroup means is approximately bell shaped. This property is known as the central limit theorem and was clearly illustrated in Figure 5.25b.

A reason that this property is important is that the $\bar{X}$ control chart constants were developed by assuming a normal distribution for the subgroup means. Thus, property 1 ensures that the $\bar{X}$ control chart is widely adaptable to a variety of processes. The R chart constants assume that the individual measurements have a normal distribution, which may not always be true. Fortunately, the R chart is still very useful in indicating process problems for a wide variety of situations.

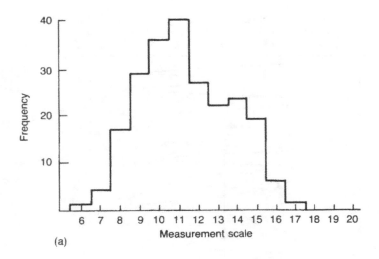
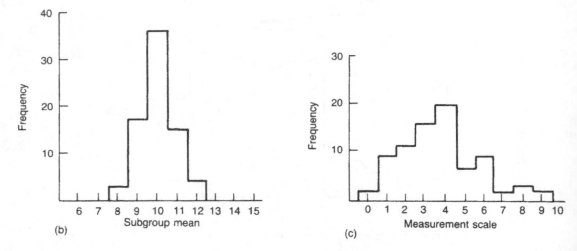

Figure 5.25 Examples of (a) individuals ($N = 225$), (b) means ($N = 75$, $n = 3$), and (c) ranges histograms.

Property 2. The standard deviation σ of individual measurements is related to the standard deviation of the subgroup means $\sigma_{\bar{x}}$ by

$$\sigma_{\bar{x}} = \frac{\sigma}{\sqrt{n}}$$

This implies that the spread of the individual measurements is wider than the spread of the subgroup means. Recall that there are two methods to estimate σ, the standard deviation s or using the average range $\bar{R}/d_2$, as shown in Procedure 3.2:

5.6 Histograms from Control Charts

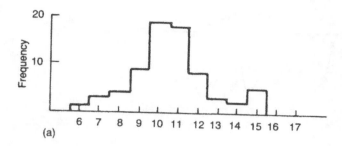

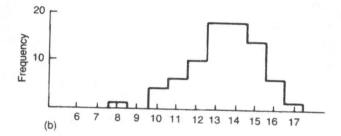

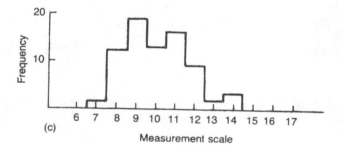

Figure 5.26 Histograms of three sources of variability ($N = 75$): (a) spindle 1; (b) spindle 2; (c) spindle 3.

$$\hat{\sigma} = s$$

$$\hat{\sigma} = \frac{\bar{R}}{d_2}$$

However, if the individuals histogram is not reasonably bell shaped, the two methods do not agree since the constant d_2 assumes a normal distribution. For non–bell-shaped patterns, s provides the better estimate.

Property 3. The A, B, and C zones on the control chart are the same as the zones on a means histogram and can be related to the zones on an individuals histogram using property 2. The zones for either an individuals or a means histogram simply divide the process

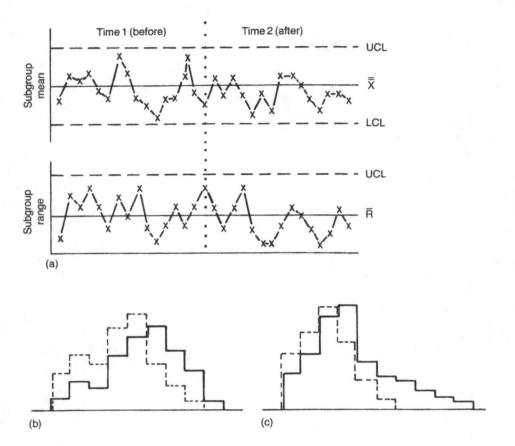

Figure 5.27 Dividing control charts into time segments: (a) control charts; (b) individuals histogram; (c) ranges histogram. In b and c, (——) before, (---) after.

spread into six equal regions corresponding to one standard deviation intervals, as shown here:

Zone	Description	Individuals histogram	Means histogram ($\bar{X}$ chart)
A	Mean + 3 standard deviations	$\bar{\bar{X}} + 3\hat{\sigma}$	$\bar{\bar{X}} + 3\hat{\sigma}/\sqrt{n}$
B	Mean + 2 standard deviations	$\bar{\bar{X}} + 2\hat{\sigma}$	$\bar{\bar{X}} + 2\hat{\sigma}/\sqrt{n}$
C	Mean + 1 standard deviation	$\bar{\bar{X}} + \hat{\sigma}$	$\bar{\bar{X}} + \hat{\sigma}/\sqrt{n}$
C	Mean	$\bar{\bar{X}}$	$\bar{\bar{X}}$
B	Mean − 1 standard deviation	$\bar{\bar{X}} - \hat{\sigma}$	$\bar{\bar{X}} - \hat{\sigma}/\sqrt{n}$
A	Mean − 2 standard deviations	$\bar{\bar{X}} - 2\hat{\sigma}$	$\bar{\bar{X}} - 2\hat{\sigma}/\sqrt{n}$
	Mean − 3 standard deviations	$\bar{\bar{X}} - 3\hat{\sigma}$	$\bar{\bar{X}} - 3\hat{\sigma}/\sqrt{n}$

5.6 Histograms from Control Charts

The means histogram can be directly related to the control chart values, as shown in Figure 5.28. Either method of finding $\hat{\sigma}$ discussed earlier could be used.

Example 5.6 Suppose that $\hat{\sigma} = 2$ with $n = 4$ and $\bar{\bar{X}} = 10$. The standard deviation of the mean is

$$\hat{\sigma}_{\bar{X}} = \frac{2}{\sqrt{4}} = 1$$

so the zones for the individuals and means histogram can be calculated:

Individuals histogram		Means histogram	
$10 + 3\times 2 = 16$	⎫ A	$10 + 3\times 1 = 13$	⎫ A
$10 + 2\times 2 = 14$	⎬ B	$10 + 2\times 1 = 12$	⎬ B
$10 + 1\times 2 = 12$	⎬ C	$10 + 1\times 1 = 11$	⎬ C
10	⎬ C	10	⎬ C
$10 - 1\times 2 = 8$	⎬ B	$10 - 1\times 1 = 9$	⎬ B
$10 - 2\times 2 = 6$	⎬ A	$10 - 2\times 1 = 8$	⎬ A
$10 - 3\times 2 = 4$	⎭	$10 - 3\times 1 = 7$	⎭

Example 5.6 shows that the spread of the individual measurements is appreciably larger than the spread of the subgroup means. In fact, from Procedure 5.4,

$$\text{Control chart spread} = \text{UCL} - \text{LCL}$$
$$= 6\hat{\sigma}_{\bar{X}}$$
$$= \frac{6\hat{\sigma}}{\sqrt{n}}$$

so that

$$\text{Process spread} = 6\hat{\sigma}$$
$$= 6\sqrt{n}\,\hat{\sigma}_{\bar{X}}$$
$$= \sqrt{n}(\text{UCL} - \text{LCL})$$

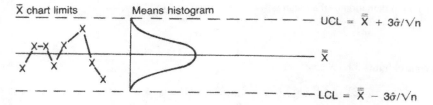

Figure 5.28 Control chart and histogram zones.

Procedure 5.4 Process Spread Zones

Assume a stable process in which the standard control chart values $\bar{\bar{X}}$ and $\bar{R}$ have been computed with subgroups of size n. The standard deviation of individual measurements can be estimated using

$$\hat{\sigma} = s$$

$$\hat{\sigma} = \frac{\bar{R}}{d_2}$$

as discussed in Procedures 3.1 and 3.2. Generally, the two approaches give similar results. Differences are often due to a significantly nonnormal distribution of individual measurements. The first method provides a better estimate in this situation.

Individuals Histogram

The A, B, and C zones on the histogram can be represented as

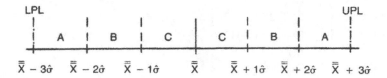

Each zone has a length $\hat{\sigma}$. The limits for the zones are generally computed using $\hat{\sigma} = s$. Assuming a stable process, the process spread in which almost all (99.73%) measurements should occur is

Upper process limit (UPL) = $\bar{\bar{X}} + 3\hat{\sigma}$
Lower process limit (LPL) = $\bar{\bar{X}} - 3\hat{\sigma}$
Process spread = $6\hat{\sigma}$

Means Histogram (or $\bar{X}$ Chart)

The A, B, and C zones for a means histogram correspond to the zones for an $\bar{X}$ chart and can be represented as

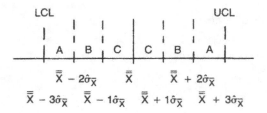

where the standard deviation of the mean is

$$\hat{\sigma}_{\bar{X}} = \frac{\hat{\sigma}}{\sqrt{n}}$$

The lower control limits for a $\bar{X}$ chart are

$$LCL = \bar{\bar{X}} - 3\hat{\sigma}_{\bar{X}} = \bar{\bar{X}} - 3\frac{\hat{\sigma}}{\sqrt{n}} = \bar{\bar{X}} - \frac{3}{\sqrt{n}}\frac{\bar{R}}{d_2} = \bar{\bar{X}} - A_2\bar{R}$$

and similarly for UCL.

5.6 Histograms from Control Charts

A quick method of calculating zones directly from the $\bar{X}$ chart control limits is to find the control chart spread:

Control chart spread = UCL − LCL

$$= 6\hat{\sigma}_{\bar{x}}$$

and divide the spread into six equal intervals of width (UCL − LCL)/6.

Percentage of Measurements in Zones

Assuming the process is stable, a normal distribution predicts the following percentage of measurements in each zone:

Zone	Percentage
A	2
B	14
C	34

Only about 0.3% of the measurement is predicted beyond zone A.

Thus, the predicted process spread for a stable process is

Subgroup size n	Process spread
3	1.7 (UCL − LCL)
4	2.0 (UCL − LCL)
5	2.2 (UCL − LCL)

The spread of individual process measurements is two times the width of the control limits for a subgroup size of $n = 4$.

5.6.2 Why Use Histograms with Control Charts?

It is good practice to periodically plot the individuals histogram from control chart measurements. This practice has the following benefits:

Multiple populations: It is sometimes difficult to detect the presence of multiple populations using control charts. Unfortunately, the process may appear stable when in fact multiple populations exist. Histograms often make detection quite easy (see Sec. 5.7).

Measurement problems: Control charts are generally not effective in detecting measurement problems. Situations with clifflike histograms are not easily uncovered with control charts. Again, control charts make identification of this problem easy (Prob. 5.1).

Shape and smoothness: It is not possible to evaluate the shape and smoothness of the collective measurements using control charts. Histograms can uncover unusual patterns that may be caused by process problems. Also, the bell-shaped normal distribution can be used as a benchmark to evaluate the process output (Prob. 5.8).

Stratified control charts: Using a stratified control chart, it is possible for some changes in one of the sources of variability forming the rows of the control chart to go unnoticed. Plotting separate histograms for each source of variability allows direct comparison of the different sources (Prob. 5.4).

Zonal evaluation: Plotting the histogram of individuals or means allows evaluation of the number of measurements in different zones. As we have seen in the three spindle example, this can indicate a process problem (Prob. 5.9).

Histograms do have limitations, the most significant being the lack of any evaluation of process stability. As we have discussed, stability means predictability, which is desirable for any process. A bell-shaped histogram is not a guarantee of process stability.

A smooth bell-shaped histogram does not ensure a stable process

Example 5.7 Consider the perfectly bell-shaped histogram in Figure 5.29. Different arrangements of the values in a subgroup are possible. Four cases for $n = 4$ are shown in Table 5.1. The $\bar{X}$ chart is plotted in Figure 5.30. Note that the completely different $\bar{X}$ chart patterns produce the same histogram.

5.7 Case Studies

5.7.1 Two Suppliers

Meeting the in-process specification of ± 10 on a dimensional characteristic of a gear case was a continual problem and caused many downstream operations to produce poor-quality parts. The first-line supervisor decided to start a $\bar{X}$ chart on the characteristic and collected

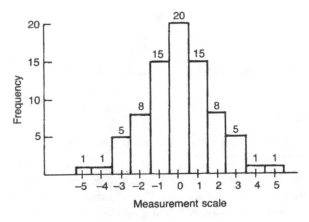

Figure 5.29 Histogram for Example 5.7.

5.7 Case Studies

Table 5.1 Four Control Chart Arrangements of the Histogram Results for Example 5.7

No.	1	2	3	4	5	6	7	8	9	10	11	12	13	14	15	16	17	18	19	20
									Stable process											
1	-3	1	2	-1	0	-2	-1	1	-3	-1	0	-3	-1	-2	-1	-2	0	-2	-1	-1
2	-1	-2	0	-1	-3	-1	1	-2	4	1	-2	0	1	-3	3	0	3	-5	2	0
3	-1	-1	-1	3	-2	1	0	2	-1	2	5	0	0	-4	0	-1	-1	0	3	2
4	2	0	3	0	0	2	-1	1	0	1	-1	0	2	0	0	0	0	0	3	-1
$\bar{X}$	-.25	-.50	1.50	.25	-1.25	.00	-.25	.50	.00	.75	.50	-.50	.50	-2.25	.50	-.25	.50	-1.50	1.25	.50
R	5	3	3	4	3	4	2	4	7	3	7	4	3	4	4	3	4	6	4	3
									Linear trend											
1	-5	-4	-3	-3	-3	-3	-2	-2	-1	-1	-1	-1	2	2	3	3	3	3	4	5
2	-2	-2	-3	-2	-1	-1	-2	-1	-1	0	0	-1	-1	2	-1	-1	2	3	2	2
3	-2	-2	-1	-1	-1	-1	0	0	0	0	0	0	0	2	-1	-1	2	-1	2	2
4	-1	-1	-1	-1	-1	-1	0	0	0	0	0	0	0	0	0	1	1	1	-1	-1
$\bar{X}$	-2.50	-2.25	-2.00	-1.75	-1.50	-1.25	-1.00	-.75	-.50	-.25	.25	.50	.75	1.00	1.25	1.50	1.75	2.00	2.25	2.50
R	4	3	2	2	2	3	2	2	1	1	1	1	2	2	3	2	2	2	3	4
									Quadratic trend											
1	-5	-2	-1	-1	1	-1	1	2	3	5	4	3	3	2	-1	1	3	-1	-2	-4
2	-3	-2	-1	-1	0	1	1	2	3	1	1	3	2	2	1	0	2	-1	-2	-3
3	-3	-2	-1	1	0	2	0	1	0	0	2	0	0	0	2	0	1	-1	-2	-3
4	-2	-2	-1	0	0	0	0	0	0	0	0	0	0	0	1	0	0	-1	-1	-3
$\bar{X}$	-3.25	-2.00	-1.00	-.25	.25	.75	1.00	1.50	1.50	1.50	1.50	1.50	1.25	1.00	.75	.25	1.75	-.25	-1.75	-3.25
R	3	0	0	2	1	3	2	1	3	5	4	3	3	2	3	2	2	0	1	1
									Up-and-down trend											
1	5	-5	4	-4	3	-3	3	-3	2	-2	3	-3	2	-2	2	-2	3	-3	3	-3
2	1	-1	2	-2	1	-1	1	-1	2	-2	2	-2	2	-2	1	-1	1	-1	1	-1
3	0	0	0	0	-1	-1	1	-1	2	-2	0	0	1	0	1	-1	0	-1	1	-1
4	0	0	0	0	-1	-1	1	-1	-1	1	0	0	0	0	0	-1	1	0	0	0
$\bar{X}$	1.50	-1.50	1.50	-1.50	.50	-1.50	1.50	-1.50	1.50	-1.50	1.25	-1.25	1.25	-1.25	1.25	-1.25	1.25	-1.25	1.25	-1.25
R	5	5	4	4	2	2	2	2	2	2	3	3	2	2	1	1	3	3	3	3

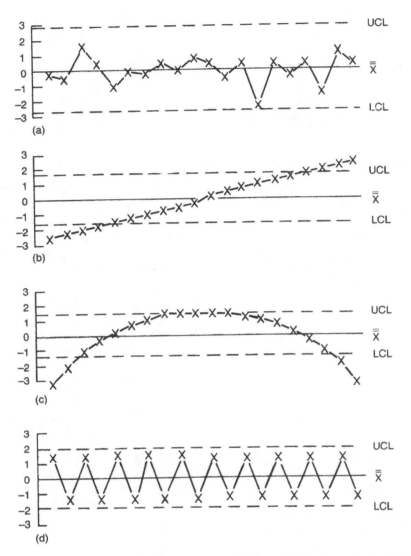

Figure 5.30 Varying $\bar{X}$ chart patterns that produce the same individuals histogram: (a) stable process; (b) linear trend; (c) quadratic trend; (d) up-and-down trend.

$n = 5$ samples per shift. As expected, the chart indicated the process was not stable (Chart 5.1), but what was he to do now? Unfortunately, the sampling plan he selected for troubleshooting the problem spread the samples out too far over time—detailed analysis was not possible. A better plan would have been to use consecutive piece sampling for troubleshooting and then change to a monitoring sampling plan once detailed information concerning special causes had been obtained. However, he decided to use the data he had collected and plot a histogram. The histogram in Figure 5.31 did suggest a possible problem. The supervisor noticed there might be two (or more) groups represented by the

5.7 Case Studies

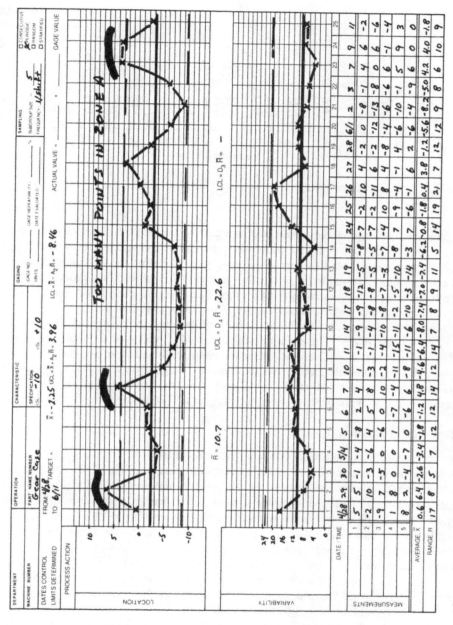

Chart 5.1 Control chart for variable data with $n = 5$ samples per shift: the process is not stable, but the sampling plan does not allow detailed analysis.

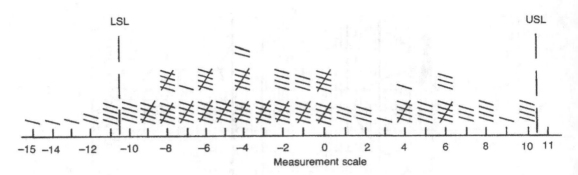

Figure 5.31 Histogram of gear case dimensional characteristics.

histogram. He then recalled there were two gear case casting sources. Checks were made for the castings, and it was found there were several critical differences between the castings. In most cases, both suppliers were meeting specifications but with differing mean levels and variability. To minimize production problems, he initiated long production runs of each supplier rather than many alternating short runs. Process stability was established, and overall quality was improved. Eventually, a single supplier was used.

Deming (1982, Chap. 2) cites the importance of reducing the number of suppliers in order to reduce variability. The example just discussed is typical of the problems introduced by multiple sourcing. Even with the most well-intentioned suppliers, their products differ, which will ultimately lead to unwanted process variability. If production capacity constraints demand multiple suppliers, long batch runs from each supplier are desirable. Alternatively, it may be possible to separately target the flow of each supplier's material.

5.7.2 Robot Evaluation

Robots are used in many industrial applications for repetitive tasks. Performing the work task typically requires the robot to move between different locations. The ability of the robot to return to a desired location is critical. Prior to the purchase of a robot, it was decided to evaluate a robot's performance using $\bar{X}$ and R charts along with an individuals histogram. Ideally the robot should provide a stable process in its ability to return to specific locations. The test consisted of cycling the robot using the stated cycle and dwell times between a home and target position. Subgroups are formed using $n = 5$ consecutive position measurements. To make measurement points comparable, the linear deviation ($1 = .001$ inch) from the target position was used to evaluate movement accuracy. The tests consisted of varying the load the robot was transporting, the speed to move between locations, and the dwell time at different locations.

Selected $\bar{X}$ and R charts appear in Figures 5.32 through 5.36. These charts show many robot characteristics exhibiting a lack of process stability. Figure 5.32 indicates instability since a large number of values are in zone A. Also, notice that around cycle 620 the range chart exhibits reduced variability, possibly due to machine warm-up. Figure 5.33 is a typical example in which a linear $\bar{X}$ chart trend can produce a bell-shaped histogram. Figure 5.34 shows a decreasing linear trend but with excessive variability indicated on the R chart and the non–bell-shaped histogram. Figure 5.35 shows two time trends with a shift in process location around cycle 260. Figure 5.36 is an example in which a process can appear stable on control charts when multiple populations are present.

5.7 Case Studies

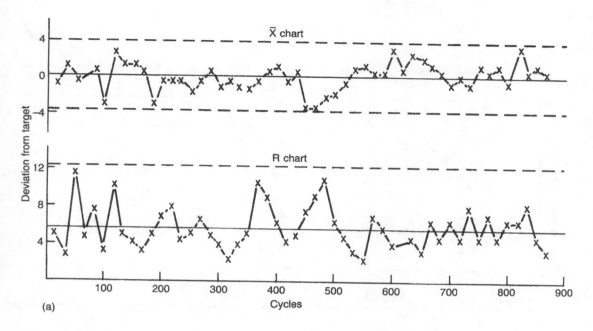

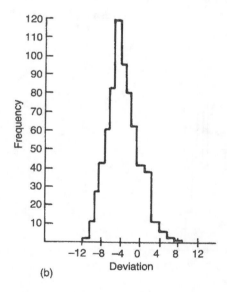

Figure 5.32 Robot evaluation indicating varying performance: (a) $\bar{X}$ and R charts; (b) individuals histogram. Load, 55 pounds; speed, 4 inches/second; cycle time, 15.3 seconds; dwell time, 3.0 seconds.

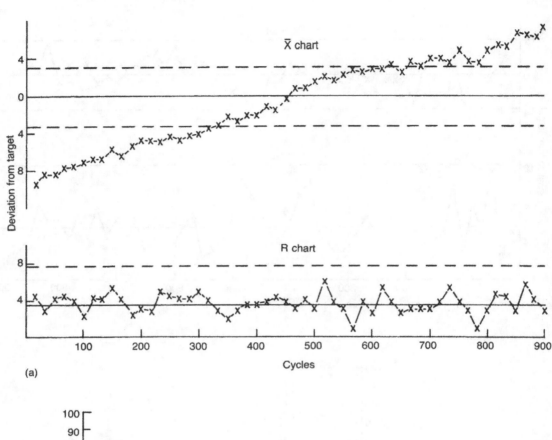

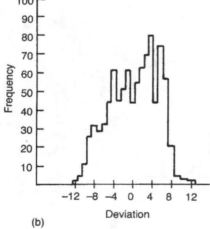

Figure 5.33 Robot evaluation indicating linear trend: (a) $\bar{X}$ and R charts; (b) individuals histogram. Load, 55 pounds; speed, 8 inches/second; cycle time, 3.6 seconds; dwell time, 0 seconds.

5.7 Case Studies

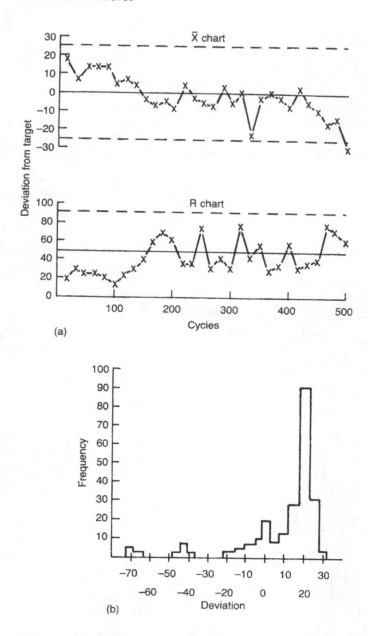

Figure 5.34 Robot evaluation indicating linear trend with outliers: (a) $\bar{X}$ and R charts; (b) individuals histogram. Load, 27 pounds; speed, 4 inches/second; cycle time, 11.9 seconds; dwell time, 0 seconds.

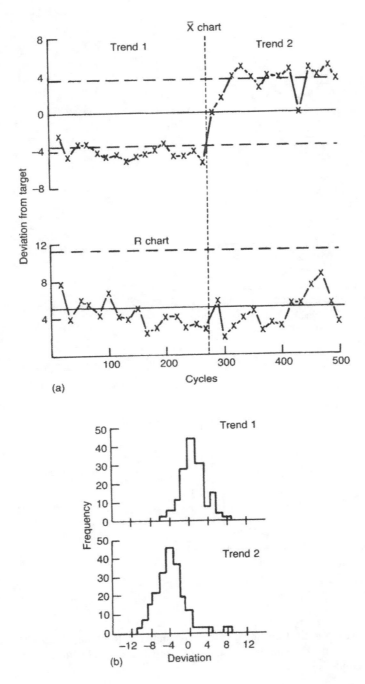

Figure 5.35 Robot evaluation indicating location shift: (a) $\bar{X}$ and R charts; (b) individuals histograms. Load, 27 pounds; speed, 8 inches/second; cycle time, 10.0 seconds; dwell time, 3.0 seconds.

5.7 Case Studies

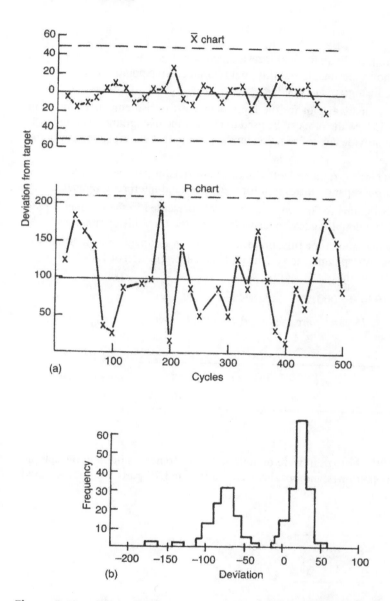

Figure 5.36 Robot evaluation indicating two populations: (a) $\bar{X}$ and R charts; (b) individuals histograms. Load, 27 pounds; speed, 8 inches/second; cycle time, 6.7 seconds; dwell time, 0 seconds.

Problems

5.1 Prepare a histogram of the individual measurements for the crankshaft case study in Chart 3.2. The chart shows reasonable stability, but two successive points are in the lower A zone. What does the histogram suggest?

5.2 (a) Prepare a stem and leaf plot for the individual chart measurements of micro in Problem 3.8. (b) Explain how the plot can be used to prepare the histograms in Figure 5.5. (c) Does the histogram provide any information on the process that was not apparent in the control chart?

5.3 Control chart data for the right and left side of a lathe machining operation was given in Problem 3.7. Prepare a separate histogram for each side, and interpret the results.

5.4 Prepare three histograms for the control chart data presented in the Chapter 4 case study on the eight spindle double-index chucker. Interpret the three histograms.

5.5 A counterboring operation for a final drive sun gear in an automatic transmission was producing quality parts. A step dimension on the part had a specification of 1.1998–1.1984 mm. The data are coded so that

Actual value = $1.1984 + .0001 \times$ coded value

To evaluate the process, 75 parts were measured with the following results:

Value	0	5	6	7	8	9	10	11	12	13	14
Frequency	1	3	13	15	5	1	5	2	1	16	13

The multiple peaks of the histogram were thought to be due to heat buildup in the spindle bridge. Additional coolant lines and nozzles were added, and 75 parts were remeasured with the following results:

Value	6	7	8	9	10
Frequency	6	29	20	12	8

(a) Prepare a histogram for the before and after studies, and interpret the results. (b) Calculate the mean standard deviation and upper and lower process limits for the after study. Convert the results to the actual value scale. Is it meaningful to perform the same calculation for the before study?

5.6 (a) Prepare separate before and after histograms for Problem 3.9. Comment on the improvement. (b) Perform the same analysis for Problem 3.10.

5.7 Consider the artificial data that follow from a normal bell-shaped distribution. The three groups could represent the output of three machines producing the same part.

(a) Prepare a histogram ($N = 50$) for the individual machines. (b) Combine the histograms for machines 1 and 2. The combined mean is $\bar{\bar{X}} = 11.2$. Are the differences in the two machines apparent? (c) Combine the histograms for machines 2 and 3. The combined mean is $\bar{\bar{X}} = 15.4$. Are the differences in the two machines apparent? (d) Combine the histograms for machines 1 and 3. The combined mean is $\bar{\bar{X}} = 13.8$. Are the differences in the two machines apparent? (e) Combine the histograms for machines 1, 2, and 3. The combined mean is $\bar{\bar{X}} = 13.5$. Are the differences in the three machines apparent? The frequency of each measurement value follows:

	Measurement scale																	
	5	6	7	8	9	10	11	12	13	14	15	16	17	18	19	20	21	22
Machine 1	1	2	4	4	11	10	10	7	1									
Machine 2				1	2	2	10	10	6	7	7	2	2	1				
Machine 3									1	1	9	10	11	5	9	3	1	

5.8 Prepare a histogram from the eight spindle screw machine concentricity measurements in Problem 4.1. Interpret the histogram considering the results of Problem 5.7.

5.9 In Problem 3.11, there were 25 subgroups of size $n = 5$. The analysis of this control chart data appears in Chart 4.1. This analysis indicated a possible "hugging the centerline" condition resulting from a mixture of populations. However, the control chart analysis had only 25 subgroups and was not conclusive. (a) Prepare a histogram for the $N = 125$ values used for the control chart. (b) Compare the histogram in part a to the histogram for the complete $N = 225$ values in Figure 5.25. (c) Compute the histogram zones using the values computed from the control chart $\bar{\bar{X}} = 11.6$ and $\hat{\sigma} = 2.4$. (d) Compute $\bar{X}$ and s from the histogram frequencies. Would we expect $\hat{\sigma}$ and s to agree very well in this example? (e) Compare the actual and expected number of points in the histogram zones.

5.10 Brinell hardness is measured by impacting a metal surface with a ball (typically a 10 mm ball with 3000 kg load). The diameter of the impact hole is a measure of the metal's

Day	Brinell hardness	Day	Brinell hardness
1	179, 179, 170, 170, 170	24	179, 197, 187, 179, 179
2	170, 170, 163, 163, 179		170, 170, 170, 179, 170
	170, 179, 170, 163, 163	32	197, 187, 187, 179, 187
3	156, 163, 170, 163, 170		187, 187, 187, 187, 187
5	163, 170, 179, 187, 187	39	170, 170, 163, 163, 163
15	179, 179, 179, 179, 163		163, 163, 170, 163, 170
	163, 163, 179, 170, 163		163, 170, 170, 163, 170
16	179, 170, 170, 170, 170		163, 163, 179, 179, 170
17	179, 170, 170, 170, 170		170, 163, 163, 179, 170
23	179, 170, 187, 187, 179		179, 163, 163, 163, 170

hardness. The Brinell hardness number for different diameters is given in Appendix III. The data below were collected on the hardness of an individual part. (a) Prepare an $\bar{X}$ and R control chart for both the diameter of the impact hole and the Brinell hardness number. Is the process stable? (b) Prepare a histogram of individual measurements for both scales. Which scale seems most appropriate for analysis? (c) The sampling plan consisted of obtaining $n = 5$ parts randomly from a 2 hour run. Owing to varying production requirements, the foundry ran batches of parts. The specification limits were 143–217. Can you suggest a better sampling plan? Comment on the performance of the process.

6
Interpretation of Control Charts

Control charts provide a method to evaluate the stability of a process. Chapter 3 provided basic computational procedures for several types of charts. Chapter 4 discussed how the sampling plan used to collect measurements to form a subgroup can alter the appearance, use, and interpretation of a chart. Chapter 5 related control charts to histograms and the normal, bell-shaped curve. This chapter relates process stability and lack of stability, including out-of-control signals, to histograms. Throughout this chapter a normal, bell-shaped distribution of measurements is assumed.

6.1 Process Stability

In Chapter 3 a process was said to be stable if the process location and variability does not change over time more than would be expected from the short-term fluctuation quantified by $\bar{R}$. Figure 6.1 shows histograms and the bell-shaped curves that illustrate this concept. The key element in establishing stability is that the five process elements: men, methods, machines, materials, and measurements (Fig. 2.2), vary predictably. These elements are not constant; they vary as a result of normal operations. However, process output, owing to the absence of special causes of variation, varies predictably. In a stable process, only common causes of variation are present, resulting in random process variation.

As an example of normal process variation, suppose two groups of $n = 5$ measurements were collected from a process at two different points in time. Figure 6.2 shows two groups of five possible measurements that could be observed. The histograms represent the process output at each time point that could have been obtained by making a large number of measurements (say, $N = 1000$ or more). Notice that the histograms have not changed position so the process is stable, but the two subgroups have a different mean and range. Also, the two groups of measurements differ from the process mean $\bar{\bar{X}} = 17$ and variability (expected $\bar{R} = 6$) quantified by the histogram.

The two subgroups and histograms can be related to a standard control chart. Figure 6.3a shows the same measurements as Figure 6.2, but arranged in a multiple-point run chart format. The individuals histogram could be obtained by standing the run chart ''on its end'' and forming a histogram. The spread of the process ($6s$) is indicated by the extremes of the histogram. Figure 6.3b shows the same measurements, but in a standard $\bar{X}$ chart

214   6 Interpretation of Control Charts

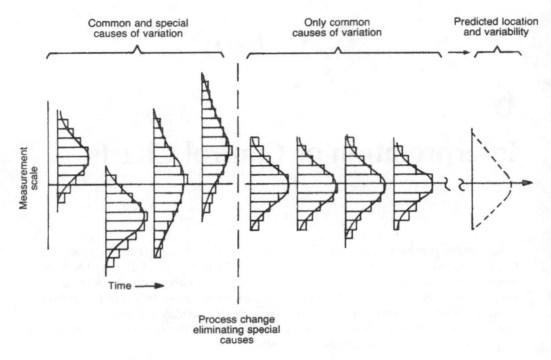

Figure 6.1 Histogram illustrating special and common causes of variation.

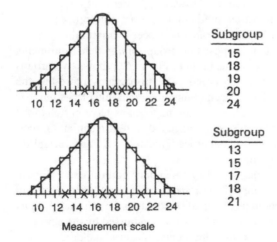

Figure 6.2 Variability in a stable process: (top) time 1, $\bar{X} = 19.2$, $R = 9$; (bottom) time 2, $\bar{X} = 16.8$, $R = 8$.

6.1 Process Stability

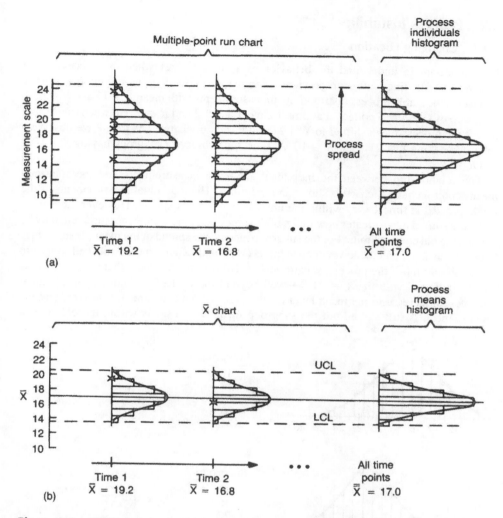

Figure 6.3 Control charts and histograms for a stable process: (a) individual measurements; (b) subgroup means.

format. As discussed in Chapter 5, the extremes of the means histogram correspond to the control limits.

Figure 6.3 shows that the location and spread of the process individuals and means histogram does not change for a stable process. The combined distribution of measurements is unchanged in a stable process, even though the individual measurements and subgroup means vary. The essence of process stability is that these measurements vary predictably between the control limits.

Stable processes have predictable variability

6.2 Process Instability

6.2.1 Changing Location

It is important to understand the behavior of a control chart when the process mean changes. Continuing with our previous example, suppose that in Figure 6.4 time 1 represents the stable process distribution of measurements with mean $\bar{\bar{X}} = 17$ and $\bar{R} = 6$. A subgroup of size $n = 5$ collected at time 1 gave $\bar{X} = 19.2$ and $R = 9$. Between time 1 and time 2 the process mean shifted to $\bar{\bar{X}} = 21$ and $\bar{R}$ did not change. At time 2 we observe a subgroup having $\bar{X} = 21.8$ and $\bar{R} = 10$. At time 3 we observe a subgroup having $\bar{X} = 18.4$ and $R = 7$.

The relationship between the individual subgroup measurements and the subgroup means control chart ($\bar{X}$ chart) is shown in Figure 6.5. The individual measurements of the stable process at time 1 vary within the limits of the process spread. However, at times 2 and 3 the shifted mean creates new limits of the process spread. Some measurement may be beyond the old control limits. For the subgroup means, the spread of the means histogram is between the upper and the lower control limits (UCL and LCL), where almost all subgroup means should fall if the process remains stable. Notice now at time 2 that shifted mean is clearly indicated with the $\bar{X} = 21.8$—well beyond the UCL. This out-of-control signal indicates an increase in the mean level for the process. Note at time 3, however, that the process mean is still shifted but the subgroup mean $\bar{X} = 18.4$ is within the old control

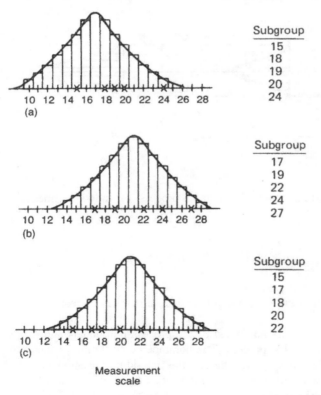

Figure 6.4 Histograms for changing process location: (a) time 1, $\bar{X} = 19.2$, $R = 9$; (b) time 2, $\bar{X} = 21.8$, $R = 10$; (c) time 3, $\bar{X} = 18.4$, $R = 7$.

6.2 Process Instability

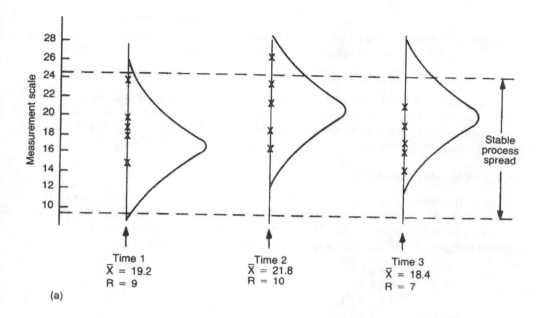

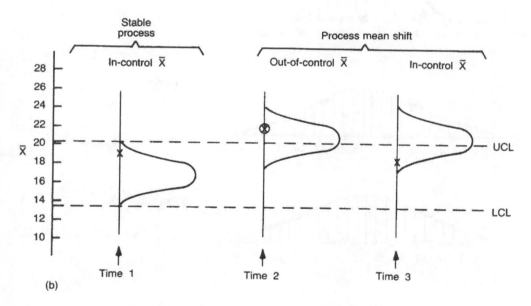

Figure 6.5 Control charts for changing process location: (a) individual measurements; (b) subgroup means ($\bar{X}$ chart).

limits. Control chart users sometimes mistakenly believe any change in the process is indicated on the control chart. Unless the shift in mean level is very large (say, out to $\bar{\bar{X}} = 23$ in our example), the control chart does not indicate a process change for all subgroups.

Process changes may not be immediately indicated by control charts

A practice sometimes used when an out-of-control subgroup is obtained is to collect additional subgroups (see Chap. 4). Clearly, from the time 3 example all the additional subgroup means may not be beyond the control limits. Generally, it is good practice to collect several subgroups (say, six or more) if any verification is necessary.

6.2.2 Changing Variability

Control charts may also be used to indicate changes in process variability. Using the example discussed previously, suppose between times 1 and 2 the variability of the process increased from $\bar{R} = 6$ to $\bar{R} = 10$. The subgroup means and ranges for times 2, 3, and 4 appear in Figure 6.6. Notice the process spread has increased. Time 1:

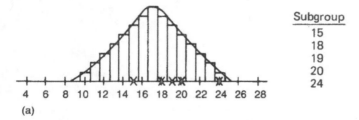

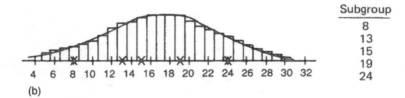

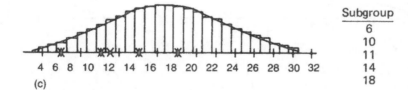

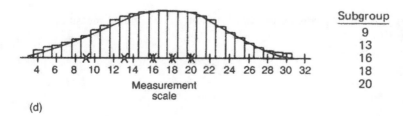

Figure 6.6 Histograms for changing process variability: (a) time 1, $\bar{X} = 19.2, R = 9$; (b) time 2, $\bar{X} = 15.8, R = 16$; (c) time 3, $\bar{X} = 11.8, R = 12$; (d) time 4, $\bar{X} = 15.2, R = 11$.

6.2 Process Instability

$$\text{Process spread} = 6\hat{\sigma} = \frac{6\bar{R}}{d_2} = \frac{6 \times 6}{2.33} = 15.5$$

Times 2, 3, and 4:

$$\text{Process spread} = 6\hat{\sigma} = \frac{6\bar{R}}{d_2} = \frac{6 \times 10}{2.33} = 25.8$$

This increase in spread is apparent in the individuals histograms in Figure 6.6.

The relationship between the individual measurements and the subgroup range control chart appears in Figure 6.7. As expected, the range $R = 9$ at time 1 is within the range

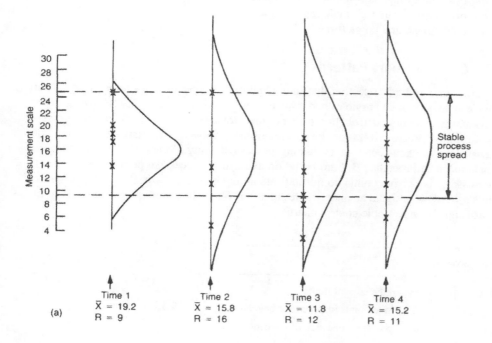

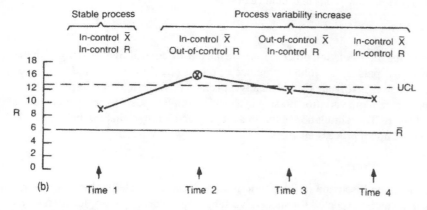

Figure 6.7 Control charts for changing process variability: (a) individual measurements; (b) subgroup ranges (R chart).

UCL. Also, the increase in variability is clearly indicated at time 2 with the range $R = 16$ beyond UCL. However, at time 3 the individual measurements randomly occurred on the low side of the process having increased variability. This resulted in a low range $R = 12$ that was within UCL, but a value of the subgroup mean $\bar{X} = 11.8$ below the $\bar{X}$ chart LCL in Figure 6.5b. Thus, an increase in process variability can result in an out-of-control subgroup mean. At time 4 the individual measurements were generally with the stable process spread, which results in a subgroup mean and range that are both within control limits. In summary, an increase in variability can be difficult to interpret from a control chart since the following cases may occur:

An out-of-control range and stable mean (time 2)
An out-of-control mean and stable range (time 3)
Both a stable mean and range (time 4)

6.3 Control Chart Patterns

We have seen in the last section that individual out-of-control points on an $\bar{X}$ or R chart provide an indication of possible changes in process location or variability. However, using individual points on a control chart does not take advantage of a key feature of a chart—monitoring the process over time. It is the relationship between successive groups of points that is often the key factor in evaluating process stability. Many of the out-of-control signals listed in Procedure 3.7 are based on evaluating successive points (i.e., runs). In other cases, successive points do not indicate an out-of-control signal, but the pattern of points nevertheless signal an undesirable situation. This section interprets standard out-of-control signals on example control charts:

No.	Signal
1	1 point beyond the control limits
2	2 of 3 consecutive points in zone A or beyond
3	8 consecutive points on one side of the centerline
4	6 consecutive points all increasing or decreasing
5	Too many or too few points in control chart zones

The objective of the interpretation process is to use the out-of-control signals along with the overall control chart pattern to evaluate whether a process change has occurred and, if so, what type of change is involved. Unfortunately, it is also possible to overinterpret control chart patterns. The random variation present in a process often appears to be a pattern to the inexperienced user. The simulation exercises at the end of the chapter help provide experience using known process changes.

6.3.1 Shift in Location

The changing location discussed in the previous section can be related to the five control chart signals that can be expected for successive subgroups. Three types of location shifts

6.3 Control Chart Patterns

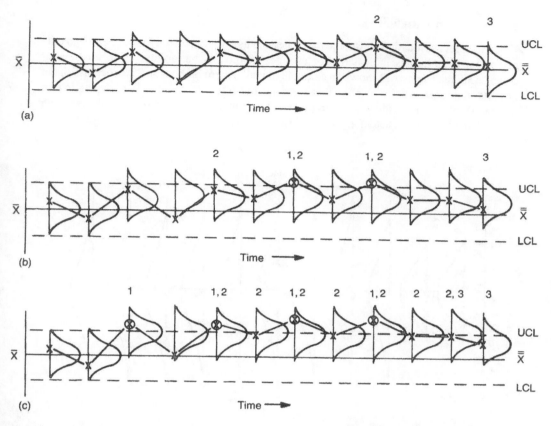

Figure 6.8 Out-of-control signals for shift in process location: (a) minor shift; (b) moderate shift; (c) Major shift. Number denotes type of out-of-control signal.

are discussed: minor, moderate, and major. Figure 6.8 shows an example pattern for each type of process shift. For a minor shift, a point beyond the control (signal 1) limits may not be observed as quickly as signals 2 or 3. It would not be unusual to wait for a fairly large number of subgroups before a point beyond a control limit occurs. Use of all the out-of-control signals improves the chance of detecting a small process shift condition. It should be noted the one-σ shift shown in Figure 6.8a may not have a "minor" impact on the process. For example, in the previous section $\bar{\bar{X}} = 17$, $\bar{R} = 6$ with $n = 5$ so $\sigma = 2.6$, which implies a shift to $\bar{X} = 19.6$, which is a change of 2.6 units.

Use of all out-of-control signals results in earlier detection of process changes

For a moderate shift in the mean, the chance of observing an out-of-control signal increases, and for a major shift, a point beyond a control limit is likely. Thus, as the magnitude of shift increases, the earlier an out-of-control signal is likely to be observed. In all the cases with only a shift in process location, no change in the behavior of the R chart is expected.

6.3.2 Shift in Variability

When the process location shifts, the major issue is how long we need to wait before an out-of-control condition is indicated. For changes in process variability, there are several considerations. Figure 6.9 shows an example in which there is about an 85% increase in the process standard deviation. Notice that the range out-of-control signals vary depending on the range of the individual measurements. The range out-of-control signals generate somewhat of an inconsistent pattern. This inconsistency would be greater for smaller

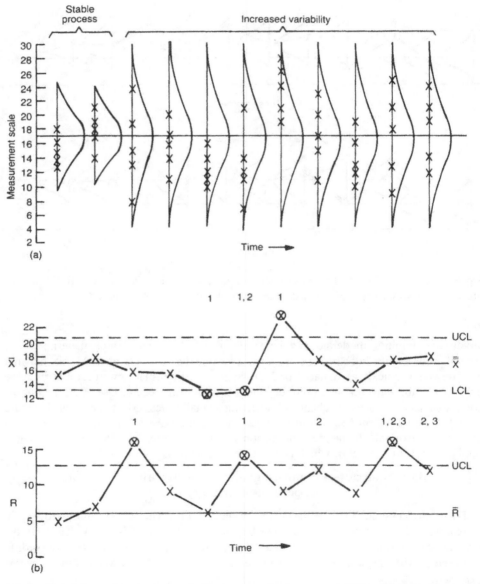

Figure 6.9 Out-of-control signals for increased process variability: (a) individual measurements; (b) $\bar{X}$ and R chart. Number denotes type of out-of-control signal.

6.3 Control Chart Patterns

increases in the process variability. Perhaps more unusual is that varying out-of-control signals appear on the $\bar{X}$ chart due only to changes in variability. Some control chart users mistakenly assume that an out-of-control signal on the $\bar{X}$ chart means a change in location has occurred: this is obviously not true. Notice that one key for interpretation is that both high and low out-of-control signals may appear on the $\bar{X}$ chart if the variability is increased. This behavior does not occur if only the process mean shifts. Finally, again notice that the use of all the out-of-control signals increases the likelihood of indicating an out-of-control condition. However, a subgroup may still not indicate an out-of-control condition when the process is unstable (see point 4).

> Out-of-control signals 1 and 2 on the $\bar{X}$ chart
> can be caused by increased process variability

6.3.3 Shift in Location and Variability

Suppose that a process shifts location and increases its variability. Figure 6.10 shows an example of this case. A variety of out-of-control signals is possible on both the $\bar{X}$ and R charts. The signals on the R chart are the same as if only an increase in variability occurred. As discussed previously, these signals produce a pattern of signals that can be somewhat inconsistent if the increase in process variability is not too large. The out-of-control signals on the $\bar{X}$ chart may also be inconsistent since the subgroup means are more variable owing to the increased variability of the individual measurements. Comparing Figures 6.9 and 6.10, it is apparent that the interpretation of process changes can be difficult and may require a number of successive subgroups to clearly establish a pattern.

6.3.4 Trend in Location

In some cases the mean does not shift abruptly, but changes gradually. Different process problems can be associated with this out-of-control signal:

Shift	Chipped tool
	New raw material
	New operator
	Machine overhauled
Trend	Machine component or tool wearing
	Gradual contamination of machine coolant
	Operator fatigue
	Machine warm-up

Determining whether the process mean is gradually increasing or has shifted can lead to identifying the cause of a problem. Figure 6.11 is an example in which the mean is increasing. Notice that signal 4, with six successive points increasing, indicates the increasing process mean.

In practice it is often not practical to wait for a large number of subgroups to clearly indicate whether a location shift or trend, variability increase, or both have occurred. As noted in Chapter 3, it is important to react to an out-of-control condition once *any* signal occurs. Generally, two courses of action are possible:

1. Use process knowledge. Employees working in an area are familiar with many factors that could influence process stability. Their knowledge can often lead to identification of factors that may cause the out-of-control condition. Detailed interpretation from the control charts is not required.

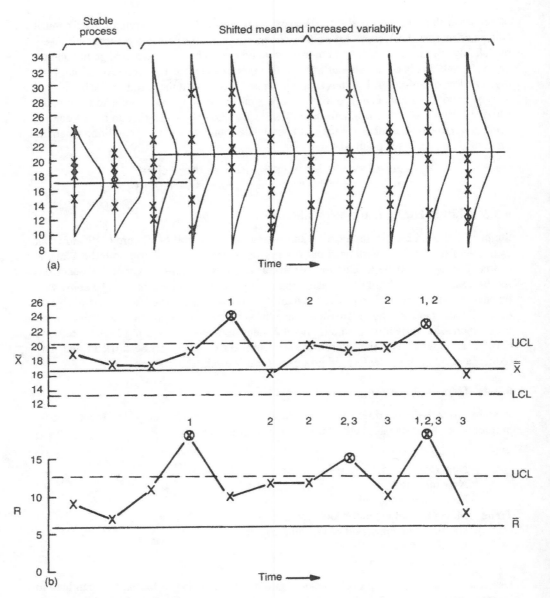

Figure 6.10 Out-of-control signals from a shifted process location with increased variability: (a) individual measurements; (b) $\bar{X}$ and R chart. Number denotes type of out-of-control signal.

6.3 Control Chart Patterns

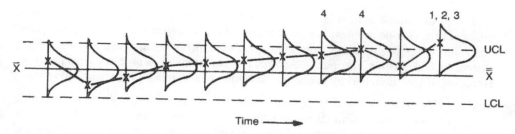

Figure 6.11 $\bar{X}$ chart out-of-control signals for a trend in the process location. Number denotes type of out-of-control signal.

2. Collect additional subgroups. When process knowledge cannot determine the causal factor, collecting additional subgroups often establishes the definition of the problem. Using additional subgroups makes the control chart patterns clearer, which can lead to isolating problem causes.

6.3.5 Process Intervention

Several control chart patterns result from operator intervention in the production process or the measurement system. This intervention alters in some way the natural random variation that would be expected in the process output. The most common example is operator overcontrol (Chap. 4), in which an up-and-down cycle repeats or there are systematic shifts in the process, as shown in Figure 6.12. Typical out-of-control signals indicating these problems are listed in Figure 6.12. However, it is possible that there is no out-of-control signal given in these cases if the control limits are wider than the adjustment level. In these cases, the up-and-down cycles indicate operator intervention.

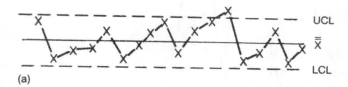

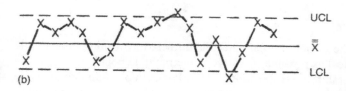

Figure 6.12 Operator overcontrol patterns: (a) up-and-down pattern, consistent operator intervention; (b) up-and-down systematic shift, periodic operator intervention. Out-of-control signals: too many points in zone A; two of three successive points in zone A; one point beyond control limits.

The measurement system is part of the overall production process monitored by a control chart. Gaging problems are sometimes difficult to detect since the original control chart limits use data obtained from the gaging system. Procedures to study the gaging system are addressed in Appendix I. Another process intervention problem is the improper recording of measurements by operators. Examples of recording error are as follows:

Few negative values. In some cases, where several operators maintain a control chart, one operator may fail to record the minus sign on measurements. This error is due to a lack of understanding by the operator. Depending on the number of successive subgroups recorded by each operator, this error can be difficult to detect.

No extreme values. Operators are sometimes reluctant to record extreme values, especially if these values are beyond specification limits. This practice can be difficult to detect using only control charts. The easiest way to detect this problem is to plot a histogram of individual measurements (see Chap. 5, Sec. 5.6.2). A clifflike pattern indicates the problem.

Many recording errors are difficult to detect. It is important that employees be trained on the correct measurement and recording procedures.

6.3.6 Mixtures

In Chapter 5 we discussed how multiple populations influence the appearance of histograms. These combinations of multiple populations (generally called "mixtures") also impact on the control chart patterns that are observed. Mixtures are a common problem in a manufacturing operation. Three populations are shown here, of machines, shifts, operators, lots of material, suppliers, and spindles, for example.

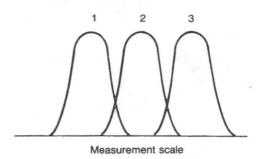

Measurement scale

The objective of reducing the overall process spread is furthered by making the multiple populations as nearly identical as possible. Clearly, as the multiple populations differ in mean or variability, the overall process spread increases. For example, suppose there are 10 spindles on a machine producing the same part and the combined process appears as follows:

6.3 Control Chart Patterns

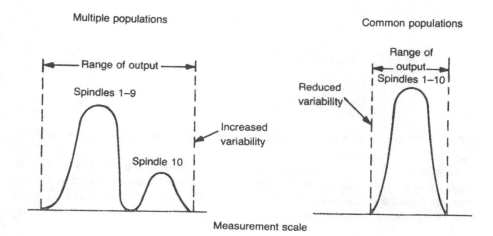

Unfortunately, it is sometimes difficult to detect the presence of multiple populations when many potential populations are present. For example, there may be several hundred machining pallets in a transfer line process. The potential differences between pallets is difficult to study.

The control chart patterns obtained from a mixture depend on two factors:

1. How different are the multiple populations?
2. How are the control chart subgroups formed from the multiple populations?

In general, the greater the population differences, the more extreme the three patterns discussed later will be. However, the sampling methods used to form the control chart subgroups completely change the out-of-control pattern. It is therefore important to understand the different sampling possibilities.

There are three ways in which subgroups may be formed from multiple populations:

1. Stratified sampling. A subgroup contains every stratum (population). If there are three spindles on a machine, each spindle is sampled for a subgroup of size $n = 3$ to form a stratified control chart. Because each subgroup contains a sample from each stratum, there cannot be too many strata.
2. Random sampling. A subgroup may or may not contain a particular population. If there are 12 populations (say, machine fixtures) and $n = 4$, then there is a chance of $4/12$ or 33% that a particular population is sampled in a subgroup. Often there are many populations (e.g., hundreds of machining pallets) in the process being charted and the presence of a particular population in a subgroup occurs randomly.
3. Cluster sampling. A subgroup contains samples only from a single population. Different subgroups contain varying populations. Suppose there are three operators for a process being charted; any subgroup will contain only the measurements made by a particular operator.

To illustrate the three sampling methods, consider an operation that has three machines producing the same part. A partial process flow diagram follows with a control chart at I:

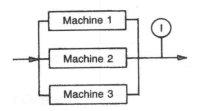

Stratified sampling. Every subgroup contains a part measurement from each of the three machines. The subgroup size is $n = 3$.

Random sampling. Every subgroup contains a random mixture of the three machines. The subgroup size may be any value. Suppose that the control chart (I) is kept at a place removed from the machines and the parts are not marked to indicate which machine was used. Essentially, a random sampling of machines to form subgroups would be the result.

Cluster sampling. Every subgroup contains samples from only one machine. The subgroup size may be any value. Suppose the sampling plan was to collect $n = 5$ samples from a machine every hour and alternate which machine was used. If these data were combined on a single chart, a cluster sampling plan would be employed.

The interpretation of any control chart pattern depends on understanding the sampling plan being used to study the various sources of process variability. Using a process flow diagram, it is possible to relate the control chart sampling plan to the process. It will then be possible to meaningfully interpret control chart patterns. There are three typical patterns that arise from the three subgroup sampling methods.

Hugging the Centerline. An example of hugging the centerline was given in Chapter 4. This control chart pattern is easy to detect in the extreme case since there are few, if any, points in zone A, as shown in Figure 6.13. This pattern results from stratified sampling in which one or more of the populations is different from the rest. Other sampling plans can also produce this pattern if the extreme population is present in most subgroups. The pattern results from the control chart range being too large owing to the continued presence of the extreme population. Since $\bar{R}$ is too large, the $\bar{X}$ chart control width $\pm A_2\bar{R}$ is too large, causing a pattern in which there are too few extreme points. A plot of the individuals histogram shows the presence of multiple populations if the separation between populations is sufficiently large (see Figs. 5.25 and 5.26).

Hugging the centerline can be caused by other factors not related to multiple populations:

Error in control limit calculations. This pattern is obtained if a computational error is made in the control limits, causing the $\bar{X}$ chart limits to be too wide.

Old control limits. This pattern is obtained if the process variability has been reduced and new control limits were not calculated. Examination of the range chart should indicate whether variability has changed.

Nonrecording of extreme values. This pattern is observed if control limits were established using all measurements and the control chart is maintained by excluding extreme values. The range chart should show the artificially reduced variability, and an individuals histogram should show the clifflike pattern.

6.3 Control Chart Patterns

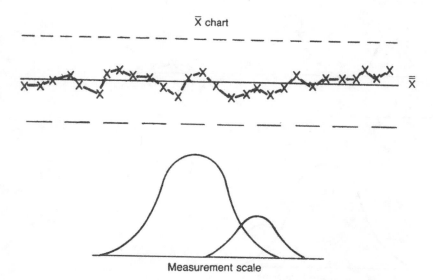

Figure 6.13 Hugging the centerline. Control chart subgroups contain each population, making the range too wide. Out-of-control signal: too many points in zone C.

Inconsistent Instability. It may be expected that instability is inherently inconsistent, but this is not true. Consider the process mean shift in Figure 6.8. The mean shift results in the out-of-control signals indicating process instability. However, since the mean shift is constant and continuing, a consistent pattern of process instability emerges. A common problem in manufacturing that it is sometimes difficult to detect the cause of an out-of-control condition since the signals seem to occur randomly and cannot be easily repeated. However, this pattern is exactly what would be expected in a transfer line with several hundred pallets only a few of which improperly locate parts to be machined. Every time a bad part machined from a bad pallet is measured, inconsistent results are obtained. If pallet numbers are not recorded with part measurements, the inconsistent instability continues indefinitely. Another example involves the random presence of machining chips on a machine locating surface. When this occurs, an out-of-control signal is likely, but it will not repeat since the chip moves when the next part enters the machining station.

The pattern of inconsistent instability is shown in Figure 6.14. The unusual characteristic of this pattern is not its appearance, but the random, nonrepeatable occurrence of the out-of-control signals. The control chart subgroup randomly contains a sample from the extreme population. When this occurs, an unusual $\bar{X}$ or R may result. Determining the cause of the problem is sometimes possible by intensively sampling from the process, maintaining a control chart, and recording all known process sources of variation.

Bunching. When some subgroups contain samples from the extreme population, an out-of-control condition frequently results. Two patterns may result from this cluster sampling, as shown in Figure 6.15. If the timing of subgroup sampling is such that several consecutive extreme subgroups are obtained, an extreme pattern of control chart points results. Examples may involve machine warm-up or one operator's not using proper procedures.

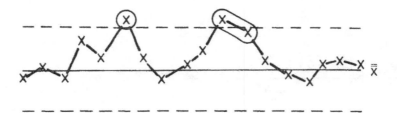

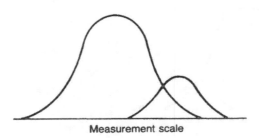
Measurement scale

Figure 6.14 Inconsistent instability. Control chart subgroups randomly contain some samples from the extreme populations. Out-of-control signals: one point beyond the control limits; two of three consecutive points in zone A or above; too many points in zone A.

Alternatively, the cluster of points from the extreme subgroup may be obtained randomly. In this case, the pattern is the same as inconsistent instability if the variability of the extreme population is the same as the main population. If the range is different from the main population, the $\bar{X}$ chart and R chart out-of-control conditions tend to follow similar patterns.

6.3.7 Control Chart Sensitivity to Shifts in Location*

This chapter emphasizes that a change in the process is reflected in a control chart in different patterns. Also, it may be necessary to wait for several subgroups after a process change has occurred before an out-of-control signal is given. The chance of detecting a process shift in location was computed by Wheeler (1983) using out-of-control signals similar to those given in Procedure 3.7. The chance of detecting a shift within a certain number of subgroups depends on the

Size of the process shift
Subgroup size n

Figure 6.16 is a plot of some of the values computed by Wheeler. For a shift of one σ of the process mean, there is about a 10% chance of detecting the shift one subgroup after the shift if $n = 3$. The chance increases to about 20% if $n = 5$. After four subgroups, the chance of detecting the shift increases to about 50% ($n = 3$) and 85% ($n = 5$). The chance of detecting a 2σ shift after one subgroup is about 65% ($n = 3$) and 90% ($n = 5$). After four subgroups, the chance increases to close to 100% for both $n = 3$ and $n = 5$. In general, it is

6.4 Stratified Control Charts

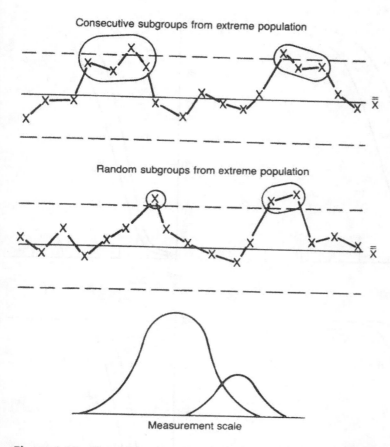

Figure 6.15 Bunching. Some control chart subgroups contain all samples from the extreme population. Out-of-control signals: one point beyond the control limits; two of three consecutive points in zone A or beyond; too many points in zone A.

difficult to detect a one σ shift quickly; frequently four or more subgroups are required. Conversely, a 2σ or 3σ shift is usually detected within one or two subgroups. This delay time must be considered when the time between subgroups is established in the control chart sampling plan. The consequence of a special cause going undetected must be evaluated when establishing the sampling intensity.

6.4 Interpretation of Stratified Control Charts

Stratified control charts were introduced in Chapter 4 as a way in which multiple sources of variability can be combined on a single control chart. Procedure 4.1 warns about the need to first establish a baseline for each unit prior to combining the units on a single chart. The previous section showed that if stratified sampling is used and the units differ, hugging the centerline would result. This section discusses the control chart patterns that result from first establishing a common baseline for all units and later having one or more of the units change location or variability.

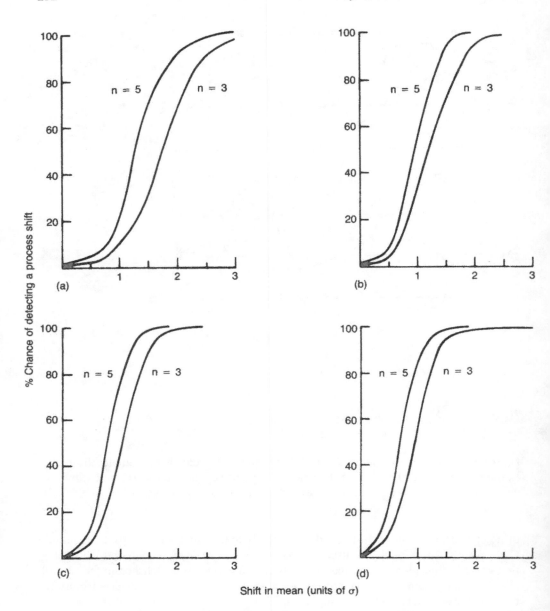

Figure 6.16 Chance of detecting a shift in the process mean for varying mean shifts and subgroup sizes: (a) one subgroup after shift; (b) within two subgroups after shift; (c) within three subgroups after shift; (d) within four subgroups after shift.

6.4 Stratified Control Charts

Consider an example in which a baseline is established for three spindles on a machine at $\bar{\bar{X}} = 17$ and $\sigma = 2.58$, which corresponds to the same population used earlier in this chapter. A stratified control chart can be established with $n = 3$. Consider the following subgroups as examples of various out-of-control conditions:

Unit	\multicolumn{11}{c}{Subgroup}										
	1	2	3	4	5	6	7	8	9	10	11
Spindle 1	16	21	15	12	22	23	12	14	23	12	19
Spindle 2	14	18	12	20	14	20	22	12	15	19	18
Spindle 3	18	16	19	23	26	25	19	17	24	26	22

One of the benefits of a stratified control chart is the ability to scan the rows of the chart to look for any obvious differences in the units. Although 11 subgroups is not enough to establish a meaningful pattern, there is some suggestion that spindle 3 has an increased mean. The control charts for these data appear in Figure 6.17. Notice that the range chart exhibits inconsistent out-of-control signals. However, the only special cause present is the spindle 3 mean increasing from 17 to 21. Because the subgroups are formed by combining the different spindles, an increase in the mean of a spindle causes the subgroup ranges to increase, as follows:

Unit	Subgroup	
Spindle 1	$\bar{\bar{X}} = 17$	⎫ ← Increased subgroup *range*
Spindle 2	$\bar{\bar{X}} = 17$	⎬ due to increased *mean*
Spindle 3	$\bar{\bar{X}} = 21$	⎭ ← of spindle 3

As expected, Figure 6.17 also shows that the $\bar{X}$ chart occasionally indicates an out-of-control subgroup mean. However, the subgroup mean is not as sensitive to an out-of-control condition since spindle 3 with the shifted mean has values that are averaged with the other spindle values. Conversely, the range uses the extremes, so spindle 3 can directly influence the individual subgroup range values.

An out-of-control range can indicate a unit mean shift on a stratified control chart

Consider the same stratified sampling situation but with a new set of subgroups:

Unit	\multicolumn{11}{c}{Subgroup}										
	1	2	3	4	5	6	7	8	9	10	11
Spindle 1	16	21	21	19	19	13	14	21	22	21	19
Spindle 2	14	18	17	15	13	20	18	16	18	18	14
Spindle 3	18	16	26	17	26	7	8	26	14	24	11

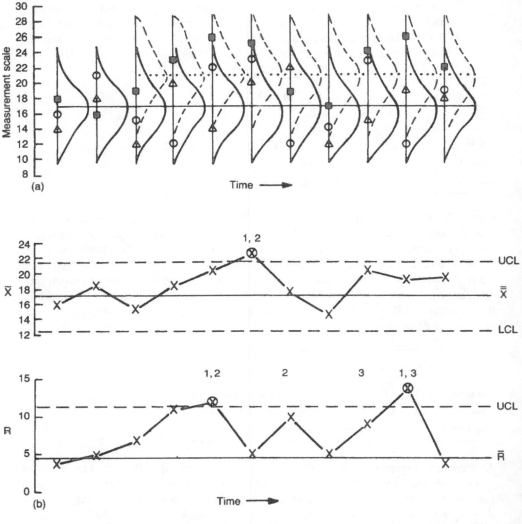

Figure 6.17 Out-of-control signals from a shifted process location in a stratified control chart: (a) individual measurements; (b) $\bar{X}$ and R chart; ○, spindle 1; △, spindle 2; ■, spindle 3. Number denotes type of out-of-control signal.

The pattern for spindle 3 is less clear, but increased variability is a possibility. The process variability for spindle 3 has increased from $\sigma = 2.58$ to 4.33. The control chart for the above data appears in Figure 6.18. Again we see the range chart indicating several out-of-control situations. However, in this case the $\bar{X}$ chart varies from high to low owing to the high and low values from spindle 3. Again we see that it is necessary to examine the $\bar{X}$ and R charts together to determine whether a mean shift or increase in variability has occurred.

This procedure uses the data plotted on the $\bar{X}$ and R chart for interpretation. It is possible to rearrange the data in a convenient manner to make interpretation easier. Three methods are discussed here.

6.4 Stratified Control Charts

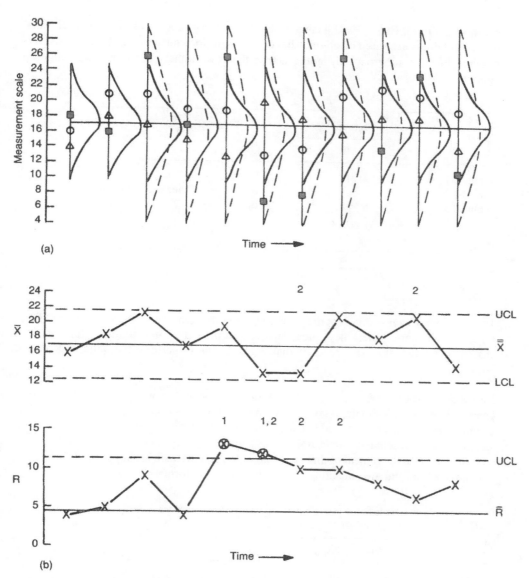

Figure 6.18 Out-of-control signals for increased process variability on a stratified control chart: (a) individual measurements; (b) $\bar{X}$ and R chart; ○, spindle 1; △, spindle 2; ■, spindle 3. Number denotes type of out-of-control signal.

Method 1: Visual Test. In many cases it is possible simply to scan the rows and detect a problem. As with any control chart, the pattern of values should be used rather than one or two measurements. The multiple spindle case study in Chapter 4 shows that scanning the rows of a chart (see Chart 4.5) makes some process problems obvious. Had the data not been organized in the row "stratified" format, some of these problems would go undetected. It is generally easier to detect a change in mean level than a change in variability.

Method 2: Quick Test. Establishing the baseline described in Procedure 4.1 ensures that the control limits are based on all units having the same location and variability. Thus, it is possible to form subgroups by going across the rows rather than down the subgroup column.

		Subgroup			$\bar{X}$	R	
Unit	...	9	10	11			
Spindle 1	...	23	12	19	18.0	11	←
Spindle 2	...	15	19	18	17.3	4	← Three new subgroups
Spindle 3	...	24	26	22	24.0	4	←
$\bar{X}$		20.7	19.0	19.7			
R		9	14	4			

↑_____↑
Standard subgroups

The subgroup size $n = 3$ is kept the same as used in the standard chart to enable use of the same control limits. These new subgroups can be plotted on the stratified chart. A new chart is not necessary. This practice makes it possible to perform the quick test on the production floor.

Method 3: Adjusted Control Limits. When five or more sources are combined on a chart, it may be undesirable to go back in time too long a period using the quick test. For example, if a single chart is used to control an eight spindle machine and an out-of-control signal is detected, it may be undesirable to use eight subgroups to compute $\bar{X}$ and R values. Using too large an increment of time may make the chart insensitive. Suppose we have a six spindle machine and are collecting data in the following format:

		Subgroup					
Unit	...	64	65	66	67	68	69
Spindle 1		0	−1	2	−1	−2	1
Spindle 2		−2	0	1	0	−1	−1
Spindle 3	...	−2	1	−1	2	4	3
Spindle 4		−1	0	1	1	3	−3
Spindle 5		2	−1	1	3	−1	0
Spindle 6		0	−1	2	−2	1	1
$\bar{X}$	...	−.5	−.33	1.0	.5	.67	.17
R		4	2	3	5	6	6

It may be desirable to use $n = 3$ (rather than $n = 6$ as would be necessary for the quick test) to determine which spindle is a problem. The following procedure can be used to compute the new ($n = 3$) limits from the original ($n = 6$) limits. Original control chart values, $n = 6$:

6.4 Stratified Control Charts

$\bar{R}_{\text{orig}} = 3.1$

$\bar{\bar{X}}_{\text{orig}} = .5$

$\hat{\sigma} = \dfrac{\bar{R}_{\text{orig}}}{d_2} = \dfrac{3.1}{2.53} = 1.23 \qquad d_2 \text{ for } n = 6$

New control chart values, $n = 3$:

$\bar{R}_{\text{new}} = d_2 \hat{\sigma} = 1.69 \times 1.23 = 2.08 \qquad d_2 \text{ for } n = 3$

$\text{UCL}_R = D_4 \bar{R}_{\text{new}} = 2.58 \times 2.08 = 5.37 \qquad D_4 \text{ for } n = 3$

$\bar{\bar{X}}_{\text{new}} = \bar{\bar{X}}_{\text{orig}} = .5$

$\text{UCL}_{\bar{X}} = \bar{\bar{X}}_{\text{new}} + A_2 \bar{R}_{\text{new}} = .5 + (1.02 \times 2.08) = 2.62$

$\text{LCL}_{\bar{X}} = \bar{\bar{X}}_{\text{new}} - A_2 \bar{R}_{\text{new}} = 0.5 - (1.02 \times 2.08) = 1.62$

The new rearranged subgroups are as follows:

	Subgroup 64–66, spindle						Subgroup 67–69, spindle					
	1	2	3	4	5	6	1	2	3	4	5	6
	0	−2	−2	−1	2	0	−1	0	2	1	3	−2
	−1	0	1	0	−1	−1	−2	−1	4	3	−1	1
	2	1	−1	1	1	2	1	−1	3	−3	0	1
$\bar{X}$	.33	−.33	−.67	0	.67	.33	−.67	−.67	3	.33	.67	0
R	3	3	3	2	3	3	3	1	2	6	4	3

It appears that spindle 3 went out of control by increasing its mean level and spindle 4 went out of control by increasing its variability in subgroups 67–69. These calculations need to be performed only when control limits change. The new $n = 3$ control limits could be written at the top of a chart to enable review of the charts (using $n = 3$) on the floor. An example of a format that can be used for plotting is given in a case study (Chart 6.7).

Considering the two previous examples, the means and ranges are as follows:

	Subgroup 3–5, spindle			Subgroup 6–8, spindle			Subgroup 9–11, spindle		
	1	2	3	1	2	3	1	2	3
Shifted mean									
$\bar{X}$	16.3	15.3	22.7	16.3	18.0	20.3	18.0	17.3	24.0
R	10	8	7	11	10	8	11	4	4
Increased variability									
$\bar{X}$	19.7	15.0	23.0	16.0	18.0	13.7	20.7	16.7	16.3
R	2	4	9	8	4	19	3	4	13

These subgroup means and ranges could be plotted directly on the $\bar{X}$ and R charts in Figures 6.17 and 6.18. This procedure gives a reasonable indication of the problem.

When using stratified control charts, it is sometimes difficult to determine what units may have changed. This is particularly true when only a few unusual subgroups have been observed. In these cases, the best approach is to use the methods described in Chapter 11 to isolate any changes. Using these procedures, a number of new measurements would be made on each unit and the data analyzed by one of the comparison methods.

6.5 Control Chart Simulation

Interpretation of control chart patterns requires practice in observing how various process changes appear on a control chart. If real manufacturing data are used for practice, there is always a question of what change really did occur. As we have seen, control chart patterns may not be consistent and precise interpretation can be difficult. Thus, the best way to gain experience is to simulate a control chart. Using simulation methods, it is possible to make known changes the process mean or variability at known times and examine the resulting pattern. This approach makes the type of change and the time of the change known. Only the way in which the control chart pattern reflects the change is subject to interpretation.

The control chart simulation uses eight approximately normal, bell-shaped populations shown in Figure 6.19. Population A is generally considered the base population, with $\bar{\bar{X}} = 0$ and $\sigma = 1.74$. Thus for a subgroup size of $n = 5$, the standard control chart values are as follows:

$$\bar{R} = d_2 \sigma \qquad d_2 = 2.33 \text{ for } n = 5$$
$$= 2.33 \times 1.74$$
$$= 4.1$$
$$\text{UCL}_R = D_4 \bar{R} \qquad D_4 = 2.12 \text{ for } n = 5$$
$$= 2.12 \times 4.1$$
$$= 8.7$$
$$\text{UCL}_{\bar{X}} = \bar{\bar{X}} + A_2 \bar{R} \qquad A_2 = 0.58 \text{ for } n = 5$$
$$= 2.4$$
$$\text{LCL}_{\bar{X}} = \bar{\bar{X}} - A_2 \bar{R}$$
$$= -2.4$$

Simulating 20–25 subgroups and computing the control chart values should produce values close to the values just calculated. The following table is convenient to check control chart values for population A for common subgroup sizes:

Subgroup size n	$\bar{R}$	UCL$_R$	UCL$_{\bar{X}}$	$\bar{\bar{X}}$	LCL$_{\bar{X}}$
3	2.9	7.5	3.0	0	−3.0
4	3.6	8.2	2.6	0	−2.6
5	4.1	8.7	2.4	0	−2.4
6	4.4	8.8	2.1	0	−2.1

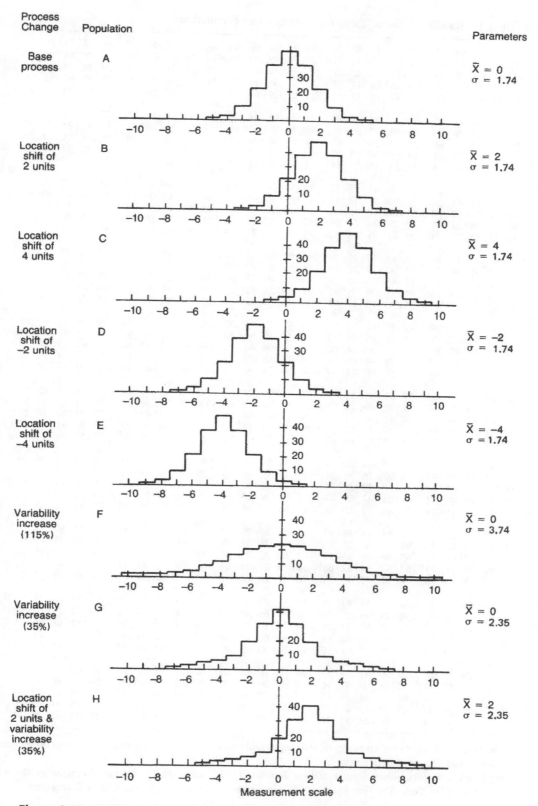

Figure 6.19 Eight populations used for control chart simulation.

Table 6.1   Random Number Table for Control Chart Simulation[a]

94	01	54	68	74	32	44	44	82	77	59	82	09	61	63	64	65	42	58	43	41	14	54	28	20
74	10	88	82	22	88	57	07	40	15	25	70	49	10	35	01	75	51	47	50	48	96	83	86	03
62	88	08	78	73	95	16	05	92	21	22	30	49	03	14	72	87	71	73	34	39	28	30	41	49
11	74	81	21	02	80	58	04	18	67	17	71	05	96	21	06	55	40	78	50	73	95	07	95	52
17	94	40	56	00	60	47	80	33	43	25	85	25	89	05	57	21	63	96	18	49	85	69	93	26
66	06	74	27	92	95	04	35	26	80	46	78	05	64	87	09	97	15	94	81	37	00	62	21	86
54	24	49	10	30	45	54	77	08	18	59	84	99	61	69	61	45	92	16	47	87	41	71	71	98
30	94	55	75	89	31	73	25	72	60	47	67	00	76	54	46	37	62	53	66	94	74	64	95	80
69	17	03	74	03	86	99	59	03	07	94	30	47	18	03	26	82	50	55	11	12	45	99	13	14
08	34	58	89	75	35	84	18	57	71	08	10	55	99	87	87	11	22	14	76	14	71	37	11	81
27	76	74	35	84	85	30	18	89	77	29	49	06	97	14	73	03	54	12	07	74	69	90	93	10
13	02	51	43	38	54	06	61	52	43	47	72	46	67	33	47	43	14	39	05	31	04	85	66	99
80	21	73	62	92	98	52	52	43	35	24	43	22	48	96	43	27	75	88	74	11	46	61	60	82
10	87	56	20	04	90	39	16	11	05	57	41	10	63	68	53	85	63	07	43	08	67	08	47	41
54	12	75	73	26	26	62	91	90	87	24	47	28	87	79	30	54	02	78	86	61	73	27	54	54
60	31	14	28	24	37	30	14	26	78	45	99	04	32	42	17	37	45	20	03	70	70	77	02	14
49	73	97	14	84	92	00	39	80	86	76	66	87	32	09	59	20	21	19	73	02	90	23	32	50
78	62	65	15	94	16	45	39	46	14	39	01	49	70	66	83	01	20	98	32	25	57	17	76	28
66	69	21	39	86	99	83	70	05	82	81	23	24	49	87	09	50	49	64	12	90	19	37	95	68
44	07	12	80	91	07	36	29	77	03	76	44	74	25	37	98	52	49	78	31	65	70	40	95	14
41	46	88	51	49	49	55	41	79	94	14	92	43	96	50	95	29	40	05	56	70	48	10	69	05
94	55	93	75	59	49	67	85	31	19	70	31	20	56	82	66	98	63	40	99	74	47	42	07	40
41	61	57	03	60	64	11	45	86	60	90	85	06	46	18	80	62	05	17	90	11	43	63	80	72
50	27	39	31	13	41	79	48	68	61	24	78	18	96	83	55	41	18	56	67	77	53	59	98	92
41	39	68	05	04	90	67	00	82	89	40	90	20	50	69	95	08	30	67	83	28	10	25	78	16
25	80	72	42	60	71	52	97	89	20	72	68	20	73	85	90	72	65	71	66	98	88	40	85	83
06	17	09	79	65	88	30	29	80	41	21	44	34	18	08	68	98	48	36	20	89	74	79	88	82
60	80	85	44	44	74	41	28	11	05	01	17	62	88	38	36	42	11	64	89	18	05	95	10	61
80	94	04	48	93	10	40	83	62	22	80	58	27	19	44	92	63	84	03	33	67	05	41	60	67
19	51	69	01	20	46	75	97	16	43	13	17	75	52	92	21	03	68	28	08	77	50	19	74	27
49	38	65	44	80	23	60	42	35	54	21	78	54	11	01	91	17	81	01	74	29	42	09	04	38
06	31	28	89	40	15	99	56	93	21	47	45	86	48	09	98	18	98	18	51	29	65	18	42	15
60	94	20	03	07	11	89	79	26	74	40	40	56	80	32	96	71	75	42	44	10	70	14	13	93
92	32	99	89	32	78	28	44	63	47	71	20	99	20	61	39	44	89	31	36	25	72	20	85	64
77	93	66	35	74	31	38	45	19	24	85	56	12	96	71	58	13	71	78	20	22	75	13	65	18
38	10	17	77	56	11	65	71	38	97	95	88	95	70	67	47	64	81	38	85	70	66	99	34	06
39	64	16	94	57	91	33	92	25	02	92	61	38	97	19	11	94	75	62	03	19	32	42	05	04
84	05	44	04	55	99	39	66	36	80	67	66	76	06	31	69	18	19	68	45	38	52	51	16	00
47	46	80	35	77	57	64	96	32	66	24	70	07	13	94	14	00	42	31	53	69	24	90	57	47
43	32	13	13	70	28	97	72	38	96	76	47	96	85	62	62	34	20	75	89	08	89	90	59	85
64	28	16	18	26	18	55	56	49	37	13	17	33	33	65	78	85	11	64	99	87	06	41	30	75
66	84	77	04	95	32	35	00	29	85	86	71	63	87	45	26	31	37	74	63	55	38	77	26	81
72	46	13	32	30	21	52	95	34	24	92	58	10	22	62	78	43	86	62	76	18	39	67	35	38
21	03	29	10	50	13	05	81	62	18	12	47	05	65	00	15	29	27	61	39	59	52	65	21	13
95	36	26	70	11	06	65	11	61	36	01	01	60	08	37	55	01	85	63	74	35	82	47	17	08
49	71	29	73	80	10	40	45	54	52	34	03	06	07	26	75	21	11	02	71	36	63	36	84	24
58	27	56	17	64	97	58	65	47	16	50	25	94	63	46	87	19	54	60	92	26	78	76	09	39
89	51	41	17	88	68	22	42	34	17	73	95	97	61	46	30	34	24	02	77	11	04	97	20	49
15	47	25	06	69	48	13	93	67	32	46	87	43	70	88	73	46	50	98	19	58	86	93	52	20
12	12	08	61	24	51	24	74	43	02	60	88	35	21	09	21	43	73	67	86	49	22	67	78	37

[a]*Note:* To select a set of random numbers, start at any table position and select numbers going up or down a column or across a row in a systematic manner.
Reproduced with permission from *A Million Random Digits with 100,000 Normal Deviates* by the RAND Corporation (New York: The Free Press, 1955). Copyright 1955 and 1983 by the RAND Corporation.

6.5 Control Chart Simulation

The procedure for simulating a control chart follows:

1. Select a population to be simulated. If the base process population A is selected, the preceding control limits may be used.
2. Select a starting position in Table 6.1. Going along a column or row, pick a random number. Do not skip any values in the row or column pattern you select.
3. For the population selected in step 1, find the interval containing the random number in Table 6.2. Select the measurement column value corresponding to the upper interval boundary. For example, assign −1 to random numbers between 38 and 19 for population A.

Table 6.2 Measurement Assignment Table for Control Chart Simulation[a]

Measurement	Population							
	A	B	C	D	E	F	G	H
10						00		
9			00			99		00
8			99			98		99
7		00	98			97	00	98
6		99	93			95	99	96
5	00	98	82			93	98	93
4	99	93	62			90	96	89
3	98	82	38	00		85	93	80
2	93	62	18	99		77	89	62
1	82	38	07	98	00	67	80	38
0	62	18	02	93	99	56	62	20
−1	38	07	00	82	98	44	38	11
−2	18	02		62	93	33	20	07
−3	07	00		38	82	23	11	04
−4	02			18	62	15	07	02
−5	01			07	38	10	04	01
−6				02	18	07	02	
−7				01	07	05	01	
−8					02	03		
−9					01	02		
−10						01		

[a] To obtain a simulated measurement value, select a two-digit random number from Table 6.1. For the desired population column, place the random number in an interval and select the measurement column value corresponding to the upper interval boundary. For example, assign −1 to random numbers between 38 and 19 for population A.

4. Repeat steps 2 and 3 to obtain the desired number of measurement values. Note that the measurement column can easily be shifted to nonnegative values by adding 10 to each measurement.

Example 6.1 Suppose we select a subgroup size of $n = 4$ and wish to examine how a process location shift of +2 units (population B) would appear on a control chart established for the base process with $\bar{X} = 0$ (population A). Control limits can be established directly for the base process from the table given previously in this section:

$\bar{R} = 3.6$
$\text{UCL}_R = 8.2$
$\bar{\bar{X}} = 0$
$\text{LCL}_{\bar{X}} = -2.6$
$\text{UCL}_{\bar{X}} = 2.6$

Normally, we would generate 20–25 subgroups, but for this example, consider only two subgroups. Following the steps,

1. Select population B.
2. Start at row 1, column 4 in Table 6.1. The first eight random numbers are 68, 82, 78, 21, 56, 27, 10, and 75.
3. For population B, the measurement values are as follows:

Random number	Measurement	Control chart values
68	3	
82	3	$\bar{X} = 2.5$
78	3	$R = 2$
21	1	
56	2	
27	1	$\bar{X} = 1.5$
10	0	$R = 3$
75	3	

The control chart values would be plotted on a control chart and the pattern examined.

The simulation procedure in Example 6.1 makes it possible to examine the patterns for a variety of location and variability changes. The chapter case studies illustrate the simulation procedure for a standard and a stratified control chart.

6.6 Summary

Competent interpretation of control charts is an important part of the training required to establish a defect prevention system (DPS). Procedure 6.1 provides questions to encourage some useful practices when an unusual pattern is suspected. The control chart patterns discussed in this chapter can serve only as guides about possible problems. The overriding

6.6 Summary

Procedure 6.1 Checklist for Interpretation of Control Chart Signals

The following list of questions can be used to guide the interpretation of control charts. Ideally, a group of individuals familiar with the process can use this list to guide the discussion concerning a problem identified by a control chart.

1. Check the obvious
   Are there any calculation errors for the $\bar{X}$ and R values?
   Are the $\bar{X}$ and R values plotted correctly?
   Are the control limits calculated correctly?
   How old are the control limits?
   Can unusual measurements be rechecked?
2. Understand the sampling plan
   How are the samples forming a subgroup collected over time?
   Where in the process are parts collected?
   Is the sampling plan appropriate to evaluate the sources of variability identified on the process flow diagram?
3. Use all out-of-control signals
   Have all five out-of-control signals been evaluated?
   Are all out-of-control signals marked on both the $\bar{X}$ and R charts?
   Are the "patterns" on the chart being overinterpreted? Could they be due to random variation?
4. Use $\bar{X}$ and R charts jointly
   Can the patterns on the $\bar{X}$ and R charts be related?
   Can an out-of-control mean be due to increased variability?
5. Evaluate all approaches
   Can you take advantage of process knowledge to suggest possible problem causes?
   What has recently changed in the process?
   Has a histogram of individual measurements been evaluated?
   Are the upstream sources of variability on the process flow diagram stable?
   Has the measurement system been evaluated?
   Does the variation or special causes diagram indicate possible problem causes?
6. Collect additional data
   Does additional subgroup sampling verify an ongoing problem?
   Can a comparison study (Chap. 11) be performed to isolate problem causes?
7. Do something
   Are the appropriate people reacting to solve the problem?
   Should a team be established (Chap. 13)?

---

concern must be to react to problems identified by the charts. Correct interpretation of the charts is important, but interpretation without action is meaningless.

A top company official on one of his manufacturing plant tours stopped at a control chart display in an aisle. He noticed that there were several out-of-control signals but no points beyond the control limits. He discussed his concern with several managers on the tour. Later he stopped to review another chart and noted several out-of-control points. The employee maintaining the chart said that they were aware of the signals but really were not sure of the cause. Also, the signal seemed to go away. In a later discussion with the plant manager, the company official asked if he could review the plant's plans to revitalize their use of control charts.

6.7 Case Studies

6.7.1 Simulating Process Changes on an $\bar{X}$ and R Chart

Consider a simulation exercise in which a single machine is studied using a standard control chart with $n = 5$ consecutive parts measured about every 2 hours. The process flow diagram, variation diagram, special causes diagram, and data collection format follow. A control chart is located at I with no upstream sources of variability.

Process flow diagram:

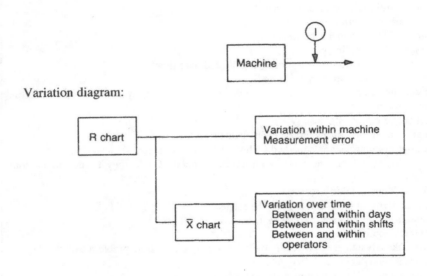

Variation diagram:

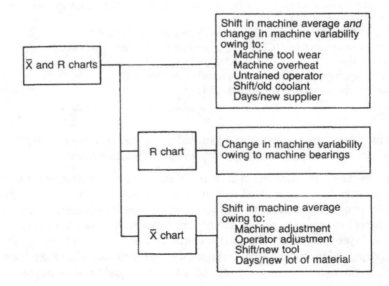

Special causes diagram:

6.7 Case Studies

Data collection format:

	Day 10–25, hour			
Piece	6:05	8:00	10:15	12:30
1				
2				
3				
4				
5				
Sum				
$\bar{X}$				
R				

These tools illustrate the planning that can be used to make better use of a control chart on the production floor. The following simulation examples can be used to study control chart behavior. Attempt to determine what type of process change occurred and when it occurred.

Chart 6.1. The base process control chart was generated by simulation from population A in Table 6.2. Notice that the actual and predicted control chart values agree rather closely:

	$\bar{R}$	UCL_R	$LCL_{\bar{X}}$	$\bar{\bar{X}}$	$UCL_{\bar{X}}$
Predicted	3.6	8.2	−2.6	0	2.6
Actual	3.56	8.1	−2.64	−.04	2.56

Using 25 subgroups makes the control chart values agree closely with the values predicted from the base process histogram in Figure 6.19. The R chart appears stable. The pattern on the $\bar{X}$ chart appears to be hugging the centerline. The limits of zone C are $(-0.91, 0.83)$, resulting in only 4 of 25 subgroups (16%) in zone B. From Procedure 3.7, 32% of the subgroups should be in zones B and A. The following plot of the individuals histogram can be used to further evaluate the distribution of measurements:

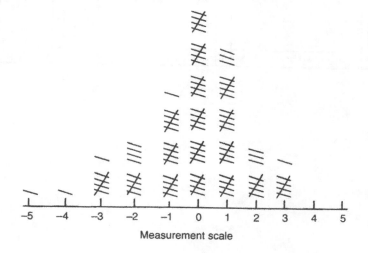

Measurement scale

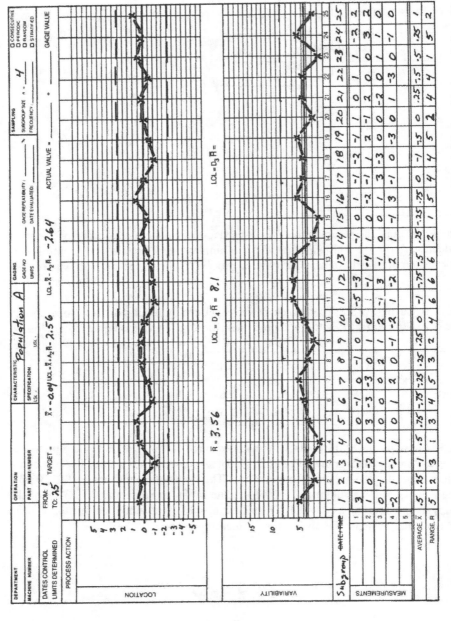

Chart 6.1 Control chart for variable data. Base process control chart generated by simulation from population A in Table 6.2.

6.7 Case Studies

The histogram appears to be reasonably bell-shaped. The zones for the individuals histogram follow, along with the number and percentages of points in each zone. The calculations are based on an estimate of the standard deviation $\hat{\sigma} = \bar{R}/d_2 = 3.56/2.06 = 1.73$.

Zone	Boundary line	Number in zone	Actual % in zone	Expected % in zone
A	$\bar{\bar{X}} + 3\hat{\sigma} = -.04 + (3 \times 1.73) = 5.15$	0	0	2
B	$\bar{\bar{X}} + 2\hat{\sigma} = -.04 + (2 \times 1.73) = 3.42$	14	14	14
C	$\bar{\bar{X}} + 1\hat{\sigma} = -.04 + 1.73 = 1.69$	53	53	34
C	$\bar{\bar{X}} = -.04$	16	16	34
B	$\bar{\bar{X}} - 1\hat{\sigma} = -.04 - 1.73 = -1.77$	15	15	14
A	$\bar{\bar{X}} - 2\hat{\sigma} = -.04 - (2 \times 1.73) = -3.50$	2	2	2
	$\bar{\bar{X}} - 3\hat{\sigma} = -.04 - (3 \times 1.73) = -5.23$			

The actual and expected percentages agree reasonably well, except in zone C, where all the 0 values were assigned to the upper part of zone C. Collectively for zone C, the actual percentage was 69% but the expected percentage was 68%. The individuals histogram does not confirm the existence of multiple populations that may have been suggested by hugging of the centerline. Thus, any lack of extreme points in Chart 6.1 must be due to random variability.

Chart 6.2. The R chart appears stable, with the exception of a run of seven points below $\bar{R}$ that is not sustained so no overall decrease in process variability seems possible. On the $\bar{X}$ chart, subgroups 31–44 indicate a run of points above the centerline, indicating an increase in process location. Subgroups 47–50 indicate a decrease in process location.

The actual simulation used the following populations:

Subgroups	Population	Mean
26–30	A	0
31–38	B	2
39–43	C	4
44–46	A	0
47–50	E	−4

The conclusions from the control chart agree reasonably well with the populations used in simulation.

Chart 6.3. The range chart appears generally unstable. Subgroups 54–61 indicate an increase in process variability. However, the remaining part of the chart provides an irregular pattern with two out-of-control subgroups. The $\bar{X}$ chart shows subgroup 57 to be

248 6 Interpretation of Control Charts

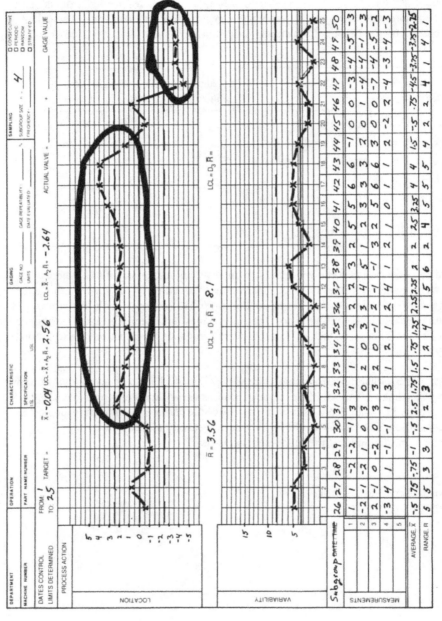

Chart 6.2 Control chart for variable data. The conclusions agree reasonably well with the populations used in simulation.

6.7 Case Studies

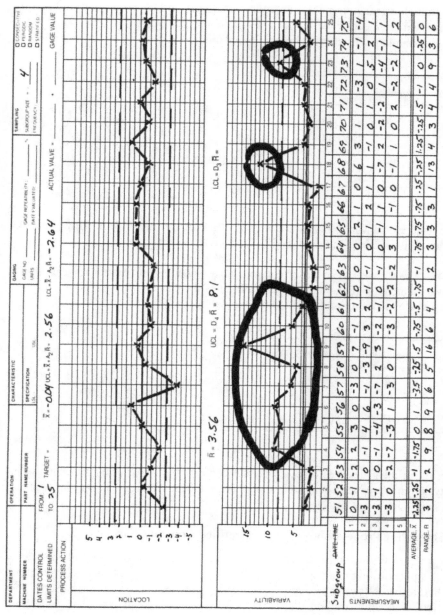

Chart 6.3 Control chart for variable data. The large increase in variability was correctly detected by the R chart, but the smaller variability increase in population G produced an inconsistent pattern.

250   6   Interpretation of Control Charts

out of control. However, this single subgroup corresponds to the out-of-control range, so we conclude that the mean is stable.

The actual simulation used the following populations:

Subgroups	Population	SD σ
51–53	A	1.74
54–60	F	3.74
61–64	A	1.74
65–75	G	2.35

The large increase in variability was correctly detected by the R chart. However, the smaller variability increase in population G produced an inconsistent pattern.

Chart 6.4. The R chart has an out-of-control signal, but no overall pattern is clear. The $\bar{X}$ chart shows an increase in the mean level for most of the chart.

The actual simulation used the following populations:

Subgroups	Population	Mean	SD σ
76–82	B	2	1.74
83–96	H	2	2.35
97–100	F	0	3.74

The increase in mean level is detected by the control chart. However, the increase in variability does not produce a consistent signal. Also, the large increase in variability of the last four subgroups is not detected.

6.7.2 Simulating Process Changes on a Stratified Control Chart

Consider a simulation exercise in which a stratified control chart is used to monitor a process. A single drilling machine with four fixtures is monitored with an $n = 4$ stratified control chart for which two subgroups of consecutive parts are collected during a shift. The process flow diagram, variation diagram, special causes diagram, and data collection format follow. Control charts are located at I and II.

Process flow diagram:

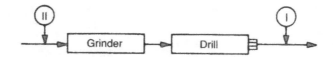

6.7 Case Studies

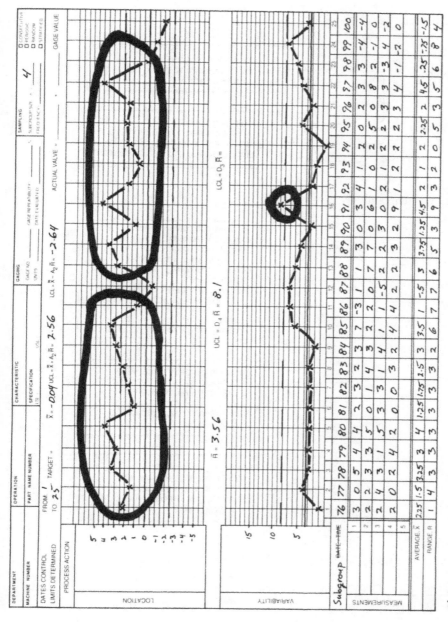

Chart 6.4 Control chart for variable data. The increase in mean level is detected by the control chart, but the increase in variability does not produce a consistent signal and the large increase in variability of the last four subgroups is not detected.

252   6 Interpretation of Control Charts

Variation diagram:

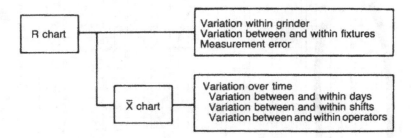

Special causes diagram:

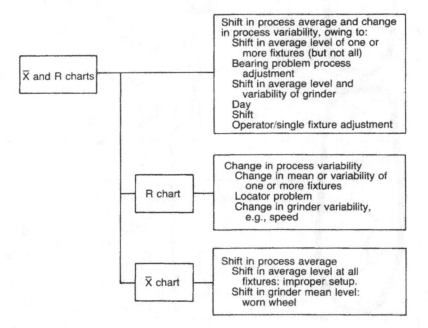

Data collection format:

	Shift 1		Shift 2		Shift 1	
Fixture	9:00	2:05	4:35	7:02	8:45	2:35
1						
2						
3						
4						

6.7 Case Studies

A baseline study was conducted on the four fixtures using separate control charts. The means and ranges were analyzed and the process improved until the following results were obtained. Note that the grinder was kept in statistical control during the study.

Quantity	Fixture			
	1	2	3	4
$\bar{\bar{X}}$	$-.02$	$.10$	$-.14$	$.06$
$\bar{R}$	3.2	2.5	3.1	2.7
n	3	3	3	3

Baseline control chart values ($n = 3$):

$$\bar{R}_{base} = \frac{\bar{R}_1 + \bar{R}_2 + \bar{R}_3 + \bar{R}_4}{4} = \frac{3.2 + 2.5 + 3.1 + 2.7}{4} = 2.875$$

$$\bar{\bar{X}}_{base} = \frac{\bar{\bar{X}}_1 + \bar{\bar{X}}_2 + \bar{\bar{X}}_3 + \bar{\bar{X}}_4}{4} = \frac{-.02 + .10 + -.14 + .06}{4} = 0$$

$$\hat{\sigma} = \frac{\bar{R}_{base}}{d_2} = \frac{2.875}{1.69} = 1.70 \ (d_2 \text{ for } n = 3)$$

New stratified control chart values:

$\bar{R}_{new} = d_2 \hat{\sigma} = 2.06 \times 1.70 = 3.50$ $\quad d_2$ for $n = 4$
$UCL_R = D_4 \bar{R}_{new} = 2.28 \times 3.50 = 7.98$ $\quad D_4$ for $n = 4$
$\bar{\bar{X}}_{new} = \bar{\bar{X}}_{base} = 0$
$UCL_{\bar{X}} = \bar{\bar{X}}_{new} + A_2 \bar{R}_{new} = 0 + (.73 \times 3.50) = 2.56$
$LCL_{\bar{X}} = \bar{\bar{X}}_{new} - A_2 \bar{R}_{new} = 0 - (.73 \times 3.50) = -2.56$

The following simulation control charts are more difficult to interpret. Not only must the type of change and time it occurred be identified, but the fixtures involved must be determined.

Chart 6.5. The new stratified control chart appears stable, as expected from the baseline study.

Charts 6.6 and 6.7. Several out-of-control subgroups appear on the R chart. However, these signals are likely due to the obviously elevated mean level for various fixtures. A visual scan of the rows suggests that fixture 4 is high for subgroups 26–36 and fixture 2 for subgroups 42–50. These high mean values would cause an out-of-control range on a stratified control chart. The means and ranges for subgroups of size $n = 4$ are plotted by fixture in Chart 6.7. This quick test indicates a stable range and high values for fixture 4 (subgroups 31–38) and fixture 2 (subgroups 43–50).

254 6 Interpretation of Control Charts

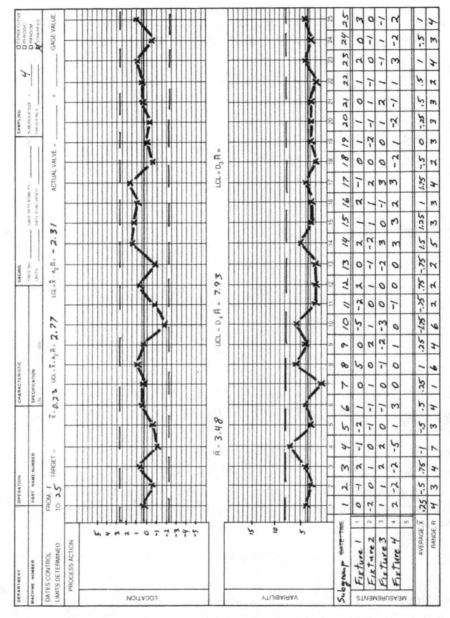

Chart 6.5 Control chart for variable data. This new stratified control chart appears stable, as expected from the baseline study.

6.7 Case Studies

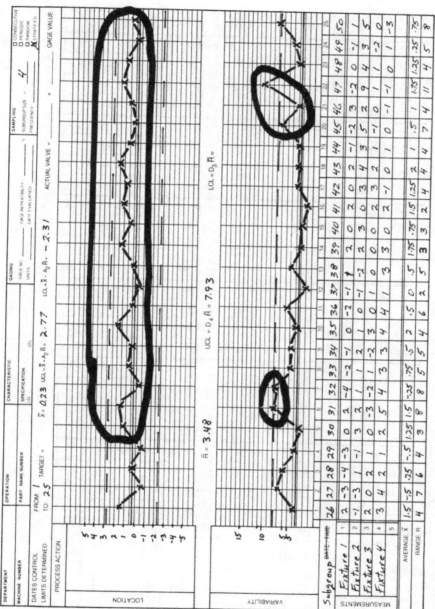

Chart 6.6 Control chart for variable data. Several out-of-control subgroups appear on the R chart, but these are probably due to the elevated mean level for various fixtures.

256   6 Interpretation of Control Charts

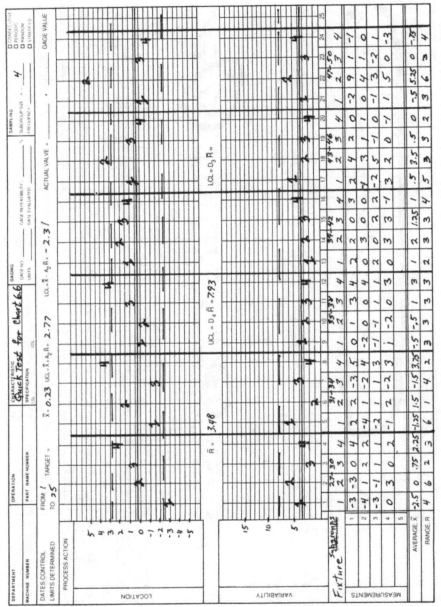

Chart 6.7 Control chart for variable data. The means and ranges for subgroups size $n = 4$ are plotted by fixture, indicating a stable range and high values for fixtures 2 and 4.

6.7 Case Studies

The actual simulation used the following populations:

Subgroup	Fixture	Population	Mean
26–41	1	A	0
	2	A	0
	3	A	0
	4	B	2
42–50	1	A	0
	2	C	4
	3	A	0
	4	A	0

The control chart analysis correctly indicated the simulated process changes. However, the exact timing of the changes could not be established correctly.

Charts 6.8 and 6.9. The R chart shows an out-of-control condition in the last half of the chart. However, the $\bar{X}$ chart shows a total lack of stability since there are too many points in zones A and B throughout this chart. Notice, however, that both the high and low zones are involved. (Operator overcontrol does not apply here since there are no routine adjustments of the machining fixtures.) This behavior could be caused by increased variability of the fixtures. Visually scanning the rows suggests that fixture 1 has increased variability for subgroups 66–75 and fixture 3 for subgroups 63–75. The quick test in Chart 6.9 indicates increased variability for fixtures 1 and 3 for several subgroups. Note that the mean level appears stable.

The actual simulation used the following populations:

Subgroup	Fixture	Population	SD σ
51–58	1	G	2.35
	2	A	1.74
	3	A	1.74
	4	A	1.74
59–75	1	F	3.74
	2	A	1.74
	3	F	3.74
	4	F	3.74

The control chart correctly indicated increased variability for fixtures 1 and 3, although again the timing was not accurate. The chart missed the moderate increase in variability of fixture 1 for subgroups 51–58 and the large increase in variability of fixture 4 for subgroups 59–75.

258   6 Interpretation of Control Charts

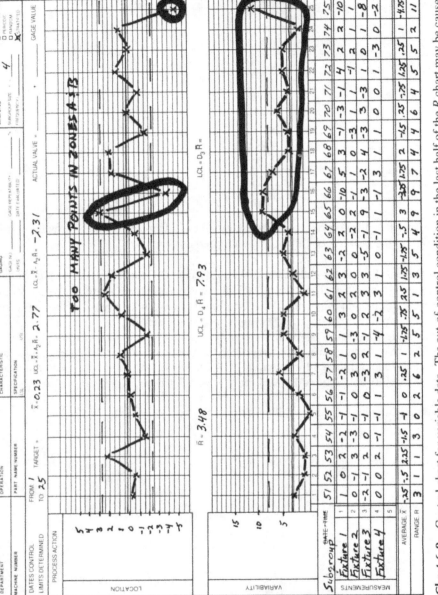

Chart 6.8 Control chart for variable data. The out-of-control condition in the last half of the *R* chart may be caused by increased variability of the fixtures.

6.7 Case Studies

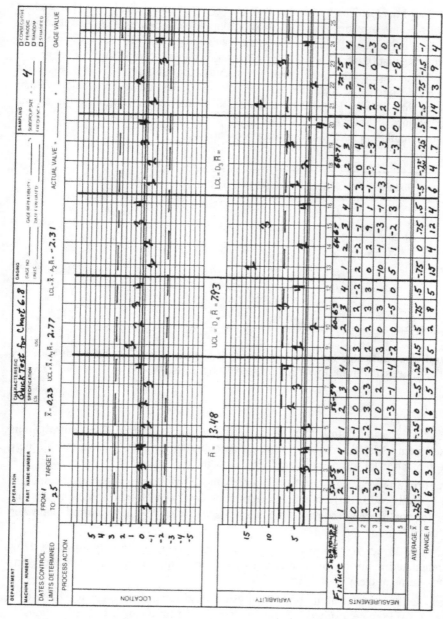

Chart 6.9 Control chart for variable data. The quick test indicates increased variability for fixtures 1 and 3 for several subgroups.

Charts 6.10 and 6.11. The R chart has two groups of out-of-control points. The $\bar{X}$ chart has too many points in zones A and B for subgroup 76–93. However, notice that most points are in the upper zones, which could be the result of a process shift. Visually scanning the rows suggests that fixtures 1 and 2 have an increased mean and variability for subgroups 76–96. Fixture 3 appears to have increased variability for subgroups 92–100. Fixture 4 appears to have a decreased mean for subgroups 93–100. The quick test results in Chart 6.11 indicate that fixtures 1 and 2 have increased locations and variability for several subgroups. Fixture 4 shows a decreased location.

The actual simulation used the following populations:

Subgroups	Fixture	Population	Mean	SD σ
76–89	1	H	2	2.35
	2	H	2	2.35
	3	A	0	1.74
	4	A	0	1.74
90–100	1	H	2	2.35
	2	H	2	2.35
	3	A	0	1.74
	4	F	0	3.74

The control chart indicated the increased mean of fixtures 1 and 2 but did not correctly identify the increased variability of fixture 4 for subgroups 90–100.

Problems

6.1 Chart 6.12 is a control chart of data from a single source. The control limits determined from a base process are correct for the subgroup size $n = 4$. Determine if there is a shift in mean level, change in variability, or both. When did these changes occur?

6.2 Chart 6.13 is a control chart from a three spindle machine in which each spindle is sampled in a subgroup. Perform a quick test on the data, and plot the spindle means and ranges on the control chart (no new chart is necessary). What changes occurred, and when did they occur?

6.3 Generate 25 subgroups for $n = 3$ and $n = 5$ from population A in Figure 6.18. How do the calculated control chart values agree with the expected values? Are there any apparent patterns in the $\bar{X}$ or R chart?

6.4 Assume that control limits have been established for population A in Figure 6.18 using $n = 3$. (a) Prepare a simulated control chart for the following set of populations:

Subgroup	Population	Mean	SD
1–10	H	2	2.35
11–20	G	0	2.35
21–25	A	0	1.74

Problems

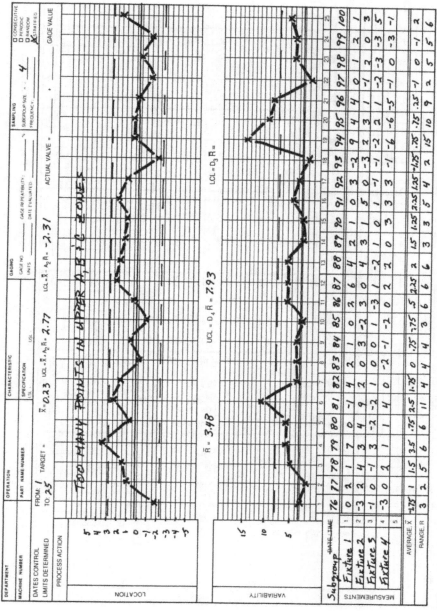

Chart 6.10 Control chart for variable data. The R chart has two groups of out-of-control points.

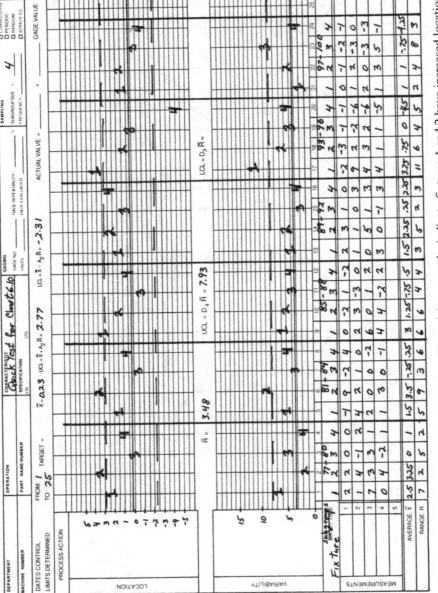

Chart 6.11 Control chart for variable data. The quick test results indicate fixtures 1 and 2 have increased locations and variability for several subgroups. Fixture 4 shows a decreased location.

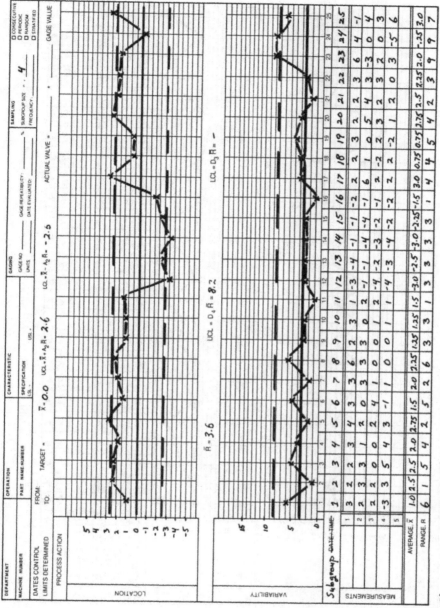

Chart 6.12 Control chart for variable data. Data are from a single source (see Prob. 6.1).

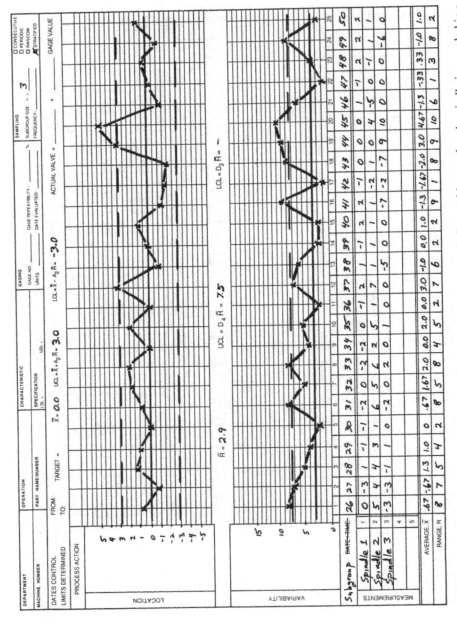

Chart 6.13 Control chart for variable data. Data are from a three spindle machine, and each spindle is sampled in a subgroup (see Prob. 6.2).

Problems

(b) Mark the out-of-control subgroups. Can you predict the type of change and when it occurred?

6.5 Assume that control limits have been established for a stratified control chart for four spindles with $n = 4$ using population A in Figure 6.19. (a) Prepare a simulated control chart for the following set of populations.

Subgroup	Spindle	Population	Mean	SD
1–13	1	A	0	1.74
	2	D	−2	1.74
	3	A	0	1.74
	4	H	2	2.35
14–25	1	E	−4	1.74
	2	A	0	1.74
	3	A	0	1.74
	4	F	0	3.74

(b) Mark the out-of-control subgroups. Can you predict the type of change and when it occurred?

6.6 Class exercise: Divide into teams, and prepare a simulated control chart assuming only one source of variability with 25 subgroups of size $n = 4$. Use population A control chart values for the base process. Use three or fewer times in which the process changes. Allow each team to present its chart, with the remainder of the class interpreting the results.

6.7 Class exercise: Using the same format as Problem 6.6, prepare a stratified control chart.

7
Process Capability

The emphasis of previous chapters has been on defining the work process, selecting an appropriate control chart, determining a meaningful sampling plan, and then obtaining stability of the process. This stability provides predictability of the process output. However, as noted in Figure 3.1 it is possible for a process with predictable, controlled variation to produce parts beyond the specification limits. This chapter provides a system of five simple indices that quantify the capability of a stable process to produce parts within the specification limits. These indices, which follow, collectively measure "process capability":

C_p = process potential index
CPU = upper process performance index
CPL = lower process performance index
k = process centering index
C_{pk} = process performance index

The five indices provide a common "language" for machine operators, process engineers, design engineers, and managers to discuss the performance of manufacturing processes. As importantly, these indices provide a system to measure and encourage continuing process improvement, as illustrated in a case study at the end of this chapter. The automotive industry (Kane, 1986) has used these five measures to encourage their plants as well as outside suppliers to improve processes to attain continuously increasing capability.

> Continuous improvement requires constantly improving process capability

7.1 General Concepts

A direct way to evaluate the performance of a stable process is to make measurements from the process and relate the resulting histogram to the upper specification limit (USL) and lower specification limit (LSL).

7.1 General Concepts

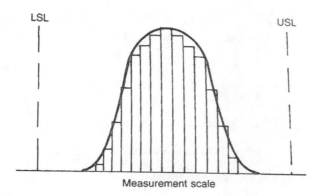

The bell-shaped curve approximating the histogram is used in the following capability discussions similarly to its use in Chapter 5. The bell-shaped curve graphically represents the capability of the process. Traditionally, a process has been called "capable" if the process spread ($6\hat{\sigma}$) was equal to the width of the specification limits.

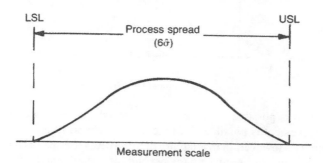

If the assumption of a normal, bell-shaped distribution is exactly correct for a capable process, virtually no parts are produced beyond the specification limits (less than 0.27%), assuming the process remains stable. Obviously, if special causes of variation are introduced into the process, the predictability of the process output is lost and parts beyond the specification limits are possible. Reacting to out-of-control signals on control charts and eliminating special causes is the best method to maintain a capable process.

There are three ways in which a process can be judged not capable:

1. The process is not stable.
2. The process is centered too close to a specification limit.
3. The process variability is excessive.

The only way to assess process stability is by using control charts. It is then meaningful to evaluate location and variability. A process can be producing parts beyond the specification limits if the mean $\bar{X}$ is located too close to one of the specification limits, as shown in Figure 7.1. Making the process capable is easy if the part characteristic is adjustable (Chap. 4). However, a nonadjustable characteristic may require significant process changes. We should note that a perfectly centered location is not necessary for a capable process. In many cases, the process mean is targeted at a noncentral value to make downstream processing or assembly easier.

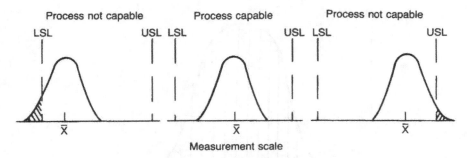

Figure 7.1 Influence of location on capability.

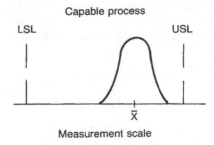

In Figure 7.1, it is possible to move the process location and be either capable or not capable. This movement is possible since the process spread is less than the width of the specification limits. However, the variability of the process may be too large, so a perfectly centered process would still produce parts beyond the specification limits.

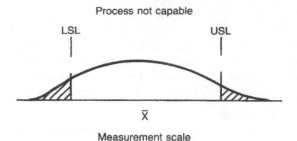

This figure demonstrates that there is no chance of obtaining a capable process if the process variability is too large. Thus, evaluation of process variability is usually the first step in assessing process capability after stability is attained. Also, for most situations a lack of capability is due to excessive process variability. Obtaining a capable process by targeting $\bar{X}$ once variability is sufficiently small is usually much easier than reducing variability.

7.2 Process Potential Index C_p

The potential of a stable process to be capable depends only on the variability of the process. A simple method of evaluating this potential is to relate the actual

7.2 Potential Index

Process spread = $6\hat{\sigma}$

to the

Allowable process spread = USL − LSL

The width of the specification limits is the allowable process spread if the process is judged capable. These two spreads can be related to form a simple index, the process potential index:

$$C_p = \frac{\text{allowable process spread}}{\text{actual process spread}}$$

$$= \frac{\text{USL} - \text{LSL}}{6\hat{\sigma}}$$

This relationship is shown in Figure 7.2.

For a capable process the actual process spread equals the allowable process spread, so $C_p = 1$. Values of C_p greater than 1 imply a desirable process in which the actual spread is less than the allowable spread. The reverse is true if C_p is less than 1. A more conventional way to evaluate process performance is to evaluate the percentage of the specification width used by the process spread:

$$\% \text{ Specification used} = \frac{\text{actual process spread}}{\text{allowable process spread}} \times 100$$

$$= \frac{1}{C_p} 100$$

This approach is informative for evaluating process potential, but the relationship between the five indices for evaluating total process performance makes the use of C_p desirable.

Varying values of C_p are related to the process spread and the percentage of specification used in Figure 7.3. It is generally believed that a minimum C_p of 1.33 (75% of specification width) is required for most manufacturing processes. This capability allows some flexi-

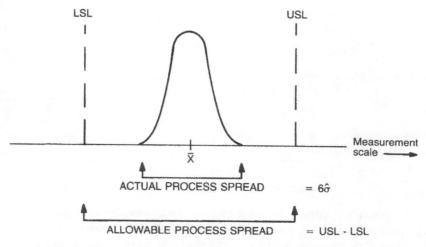

Figure 7.2 Relationship of C_p parameters.

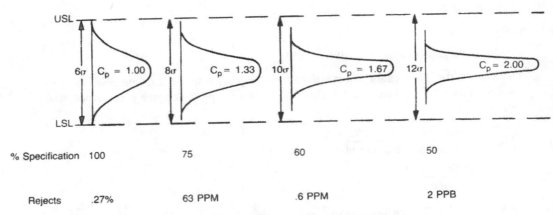

Figure 7.3 C_p indices for varying widths of the process distribution.

bility if the process is slightly off center. To assure less than a part per million (PPM) defective rate, capabilities of 1.66 or higher may be required, as shown in a later section.

Example 7.1 Suppose USL = 10, LSL = 5, and $\bar{R} = 2.5$ (with $n = 4$) in a stable process. The C_p calculations are

$$\begin{aligned}
\text{Allowable process spread} &= \text{USL} - \text{LSL} \\
&= 10 - 5 \\
&= 5 \\
\text{Standard deviation} &= \hat{\sigma} \\
&= \frac{\bar{R}}{d_2} \\
&= \frac{2.5}{2.06} \\
&= 1.21 \\
\text{Actual process spread} &= 6\hat{\sigma} \\
&= 6 \times 1.21 \\
&= 7.26 \\
\text{Process potential} &= C_p \\
&= \frac{5}{7.26} \\
&= 0.69
\end{aligned}$$

Since C_p is considerably less than 1, the stable process can be expected to produce parts beyond the specification limits. Any absence of process stability further contributes to out-of-specification parts. Improvement must focus on reducing process variability.

7.3 Process Performance Index C_{pk}

The performance of a process must relate the process potential (quantifying variability) to the location, measured by $\bar{X}$. Consider the simple situation in which only a unilateral tolerance exists so the process parameters ($\bar{X}$, $\hat{\sigma}$) must be related to a single specification limit. Following the same format as C_p, we relate the actual spread to the allowable spread for a process with only an upper specification limit

$$\text{Actual upper process spread} = \frac{1}{2} \text{ actual process spread}$$
$$= 3\hat{\sigma}$$
$$\text{Allowable upper process spread} = \text{USL} - \bar{X}$$

The allowable spread is the maximum distance the upper process spread can occupy and still have a capable process. These spreads can be related to form the upper capability index CPU (upper process performance index):

$$\text{CPU} = \frac{\text{allowable upper process spread}}{\text{actual upper process spread}}$$
$$= \frac{\text{USL} - \bar{X}}{3\hat{\sigma}}$$

These relationships are shown in Figure 7.4. Using the same approach, the actual spread can be related to the allowable spread for a process with only a lower specification limit to form a lower capability index (lower process performance index):

$$\text{CPL} = \frac{\text{allowable lower process spread}}{\text{actual lower process spread}}$$
$$= \frac{\bar{X} - \text{LSL}}{3\hat{\sigma}}$$

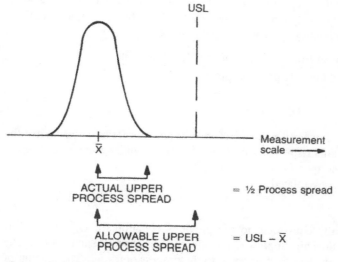

Figure 7.4 Relationship of CPU parameters.

For both CPU and CPL, we assume $\bar{X}$ is not beyond the specification limits so the indices are never negative. These indices provide a convenient measure of the performance of a process when only one specification limit is used.

The case for two-sided, bilateral specification limits is a natural extension of the single-limit case. The index C_{pk} is a measure of process performance in this case, which can be calculated simply by

$$\text{Process performance} = C_{pk} = \text{minimum (CPL, CPU)}$$

Thus, the C_{pk} index measures capability at the specification limit, which has the highest chance of a part beyond the limit.

Another equivalent approach for obtaining C_{pk} allows relating the process potential C_p to the process performance C_{pk}. Consider the midpoint of the specification limits,

$$m = \frac{\text{USL} + \text{LSL}}{2}$$

If the process is exactly centered, $\bar{X} = m$ and there are the fewest possible parts beyond the specification limits. The distance between m and $\bar{X}$ can be computed by the difference

Difference = $m - \bar{X}$

Difference = $\bar{X} - m$

We assume the difference is always positive (i.e., take the absolute value of $m - \bar{X}$ denoted by $|m - \bar{X}|$). This difference can be related to one-half the allowable spread, as with CPU and CPL, to form a centering index (process centering index):

$$k = \frac{\text{distance process mean from midpoint of specification limits}}{\text{½ allowable process spread}}$$

$$= \frac{2|m - \bar{X}|}{\text{USL} - \text{LSL}}$$

Note that when the process is exactly centered, $\bar{X} = m$, which results in $k = 0$. When the process mean is located at either specification limit, $k = 1$. Again, we are assuming the mean does not go beyond the specification limits.

The values of C_{pk} and C_p are related:

$$C_{pk} = C_p(1 - k)$$

This equation is algebraically equivalent to the previous equation for C_{pk}. Since k is between 0 and 1, this relationship implies C_{pk} is always less than or equal to C_p. The process performance is always less than or equal to the process potential. The amount of reduction in C_p depends on the distance the process is off center as measured by k. These relationships are shown in Figure 7.5.

7.3 Performance Index

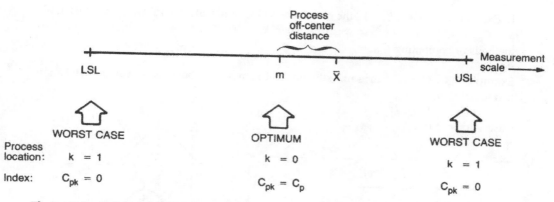

Figure 7.5 Relationship of C_{pk} parameters.

The following examples give insight into two capability indices.

Example 7.2 Suppose USL = 20, LSL = 8, and $\hat{\sigma} = 2$. The process potential is

Allowable process spread = USL − LSL
= 12

Actual process spread = $6\hat{\sigma}$
= 12

Process potential = C_p
= 1

Thus, the process is capable if $\bar{X} = 14$ but not capable for any off-center mean. Suppose the means are $\bar{X} = 12, 14,$ and 16; the performance indices are shown below.

	CPU	CPL	k		C_{pk}
Formula	$\dfrac{USL - \bar{X}}{3\hat{\sigma}}$	$\dfrac{\bar{X} - LSL}{3\hat{\sigma}}$	$\dfrac{2\|m - \bar{X}\|}{USL - LSL}$	$C_p(1 - k)$	
$\bar{X} = 12$	$\dfrac{20 - 12}{6} = 1.33$	$\dfrac{12 - 8}{6} = .67$	$\dfrac{2(14 - 12)}{12} = .33$	$1(1 - 0.33) = .67$	
$\bar{X} = 14$	$\dfrac{20 - 14}{6} = 1.0$	$\dfrac{14 - 8}{6} = 1.0$	$\dfrac{2(14 - 14)}{12} = 0$	$1(1 - 0) = 1.0$	
$\bar{X} = 16$	$\dfrac{20 - 16}{6} = .67$	$\dfrac{16 - 8}{6} = 1.33$	$\dfrac{2(16 - 14)}{12} = .33$	$1(1 - 0.33) = .67$	

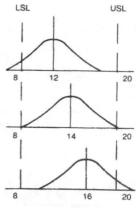

7 Process Capability

In Example 7.2, when $\bar{X} = 12$ the process was closest to LSL so CPL was lower than CPU and C_{pk} = CPL. The reverse was true when $\bar{X} = 16$. Notice the equivalence of the two methods of computing C_{pk}.

Example 7.3 Consider the same situation as Example 7.2, but now we vary both location and variability.

Case 1, $\hat{\sigma} = 2.0$:

	C_p	CPU	CPL	k	C_{pk}
$\bar{X} = 12$	1.0	1.33	.67	.33	.67
$\bar{X} = 14$	1.0	1.0	1.0	0	1.0
$\bar{X} = 16$	1.0	.67	1.33	.33	.67

Case 2, $\hat{\sigma} = 1.33$:

	C_p	CPU	CPL	k	C_{pk}
$\bar{X} = 12$	1.5	2.0	1.0	.33	1.0
$\bar{X} = 14$	1.5	1.5	1.5	0	1.5
$\bar{X} = 16$	1.5	1.0	2.0	.33	1.0

Case 3, $\hat{\sigma} = 1.0$:

	C_p	CPU	CPL	k	C_{pk}
$\bar{X} = 12$	2.0	2.67	1.33	.33	1.33
$\bar{X} = 14$	2.0	2.0	2.0	0	2.0
$\bar{X} = 16$	2.0	1.33	2.67	.33	1.33

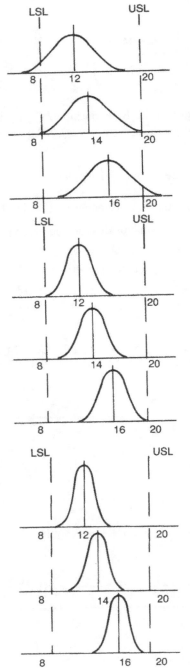

7.4 Process Capability Applications

The minimum process potential of case 1 with $C_p = 1.0$ does not allow off centering of the process. In case 2 the reduced variability enables some movement of the process mean while maintaining a capable process. In case 3 the further reduced variability allows targeting of the process mean if that is desirable.

7.4 Process Capability Applications

7.4.1 Two-Sided Specification

A capability improvement effort is discussed in a case study at the end of this chapter. The hardened metal case depth of an engine camshaft lobe was improved by a team effort. Process stability was attained, along with reduced variability. The process mean and range for the third study period were $\bar{\bar{X}} = 5.40$ mm and $\bar{R} = 1.2$ using $n = 5$, where LSL = 3.5 and USL = 10.5 mm. Calculations of the five capability indices follow:

$$\hat{\sigma} = \frac{\bar{R}}{d_2} = \frac{1.2}{2.33} = 0.52$$

$$m = \frac{10.5 + 3.5}{2} = 7$$

$$C_p = \frac{USL - LSL}{6\hat{\sigma}} = \frac{10.5 - 3.5}{6 \times 0.52} = 2.24$$

$$CPU = \frac{USL - \bar{\bar{X}}}{3\hat{\sigma}} = \frac{10.5 - 5.40}{3 \times 0.52} = 3.27$$

$$CPL = \frac{\bar{\bar{X}} - LSL}{3\hat{\sigma}} = \frac{5.40 - 3.5}{3 \times 0.52} = 1.22$$

$$k = \frac{2|m - \bar{\bar{X}}|}{USL - LSL} = \frac{2|7 - 5.40|}{10.5 - 3.5} = 0.46$$

$$C_{pk} = \text{minimum (CPL, CPU)} = 1.22$$

or

$$C_{pk} = C_p(1 - k) = 2.24(1 - 0.46) = 1.22$$

Since k is greater than 0, an off-center location is indicated. Unfortunately, it is difficult to center the process mean in this example.

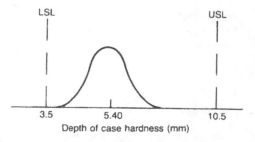
Depth of case hardness (mm)

It is always a good practice to check whether the histogram is an approximately normal, bell-shaped distribution. For this example, the histogram from the control chart appears in Figure 7.6.

As often occurs in improvement efforts, the ability of the process to produce parts within the specification limits has dramatically increased. However, further improvements are

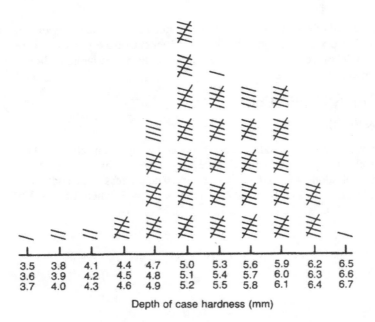

Figure 7.6 Histogram of engine camshaft depth of case hardness, third study.

necessary. Notice that the range of the process is stable, so $\hat{\sigma}$ is probably a reasonable estimate of the process standard deviation. The capability estimates are suspect owing to the lack of an approximately normal, bell-shaped histogram.

7.4.2 One-Sided Specification

The concentricity of an oil seal groove was monitored by the chart in Figure 7.7, where the data were coded so that $1 = .001$ inch. Initially, the process was stable, with $\bar{\bar{X}} = 4.2$ and

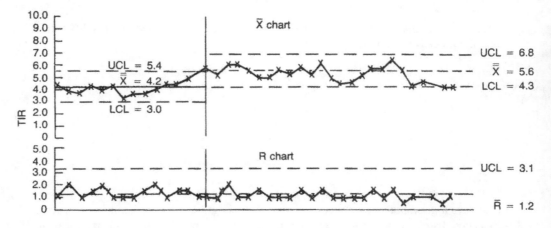

Figure 7.7 Oil seal groove concentricity showing changing capability CPU with changing location.

$\bar{R} = 1.2$ using a subgroup size of $n = 3$. The process could adequately meet the specification of .010 inch total indicator reading (TIR). The process could be run more economically at a higher mean level and not adversely impact on product function. The mean was moved to $\bar{\bar{X}} = 5.6$; the process variability remained unchanged. The process was reasonably stable at the new level, although there is an indication of a possible mean shift to a lower level at the end of the chart. Assuming the process remains stable, the capabilities before and after the planned mean shift follow:

$$\hat{\sigma} = \frac{1.2}{1.69} = 0.71$$

Before shift capability:

$$\text{CPU} = \frac{\text{USL} - \bar{X}}{3\hat{\sigma}} = \frac{10 - 4.2}{3 \times 0.71} = 2.72$$

After shift capability:

$$\text{CPU} = \frac{\text{USL} - \bar{X}}{3\hat{\sigma}} = \frac{10 - 5.6}{3 \times 0.71} = 2.06$$

The process is still highly capable after the planned shift.

7.5 Evaluation of Stability, C_p, and C_{pk}

Stability must be evaluated prior to assessment of C_p and C_{pk}. The initial phase of an improvement effort should focus on eliminating special causes that create process instability. Once stability is attained, the two capability indices can be used to direct the improvement effort. Figure 7.8 shows a diagram of how continuous improvement of process capability can be approached. This flowchart can serve as a guide of how to use stability assessment along with C_p and C_{pk} for process improvement.

7.5.1 Stability Versus Capability

Recall from the sampling plan case study in Chapter 4 that two different sampling plans were used to study the same process over a common time period:

Plan 1. One part every 30 minutes, grouped so that the subgroup size was $n = 5$ ($\hat{\sigma} = 8.2$).
Plan 2. Five consecutive parts each hour ($\hat{\sigma} = 5.1$).

The variability for plan 1 was 62% higher than that for plan 2, yet the process was sampled over the same time period! Also, recall that plan 2 showed several out-of-control points so the variability would probably be lower were the special causes eliminated. Suppose we computed the capability for both plans.

Plan 1:

$$C_p = \frac{255 - 187}{6 \times 8.2} = 1.4$$

Plan 2:

$$C_p = \frac{255 - 187}{6 \times 5.1} = 2.2$$

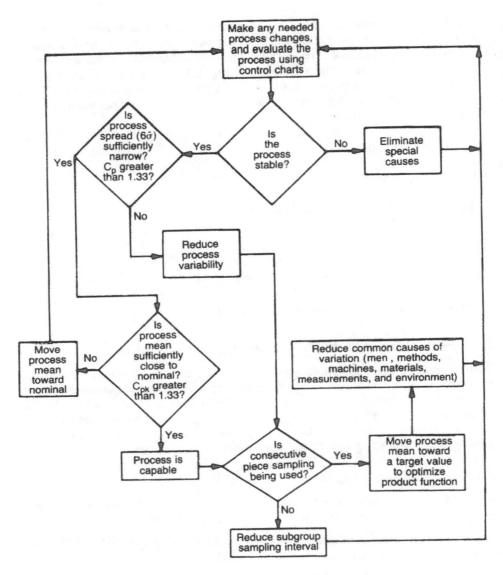

Figure 7.8 Continuous improvement in process capability.

Note that we should not compute the capability for plan 2 because of the presence of special causes. These calculations are performed only to illustrate the relationship of the sampling plan to capability. Plan 1 results in an appreciably lower capability than plan 2. Although this difference may at first seem unusual, it is entirely an expected result.

In Chapter 4, four subgroup sampling plans were compared. The variation diagram showed that increasing the time interval between parts used in a subgroup resulted in encompassing more sources of variability in the R chart:

7.5 Evaluation of Stability, C_p, and C_{pk}

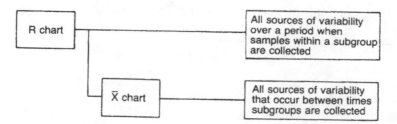

The sources of variability are reduced as the time interval shrinks over which parts in a subgroup are collected. A reduction in the sources of variability results in a lower value of $\bar{R}$. Now since $\hat{\sigma} = \bar{R}/d_2$, a lower $\bar{R}$ results in a lower $\hat{\sigma}$ and a higher capability evaluation since reducing $\hat{\sigma}$ increases C_p and C_{pk}. Thus, the following sequence explains the relationship of subgroup sampling plan and capability:

Decrease time interval for parts within a subgroup $\longrightarrow$ Decrease $\bar{R}$ $\longrightarrow$ Decrease $\hat{\sigma}$ $\longrightarrow$ Increase C_p and C_{pk}

When increasing the subgroup time interval does not include additional sources of variation, the value of $\bar{R}$ does not change. It seems that the best approach for obtaining a high capability would be to use a small time interval for collecting parts within a subgroup. However, the lower $\bar{R}$ causes a narrower width of the control limits on both the $\bar{X}$ and R charts, since

$\bar{X}$ chart width between control limits = $UCL_{\bar{X}} - LCL_{\bar{X}}$
$= \bar{\bar{X}} + A_2\bar{R} - (\bar{\bar{X}} - A_2\bar{R})$
$= 2A_2\bar{R}$

R chart width between control limits = $UCL_R - LCL_R$
$= D_4\bar{R} - D_3\bar{R}$
$= (D_4 - D_3)\bar{R}$ D_4 greater than D_3

The narrower width of the control limits means more special causes are likely to be detected. The presence of special causes makes capability evaluation not meaningful.

The evaluation of process capability is thus directly related to the subgroup sampling plan in two ways. First, the subgroup sampling plan determines the evaluation of stability required to make capability calculations meaningful. Second, the estimated process variability $\hat{\sigma}$ used to evaluate capability depends on the subgroup sampling plan. Table 7.1 presents an overview of these relationships. The results for plans 1 and 2 discussed at the beginning of this section were predictable. The consecutive piece sampling plan produced a smaller $\hat{\sigma}$, but the presence of out-of-control signals made capability evaluation not meaningful.

Identification and elimination of special causes reduce $\hat{\sigma}$ and increase the capability of the process. In Figure 7.8, a process judged capable may not have identified all the special causes in a process because the subgroup sampling plan spread subgroup samples over too large a time interval. Reducing the subgroup sampling time interval highlights the special causes and produces an opportunity for improvement. If consecutive piece sampling is used on a stable process, the common causes of variation originating from men, methods, machines, materials, measurements, and environment must be changed. Generally, it is much easier to eliminate special causes to improve process capability.

Table 7.1 Relationship Between Process Stability Evaluation and Capability

	Sampling plan 1 periodic	Sampling plan 2 consecutive
Time between samples within a subgroup	Larger	Smaller
Estimate of process variability $\hat{\sigma}$	Larger	Smaller
Width of control limits	Wider	Narrower
Possibility of out-of-control signals (instability detected)	Lower	Higher
Process capability estimates	Lower	Higher

Seek and eliminate special causes of variation to improve capability

Stating that a process is stable is not sufficient—capability must be evaluated. Stability may have been attained using a subgroup sampling plan that spanned too broad a time period (plan 1), so detecting special causes is difficult. The resulting $\bar{R}$ is higher than if consecutive piece sampling is used and often results in unacceptable low capability. A stable process need not be capable, in which case poor performance can be predicted. The best practice to evaluate true capability is to use consecutive piece subgroup sampling. It is more difficult to attain stability, but once attained, the highest capability is obtained. Other sampling plans may be more desirable for monitoring a process, as discussed in Chapter 14. In practice, it is still useful to calculate capability from a monitoring sampling plan, but the relationship between stability, capability, and the subgroup sampling plan must be realized.

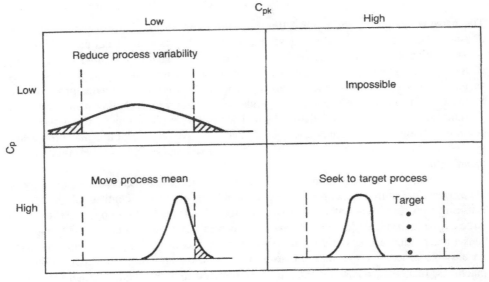

Figure 7.9 Comparison of C_p and C_{pk}.

7.5.2 C_p versus C_{pk}

After process stability is established, a comparison of C_p and C_{pk} provides direction for any required process action. Figure 7.9 provides a guide for determining the type of action. First, from Figure 7.8 we evaluate process potential C_p to determine whether the process spread is sufficiently small to enable positioning of the process spread within the specification limits. Typically, the benchmark is a C_p greater than 1.33 or 1.67. If the process variability is too high, the first objective should be to reduce variability. Any adjustment of the mean level cannot produce a capable process.

If the value of C_p is higher than the benchmark level, then it is necessary to consider total process performance C_{pk}. A low value of C_{pk} implies the process is not centered, and

Table 7.2 Predicted Number of Parts Beyond Specification Limits for a Stable Process[a]

Capability	One-sided specification (CPL or CPU)	Two-sided specification[b] (C_{pk})
.1	38.2%	76.4%
.2	27.4%	54.9%
.3	18.4%	36.8%
.4	11.5%	23.0%
.5	6.7%	13.4%
.6	3.6%	7.2%
.7	1.8%	3.6%
.8	.82%	1.6%
.9	.35%	.69%
1.0	.14%	.27%
1.1	.048%	.097%
1.2	159 PPM	318 PPM
1.3	48 PPM	96 PPM
1.33	32 PPM	63 PPM
1.4	13 PPM	27 PPM
1.5	3.4 PPM	6.8 PPM
1.6	.79 PPM	1.6 PPM
1.67	.29 PPM	.57 PPM
1.7	.17 PPM	.34 PPM
1.8	33 PPB	67 PPB
1.9	6 PPB	12 PPB
2.0	1 PPB	2 PPB

[a] PPM = parts per million; PPB = parts per billion.
[b] Process assumed centered.

capability improvement must focus on moving the process mean. However, if C_{pk} is higher than the selected benchmark level, it is possible to consider targeting the process mean at a level that optimizes product function, assembly, or some other criterion.

7.5.3 Why Excess Capability Is Necessary

If the measurements have a normal, bell-shaped distribution, Table 7.2 gives the percentage of parts that would be expected beyond the specification limits, assuming the process remains stable. Capabilities of 1.0–1.5 seem to provide adequate control. However, Table 7.2 serves only to educate our intuition. In a manufacturing situation, the results of Table 7.2 are largely meaningless for actual prediction. This table implies that seven parts in a million would be expected to be outside the specification limits based on normal random variability from a stable process with $C_p = 1.5$. However, the most likely way of generating parts beyond the specification limits from a process would be for the process to become unstable owing to the introduction of a special cause of variation. The dynamic nature of manufacturing processes make possible changes in men, methods, machines, materials, measurements, and environment.

In Chapter 6 we saw how it is possible for process changes to go undetected for a varying number of successive subgroups using a standard $\bar{X}$ and R chart. During this period, parts could be produced beyond the specification limits without out-of-control signals appearing on the charts. Of course, this is an undesirable situation. Figure 7.10 compares shifts in the

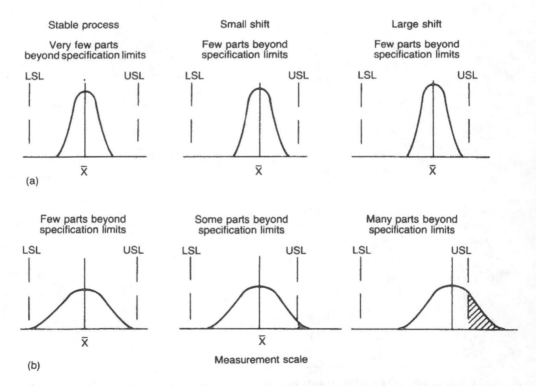

Figure 7.10 Parts beyond specification limits for varying process capabilities and mean shifts: (a) $C_{pk} = 2$; (b) $C_{pk} = 1$.

process mean for a process with $C_{pk} = 2$ and $C_{pk} = 1$. The large shift ($2\hat{\sigma}$) does not produce a problem in the process with $C_{pk} = 2$. However, for the process with $C_{pk} = 1$, the large shift produces many parts beyond the specification limits. It is undesirable for this case to go undetected for even a few subgroups. The small shift (1σ) produces a number of parts beyond the specification limits for the process with $C_{pk} = 1$. Unfortunately, from Figure 6.16 with $n = 5$ the small shift would have only about a 20% chance of being immediately detected. After four subgroups, the chance of detecting the shift increases to about 80%. The large shift would have about a 90% chance of being immediately detected. Thus, "extra" capability is necessary to protect against process changes that may go temporarily undetected.

7.6 Relating Control Chart Sampling Intensity to Capability

The previous section showed how a process with $C_{pk} = 1$ could easily produce parts beyond the specification limits if the process changed even a small amount. Correspondingly, the process with $C_{pk} = 2$ could have process changes detected well before any parts beyond the specification limits were likely. This observation suggests that process capability should be a factor in developing control chart sampling frequencies. Two considerations are necessary to develop guidelines.

First, a relatively small 1σ shift for a process with $C_{pk} = 1$ has a significant impact on process output quality, where a 1σ shift for a process with $C_{pk} = 2$ has relatively little impact, as Figure 7.10 shows. The actual percentage of parts beyond the specification limits for varying C_{pk} values and shifts in the process mean is given in Figure 7.11. A 1σ shift when $C_{pk} = 1$ results in 2.3% of the output beyond the specification limits; the corresponding percentage when $C_{pk} = 2$ is below 0.0001%.

Second, the average number of successive subgroups required to detect a process shift varies with the subgroup size and the size of the shift. Figure 6.16 shows the relationship between these factors. Example 7.4 illustrates the relationship between Figures 7.11 and 6.16.

Example 7.4 Suppose a stable process has USL = 22, LSL = 10, and $\hat{\sigma} = 1.5$. If the mean of $\bar{\bar{X}} = 16$ shifts to $\bar{\bar{X}} = 19$, the percentage of points beyond the specification limits is calculated as follows:

$$C_p = \frac{USL - LSL}{6\hat{\sigma}}$$

$$= \frac{22 - 10}{9}$$

$$= 1.33$$

Mean shift = $19 - 16 = 3$

Units of $\sigma = \frac{\text{mean shift}}{\sigma} = \frac{3}{1.5} = 2$

From Figure 7.11(b), a process with $C_p = 1.33$ having a mean shift of 2 standard deviation units results in about 2% of its output expected to be beyond the specification limits. From Figure 6.16(a), with $n = 3$ we would have about a 65% chance of detecting the shift immediately.

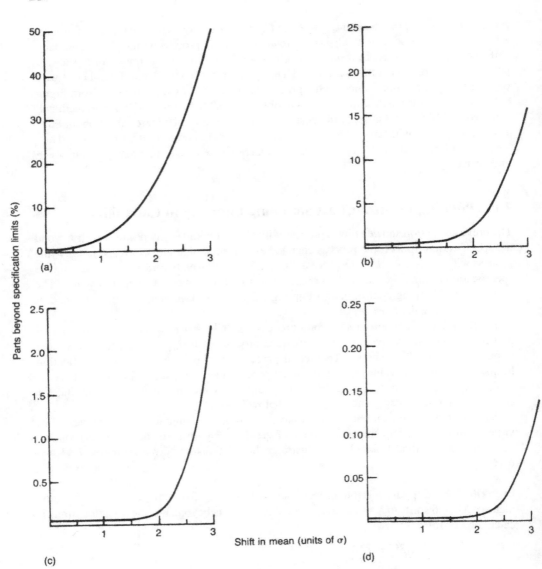

Figure 7.11 Percentage of parts beyond the specification limit for varying shifts of the mean and process capability: (a) $C_p = 1.00$; (b) $C_p = 1.33$; (c) $C_p = 1.67$; (d) $C_p = 2.00$. Process mean is assumed centered.

7.7 Computing Capability from Histograms

The average number of successive subgroups required to detect process changes is much larger than most control chart users would expect. Chapter 14 discusses factors other than process capability that must be evaluated to design a comprehensive control plan. The important consideration here is that the capability of a process is a significant factor in assessing the intensity of sampling required. The greater the capability, the lower the required sampling intensity needed to protect a given quality level.

In a discussion of the use of control charts in a plant, a former user of control charts indicated he saw no need for any charts. Further inquiry revealed that the standard sampling plan used in the plant was to collect $n = 5$ samples per day for all charts. The employee explained that his system of measuring a part every 15 minutes provided him better information than he was getting from the charts.

Relate control chart sampling intensity to process capability

7.7 Computing Capability from Histograms

Many occasions arise in manufacturing when an adjustment is made to a stable process and a new process capability is desired. These process adjustments are often suggested as a result of a control chart signal. For example, for an adjustable characteristic, the process may have shifted off target and an adjustment is necessary. To compute the process capability in this situation, it is possible to collect control chart data and compute the five indices directly. However, if a consecutive piece sampling plan is not being used, waiting for control chart data may not be convenient. Using a measurement scale check sheet (Chap. 8), it is possible to collect measurements and form a histogram from which a new process capability can be computed. Generally, a minimum of $N = 50$ measurements is used.

Computing capability from histograms is easy since once the histogram is complete the work sheet in Chapter 5 can be used to compute the process mean $\bar{X}$ and variability s. Procedure 7.1 gives the procedure to calculate all five capability indices using either the standard control chart approach or histograms. A work sheet to compute capabilities is also given.

Example 7.5 The shift effort on a car lever was produced by a stable process. A large study was performed to quantify the capability of the process. The two-sided specification was 3.0–15.3 newtons. The histogram of the $N = 300$ measurements follows:

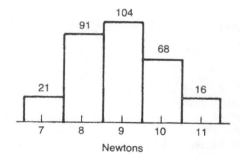

Procedure 7.1 Process Capability Indices

Capability indices are used to evaluate the ability of a stable process to produce parts within the specification limits. The stability of the process is required to ensure meaningful estimates of the process location $\bar{\bar{X}}$ and variability $\bar{R}$. Typically, the five capability indices are computed from standard $\bar{X}$ and R control charts. However, histograms can be used to estimate the process location $\bar{X}$ and variability s if the process is stable. The work sheet associated with this display can be used for all calculations. It is assumed that the process mean is located between the specification limits.

Process Potential C_p

The potential of a process to produce parts between the upper (USL) and lower (LSL) specification limits depends on the variability $\hat{\sigma}$ of the process measured by $\bar{R}/d_2$ or s and the width of the specification limits USL − LSL:

$$C_p = \frac{\text{USL} - \text{LSL}}{6\hat{\sigma}}$$

If C_p is less than 1.0, the process is considered incapable of producing most parts (99.73%) within the specification limits and the process variability must be reduced. A $C_p = 2.0$ or higher is desirable.

Upper Process Performance CPU

The ability of a process to produce parts below the upper specification limit depends on the distance the process mean is from USL and the upper half of the process variability $3\hat{\sigma}$:

$$\text{CPU} = \frac{\text{USL} - \bar{\bar{X}}}{3\hat{\sigma}}$$

If CPU is less than 1.0, the process is considered not capable. The process mean must be shifted away from USL and/or the variability must be reduced. Total process capability for a one-sided upper specification limit can be measured by CPU.

Lower Process Performance CPL

The process performance relative to the lower specification limit is

$$\text{CPL} = \frac{\bar{\bar{X}} - \text{LSL}}{3\hat{\sigma}}$$

The discussion of CPU can be directly related to CPL using the lower specification limit. The total process capability for a one-sided lower specification limit can be measured by CPL.

Process Centering k

The scaled deviation of the process mean from the midpoint of the specification limits,

$$m = \frac{\text{USL} + \text{LSL}}{2}$$

is measured by

$$k = \frac{2|m - \bar{\bar{X}}|}{\text{USL} - \text{LSL}}$$

The bars denote a positive difference. If the process is centered, $k = 0$. If $\bar{\bar{X}}$ is located at USL or LSL, $k = 1$.

Process Performance C_{pk}

The ability to produce parts within the specification limits considering both process location and variability is

C_{pk} = minimum (CPL, CPU)

or

$C_{pk} = C_p(1 - k)$

The two expressions for computing C_{pk} are equivalent. Analogously to C_p, a value of C_{pk} below 1.0 indicates the process is not capable. A minimum value of 1.33–2.0 for C_{pk} is desirable.

Process Capability Work Sheet

n	d_2
3	1.69
4	2.06
5	2.33
6	2.53
7	2.70
8	2.85
9	2.97
10	3.08

Process specifications:
  Upper specification limit   USL =
  Lower specification limit   LSL =
  Width of specification   W = USL − LSL =
  Midpoint of specification   m = (USL + LSL)/2 =
Process variability:
  Control chart
    Average range $\bar{R}$ =
    Standard deviation of individuals $\hat{\sigma} = \bar{R}/d_2$ =
  Histogram (use histogram work sheet)
    Standard deviation of individuals   s =
    Note that a histogram does not establish process stability.
  Process spread   PS = $6\hat{\sigma}$ =
                    PS = 6s =
  Note that $\bar{R}/d_2$ and s should be approximately equal for a stable normal process.
Process location:
  Process mean $\bar{\bar{X}}$ =
  Off-target difference $D = m - \bar{\bar{X}}$ =
    Note that if D is negative, ignore the negative sign.
Process capability:
  Process potential C_p = W/PS =
  Upper process performance   CPU = 2(USL − $\bar{\bar{X}}$)/PS =
  Lower process performance   CPL = 2($\bar{\bar{X}}$ − LSL)/PS =
  Process centering   k = 2D/W =
  Process performance $C_{pk} = C_p(1 - k)$
  Note that for a one-sided specification, capability is measured by CPU or CPL.

Using the work sheet in Chapter 5, $\bar{X} = 8.9$, and $s = 1.0$:

$$C_p = \frac{USL - LSL}{6\hat{\sigma}} = \frac{15.3 - 3.0}{6 \times 1.0} = 2.05$$

$$CPU = \frac{USL - \bar{X}}{3s} = \frac{15.3 - 8.9}{3 \times 1.0} = 2.13$$

$$CPL = \frac{\bar{X} - LSL}{3s} = \frac{8.9 - 3.0}{3 \times 1.0} = 1.97$$

$$k = \frac{2|m - \bar{X}|}{USL - LSL} = \frac{2|9.15 - 8.9|}{15.3 - 3.0} = .04$$

$$C_{pk} = 2.05(1 - .04)$$
$$= 1.97$$

The problem with using histograms to compute process capability is that process stability has not been established. Recall from Figure 5.30 that a variety of unstable processes can produce a bell-shaped histogram. Assuming that the capability indices predict anything about future performance in these cases is clearly not justified.

Capability indices computed from histograms assume a stable process

The importance of evaluating both stability and capability was discussed earlier in the chapter. The importance of preparing a histogram of individual control chart measurements was discussed in Chapter 5. It is possible to use a dot control chart to collect data, evaluate stability, prepare a histogram, and compute capability. On the location chart, a dot denotes an individual measurement, a circled dot a median, and an X a mean value. The following examples illustrate different applications of the chart.

Example 7.6 Consider a set of simulated measurements collected from population A in Chart 6.1. The histogram in Chart 7.1 can be obtained by counting the number of dots in the histogram intervals. The capability calculations are illustrated in the chart.

Example 7.7 The depth of case hardness measurements from the fuel pump case study are plotted in Chart 7.2. Since an odd number of measurement is used to form a subgroup, a median range control chart is convenient. The medians are circled and connected. Using either the control limits or the histogram, a lack of stability can be detected. Capability calculations should not be performed in this case.

Example 7.8 Concentricity measurements of a machined bore in a transmission housing appear in Chart 7.3. Since a median control chart was used with $n = 3$, it was not necessary to record the individual measurements. All calculations can be computed directly from the plotted points.

Example 7.9 The flatness measurements of an automatic transmission valve body discussed in Chapters 4, 7 and 11 case studies appear in Chart 7.4. There are eight fixtures on the machine, so a stratified chart is used. Instead of using dots to record individual measurements, extreme fixtures are indicated on the chart.

7.7 Computing Capability from Histograms

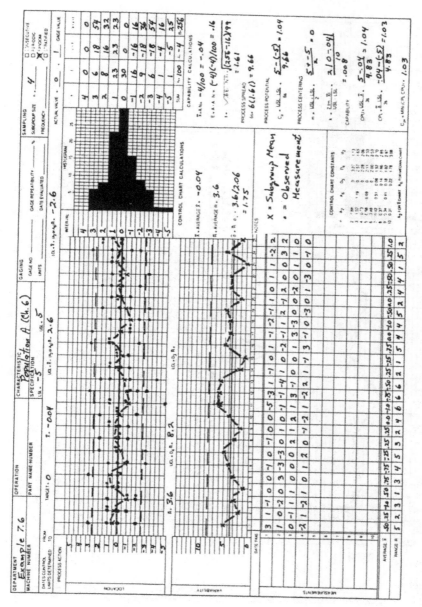

Chart 7.1 Dot control chart for the simulated measurements in Example 7.6.

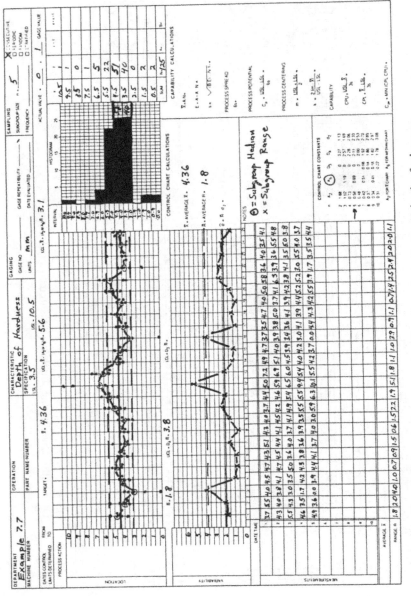

Chart 7.2 Dot control chart for the depth of hardness measurements in the fuel pump case study (see Example 7.7).

7.7 Computing Capability from Histograms

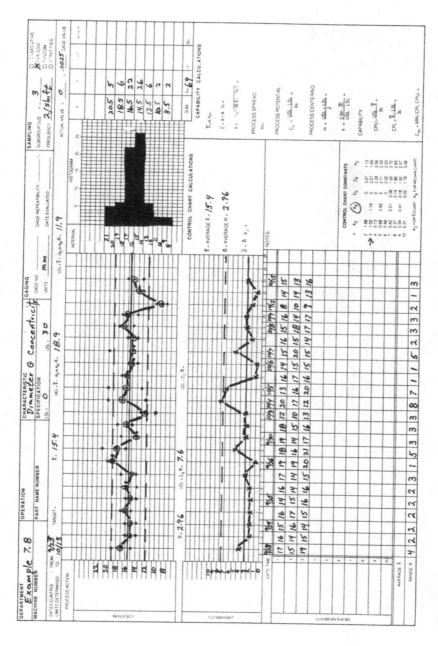

Chart 7.3 Dot control chart for concentricity measurements of a machined bore (see Example 7.8).

292   7 Process Capability

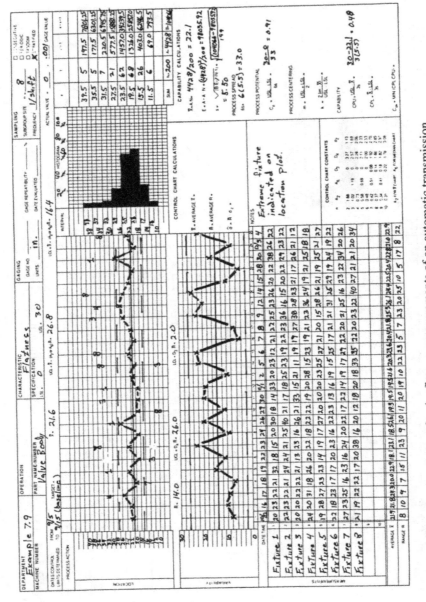

Chart 7.4 Dot control chart for the flatness measurements of an automatic transmission valve body (see Example 7.9).

7.8 New Machine Tryout*

Evaluating the capability of existing machinery should be performed during routine production. This enables all the potential sources of variability to have an opportunity to influence the process output. Normally, capability is evaluated over a several week period, but the manufacturing situation influences the time period. Unfortunately, this approach can generally not be used when evaluating new machinery located at a machine builder's plant. Typically, the purchaser of the machinery wants to evaluate a new machine prior to shipment to the plant. To accomplish this, special machine tryout studies are performed. However, it is often not possible to predict how the machine will perform in its intended manufacturing process since the machine tryout includes many different process factors (men, methods, machines, materials, measurement, and environment). These factors can work to either increase or reduce variability when compared with the variability observed in the production situation:

Factors increasing variability at a machine tryout	Factors reducing variability at a machine tryout
Machine break-in	Low production volumes
Special parts and materials to process	All new equipment
Special gaging	Specially trained operators
Nonstandard machine hookups (e.g., power, coolants)	Strict environment controls

In most cases the combined factors result in lower overall variability (i.e., higher capability) during a tryout than is experienced during normal production.

Since the process at machine tryout is quite different from the intended production process, does it make sense to evaluate capability? The realistic objective of a machine tryout should be to identify problems that would adversely impact on long-term capability (and reliability) in the normal production process. Since capability at the machine tryout is generally better than the long-term production performance, a higher benchmark capability is needed. Experience has shown that if the target production process capability is 1.0, the tryout capability should be 1.33 or greater. If the target production process capability is 1.33, the tryout capability should be 1.67 or greater, and so on.

A number of considerations should be addressed when conducting a machine tryout.

Site. It is important to conduct the tryout at the machine builder's site. This gives the opportunity to correct any deficiencies quickly. In many cases any lack of capability can be corrected by "fine-tuning" machine parameters. This process is time consuming and often requires a variety of engineering expertise. The required personnel and facilities are most accessible at the machine builder's site.

Parts and Material. As the process flow diagrams indicate, machines are often part of a sequential process of manufacturing a finished product. Machines that operate on various stages of semifinished products require parts or materials in the required semifinished stage of processing. For a new process all machinery is purchased at the same time, so it can be difficult to obtain the required parts. In many cases, these parts must be specially manufactured at a high cost. The quality of these special parts can be much better or worse than the normal production parts. Too often, too few parts are available at a tryout to adequately evaluate a new machine.

Cycle Time. The rate at which a machine produces parts is often related to the variability of the output. Thus, it is important to perform the tryout at the same cycle time used in normal production. This is a common oversight.

Tooling. Machine output is directly influenced by the type of tools used. It is important to use the tooling expected to be used in normal production. If temporary nonstandard tooling must be used, the machine cycle time and capability criteria must not change.

Machine Adjustments. During normal production, machines are sometimes adjusted to attempt to change the mean of the process output. A machine tryout should make as few adjustments as possible. For some machines, a few adjustments are a normal part of the process. However, there is a tendency to want to make many adjustments during a tryout. This is a bad (and often self-defeating) practice, as discussed in the Chapter 4, Section 4.7.2. A detailed chronological log should be kept of all process factors, including any machine adjustments. Information on cycle times, feed rates, speeds, tools, coolants, and setup, for example, should be noted for each machine run. All machine modifications, no matter how small, should be recorded.

Time. The time required to perform a tryout to obtain the required capability is generally underestimated. In many cases the fine-tuning process requires several months. Sufficient time must be part of the machine delivery schedule. A useful practice is for the machine builders to perform a preliminary capability study prior to involvement of the purchaser.

Rerun of Tryout. A good practice is to rerun the tryout immediately after the machine is installed at the purchaser's manufacturing plant. Any shipping or installation damage can then be quickly corrected.

During the tryout, the machine should be run continuously at the production cycle time. It is desirable to run the machine at least 8 hours, although 4 hour preliminary runs often identify many problems. If too few parts are available, many machines can run without parts (dry cycle) so the available parts can be used periodically during a 4 or 8 hour period. The more parts measured the better the analysis, but a minimum of 75 is often used. These parts should be collected periodically throughout the production run. Analysis of the data should be performed using the following steps.

1. Gage study. The variability of the measurement system should not be confused with machine variability. It is desirable to use the production gaging system to evaluate the machine. Generally, the special fixturing and design result in lower gaging variability. If the production gaging system is not available, a measurement system with a lower gaging error should be used. In either case, a gage study should be performed to assess the adequacy of the gaging (App. I).
2. Run chart. A run chart in the production order should be prepared to evaluate any obvious machine problems. It is good practice to number the parts in production order and to save the parts used in the final machine tryout so that they can be measured or

analyzed later when the machine is on the production floor and all production gaging is available.
3. $\bar{X}$ and R chart. An $\bar{X}$ and R chart should be prepared to evaluate the stability of the machine. All major stratification factors (e.g., fixtures, spindles, and pallets) should be identified for each part. It is often not possible to evaluate possible major stratification factors separately, so a baseline control chart cannot be established. A stratified control chart is often the best choice.
4. Histograms. A histogram of the individual measurements for all stratification factors should be prepared. Any unusual shapes can be an indication of a problem. The dot control chart used in the previous section can be used.
5. Compare major sources of variation. All stratification factors should be compared using the procedures discussed in Chapter 11. The comparison principle says that a machine having four fixtures performing the same operations should have a similar location and variability for the fixtures. Many machine problems can be identified using this approach.

7.9 Process Audits*

Traditional audits conducted by quality control departments consist of a random selection of finished parts and report the percentage of parts or characteristics within the specification limits. For most applications, we expect high percentages. It is not uncommon to find such statements as "99.8% of the dimensions are within specification" on management reports. Yet, many of these same parts are associated with customer complaints and poor-quality products. The intent of the traditional system is to provide upper management with the "evidence" to motivate (often punitively) lower level managers. The system is rarely associated with high-quality products and is largely useless and demoralizing. Deming (1982, Chap. 14) discusses the uselessness of the inspection-oriented quality audit.

Audits of the process are an essential management tool, but the purpose should not be to evaluate conformance to specification, but rather to

Motivate continuous improvement
Evaluate the defect prevention system

Three types of audits should be used to accomplish these objectives. In each case the capability indices C_p and C_{pk} should be part of the audit system. As noted in the discussion of Figure 7.8, these indices serve as a central tool to use in attaining and monitoring continuous improvement. Unfortunately, the mind-set in an organization can be "Make it to print" rather than "Make it the best you can—and then improve the process." The audit systems that encourage continuous improvement are discussed in the following subsections.

7.9.1 Management Audits

Management's responsibility is to work on the system so it is necessary to evaluate the system, not just the results. Most audits of manufacturing facilities by top management are well-planned, highly orchestrated events. Tour routes, special presentations, and interviews, for example, are all prearranged. These events may serve to indicate top management's interest but provide no evaluation of the operating systems. Worse yet, there is a large loss in productivity since many employees spend weeks making special preparations.

(The special painting and cleanup activity do serve a useful purpose.) A preferable audit system would be unannounced visits with the intent of talking with all levels of the organization. The focus of the discussions would be their individual actions directed toward continuous improvement. Control charts and other types of data could serve to initiate these discussions. Within several hours, top management would have a reasonably clear picture of whether a system existed to foster continuous improvement.

7.9.2 System Audits

The intent of the system audit is to evaluate whether an effective defect prevention system (DPS) exists. The steps discussed in Chapter 14 used to establish a DPS provide a benchmark that can be used to evaluate the prevention abilities of any system. In particular, the operation control plan table provides a basis to evaluate the system to control every operation in an entire process. The benefit of this auditing approach is that the control *system* is the focus rather than the inspection of a few parts. Clearly, the existence of an effective system ensures few parts beyond the specification limits. Conversely, if an effective system does not exist, the inspection approach may or may not find parts beyond the specification limits based on random chance. Noting system deficiencies provides clear, positive actions that can be taken to help ensure that the control system is changed.

7.9.3 Sampling Audits

An individual part may have several hundred final dimensions, only a few of which can reasonably be monitored by standard control charts. It is useful to establish an ongoing sampling audit that measures a reasonable number of critical characteristics on a periodic (e.g., weekly) basis. A typical plan might be to sample one part per day for a week and use $n = 5$ as the subgroup size. Over a period of several months, it is possible to monitor major process changes. The intent of the chart is to encourage and monitor the continuing improvement of process capability. Control charts maintained from sampling audits have proven useful for the following:

Monitor the process output to verify operation of a control system
Assess the need for additional in-process controls
Verify the proper use of control charts
Detect problems with the measurement system
Provide variable data when floor applications use go/no-go gaging

In some applications of a sampling audit system, data can be collected by sophisticated gaging (e.g., coordinate measuring machines) and maintained on a computerized system.

One of the dangers of any data collection system is that the volume of data can limit its usefulness. Fortunately, it is convenient to summarize the sampling data in a simple format using the capability indices. When combined with process results, the data can be summarized as follows:

No.	Characteristic	Target value	Process				Audit				
			Dates	N	$\bar{X}$	C_p	C_{pk}	Dates	N	$\bar{X}$	C_p	C_{pk}
1												
2												
3												
.												
.												

The capability indices can also be plotted over time. This table can be used to evaluate how well a process is targeted as well as to compare C_p and C_{pk} values. Comparison of process and audit capability values highlights problem areas.

7.10 Summary

The five capability indices C_p, CPU, CPL, k, and C_{pk} provide a unitless "language" that can be understood from the plant floor to the manager's desk. The performance of a stable process relative to specification limits is conveniently summarized. Application areas include the following:

1. Defect prevention. The use of capability indices helps to prevent defects since the indices quantify the ability of a process to produce parts within the specification limits. A benchmark of $C_{pk} = 1.33$ is often used, but a $C_{pk} = 2.0$ is often needed to ensure that moderate changes in the process do not result in parts beyond the specification limits.
2. Continuous improvement. The capability of processes must be continually improved to produce continually improving quality products. Capability indices provide a measure of process performance and can be used to monitor improving quality. This approach helps change the production mind-set from "Make it to print" to "Make it the best you can—and then improve the process."
3. Communication. The use of C_p and C_{pk} establishes a common language that is dimensionless and assesses both the potential and actual performance of production processes. Engineering and manufacturing can communicate and note those processes with high capabilities. Poor capabilities are an indication of a needed design or process change. Manufacturers can communicate with suppliers in a straightforward manner.
4. Prioritization. A simple computer printout of processes with unacceptable C_p or C_{pk} values is an aid in establishing priority for process improvements.
5. Location or variability. For each characteristic, it is meaningful to compare C_p and C_{pk}. If C_{pk} is too low, then C_p must be examined to determine whether the variability is unacceptably high. If C_p is close to C_{pk}, then process location is not a problem. The indices CPU, CPL, and k provide an assessment of how close the process mean is to a single specification limit and how far the process mean is from the target mean.
6. New machine tryout. A common basis for evaluating new machines can be established. A minimum benchmark of 1.33–1.67 is typically used.
7. Process audits. Capability indices can be used in process audits to focus attention on problem areas. These indices should be used in management, system, and sampling audits when discussing process performance.

7.11 Case Studies

7.11.1 Continuing Improvement

In the early stages of production of a new engine, there was a 9% scrap and 12% rework rate on a camshaft fuel pump eccentric. Also, there was excessive drill breakage where an oil hole was drilled near the hardened fuel pump eccentric. A cross-functional team was formed to address the high scrap and rework. As a first step, the team prepared a process flow diagram and analyzed past data to determine that the hardened metal case depth accounted for a large portion of the departmental problems. The case depth was controlled by heating the camshafts in induction coils. A cause-and-effect diagram (Chap. 13) was prepared to identify potential causes of the out-of-specification parts. Many variables were thought to affect the case hardness depth, although the impact of each was not known. These variables included the design of the induction coil, the metallurgical composition of the camshaft, positioning of the part within the coil, alternating current frequency and settings, the rate at which the part is heated, the time it is exposed to air prior to quenching, the heat cycle time to reach temperature, electrical flux patterns, the quenching solution used, and the amount of electrical power in the part.

First Study Period. The team decided that the first step would be to evaluate the stability of the case hardness produced by the induction heater. Because there was a tendency for some employees to make adjustments to the heater, a padlock was placed on the control panel. A standard $\bar{X}$ and R chart was used with a subgroup size of $n = 5$ consecutive samples. During the first day, samples of five consecutively produced pieces were recorded for every interval of 100 parts produced. After the first day, the interval was increased to 200 parts to reduce the high cost of the destructive inspection technique. A total of 30 subgroups was recorded. Also, a log of all changes, repairs, and so forth to the process was kept. The measurements follow:

Subgroup	\multicolumn{15}{c}{Piece}
	1	2	3	4	5	6	7	8	9	10	11	12	13	14	15
1	3.7	5.5	4.0	4.5	4.7	4.3	5.1	4.3	4.0	3.7	4.4	5.0	7.2	4.9	4.7
2	4.3	4.0	3.8	4.1	4.7	4.5	4.4	4.1	4.5	4.2	4.6	5.9	6.9	5.1	4.0
3	5.5	4.3	3.0	3.5	5.0	3.6	4.0	3.7	4.1	4.9	5.4	6.5	6.0	4.5	3.9
4	4.6	3.5	1.7	4.2	4.3	3.8	3.6	3.9	3.5	5.5	5.5	9.4	5.4	4.0	4.2
5	4.9	3.6	0	3.9	4.4	4.1	3.7	4.0	3.0	5.9	6.3	10.1	5.5	4.2	3.7
	16	17	18	19	20	21	22	23	24	25	26	27	28	29	30
1	3.7	3.5	4.7	4.0	5.0	5.8	3.6	4.0	3.5	4.1	6.2	5.5	4.4	4.0	3.9
2	3.9	3.8	5.0	3.7	4.1	6.3	3.9	3.6	5.5	4.8	5.1	5.0	4.0	3.6	3.5
3	3.4	3.6	4.1	3.9	4.2	3.8	4.1	3.5	5.0	3.8	5.4	3.9	3.7	3.7	3.3
4	3.0	4.1	3.9	4.4	5.2	5.2	3.0	5.5	4.0	3.7	3.9	4.2	3.9	3.5	1.7
5	0	4.4	4.3	4.2	5.5	3.9	1.7	3.5	3.5	4.4	4.7	4.1	3.6	3.7	0

7.11 Case Studies

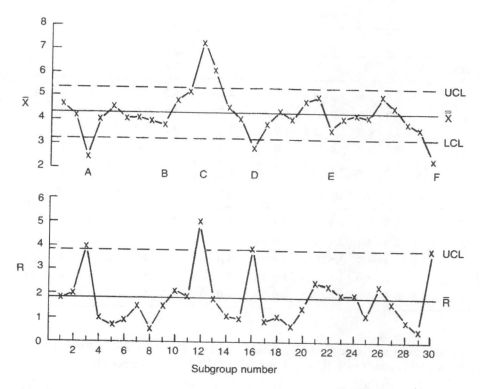

Figure 7.12 $\bar{X}$ and R chart for first study period of depth of case hardness.

Results of the $\bar{X}$ and R control chart appear in Figure 7.12. The process was obviously unstable, and a number of actions were taken during the study period:

Point A. The team increased the power on the induction heater coil from 8.2 to 9.2, and subsequent samples showed smaller ranges and more stable averages.

Point B. The team discovered a bent coil and straightened it. This started an uptrend, and at point C, the sample range and average were beyond the control limits again.

Point C. The team decreased the power back to 8.8 with the second and third subsequent samples back in control.

Point D. The coil was shorted to a part and needed to be straightened. At that point, the team devised a gage to check the coil spacing to the camshaft and instituted hourly checks of that spacing.

Point E. The team decided to decrease the air gap between the part and the coil.

Point F. The coil shorted and was replaced with a second coil of the same type.

A histogram of the measurements indicated a nonsmooth pattern (Chart 7.2) consistent with the lack of stability.

Second Study Period. After the new coil was installed, a second study period was started. The measurements from the process study follow:

	Piece														
Subgroup	1	2	3	4	5	6	7	8	9	10	11	12	13	14	15
1	5.0	5.5	3.6	5.5	4.4	3.8	5.0	4.0	3.9	5.5	5.5	3.5	5.3	4.4	5.5
2	4.1	4.0	4.3	4.0	3.5	4.4	5.1	3.7	5.0	4.0	4.4	5.3	3.9	5.1	5.0
3	4.0	3.9	4.0	3.7	5.0	4.0	3.9	5.2	5.5	3.5	3.9	4.2	3.6	4.7	4.7
4	3.7	3.6	4.0	3.9	3.7	5.0	3.7	3.5	5.2	3.7	5.0	4.7	3.5	5.3	3.9
5	3.9	4.4	5.0	4.1	5.1	5.5	5.0	3.7	5.0	3.8	3.5	5.1	5.0	5.1	4.4
	16	17	18	19	20	21	22	23	24	25	26	27	28	29	30
1	3.9	3.5	5.3	3.9	5.1	3.5	3.7	4.0	5.5	3.7	4.9	3.5	4.4	4.0	5.5
2	4.7	5.0	5.1	4.4	4.4	4.0	5.1	4.4	3.6	5.2	3.6	5.1	5.5	3.6	4.4
3	5.1	3.9	4.7	3.5	5.3	4.4	5.5	4.9	4.3	4.0	4.7	4.0	4.6	3.7	4.9
4	5.3	4.2	5.0	5.1	3.9	5.5	4.3	3.7	3.6	4.3	5.5	4.4	4.3	4.7	3.6
5	4.2	5.1	5.3	4.4	3.5	4.9	3.7	4.0	3.5	5.5	4.9	3.7	5.5	3.5	4.0

The $\bar{X}$ and R chart in Figure 7.13 was stable. Based on $\bar{\bar{X}} = 4.43$ mm and $\bar{R} = 1.6$, the process capability based on LSL = 3.5 and USL = 10.5 was unacceptable.

$$\hat{\sigma} = \frac{\bar{R}}{d_2} = \frac{1.6}{2.33} = 0.69$$

$$C_p = \frac{\text{USL} - \text{LSL}}{6\hat{\sigma}} = \frac{10.5 - 3.5}{6 \times .69} = 1.69$$

$$\text{CPU} = \frac{\text{USL} - \bar{\bar{X}}}{3\hat{\sigma}} = \frac{10.5 - 4.43}{3 \times .69} = 2.93$$

$$\text{CPL} = \frac{\bar{\bar{X}} - \text{LSL}}{3\hat{\sigma}} = \frac{4.43 - 3.5}{3 \times .69} = .45$$

$$C_{pk} = \text{minimum (CPL, CPU)} = .45$$

Since C_p was sufficiently high, it is apparent that the lack of capability is due to the mean being too close to the LSL. However, further analysis reveals an additional problem. A histogram of the measurements appears in Figure 7.14. The apparent nonsmooth pattern seems to indicate a mixture of at least two populations. The importance of using histograms with a control chart is again emphasized. In this case, the capability calculations are a poor indication of actual process performance.

The team attempted to identify the source of the two populations and sought a way to increase the mean. Unfortunately, case hardness depth mean is not easily adjusted. With assistance from the induction heater supplier, the team tried to fine-tune the process. The fine-tuning actions included leveling the machine, performing layout checks of coils before installation, and attempts to optimize machine settings for preheat, induction, delay time, and power. These actions showed no significant effect.

7.11 Case Studies

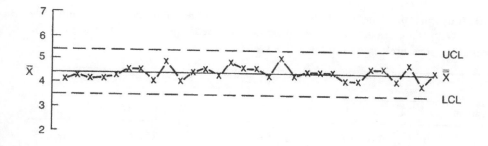

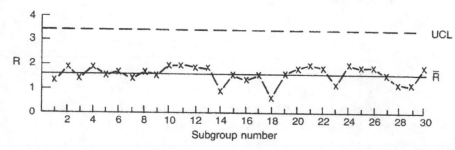

Figure 7.13 $\bar{X}$ and R chart for second study period of depth of case hardness.

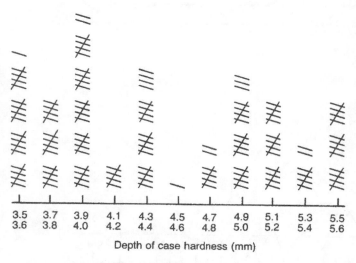

Figure 7.14 Histogram of engine camshaft depth of case hardness, second study.

Third Study Period. The first study indicated the importance of the coil in controlling the process. With the assistance of the supplier, a newly redesigned coil was placed in the process. The measurements from the process study follow:

Subgroup	Piece														
	1	2	3	4	5	6	7	8	9	10	11	12	13	14	15
1	5.0	5.5	3.6	5.5	4.4	3.8	5.0	4.0	4.1	5.5	5.2	6.2	4.9	6.1	5.3
2	4.9	5.8	5.9	5.9	6.2	6.0	5.2	5.1	5.8	5.2	4.6	5.8	4.9	6.2	5.4
3	5.5	5.5	6.0	5.9	5.9	5.7	4.6	5.8	6.1	5.4	5.4	5.1	4.9	5.9	5.4
4	5.3	5.6	5.9	6.4	5.7	5.7	5.4	6.2	5.7	5.2	6.1	5.2	4.9	4.5	5.4
5	5.6	6.3	5.8	5.3	4.9	6.3	6.1	5.9	6.5	5.8	5.2	5.4	4.6	5.6	5.2
	16	17	18	19	20	21	22	23	24	25	26	27	28	29	30
1	6.3	5.4	5.6	4.7	6.0	5.1	5.6	5.5	4.9	4.7	6.0	5.5	5.2	5.0	5.5
2	5.6	5.3	5.0	4.7	5.3	6.2	6.1	6.0	4.8	4.9	5.1	5.0	4.9	5.2	5.1
3	5.9	5.4	5.2	5.6	5.6	5.0	5.8	5.7	6.1	5.2	5.3	6.0	5.2	6.0	5.3
4	4.7	5.2	4.9	5.0	5.0	5.2	5.9	5.0	5.3	6.0	4.9	5.7	5.0	4.9	4.1
5	6.2	5.2	4.7	5.2	5.2	5.7	5.8	4.9	5.2	5.7	6.1	5.0	5.3	6.1	5.0

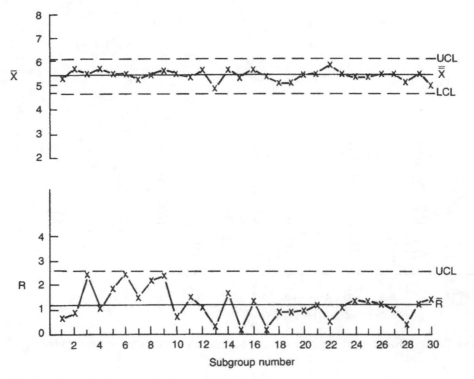

Figure 7.15 $\bar{X}$ and R chart for third study period of depth of case hardness.

7.11 Case Studies

The $\bar{X}$ and R chart in Figure 7.15 appears stable. The capability calculations are in Section 7.4, along with the histogram. The capability has been improved dramatically to $C_{pk} = 1.22$. The few low case depths indicated on the histogram are still unacceptable. The improvement process must go on—continuous improvement. However, the improvement actions have produced dramatic results. Both scrap and rejects have been eliminated at an appreciable savings to the plant. Tool use has stabilized, and the process is operating predictably.

7.11.2 Mating Parts

Manufactured parts are typically assembled together with other parts to form subassemblies that in turn are combined with other subassemblies to form a final product. The function of the final product is directly related to the dimensional relationships of mating parts. A simple example of this relationship is a bore and valve in an automatic transmission valve body. A bore generally has several steps with different diameters. The valve has corresponding steps designed to mate with the bore diameters. The interrelationship of the bore and valve dimensions is an important factor in determining the shift quality of a transmission.

A useful approach to improving the function of products is to evaluate the capability of mating parts. Typically, design engineers use historical guidelines to develop the specification limits that control the clearances between mating parts. The quality of the final product can often be improved if manufacturing and design engineers work jointly to evaluate process capabilities and target processes to optimize the function of the final product. Consider the following bore and valve relationship:

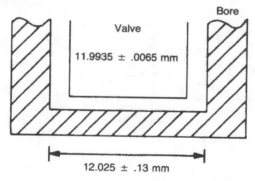

Assuming the processes are centered and have a normal, bell-shaped distribution, the relationship of the bore and valve for $C_{pk} = 1$ is shown in Figure 7.16. The designed minimum clearance is .012 mm, and the target clearance is $12.025 - 11.9935 = .0315$ mm. The transmission functions better if the clearances are consistent and centered around the target clearance. Also, if the clearance becomes too small, there are assembly problems and a possibility of damaging the valve or bore during assembly. Figure 7.17 shows several combinations of valve and bore capabilities. The importance of relating the capabilities is illustrated in Figure 7.17c, where both C_{pk} are 1.00. If the optimal product function is the distance between the two specification limit midpoints, this process produces almost no optimal parts. In this case excess pressure leakage between the valve and bore may result. Unfortunately, this case is not uncommon since the extra clearance between the valve and bore makes assembly easier. Although no parts are out of specification, a significant improvement opportunity is wasted. The customers would certainly be more pleased with the shift feel if the processes were properly targeted.

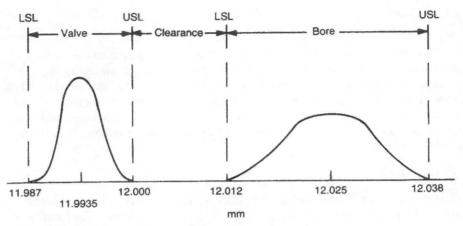

Figure 7.16 Valve and bore specifications.

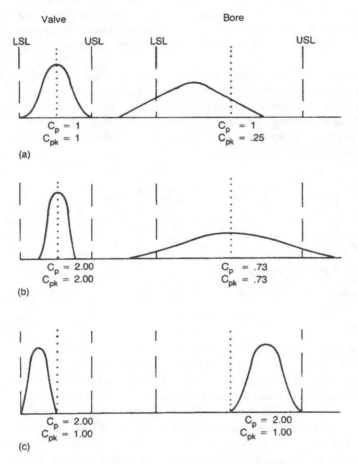

Figure 7.17 Relationship of bore and valve capabilities: (a) undersized bores result in assembly problems and poor product function, bore mean must be adjusted; (b) undersized and oversized bores result in poor product function but no assembly problems, bore variability must be reduced; (c) optimal ease of assembly but possible poor product function due to excess clearance, valve and bore means must be adjusted.

7.11 Case Studies

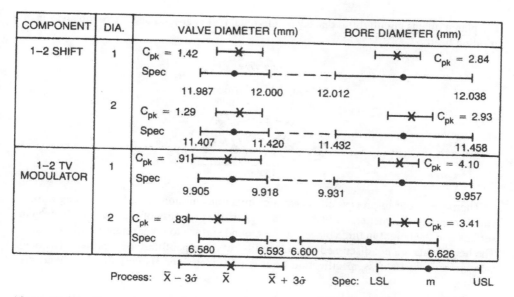

Figure 7.18 Transmission valve and bore process capabilities related to specification limits.

A month's study of the capability of a bore and valve process was performed by collecting $n = 5$ parts randomly throughout the day. The processes were stable, and the resulting capability indices are indicated in Figure 7.18. The capabilities appear acceptable for all parts except the TV modular valve. It would be useful to temporarily use a higher sampling frequency (e.g., $n = 5$ every 2 hours) to identify any special causes. Notice in Figure 7.18 that most means are targeted reasonably well, except diameter 2 on the TV modulator bore.

7.11.3 Machine Tryout

A replacement grinding machine was purchased because of the unacceptable quality performance and reliability of the old machine, discussed in the Chapter 4 case study, Section 4.9.2 (see Prob. 7.6). The new machine had a fixture arrangement similar to that of the old machine, with eight fixtures located in pairs on four tables that rotate on a large circular platen:

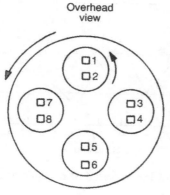

The platen rotates circularly so that each individual part is ground under a rough and finish grinding head.

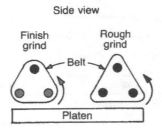

There were four in-process characteristics on a transmission valve body casting important to maintain downstream finish part characteristics: flatness of the valve body ground surface and the height in three places of the ground surface to other reference dimensions. The height data are not discussed here in detail, but the capabilities are reported. The one-sided in-process flatness specification limit is .030 inch. The flatness measurements are coded as

Actual value = .001 × coded value

so USL = 30. A gage study was performed on the production gage prior to the tryout of the machine, indicating about 15% repeatability error.

The initial study of the machine focused on reliability. The minor machine and process adjustments required 4 days, 6/18 through 6/21. The initial flatness data appear in Table 7.3, with an associated control chart shown in Figure 7.19. The lack of stability (preliminary control limits computed from subgroups 1–15) shown in the figure is somewhat expected because of the short runs of parts and numerous machine adjustments. Even in

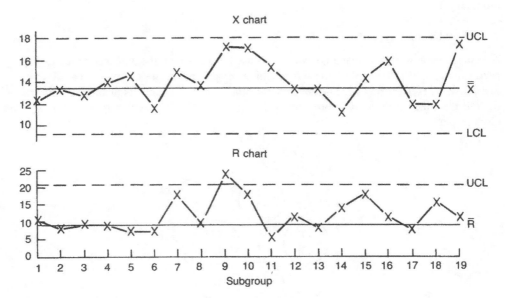

Figure 7.19 Control chart for initial period of machine tryout adjustment.

7.11 Case Studies

Table 7.3 Flatness Measurements for Initial Period of Machine Tryout

Fixture	6/18, Time					6/19, Time						6/20, Time							6/21, Time			
	6	7	8	11	12	7	8	9	10	11	12	7	8	9	10	11	12	1	10	11	12	1
1	15	12	14	18	15	13	26	15	18	28	16	14	10	10	16	14	13	22	30	24	12	14
2	17	17	15	19	14	15	12	12	32	23	18	18	9	25	20	10	12	23	10	12	10	14
3	7	17	8	10	14	9	10	14	8	13	13	14	10	15	15	15	13	13	17	20	10	10
4	12	11	14	14	18	8	28	18	22	22	14	11	13	13	20	16	14	18	17	18	12	9
5	9	10	9	10	17	11	13	15	21	12	17	16	12	13	9	9	12	18	14	11	12	11
6	7	9	10	10	14	9	10	9	11	10	14	10	8	7	18	8	4	12	16	18	13	12
7	14	17	17	17	11	13	12	12	13	17	17	12	20	21	10	16	20	22	18	16	8	13
8	17	14	15	15	14	13	10	13	14	12	14	8	6	11	10	8	8	13	18	15	10	10
Subgroup $\bar{X}$	12.3	13.4	12.8	14.1	14.6	11.4	15.1	13.5	17.4	17.1	15.4	13.4	11.0	14.4	16.0	12.0	12.0	17.6	17.5	16.8	10.8	11.6
R	10	8	9	9	7	7	18	9	24	18	5	8	14	18	11	8	16	11	20	13	5	5

this initial test phase, it is good practice to maintain data in a control chart format. Especially important in this phase is maintaining an accurate log of all machine modifications and adjustments.

The process of attaining stability and then capability required a number of machine modifications. Some of the variables addressed were

Grinding belt parameters, such as grit size, type of backing and splicing, grit material and width
Dwell time on rough and finish belts
Part fixturing adjustments
Coolant pressure and distribution
Speed of tables and platen
Amount of rough versus finish stock removed

The unfortunate part of these investigations is that there is a tendency for individuals close to the machine development to believe that "one more change will provide the necessary breakthrough." Fortunately, a disciplined approach was used to compare the fixtures (Chap. 11) and conduct small designed experiments on several parameters. It was found that the amount of grinding dwell time, width of the belt, and amount of stock removed from the rough to the finished part were all critical parameters for improving machine performance.

It was difficult to obtain a large number of parts, so subgroups of size $n = 4$ were used in which parts from fixtures 1, 2, 3, and 4 and 5, 6, 7, and 8 formed two separate subgroups. A control chart for the 6/24 trial appears in Chart 7.5. The odd-numbered subgroups are fixtures 1, 2, 3, and 4 and the even-numbered subgroups are fixtures 5, 6, 7, and 8. The control chart indicates an out-of-control process.

Approximately 10 days after the initial machine runs, process stability and capability were obtained in run 4 for flatness as shown in Table 7.4. Unfortunately, the height characteristics were measured and found not capable using the same analysis approach. Development of the machine continued for another 3 weeks. In all cases no major flaws were discovered. Improvement of machine capability involved a number of minor changes that collectively improved the machine's performance dramatically. The final run (run 7) at the machine builder's site produced the following flatness results:

Subgroup	Fixture							
	1	2	3	4	5	6	7	8
1	14	11	13	14	12	12	14	12
2	11	12	13	11	14	8	11	11
3	15	10	11	11	13	12	13	10
4	15	10	9	10	9	9	11	10
5	14	11	14	11	10	12	13	11
6	15	12	12	14	13	12	12	11
7	16	13	12	11	12	11	10	12
8	15	13	13	11	12	11	10	12
9	12	12	14	13	11	10	12	11
10	16	13	13	14	13	11	12	13

7.11 Case Studies

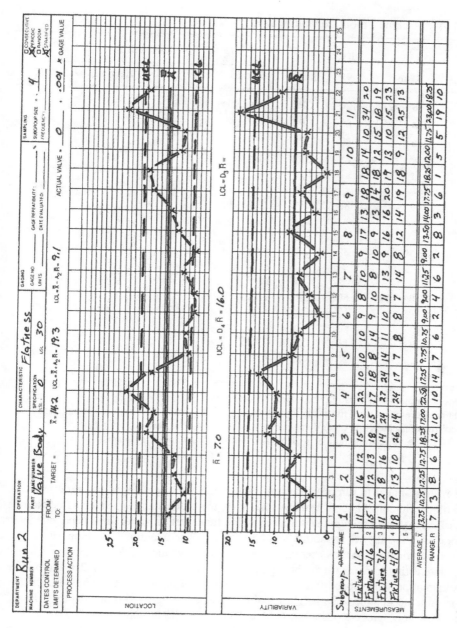

Chart 7.5 Control chart for variable data, the 6/24 trial.

Table 7.4 Machine Tryout Capability Progress

					Characteristics		
Run	Date	N	Index	Flatness	Height 1	Height 2	Height 3
1	6/18–6/21	120	CPU	Unstable	—[b]	—	—
2	6/24	88	CPU	1.01[a]	—	—	—
3	6/26	80	CPU	.64[a]	—	—	—
4	6/28	48	CPU	1.42			
			C_p		.52	.98	1.11
			C_{pk}		.39	.71	.66
5	6/29	80	CPU	2.24			
			C_p		.79	.37	1.13
			C_{pk}		.79	.26	.96
6	7/15[c]	80	CPU	7.85			
			C_p		1.39	1.28	1.19
			C_{pk}		1.06	1.18	1.06
7	7/19	80	CPU	3.63			
			C_p		1.55	1.69	1.70
			C_{pk}		1.21	1.65	1.44
8	8/21[d]	80	CPU	1.52			
			C_p		1.40	1.53	1.68
			C_{pk}		1.34	1.30	1.62

[a]Process was not stable; capability reported for reference only.
[b]Height measurements were not made initially since initially flatness was the major concern.
[c]Parts all run consecutively; other dates run over at least a 4 hour period.
[d]Runs 1–7 were at machine builder's plant. Run 8 was at manufacturing plant.

These results appear in run chart form in Figure 7.20, where no apparent patterns exist. When plotted in control chart format with $n = 4$, the process appears stable. It should be emphasized that even though the 4 hour machine runs were capable, there is no guarantee that long-term stability or capability will be obtained. The intent of a machine tryout is to discover significant problems needing correction at a machine builder's site.

It is interesting to compare the histograms for runs 3 and 7 in Figure 7.21a and b. The progress in improving capability is apparent. The flatness capability has improved: CPU = 3.63 is quite acceptable. The height capability has also improved dramatically. The C_p values show good process potential. Only height 1 needs to be centered.

It is interesting to note that run 6 was an 80 consecutive piece run resulting in a high capability of CPU = 7.85. It is not unusual for the minimal variability introduced in consecutive samples to result in minimal variation and correspondingly high capability. There were essentially no changes to the machine between runs 6 and 7. Run 6 was completed in about 45 minutes, but run 7 required 5 hours with an eight piece subgroup collected about every 30 minutes. It is important to introduce time variability in a machine tryout. Because of the unavailability of parts, consecutive subgroups are often collected. Even if a machine cycles without parts, as is often possible, a 4–8 hour trial is essential.

7.11 Case Studies

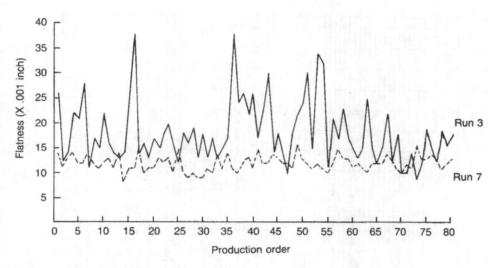

Figure 7.20 Run chart for runs 3 and 7.

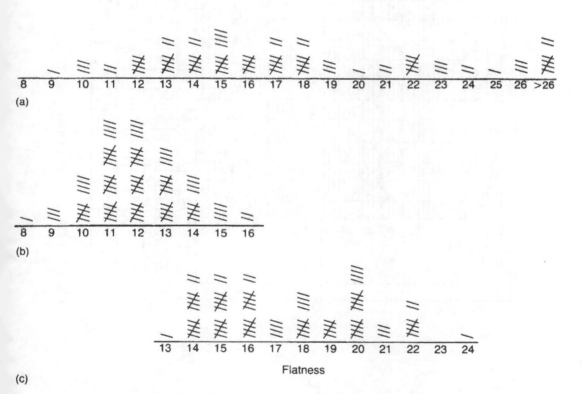

Figure 7.21 Histograms of flatness for different runs: (a) run 3, 6/26; (b) run 7, 7/19; (c) run 8, 8/21.

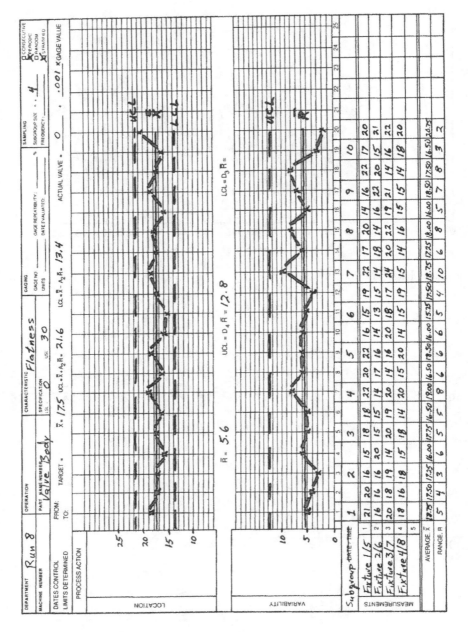

Chart 7.6 Control chart for variable data. The process appears stable.

This example illustrates why many machine tryouts result in poor predictability of how a machine performs at a manufacturer's plant.

After the grinding machine was installed at the manufacturer's site, a large number of parts were run on the machine to evaluate the overall machine performance. Run 8 was made over a 5 hour period. The 80 parts measured were collected and measured using the same procedure as for run 7. The measurements appear in control chart form in Chart 7.6, where the process appears stable. The flatness capability was CPU = 1.52 and all height capabilities were above 1.33, except height 2, which had $C_{pk} = 1.30$. The height characteristic is easily adjustable, and it was found that one fixture needed an adjustment. The process potential C_p was adequate in all cases. It is interesting to note that there was a significant reduction in flatness capability from the machine builder's site, where CPU = 3.63, to the manufacturing plant, where CPU = 1.52. A decrease in capability is not unusual, although a reduction of over 50% is greater than expected. The height capability did increase, possibly owing to a better gaging system. (The production gage was not available for runs 1–7.)

A histogram for run 8 appears in Figure 7.21c. Recall from Chart 7.6 that the process was judged stable using a control chart. The histogram suggests multiple populations are present and illustrates why histograms must be used with control charts. One of the essential machine tryout procedures involves a comparison of common sources of variation, such as machine fixtures. The analysis of these data appears in a Chapter 11 case study and shows fixture differences that explain the decrease in capability (see Prob. 7.7).

The improvement process during machine tryout required about 6 weeks. Typically, 1–2 months is required, depending on the time required for procurement of any material needed. Unfortunately, machine delivery schedules often do not allow this development time. When time is initially scheduled, delays during machine development are absorbed by shortening the tryout time. As this case study illustrates, the tryout or improvement process is an essential element of producing high-quality parts. The capability improvement process should be included as a required part of the machine builder's schedule. Once a machine is ready to run parts, a detailed log should be kept of any process changes and the data analyzed using the four analysis methods (run chart, control chart, histogram, and comparison plot). Failure to follow these practices jeopardizes a smooth production launch from both the quality and productivity standpoint.

Problems

7.1 In Figure 6.19 there are eight different populations. Assuming process stability, calculate the five capability indices where LSL = -10 and USL = 10.

7.2 Suppose you have the following stable process (Process Spread = 6):

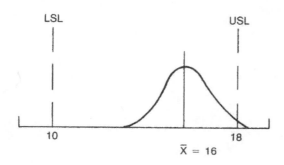

Compute the C_p and C_{pk} indices. What new mean would be required to make the process "capable" ($C_{pk} = 1$)? What new standard deviation would be required to make the process capable without changing the mean?

7.3 Chart 6.1 is a simulated sample from population A. Recall that

$$\bar{\bar{X}} = -.04$$
$$\bar{R} = 3.56$$
$$\hat{\sigma} = \frac{\bar{R}}{d_2} = \frac{3.56}{2.06} = 1.73$$

Make a histogram of the data, and compute s from the histogram data. Note that the value of s is close to $\hat{\sigma}$. Why? Assume that USL = 6 and LSL = −6, use the process capability work sheet and compute C_p and C_{pk} for the histogram data.

7.4 Assuming the process is stable, calculate and interpret the five capability indices for the engine crankshaft example in the Chapter 3 case study (Chart 3.2). How do the control chart results compare with the capabilities computed using the histogram prepared in Problem 5.1? What might explain any differences?

7.5 Compute the capability of the "after" anodizing process in Problem 3.9b. How do the control chart results compare with the capabilities computed using the histogram prepared in Problem 5.6a?

7.6 Calculate the capability of the machine with the eight fixtures discussed in the case study of Chapter 4 (Chart 7.4).

7.7 Assume the process for each machine is stable in Problem 5.7. (a) What is the capability for each machine if LSL = 5 and USL = 23? (b) Suppose the process output is combined, what is the capability of all three machines combined? (c) Suppose the process is centered, on all machines, what would be its capability? (d) Comment on how mean differences of stratification factors (e.g., machines) influence capability evaluations.

7.8 A finishing operation was having difficulty meeting capability requirements, so it was decided to study an upstream operation that had an in-process hole diameter specification of ±3. Since machining was performed on two spindles, both spindles were evaluated using 20 consecutive parts. Also of concern was possible variability in the diameter, so the minimum and maximum diameter were recorded for each part. (a) Comment on the design of the study. How could it be improved? (b) For each spindle, compute the capability of the minimum diameter using CPL and the maximum diameter using CPU. Is this calculation meaningful? Evaluate the capability results. (c) Is the difference between the minimum and maximum diameter of concern?

| | Spindle 1 | | Spindle 2 | |
Part	Minimum	Maximum	Minimum	Maximum
1	6.5	7.5	2.5	3.5
2	7.0	8.0	3.0	4.5
3	4.5	5.5	.0	1.0
4	4.0	5.0	.5	2.0
5	3.5	4.5	−1.0	.0
6	3.5	4.5	.0	1.0
7	3.5	4.5	−1.5	−.5
8	5.0	6.5	.5	2.5
9	4.5	5.5	.0	1.5
10	3.0	5.0	.5	3.0
11	5.0	6.0	.0	1.5
12	5.0	6.0	.0	2.0
13	5.0	6.5	−.5	.0
14	4.5	5.0	−1.0	.0
15	2.5	3.5	−.5	1.0
16	3.0	4.5	.0	2.0
17	1.5	3.0	.0	1.5
18	3.5	4.5	−.5	1.5
19	5.0	6.0	−1.0	1.0
20	2.5	4.0	.0	2.0

Part II
Simple Problem Analysis Tools

8
Check Sheets

8.1 The Role of Data

In most modern work settings there is an abundance of data already collected or readily available on work processes. Unfortunately, many of the numerical "figures" are of limited use since all the following are not present:

The work process in which the data are collected is not clearly defined and understood (process flow diagram, Chap. 2).

The sampling method for collecting parts from the process is not identified (e.g., consecutive, periodic, random, or stratified, Chap. 4).

The data are collected in a haphazard manner without clear operational definitions of the measurement intent (check sheets, Chap. 8).

The bias and repeatability of the measurement system is not understood (gage evaluation, App. I).

The data are not conveniently summarized in easily understood graphs.

Previous chapters showed how a relatively small amount of data displayed graphically using control charts and histograms provides a great deal of information about a process. Later chapters provide additional graphic displays that have proven useful in industrial settings. This chapter discusses the use of check sheets that can be used to collect almost any type of data. A well-designed check sheet is the launching point for an effective analysis in which numerical figures become meaningful data.

It is unfortunate that data is so readily available but so little used. Chapter 1 emphasized how data can form a basis for effective, unemotional communication, which is the first step in solving many problems. For example, a machine operator may be concerned about a problem part characteristic so he or she measures 1 part in every 10, yet does not record any data. For both control charts and histograms, the *pattern* of measurements is important in understanding the process. The only method to assess patterns is to record the data in a convenient format. The operator who measures 1 in 10 parts for conformance to specification is missing the opportunity to combine the measurements in an appropriate graphic display to help identify a process problem. It is possible to develop well-designed check sheets for most applications so that only a single pencil stroke is necessary to record data (e.g., $\sqrt{}$, /, 0). The excuse that data collection requires too much time is then unfounded.

Speak with data

It should be stressed that data collection check sheets are most commonly used to provide a simple, orderly data collection format for later analysis. A check sheet is not typically used to provide a basis for immediate process action as is a control chart. However, a check sheet can play an important role in a defect prevention system (DPS) since it is not logistically possible to have all part characteristics monitored with control charts.

8.2 Types of Check Sheets

Five different types of check sheets are discussed here. In practice a check sheet is developed for each application to make data collection both easy and thorough. Combinations of different types of check sheets are possible to obtain this objective. The five basic types of check sheets are as follows:

1. Classification. A trait such as a defect or failure mode must be classified into a category.
2. Location. The physical location of a trait is indicated on a picture of a part or item being evaluated.
3. Frequency. The presence or absence of a trait or combination of traits is indicated. Also, the number of occurrences of a trait on a part can be indicated.
4. Measurement scale. A measurement scale is divided into intervals, and measurements are indicated by checking an appropriate interval.
5. Check list. The items to be performed for a task are listed so that, as each is accomplished, it can be indicated as having been completed.

8.2.1 Classification Check Sheet

A classification check sheet is used to subdivide a trait into categories. The most common example is classifying a defect into categories. For example, consider a gear

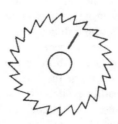

with defect categories

Inner diameter (ID) size
Outer diameter (OD) size
Nicks
Burrs
Tooth geometry

(In real applications there may be 20 or more defect categories.) A simple check sheet is shown below.

8.2 Types of Check Sheets

Defect category	Number	Total
ID size	⊬⊬	5
OD size	/ /	2
Nicks	⊬⊬ ⊬⊬ / / /	13
Burrs	/ / /	3
Tooth geometry	⊬⊬ ⊬⊬ /	11
Other (identify)		0
Total examined	⊬⊬ ⊬⊬ ⊬⊬ ...	120

Completion of the check sheet is easy. A simple slash is used to record each occurrence of a defect. It is always a good practice to leave room for an "other" category. Instructions should state when an unusual defect is found, the defect should be written on the check sheet (e.g., "porosity"). Also, it is always important to identify the total number of parts examined.

The classification check sheet can be used to collect many different types of information. Some common examples in a manufacturing setting follow:

Defects. Parts are examined for defects, and defects are categorized into meaningful subsets. The purpose of this information is to identify the most common problems and then eliminate the cause.

Machine failure modes. Each cause of a machine failure is classified. Again, the purpose is to accumulate data on causes (and duration) of downtime and then develop preventive maintenance requirements.

Lost production time. Any time a machine is not producing parts there is a reason, such as no stock, awaiting repair, tool change, or operator break. Accumulating the causes (and duration) of lost time allows productivity improvements to be made.

Testing rejects. If a subassembly or final assembly is functionally tested, the cause of any failures should be identified. Accumulating categories of the causes of rejects enables the most important problems to be identified.

Customer complaints. The customer may be the end user of a product or the next processing operation. In either case, classification of complaints is an important first step in ensuring customer problems are addressed and eventually eliminated.

The previous gear defect check sheet used a single factor to classify a defect. It can be important to use several factors. For example, time is often an important factor.

Two-Factor Check Sheet. The following chart shows that burrs may be associated with a start-up problem.

Defect category	7 a.m.	8 a.m.	9 a.m.	10 a.m.	11 a.m.	12 p.m.	1 p.m.	2 p.m.	3 p.m.	4 p.m.	5 p.m.	Total
ID size	/			/	/					/		4
OD size	/ /						/					3
Nicks									/			1
Burrs	⊬⊬ ⊬⊬			/	/	/	⊬⊬					18
Tooth geometry	/		⊬⊬					/		/		8
Other	/	/		/			/ / /		/		/	8
Total examined	35	40	31	17	22	5	31	45	23	17	5	271

Three- and Four-Factor Check Sheets. The amount of information (and complexity) obtained from a check sheet can be increased by considering more factors. Suppose we wish to evaluate the downtime of a machining operation with two machines, each having three spindles:

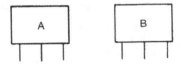

A three-factor check sheet:

Downtime cause	Beginning hour											Total	
	7 a.m.	8 a.m.	9 a.m.	10 a.m.	11 a.m.	12 p.m.	1 p.m.	2 p.m.	3 p.m.	4 p.m.	5 p.m.	A	B
Spindle bearing	A	A	A						B	B		3	2
Coolant line		B	B										2
.													
.													
.													
Switch	B	B			A		A					2	2
Total	2	3	2		1		1		1	1		5	6

Note: If a machine is not operating at the beginning of an hour, the machine letter indicates an inoperable condition.

A four-factor check sheet:

Time down	Machine A				Machine B				Total
	Spin 1	Spin 2	Spin 3	Other	Spin 1	Spin 2	Spin 3	Other	
0–30 minutes		T						E	2
30 minutes to 1 hour								T	1
1–2 hours				C					1
2–3 hours									
3–4 hours						M			1
4–8 hours									
Frequency		1		1		1		2	5

where

    M = mechanical
    E = electrical
    C = coolant
    T = tools

The factors in each check sheet differ.

8.2 Types of Check Sheets

Three factor example
  Specific downtime cause
  Hour of day or duration of downtime
  Machine
Four factor example
  Duration of downtime
  Spindle
  Machine
  General downtime cause

The benefits of each approach would need to be evaluated for the problem of concern. For example, if spindle identification was not of great interest, the three factor example may be appropriate. Alternatively, the four factor example focuses on quantifying the amount, cause, and duration of downtime. Unfortunately, typical production logs of this information are of little use since the data are largely incomplete. A check sheet approach is preferable since a disciplined system for data collection is thus established.

A common problem with classification check sheets is that a complete evaluation of all traits is often not performed. For example, in the defective gear example, the presence of burr would likely be evaluated first, and if present, the remainder of the defect categories would not be evaluated. The operator would assume the gear is defective if any defect is found, so why not evaluate the easiest traits first? Unfortunately, this practice can lead to improper evaluation of which defects are the most common. When multiple defects are present, only the defect easiest to find is typically reported. Also, when several operators are involved, each may look for defects in a different order. An appropriate but often difficult to implement practice is to evaluate all defect categories. Thus it is possible for the total number of defects to be greater than the total number of parts inspected.

8.2.2 Location Check Sheet

A location check sheet identifies the physical location of a trait. For example, suppose the location of the burrs on a defective gear were of interest. A picture of the gear could be used to identify the burr locations.

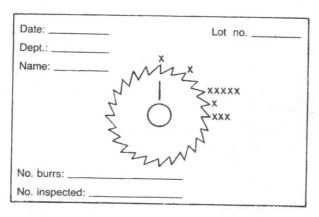

It is important to identify a part feature (such as the line in this check sheet) so that the orientation of the gear being evaluated is always the same. For symmetrical parts, the machining position can be indicated to obtain a consistent orientation. Note that it may be necessary to identify the location of the defect in a third dimension. For example, the location of a burr along a gear tooth may be of interest so another part view may be necessary.

A location check sheet is very useful to identify visual defects that are generally not measurable with a gage, for example,

Appearance rejects. Customers are intolerant of bubbles, scratches, and dents, for example, in finished products. Often the appearance characteristics of paint, windshields, and upholstery are routinely evaluated. The location of defects on the product is critical in resolving a problem.

Machining visual defects. Improper machine design or adjustments result in nicks, burrs, abrasions, tool marks, clamping marks, and other defects in machined components. Location of the problem on the part assists in resolution of the problem.

Material defects. The presence of porosity, flash, or metal defects is an indication of casting problems. The location of the problem is critical to revising metal flows or altering dies.

Leaks. The position of leaks of oil, air, or gas, for example, is important for problem identification.

A location check sheet is sometimes combined with a classification check sheet. Suppose we are evaluating the paint finish on a car hood. The type of defect as well as the location is of interest.

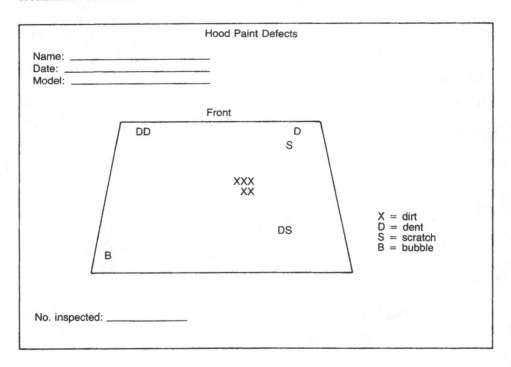

A location check sheet can also be used to collect variable measurements. The position in which a measurement is made can best be indicated using a picture. For example,

8.2 Types of Check Sheets

recording the torque readings of five plugs on a transmission can be performed using a location check sheet.

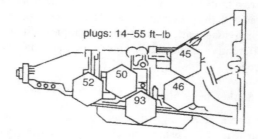

The check sheet enables clear identification of the bolt position so that data recording errors are minimized.

8.2.3 Frequency Check Sheet

A frequency check sheet is similar to a classification check sheet but focuses on a few select traits by quantifying the presence of absence of a trait or by counting the number of occurrences of a trait on an individual part. The following examples illustrate the two approaches.

Presence or absence
    One factor (X = sample of size 10):

Problem	Number	Total
Nicks	﹄﹄﹄ /	6
Number examined	XXXXXXXX	80

Two factors (X = 10 units):

Problem	Cuts	No cuts	Total
Nicks	﹄﹄﹄ ///	///	11
No nicks	﹄﹄﹄ ﹄﹄﹄ /	XXXXXXXX	91
Total	19	83	102

Number of occurrences:

Problem	0	1	2	3	4	5
Nicks	﹄﹄﹄ ﹄﹄﹄ ﹄﹄﹄ ﹄﹄﹄ //	//	/	///		
Cuts	﹄﹄﹄ ﹄﹄﹄ ﹄﹄﹄		﹄﹄﹄	﹄﹄﹄ /	/	///	/
		7	7	4	3	1

The first example is analogous to a classification check sheet with which nicks are examined. Focusing only on nicks helps to ensure a proper assessment is obtained. The second example evaluates two traits, nicks and cuts, and seeks to evaluate whether the presence of nicks is associated with the presence of cuts. The last example evaluates the number of nicks and cuts on an individual part. Each of these examples addresses the collection of information slightly differently. Again the work situation and problem being addressed determine the best approach.

8.2.4 Measurement Scale Check Sheet

The measurement scale check sheet simply divides the measurement scale into convenient intervals so that data can be recorded using a checkoff system. This check sheet results in a histogram that can be used in the same way as described in Chapter 5. An example of the basic format follows:

Frequency													
20							≢	=					
15							≢	=					
10						\	≢	≢					
5						≢	≢	≢	≢				
			=	⫤	≢	≢	≢	≢	≢	⫤	\		
Size	5	6	7	8	9	10	11	12	13	14	15	16	17
Total			2	3	5	11	20	17	10	3	1		

Recall from Procedure 5.2 and the associated work sheet that the process mean and standard deviation can be calculated from these frequencies. In this example, $\bar{X} = 11.2$ and $s = 1.6$. From Chapter 3,

$$\text{Upper process limit (UPL)} = \bar{X} + 3s$$
$$= 11.2 + (3 \times 1.6)$$
$$= 16.0$$
$$\text{Lower process limit (LPL)} = \bar{X} - 3s$$
$$= 11.2 - (3 \times 1.6)$$
$$= 6.4$$
$$\text{Process spread} = \text{UPL} - \text{LPL}$$
$$= 16.0 - 6.4$$
$$= 9.6$$

The process spread can be indicated on the check sheet. These calculations assume a stable process.

The measurement scale check sheet can be used for a variety of purposes:

8.2 Types of Check Sheets

Record normal operator measurement data.
Provide a training tool prior to SPC training for understanding process variability.
Assess the need for control charts.
Monitor a process once stability has been established using a control chart.
Monitor secondary characteristics.

It must be emphasized that this check sheet is not a substitute for a control chart. Recall from Chapter 5 that it is possible for an unstable process to generate a standard, bell-shaped histogram. However, a histogram can provide a great deal of valuable information about a process. Since the number of control charts in a process must be limited for logistical reasons, a measurement scale check sheet is a useful alternative for providing meaningful information in a simple, easy-to-obtain manner.

The exact format for a measurement check sheet can be varied. Often placing the histogram on its side makes completion of the form easier:

Measurement	Frequency (5, 10, 15, 20, 25)	Total
0		
1	/	1
2	///	3
3	⊪ //	7
4	⊪ ⊪ ⊪ //	17
5	⊪ ⊪ /	11
6	⊪ ⊪ ⊪ ⊪ /	21
7	⊪ ///	8
8	////	4
9	//	2
10	/	1
11	/	1
12		

(Process spread indicated on right side)

In some cases it is convenient not to use paper forms. Operators can simply mark a chalkboard or plexiglass board with a permanent measurement scale check sheet layout. The performance of the process is quickly evident to anyone who sees the board.

We see in Chapter 11 that one of the best methods for troubleshooting is to use the comparison principle, in which similar machines, fixtures, or spindles, for example, are compared. Using different colors or symbols, it is possible to use a measurement scale check sheet to visually compare different sources of variation. Figure 8.1 shows it is easy to visually detect some spindle differences when the check sheets are aligned vertically. It is

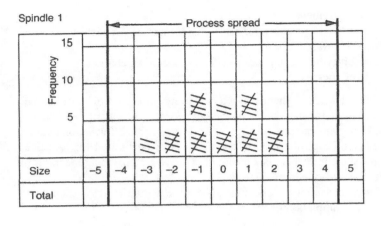
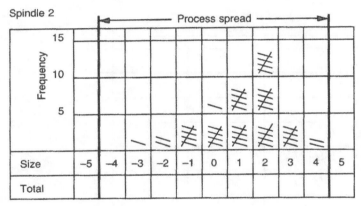
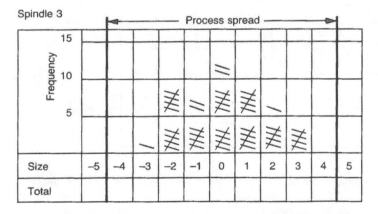

Figure 8.1 Measurement check sheet for a three spindle machine.

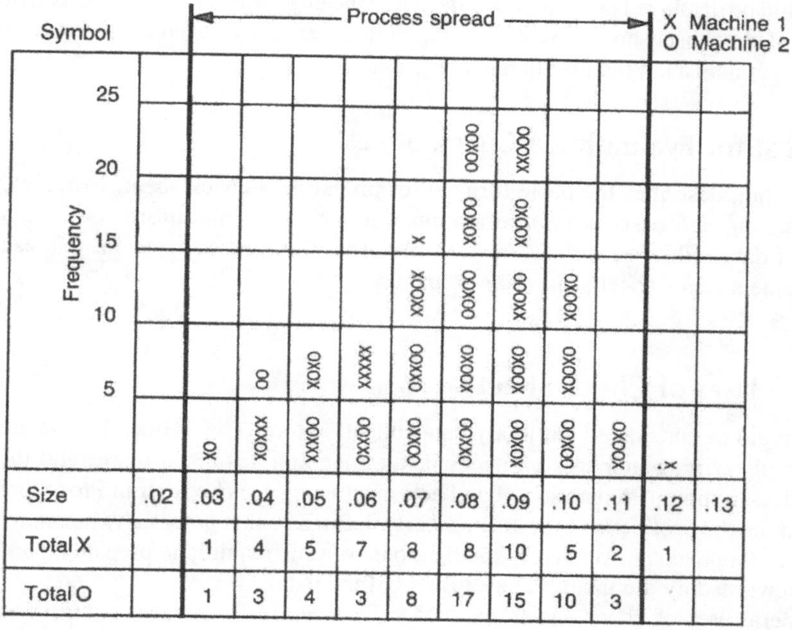

8.2 Types of Check Sheets

Figure 8.2 Combined measurement check sheet for two machines.

Symbol		Process spread											X Machine 1 / O Machine 2
Frequency 25													
20						OOXOO	XXOXO						
15					X	XOXOO	XOXOO						
10						XXOOX	OOXOO	XXOXO	XOOXO				
5			OO	XOXO	XXXXX	OXXOO	OXOOO	OXXOO	XOOXO				
		XO	XOXXX	XXXOO	OXXOO	OOOXO	OXXOO	OXOXO	OOXOO	XOOXO	.X		
Size		.02	.03	.04	.05	.06	.07	.08	.09	.10	.11	.12	.13
Total X			1	4	5	7	8	8	10	5	2	1	
Total O			1	3	4	3	8	17	15	10	3		

apparent that spindle 2 has shifted to the right in Figure 8.1. In Figure 8.2 the output of two machines is combined on a single check sheet. This arrangement of the data does not enable direct visual comparison, so it is less desirable than Figure 8.1.

8.2.5 Checklist Check Sheet

A common check sheet is a checklist of items to be performed to accomplish a task. These check sheets are used for

Machine operating procedures
Gaging instructions
Preventive maintenance items
Machine diagnosis and repair
Test procedures

Anything that can be performed in a repetitive manner can be put in a checklist format. A checklist provides a convenient format for writing an operating procedure.

A common mistake made in preparing most checklists is in not having potential users operationally test the listed instructions. The author of a checklist is often quite familiar with the correct procedure so that the checklist unintentionally omits minor details. A new checklist user is unfamiliar with the procedure and is likely not to understand all the details. A good practice is to have a trial run of a checklist with an individual unfamiliar with the process. This always results in an improved checklist.

Another common mistake is to place checklists in file drawers or notebooks rather than to display them prominently in an area where they will be used. For example, gaging instructions should be displayed prominently adjacent to the gage where measurements will be made. Finally, checklists should use diagrams, pictures, or other visual cues. Pictures are much easier to understand than written explanations.

8.3 Checklist for Evaluating a Check Sheet

The previous section describes the basic formats for preparing a check sheet. However, much of the heading and descriptive information was omitted. This information is an important part of data collection. Procedure 8.1 provides a checklist of questions that can be used to evaluate a check sheet prior to its actual use.

8.4 Primary Uses of Check Sheets

Check sheets should be considered a primary data collection tool. Too frequently data are collected by simply writing numbers on paper. Little planning has taken place, and the results are predictably poor. Preparation of a check sheet using the checklist in Procedure 8.1 requires that the data collection process be clearly defined so that potentially meaningful information is obtained. The relatively small amount of time required to prepare a good check sheet is rewarded by the increased usefulness of the data.

The two general uses of check sheets for data collection are for process control and problem analysis (Fig. 1.2). It is important to monitor the work process by collecting data that can be used to evaluate the process performance and develop improvement actions. Also, when problems arise, we have seen (see Sec. 8.6.1) that one of the primary initial objectives is to obtain meaningful data. The use of check sheets is thus a central part of building a DPS since the data become the basis for action by addressing the following considerations:

Process control. Every work process has a number of indicators on how the process is performing. Collection and later analysis of this information is an important part of monitoring a process. All types of check sheets can be used to collect information on key process indicators. Examples include defect identification or definition, machine downtime and its cause, and routine measurement checks. A check sheet cannot evaluate the stability of a process but can be used to monitor process characteristics as part of an overall operation control plan (Chap. 14).

Problem analysis. Troubleshooting to determine the root cause of a problem requires detailed information to clearly define the problem. Check sheets can be used to answer who, what, where, when, why, how, and how many. Typically, check sheets are used for a short period of time to collect detailed information. Also, check sheets are utilized to verify that a problem has been solved when a root cause has been eliminated.

A key element in both process control and problem analysis is that the data obtained using check sheets must be summarized in a convenient, often graphic form. Chapters 9 and 10 discuss several common approaches.

8.5 Summary

Check sheets can be used in any industrial setting to collect data for either process control or problem analysis. Some of the benefits of using check sheets are as follows:

8.2 Types of Check Sheets

Procedure 8.1 Checklist for Evaluating a Check Sheet

The following list of questions can be used to evaluate a proposed check sheet. All questions do not necessarily apply to every check sheet.

Check if true	No.	Question
☐	1.	Is an appropriate title given? Are department or operation names and numbers specified? Is the part name and number given?
☐	2.	Is the individual who completed the check sheet identified (who)? If more than one person completed the check sheet, is each person identified?
☐	3.	Is the appropriate method of collecting the data understood (what)? Are defects and gaging methods, for example, operationally defined? Are brief instructions for completing the check sheet written on the collection form?
☐	4.	Is the process location where parts are collected identified (where)? Is the sampling method defined and understood?
☐	5.	Is the time in which parts are sampled identified (when)?
☐	6.	Do individuals who are collecting the data understand why the data are being obtained (why)? Has their input for improving the check sheet been obtained?
☐	7.	Is the evaluation or gaging method understood (how)? Have gage studies been performed to evaluate the bias and repeatability of gages? Is a gage identification number required?
☐	8.	Are all stratification variables identified, such as machine, spindle, pallet, and fixture? Are such factors as vendor, test stand, furnace, die, tool, and material identified? Are any lot identification numbers recorded?
☐	9.	Will the total number of parts evaluated be recorded (how many)? Is the number of parts produced during the evaluation period indicated?
☐	10.	Do individuals understand how to record multiple defects? Will all defects on a part be evaluated? Do individuals understand how multiple occurrences of the same defect on a unit will be recorded (e.g., is 1 missing screw or 10 missing screws on a unit recorded as a single defect)?
☐	11.	Are data recorded on check sheets in a manner analogous to the gage measurement scales? Is there a minimal recording of numbers? Are there appropriate "Totals" columns for summarizing the data? Is there a space to indicate measurements beyond the scale on the check sheet (use "over" or "under" column)?
☐	12.	Is coded gage data used if appropriate? Is the relationship between the coded gage data and the actual measurement stated as Actual value = offset + scale × gage value
☐	13.	Has a trial run of the check sheet been attempted? Has a trial analysis of the data been attempted?
☐	14.	Is there sufficient space for the user of the check sheet to write in and make comments concerning unusual occurrences?

Preparation of a check sheet using Procedure 8.1 provides a disciplined approach to data collection that results in meaningful data analyses.

Many measurements are typically already being made in a manufacturing setting. Check sheets provide a mechanism for quickly recording this information.

A usable historical record of process performance is available from past check sheets.

Minimal training is required to use a check sheet. Measurement scale check sheets can be used prior to control chart training to make operators familiar with process variability and histograms using data derived from their own work setting.

Measurement scale check sheets provide a logical alternative to maintaining control charts for all characteristics. Critical characteristics or characteristics with low process capabilities should be monitored by control charts. However, other characteristics can be monitored using check sheets.

When problems are encountered in a process, the number of measurements can be easily increased on a measurement scale check sheet.

The use of check sheets is only a first step in monitoring a process or solving a problem. After an effective analysis of the data has been performed, it is imperative that appropriate corrective actions be taken. If the data collection is seen as not serving any useful purpose, the care and discipline required to collect meaningful data will soon be absent.

> Data collection does not solve a problem—
> analysis and action are needed

8.6 Case Studies

8.6.1 "Who Done It?"

A common manufacturing situation is to find a quality defect at the end of a production line or during the assembly of components and not know the cause of the problem. What usually transpires is a search for obvious causes with little data collection or evaluation using the graphic procedures discussed in this text. If this approach fails, meetings with different groups take place with the hope of utilizing the group's experience to resolve the problem. These meetings can become heated since frustration leads to such statements as, "If only you would do (or would have done) *xyz*, we would not have this problem!" Since this problem-solving approach is not based on meaningful data, emotional confrontation is not uncommon. The who done it? situation discussed here seeks to determine the root cause of a problem. Typically, elimination of the problem is then relatively easy. Chapter 13 develops a complete problem-solving approach based on data often collected from check sheets.

A problem occurred with the purchase of a new grinder for a transmission valve body. The process flow is shown in Figure 8.3. After the installation of the new machine, both the new and old machines were temporarily run. The old machine was scheduled for removal after the new machine proved capable in the production setting. However, after the new machine was installed, the incidence of defective valve bodies increased to the point at which 25% of production was being affected. The condition that resulted in the rejections was a no-cleanup (NCU) condition in the finished bored holes, where the finish reaming tool did not clean up the entire surface area of the bore. After color coding the parts from the new and old machines, it was discovered that the rejects were mostly from the new machine. This was a surprise to the manufacturing engineers since the capability of the new machine was appreciably better than that of the old machine.

8.6 Case Studies

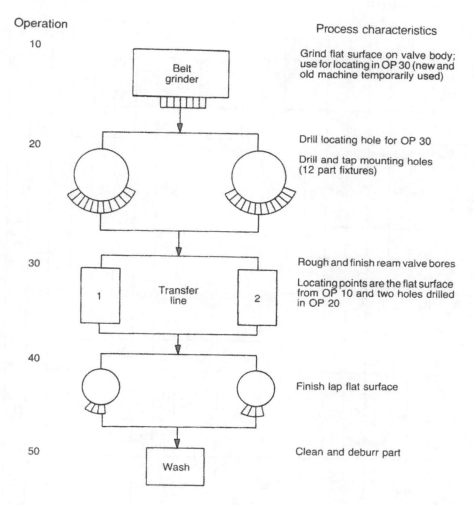

Figure 8.3 Process flow diagram for a transmission valve body.

The meeting that ensued between production personnel and the manufacturing engineers was emotional since there appeared to be equally clear positions on both sides. The production personnel were convinced that the new machine was causing the rejects. The engineers were equally sure that the new machine was superior to the old machine. After a period of discussion on various opinions on the source of the problem, an observer asked several questions involving the defects, including

Which bores had the defect?
What was the angular (i.e., clock) position in the bore of the defect?

A number of other questions were asked to characterize the defect (i.e., what, where, when, how, and how many). Although the high defect rate had been present for several days, the definition of the problem was still not clear.

It was decided that Check Sheet 8.1 would be placed in the department so that personnel

Inspector: Jones
Date: 10/19

Transfer line: 1
Shift: 2

Grinder N=new O=old	Die cavity	Ejector no. No. of dots	Pallet no.	Bore no.	NCU position	Bore no.	NCU position	Bore no.	NCU position	Bore no.	NCU position
N	4	1	42	47	⊖						
N	4	2	21	47	⊖	44					
N	5	2	7	47	⊖	44					
O	3	3	9	47	⊖						
N	2	3	17		○	44	⊖				
O	2	3	4		⊖	44	⊖	41	⊖		
N	2	1	10	47	⊖	44	⊖	51	⊖		
O	1	2	43	47	⊖		○	49	○		
N	5	2	22	47	⊖	44	⊖	53	⊖		
N	3	1	42	47	⊖		⊖	45	⊖		
O	4	1	7	47	⊖	44	⊖	46	⊖		
O	2	1	31	47	⊖		⊖	53	⊖		
N	1	3	35		⊖	44	⊖	49	⊖	46	
N	1	1	41	47	⊖	44	⊖		⊖	45	
N	2	2	8	47	⊖	44	⊖	53	⊖		
O	1	1	19	47	⊖		⊖		⊖		
N	5	1	25	47	⊖		○		○		
O	1	1	9	47	⊖	44	⊖		○		
O	3	1	44	47	⊖	44	⊖		○		
N	4	3	3		○	44	⊖		○		

Check Sheet 8.1 Valve body department final inspection: no cleanup.

8.6 Case Studies

inspecting the parts could help to fully define the problem. Unfortunately, only 2 of the 15 bores were routinely inspected (bores 47 and 44). The problem of selective inspection is common and makes troubleshooting more difficult. Inspecting all 15 bores on a part requires a significant effort, so once a part is determined to be bad, additional inspection is not routinely performed. This practice makes it impossible to determine whether some combination of bore rejects occurs at the same time (and thus indicating a common cause). However, even with the lack of complete inspection, the completed check sheets can be used to help start the problem-solving process.

A summary of the results for a 3 day period appears in Table 8.1. These data were summarized in graph form (Prob. 9.5), and the group met to evaluate the results. It was obvious to everyone that bore 47 had most rejects in the 9 and 12 o'clock positions. Similarly, for bores 44 and 49 the 9 and 12 o'clock positions were dominant. For bore 53 the 6 and 9 o'clock positions were dominant. The consistency of the position of the reject suggested to the group two possible causes for the rejects:

1. The rough and finish tools were not aligned.
2. The locating holes drilled in operation 20 were off location, causing misalignment in operation 30 stations.

The group was confused by the rejection of mostly new grinder parts. However, the check sheet data provided a basis for a rational discussion of what steps should be taken next. Three actions were decided upon to help identify the problem:

1. Alignment data were to be gathered from the machining stations for bores 47 and 44.
2. A control chart was started in operation 20 for the operation locating holes.
3. A measurement scale check sheet (rather than go/no-go data) was started for both the new and old grinders on a critical height dimension.

Table 8.1 Summary of the Valve Body No-Cleanup Defects Data

Day	Clock position	Bore 47	44	53	41	49	50	45	42	51	46	40
10/19	3	1	18			2			1			
	6			8								
	9	73	2	4	1		1	1			4	1
	12	4	12					1		1		
10/20	3	5					1					
	6						2	1				
	9	15	43			1					6	
	12	3	7			1						2
10/23	3											
	6											
	9	55	8			6						
	12	25	36			14						

The group assembled the results after a few days and determined:

The machining stations for bores 47 and 44 were in need of alignment.
No problem for locating holes existed in operation 20.
The old grinder was targeted to the high side of the specification limit, but the new grinder was targeted at the midpoint.

Thus, the group was able to solve the problem by aligning the appropriate stations and targeting the new grinder. It was determined that the difference in the height targeting on the grinders was sufficient that an operation 30 station with a marginal alignment condition could produce a NCU reject.

This example illustrates how data collection, summarization, and interpretation can be used to focus a group's effort to solve a problem. Also, several iterations are generally required before the root cause of a problem can be defined. A disciplined problem-solving approach is discussed in Chapter 13.

It should be noted that a better check sheet could have been developed. The problems with Check Sheet 8.1 include the following:

Multiple defective bores were not necessarily indicated.
There was no coding of the operation 20 and 40 machines.
There was no clear definition of "no cleanup."
The total number of new and old grinder parts was not recorded.
The form does not have brief instructions for completion.

The design of the check sheet should consider the process flow diagram in Figure 8.3, where upstream operations can be related to the operation being studied. For example, for the period of study, color coding of parts for drills 1 and 2 in operation 20 and lappers 1 and 2 in operation 40 would have enhanced the information.

8.6.2 Selected Examples

The following examples illustrate several types of check sheets. Most of the examples address some of the checklist questions in Procedure 8.1 but could be improved by addressing all the questions more thoroughly.

Assembly Repair. Use of a check sheet for repair operations is a good practice since the objective of a DPS is to eliminate all repair operations. There are two reasons this objective should be pursued. First, all units in need of repair cannot always be detected and some defective products are thus delivered to customers. Elimination of the cause of a repair reduces the risk that customers receive a defective product. Second, repaired units typically have a higher rate of defects than units produced through the normal production process. Often the cause of the repair is eliminated but the repair process produces another defect. Reducing the need for repairs reduces the risk of a faulty repair. Addressing problems identified by repair check sheets is an effective way to improve quality.

Check Sheet 8.2 is used for a car air-conditioning evaporator repair process. Notice that completion of the sheet by the repairman is quite easy. For most repairs only two / marks are required: one / to indicate the "units repaired" so that the frequency of repairs can be determined, and another / required to indicate the defect that was repaired. If there were several defects, a / would be required for each defect. A number of other features are illustrated in Check Sheet 8.2, but the most critical feature is ease of completion. The data

8.6 Case Studies

Name: _____
Shift: _____
Date: _____

Dept.: _____
Supervisor: _____
Units assembled: _____

Units repaired	Hour 1	Hour 2	Hour 3	Hour 4	Hour 5	Hour 6	Hour 7	Hour 8	Total
Total	卌 卌 卌	卌 卌 卌	卌 卌 ///	////	卌 卌 卌 //	卌 卌 /	卌 卌 /	卌 卌 卌 ////	105
	15	15	13	4	17	11	11	19	
Defects									
Def. housings	////	/	卌 卌	////	///	////	///		29
Foreign material									
Def. welding									
Missing screws	卌 卌 卌 ////	卌 ///	卌 卌 //	///	卌 卌	卌 卌	卌 卌 卌 /	卌 卌 卌 ///	96
Missing doors		/							2
Cams loose/missing									
Missing spring	//		/						2
Cam screws loose/missing	/	//		/	/				12
Missing/misal. gasket						////	//	//	
Missing door arm	/		/		///	///	///	/	13
Missing/inc. studs	/	/	/			//	/		2
Door shaft arm not connected	////	/	卌 ////	/	//		/	//	6
Missing rectrc duct	卌 ///	卌	///		/				17
Scrapped entire unit									18
Motor wire not connected									

Procedure:
1. Record / for every unit completely repaired each hour in top area of check sheet.
2. Record / for the cause of every repair. If multiple causes exist, record / for each cause.
3. Record the total number of assembled units during the shift, including any repaired assemblies.
4. Turn in sheet to supervisor at the end of the shift. Record any unusual occurrences in Comments section below.

Comments: _____

Check Sheet 8.2 Assembly repair area: A/C evaporator.

presented in Check Sheets 8.2 are analyzed in Chapter 9, but there are several obvious conclusions. Missing screws are an important problem and very likely not totally detected by the inspection process. Thus, evaporators with missing screws are likely a customer use problem. In many cases it is useful to develop a special check sheet to obtain more detailed information. It is apparent that each missing screw was recorded with a /. It would be useful to develop a check sheet to determine the number and location of missing screws on rejected units. The current approach of recording each missing screw does not make it possible to use a P chart of the percentage of units with missing screws.

Functional Test Results. Functional testing of products is often performed in manufacturing. The testing may range from automated testing of 100% of the products to manual selective testing. In any case it is often useful to utilize control charts for test parameters. The intent of the control chart is not to control the process since no process adjustments are possible at the test station. Rather the intent of the control chart is to

Monitor the product the customer is receiving
Evaluate the stability and capability of the production process
Provide a benchmark for evaluating process improvements
Evaluate long-term process trends

In most cases test equipment is used in a go/no-go manner even though variable measurements are often available. Since immediate process adjustments are not of the intent, a check sheet can be used to collect data that can be used for later construction of control charts. This approach is particularly useful for automated test equipment for which dedicated personnel are not available. Appendix I discusses a control chart procedure that can be used to evaluate the measurement process in this situation.

Check Sheet 8.3 is used to evaluate the leak rate (CFM) of a car air-conditioner evaporator and blower assembly. Completion of the check sheet requires minimal effort. One / is recorded for every part measured. The sheet is organized so the "Total" column provides a convenient histogramlike summary of the day's results.

Measurement Scale Check Sheet Examples. Check Sheets 8.4 through 8.7 are from actual manufacturing operations in which variable measurements were recorded. These check sheets illustrate a variety of techniques for quickly collecting data. Most of the check sheets could be improved by addressing the questions in Procedure 8.1 more thoroughly and by not assuming that the operator understands how to complete the information.

8.6 Case Studies

Inspector: _____
Shift: _____
Date: _____

Part name: A/C evaporator and blower assembly
Part no.: 1234

Spec.: 5.0 max.

CFM	GAGE VALUE	Hour 1	Hour 2	Hour 3	Hour 4	Hour 5	Hour 6	Hour 7	Hour 8	Total
1.3	.5									
1.6	.6									
1.9	.7									
2.2	.8									
2.4	.9									
2.7	1.0									
3.0	1.1									
3.3	1.2									
3.6	1.3									
3.8	1.4									
4.1	1.5									
4.4	1.6									
4.6	1.7									
4.9	1.8									
5.2	1.9									
5.5	2.0									
5.8	2.1									
6.0	2.2									
6.3	2.3									
6.6	2.4									
6.8	2.5									
OVER 2.5 WRITE IN										

Procedure:

1. Record results from five consecutive assemblies from production line every one-half hour.
2. Measure each assembly on leak rate test gage (#12-7G4) using standard procedure (ZP-176). See display board for brief instructions.
3. Turn in completed check sheet to supervisor at the end of the shift.
4. Record any unusual occurrences in the Comments section below.

Comments: _____
_____
_____

Check Sheet 8.3 Leak rate.

340   8   Check Sheets

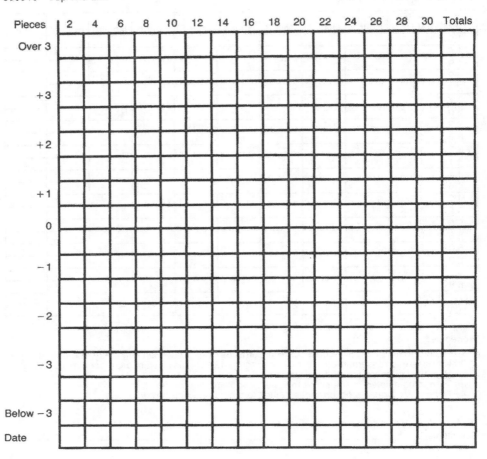

Machine: Circle one
660915   Top Mid Bot
660916   Top Mid Bot

Date:
Gage: 60169
Spec.: 4.0079/4.0082 ($\pm$ 1.5)

Procedure:
1. Record normal operator inspection data.
2. Check 30 pieces between tool changes and plot in squares (two plots per square).
3. Calculate actual dimensions: actual dimensions = 4.0080 + (.0001 x gage value).
4. If reading is over 3 or below -3, write in actual reading.

Comments: _____

Check Sheet 8.4   Gear department, operation 40, length.

8.6 Case Studies

Machine: XYZ drills
Part no.: 1234
Gage no.: 1382

East ☐   West ☐
661035   661579
Spec. 0/.014 TIR

Fixture no.	Hole 1	Hole 2	Comments
1			
2			
3			
4			
5			
6			
7			
8			
9			
10			
11			
12			

Procedure:

1. On Wednesday of each week supervisor or available qualified operator is to work with the drill operator to obtain TIR gage readings off east and west drills for hole locations 1 and 2.
2. Five (5) parts from each of the twelve (12) fixtures are to be gaged and gage readings recorded on form. Use coded data: actual TIR = .001 x gage value.
3. In the event fixture is off or not working, comment is to be noted on form.
4. Care is to be taken to obtain fixture numbers for selected part.
5. Once readings are recorded, supervisor is to notify MPS to pick up information for processing.

Additional notes: _____
_____
_____

Check Sheet 8.5   Valve body department, operation 20, locating hole.

Spec. 1.5015–1.5035

	Machine	661093	661094	661095	661096	Total	Histogram
a.m. Check	1.4995						
	1.5000						
	1.5005						
	1.5010						
	1.5015						
	1.5020						
	1.5025						
	1.5030						
	1.5035						
	1.5040						

Comments: _____

	Machine	661093	661094	661095	661096	Total	Histogram
p.m. Check	1.4995						
	1.5000						
	1.5005						
	1.5010						
	1.5015						
	1.5020						
	1.5025						
	1.5030						
	1.5035						
	1.5040						

Comments: _____

Instructions:
 Sample size: Five pieces per machine
 Frequency: Twice per shift, hour 1 and hour 5
 Method: Use slash marks for readings under each machine number; record total in total column; darken accumulated totals in histogram grid

Operator: _____ Shift: _____ Date: _____

Check Sheet 8.6 Forward sun gear department, operation 80, hob quality.

8.6 Case Studies

Machine: XYZ transfer
Circle one: Hole   29   32   33   34
Circle one: Machine 660944 (south)   661484 (north)
Circle one: Shift   1   2   3
Date:
Spec.: .2765/.2785

Value		Hour 1	2	3	4	5	6	7	8	Totals
Over .2788										
.2788										
.2787										
.2786										
.2785	USL 1									
.2784										
.2783										
.2782										
.2781										
.2780										
.2779										
.2778										
.2777										
.2776										
.2775	0									
.2774		✓	✓							2
.2773		✓	✓	✓	✓	✓✓	✓	✓✓		9
.2772				✓	✓		✓		✓✓	5
.2771										
.2770										
.2769										
.2768										
.2767										
.2766										
.2765	LSL-1									
.2764										
.2763										
.2762										
.2761										
Under .2761										

Procedures:
1. Check two pieces per hour for 8 hr.
2. Place check mark in the appropriate space, which matches gage reading.

Comments: _____

_____

_____

_____

Check Sheet 8.7 Pump department, operation 20A, hole size.

9
Pareto Diagrams

9.1 General Concepts

The Pareto principle states that it is possible for many performance measures, such as scrap, machine failures, vendor problems, inventory costs, and product development time, to separate the "*vital few*" causes resulting in unacceptable performance from the "*trivial many*" causes. Historically, this concept has also become known as the 80/20 rule, which states that a performance measure can be improved 80% by eliminating only 20% of the causes of unacceptable performance.

This rule has been applied to a wide range of performance measures:

Customer complaints
Warranty repairs or cost
Quality defects
Rework
Machine downtime
Crisis maintenance activity
Material utilization
Time utilization (machines and personnel)
Injury types and causes
Energy use
Inventory costs
Clerical errors
Product development time
Sales

Performance indicators on these categories are maintained to identify problems. These problems can most easily be addressed using the Pareto principle. Consider how performance improvement efforts might address the following areas:

80% of the customer complaints result from problems with 20% of the components of a product
80% of the quality defects result from 20% of the processing stages
80% of the machine productivity losses result from 20% of the operations
80% of the vendor quality defects result from 20% of the vendors
80% of injuries result from 20% of the operations

9.2 How to Construct a Pareto Diagram

80% of the inventory cost results from 20% of the parts
80% of the product development time results from 20% of the stages
80% of sales results from 20% of the customers

Although the percentages 80% and 20% are not, of course, exactly correct, these statements initiate a thought process that can lead to meaningful improvement. Focusing effort on the vital few items results in the maximum improvement in performance.

The concept of the vital few and trivial many was first stated by Juran (see Juran et al., 1979), who named this the "Pareto principle." Vilfredo Pareto (1843–1923) was an Italian sociologist and economist who studied the unequal distribution of wealth, noting that most of the wealth resided in the hands of a few individuals. Juran was the first to note that the causes of problems had an unequal distribution and that performance could most easily be improved by focusing on the vital few causes.

9.2 How to Construct a Pareto Diagram

In earlier chapters we repeatedly see that it is useful to display data in a graphic form. Identification of the vital few items from the Pareto principle is most easily conveyed using a Pareto diagram. Consider the following defects for a machined part:

Defect	Quantity
Undersized A hole	224
Porosity	52
Nicks	149
Concentricity diameter $A-F$	46
Parallelism surface $Z-W$	58
Oversized A hole	5
Other	23

Number of parts with 1 or more defect	$M = 557$
Number inspected	$N = 2217$

It is apparent from this short list that undersized A holes are the main problem. However, real applications typically have many defect categories and many parts, all of which are monitored over time. It is convenient to represent these data graphically as in Figure 9.1. This graph can be prepared using the work sheet in Table 9.1. The defects are arranged in rank order in column 1. The number of defects appears in column 2. The percentage that each defect represents of the total number of defects appears in column 3. The cumulative percentages of column 3 appear in column 4. The percentage that each defect represents of the total number of parts inspected appears in column 5. The Pareto diagram bars (part A) are plotted using column 2 and the cumulative percentage (part B) uses column 4.

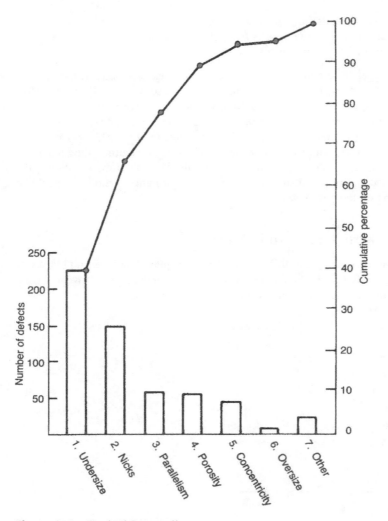

Figure 9.1 Typical Pareto diagram.

One difficulty in collecting data by such categories as undersize, nicks, and oversize is that a particular part or item being evaluated may fit into several categories. In this case the preferred approach is to mark each defect. This procedure was discussed in Chapter 8. However, in order for the total percentage of defective parts to be computed, the number of parts with one or more defects (denoted M) must be recorded. For the preceding data, $M = 557$ is equal to the sum of the individual defect categories so there are no parts with multiple defects. If $M = 493$ then $N = 2217$ parts were evaluated, 493 with one or more defects. In Table 9.1 the lower right-hand corner then becomes

$$\frac{M}{N} \times 100 = \frac{493}{2217} \times 100 = 22.2\%$$

instead of 25.1%.

9.2 How to Construct a Pareto Diagram

Table 9.1 Example Pareto Analysis Work Sheet

Column 1 Defect	2 Number of defects	3 % Composition	4 Cumulative %	5 % of Total inspected
1. Undersize	224	224/557 = 40	40	224/2217 = 10.1
2. Nicks	149	149/557 = 27	40 + 27 = 67	149/2217 = 6.7
3. Parallelism	58	58/557 = 11	67 + 11 = 78	58/2217 = 2.6
4. Porosity	52	52/557 = 9	78 + 9 = 87	52/2217 = 2.3
5. Concentricity	46	46/557 = 8	87 + 8 = 95	46/2217 = 2.1
6. Oversize	5	5/557 = 1	95 + 1 = 96	5/2217 = 0.2
7. Other	23	23/557 = 4	96 + 4 = 100	23/2217 = 1.0
Total	557	557/557 = 100	100	557/2217 = 25.1
Pareto diagram	Part A	Part A	Part B	Part A

9.2.1 Stacked Bars for Cumulative Percentages

Another method to display the cumulative percentage of the categories (column 4 in Table 9.1) is to use a stacked bar graph. This graph simply stacks the percentages as follows:

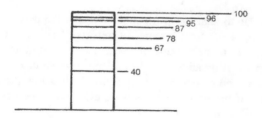

Typically, the percentage scale on the right-hand side of a standard Pareto diagram significantly increases the size of the plot. Thus, the cumulative line is sometimes not drawn. The stacked bar plot enables an arbitrary scaling to fit the size of the Pareto bars. It is convenient to use different patterns for shading each of the Pareto bars. These same patterns can be used in the stacked bar to enable easy identification of the categories, as shown in Figure 9.2.

9.2.2 Use of Percentages

In many cases it is necessary to use percentages on the left-hand scale rather than actual counts. For any analysis in which two or more Pareto diagrams are compared, it is not meaningful to compare counts if the total number of items inspected is different. Converting the counts to percentages makes direct comparison possible. Some typical cases in which the number of items inspected varies and percentages are commonly used include

Comparisons of different groups or strata
Comparisons over time
Comparisons before and after a change

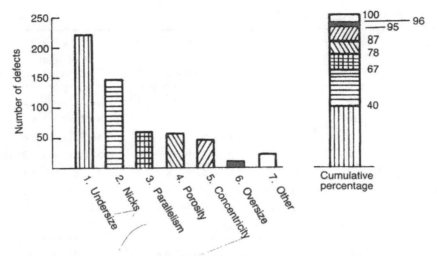

Figure 9.2 Pareto diagram with stacked bars.

Also, a percentage scale is used when attention is to be drawn to very large or small percentages. Two types of percentages can be used: percentage composition or percentage of total items inspected. Each is useful for different applications. For an individual analysis, the same ranking of problems is obtained using any of the three measurement scales shown in Figure 9.3.

Column 3 of the Pareto analysis work sheet contains the percentage composition that each defect represents of the total number of defects. These percentages are useful for evaluating the differences in the ranking of a defect among all other defects. For example, suppose an improvement effort addressed the undersize defect in Table 9.1. The percentage composition for undersized defects should decrease well below the original 40% for any time period evaluated if the improvements were effective. Another application involves

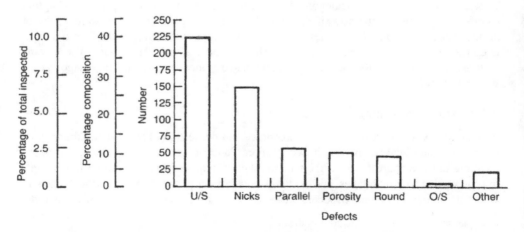

Figure 9.3 Same Pareto bars produced using three different measurement scales.

9.3 Directing Improvement Efforts

comparison of the degree of contribution to a problem by components of a stratification factor, as shown later in Figure 10.8b. Since the sum of the percentages for all defects is 100%, it is not meaningful to use percentage composition if the magnitude of defects is important. A pie chart, discussed in the next chapter, is another method of plotting this type of data.

Column 5 of the Pareto analysis work sheet contains the percentage that each defect represents of the total number of parts inspected. These percentages are often the most useful way to present Pareto analysis results. Many types of comparisons can be made using this measurement scale. P charts for individual defects can be used to provide a definitive evaluation of changes. Procedure 9.1 gives the computational procedures for different Pareto diagrams.

9.3 Directing Improvement Efforts

The use of Pareto diagrams maximizes the impact of improvement efforts by concentrating on the "vital few" most significant problems identified by the highest Pareto bars. Reducing the large bars by one-half yields a much greater improvement than reducing the small bars by one-half. Comparison of the two shaded areas in Figure 9.4 indicates 75 defects for categories 1 and 2 versus 19 in categories 3–6. Thus, making improvements that reduce categories 1 and 2 by one-half has a 400% greater impact than if categories 3–6 were addressed. Also, a larger number of separate actions is typically required to obtain improvement in the "trivial many" categories. Too often, teams do not examine the "big picture" before deciding on what problems to address.

> Pareto diagrams help focus improvement efforts
> where actions will have the largest impact

Part B of the Pareto diagram helps to determine a logical "break point" in the cumulative percentage curve. The examples in Figure 9.5 indicate both a good and a poor

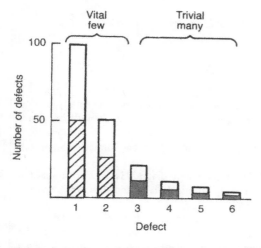

Figure 9.4 Example of different impacts of improvement efforts.

Procedure 9.1 Pareto Diagram Construction

A Pareto diagram consists of the Pareto bars (part A), which may use any of three scales on the left-side of the diagram, and the cumulative percentage (part B), which may be expressed as a line plot or stacked bar graph. The following steps are used to construct the diagram:

1. List the potential categories that can be used to meaningfully subdivide the performance indicator being considered. Prepare a check sheet using Procedure 8.1 to collect the frequency at which each category occurs. If an "other" category is used on the check sheet, each occurrence of this category should be identified completely.
2. Operationally define all categories so that all items can be classified correctly. A group or team approach should be used for steps 1 and 2.
3. Assemble the counts in which each category occurred. List the categories in the following work sheet format with the category with the highest count first, second highest next, and so on.

Column 1	2	3	4	5
Category	Count	% Composition	Cumulative %	% of Total inspected
1. XXX	C_1	$C_1/C \times 100 = P_1$	P_1	$C_1/N \times 100$
2. XXX	C_2	$C_2/C \times 100 = P_2$	$P_1 + P_2$	$C_2/N \times 100$
3. XXX	C_3	$C_3/C \times 100 = P_3$	$P_1 + P_2 + P_3$	$C_3/N \times 100$
.	.	.	.	.
.	.	.	.	.
.	.	.	.	.
Other				
Totals	Total of counts = C	100%	100%	$M/N \times 100$

M = total number of parts or items inspected with one or more defects
N = total number of parts or items inspected

4. Part A, Pareto bars: Select the most appropriate scale to plot the Pareto bars:

   Counts = column 2
   % composition = column 3
   % of total inspected = column 5

   Prepare a bar graph using the selected scale. The "other" category is generally plotted farthest to the right. The use of 6–10 bars is generally sufficient to identify important problems.

5. Part B, cumulative percentages: Select the line plot or stacked bar format to graph the cumulative percentages and plot the data in column 4.
   Line plot: The scale on the right-hand side of the Pareto diagram is selected so that the distance from 0 to the top of the first bar is P_1. The scale is extended to 100% and the values in column 4 plotted for each category.
   Stacked bar: Select a convenient scale so that the distance from 0 to 100% is about the same height as the bar for the first category. The height of individual stacked categories can be obtained from column 4. Write in the cumulative percentages. It is convenient to shade the categories with patterns corresponding to part A.

9.3 Directing Improvement Efforts

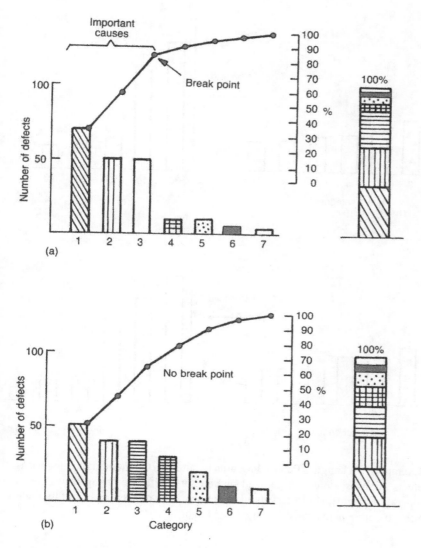

Figure 9.5 A good (a) and a poor (b) Pareto analysis.

Pareto diagram. In a good Pareto diagram, there is a break in the line plot to indicate the number of categories that should be considered. From the 80/20 rule guide, a break around the 80% level is often encountered. In a poor Pareto diagram, there is no apparent break and selection of the number of problem cause categories is somewhat arbitrary. A line plot or stacked bar can be used to select a logical break point.

How can a poor Pareto diagram arise? The most common reason is poor selection of the classification categories. For example, suppose a poor Pareto diagram is obtained for a department's quality defects. This can occur when many individual defect categories are selected. Combining categories in a logical manner or analyzing the operations generating the defects can lead to a much more meaningful analysis, as Figure 9.6a illustrates.

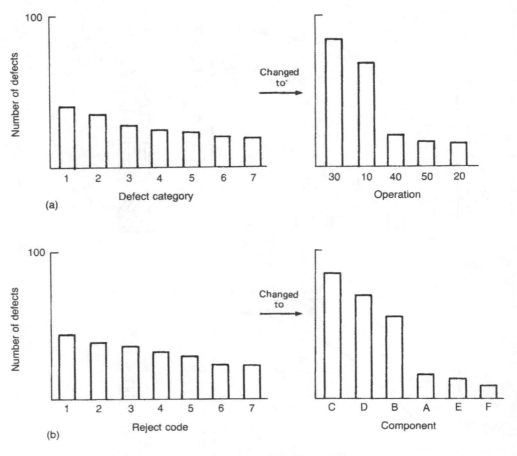

Figure 9.6 Examples of changing Pareto diagrams by altering the method of categorization: (a) in defects analysis, defects organized by category (left) are changed to defects organized by operation (right); (b) in test stand reject analysis, rejects organized by code (left) are changed to components causing the rejects (right).

Another example involves forming a Pareto diagram from reject codes from a test stand. A more useful Pareto diagram may be obtained by considering categories of components causing the reject, as shown in Figure 9.6b.

These examples illustrate the usefulness of a break point. In the first example, improvement efforts may more usefully be focused on individual operations than on individual defect categories. In practice, many defect categories are interrelated. In the second example, improvement efforts should address individual components rather than reject codes. The use of a break point helps to evaluate whether the performance indicator is subdivided into too many categories (or too few) so improvement efforts cannot be meaningfully focused. The selection of different, more meaningful, categories often produces a useful Pareto diagram that effectively targets improvement efforts.

Another problem arises when the "other" category on the right-hand side of the Pareto

9.3 Directing Improvement Efforts

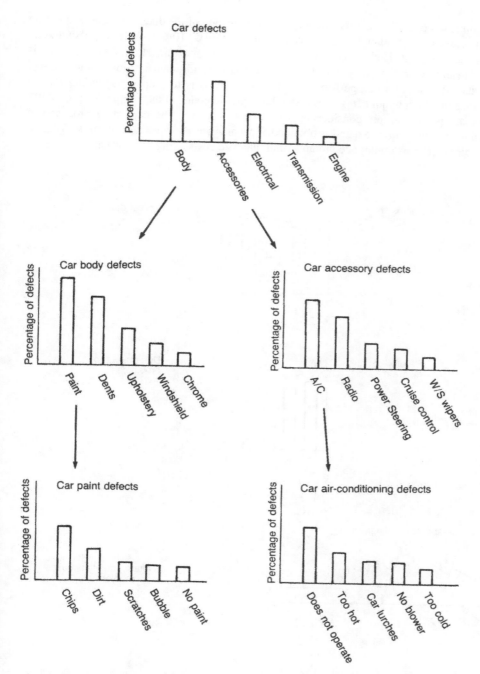

Figure 9.7 Car defects as an example of an exploding Pareto diagram.

diagram is too big. Placing this cateogry on the right side of the diagram focuses attention on this case. If the "other" category is more than about 5–10% of the total, it should be subdivided. A good check sheet practice is to write down an explanation of any item that is placed in an "other" category. It is then possible to form additional categories if necessary.

A useful practice when performing a Pareto analysis is to prepare *exploding Pareto diagrams* in which major categories at one level are exploded at the next level. Figure 9.7 shows a typical example considering the defects in a car. The exploding process can continue until the required detail is obtained. This analysis provides an overall perspective of problems. For example, accountants often analyze costs using this format.

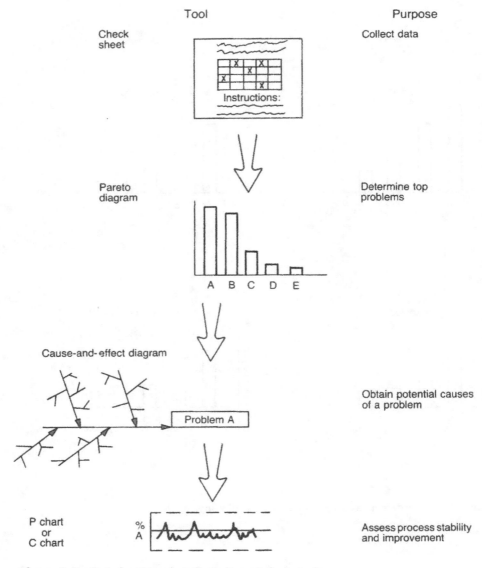

Figure 9.8 Relationship of attribute data analysis tools.

9.4 Pareto Comparisons

Pareto analyses play a pivotal role in the overall improvement process. Typically, data collection for a performance indicator starts with a classification or location check sheet. The check sheet results are then analyzed using a Pareto diagram. The vital few problems are then identified, and a cause-and-effect diagram is prepared to start the problem analysis (Chap. 13). Frequently, P charts or C charts are prepared on recurring problems. These charts provide an ongoing evaluation of progress in solving a problem and the stability of the process generating the problem. Figure 9.8 shows the relationship of the attribute data analysis tools used to improve a process.

Employee improvement teams were used in a plant for several years without much success. Meetings tended to focus on environmental or personality issues and frequently demonstrated varying degrees of hostility. The system was changed so that teams addressed only the top-quality problems in an area using data they collected. Pareto diagrams were used in the meetings and displayed prominently in work areas. Meaningful problems began to be solved. The graphic communication of problems made employees focus on what they personally might do to help eliminate a problem. The tone of the meetings changed. Quality performance steadily improved.

9.4 Pareto Comparisons

It is possible to use Pareto diagrams for different comparisons. In most cases it is necessary to use percentages on the left-hand scale since the number of items inspected N frequently differs when comparisons are made. An easy mistake to make in comparing the two diagrams in Figure 9.9 is to conclude that area A (left) is worse than area B (right). Without knowing the number of items evaluated for defects, no comparison is meaningful. If a significantly higher number of items were evaluated for area A, it would be performing better than area B. The use of percentages makes direct comparisons possible. Chapter 11 discusses the rationale for making comparisons to generate improvement. Three types of comparisons are discussed here.

9.4.1 Stratification

Comparison of different groups, units, or other types of strata often can lead to suggesting one of two types of improvement strategies. For example, suppose defects in an operation

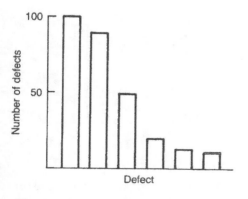
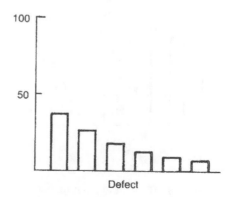

Figure 9.9 Pareto comparison of two areas.

are being studied. If there are two or more machines in the operation, performing a Pareto analysis for each machine can uncover whether a common or special cause problem exists. Figure 9.10 shows Pareto diagram patterns that indicate common and special cause problems. A common cause problem is indicated by about the same percentage of an individual defect for each machine. Improvement of the process must address a system problem common to both machines. A special cause problem is indicated by a significantly different percentage of an individual defect for a machine. In Figure 9.10, defects B on machine 1 and D on machine 2 appear due to a special cause problem. The logic is simple. For defect B, if machine 2 has a low level of the defect then why can machine 1 not attain the same level? It is convenient to use the same ordering of defects to make visual comparison easy.

An engineer turned on the water at the laboratory sink and noticed that the water pressure was much lower than usual. He then went to the bathroom sink to test the water pressure.

A danger in using Pareto comparisons is misinterpretation of differences that are really due to random variability. A precise approach would be to maintain a P chart for each machine and evaluate stability and machine differences for a common defect category. Determining common and special cause problems can then be accomplished by comparing the charts. This detail is difficult to maintain if there are a large number of machines and/or

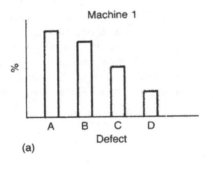
(a)

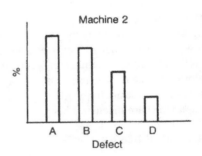

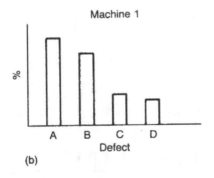
(b)

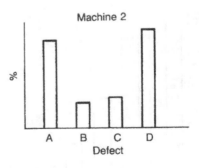

Figure 9.10 Pareto diagram patterns indicating (a) common cause problems $A–D$ and (b) special cause problems B and D.

9.4 Pareto Comparisons

defect categories. In practice, special cause problems often result in very large differences in the percentage of defects, and Pareto comparisons can be used to highlight problem areas.

Stratification Pareto comparisons can be made by considering any logical subdivision of the data. For example, the strata may be different:

Operators
Spindles, stations, or fixtures, for example, on a machine
Machines in an operation
Departments
Processes

This approach not only produces a priority ordering of the problems from the Pareto analysis, but identifies an improvement strategy that focuses on a common cause, system problem or a special cause, unique problem.

9.4.2 Time

Control charts measure the performance of a process over time. Similarly, it is of interest to perform Pareto analyses over time. Often this approach is used to monitor performance indicators. For example, 2 weeks can be compared, as shown in Figure 9.11a. These charts

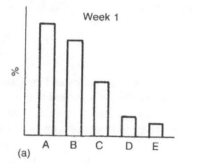
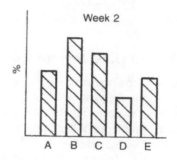

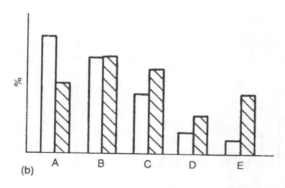

Figure 9.11 Pareto analysis over time: (a) separate weekly Pareto analyses; (b) combined weekly Pareto analyses; open bars, week 1; hatched bars, week 2.

can be more easily compared if the plots are combined in a multiple-bar Pareto diagram, as shown in Figure 9.11b. Frequently, several weeks are involved and the week 1 open bar is replaced by quarterly, monthly, or moving averages.

An interesting feature of Pareto comparisons over time is that the order of the categories should not change a great deal if the process is stable. Conversely, if the order does change dramatically, special causes are acting on the process that should be eliminated. If the highest defect changes from week to week, as in Figure 9.11b, there is insufficient control of the process.

Again, caution must be exercised in evaluating changes. Just as a P chart can be expected to vary about a mean, so do Pareto bars vary over time. Attempting to over-interpret changes in performance indicators can be misleading. If small changes are important, a P chart should be maintained. The purpose of the time Pareto comparison is to graphically display changes for a number of categories.

9.4.3 Before and After

A common application of Pareto comparisons is to assess the effectiveness of a process improvement by preparing a Pareto diagram before and after an improvement action is made. If the category of interest has decreased appreciably, the improvement is judged effective, as shown in Figure 9.12. This approach is particularly useful for employee meetings. A formal approach is to prepare a P chart for the defects of interest. If a number of defects are being addressed by a general improvement action (e.g., training or PM program), the before and after Pareto diagrams are simpler.

9.5 Pareto Diagrams for Cost

It is beneficial to associate costs with different performance indicator categories and then to prepare a Pareto diagram based on cost. It is not unusual for lower frequency items to have larger associated costs. Forming the Pareto analysis based on cost can produce a different "vital few" problems. When assigning costs, it is important to include all costs known to be associated with a category.

For a quality defect, costs should include scrap, labor rework, warranty, and so on.

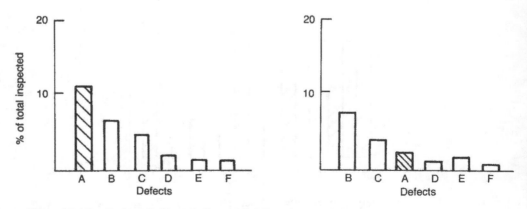

Figure 9.12 Before (left) and after (right) Pareto comparisons.

9.5 Pareto Diagrams for Cost

Consider the categories used in the first part of this chapter with the following costs per unit:

Undersize	$.25
Nicks	$.10
Parallelism	$1.25
Porosity	$.75
Concentricity	$1.25
Oversize	$1.25

A Pareto analysis for costs can be prepared by preparing cost work sheets, which obtain the cost of a defect in Tables 9.2 and 9.3 from

Cost of defect = number of defects × cost per unit for defect

Table 9.2 Pareto Analysis Cost Work Sheets: Cost per Defect Work Sheet

Defect	Number of defects	Cost per unit ($)	Cost of defect ($)	Rank
Undersize	224	.25	56.00	3
Nicks	149	.10	14.90	5
Parallelism	58	1.25	72.50	1
Porosity	52	.75	39.00	4
Concentricity	46	1.25	57.50	2
Oversize	5	1.25	6.25	6
Total	534		246.15	

Table 9.3 Pareto Analysis Cost Work Sheet

Defect	Cost of defect ($)	% Composition	Cumulative %	
Parallelism	72.50	72.50/246.15 = 29		29
Concentricity	57.50	57.50/246.15 = 23	23 + 29 =	52
Undersize	56.00	56.00/246.15 = 23	52 + 23 =	75
Porosity	39.00	39.00/246.15 = 16	75 + 16 =	91
Nicks	14.90	14.90/246.15 = 6	91 + 6 =	97
Oversize	6.25	6.25/246.15 = 3	97 + 3 =	100

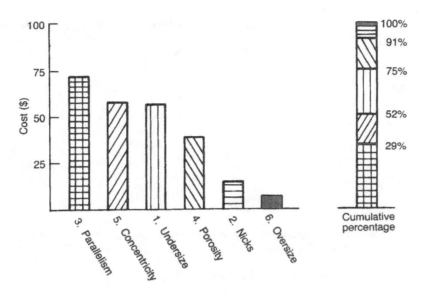

Figure 9.13 Pareto analysis for defect cost.

The costs can then be ranked and the cumulative percentages calculated as shown in Table 9.3. The Pareto diagram for cost appears in Figure 9.13. The difference ordering from that in Figure 9.2 is apparent.

How should the vital few problems be selected? Which performance indicator should be used, the number of defects or the total defect costs? There is no clear answer, but individuals working on improvements should be aware of any differences in the two indicators. Ideally, the indicator should be related to customer satisfaction, but this is often difficult to quantify. High-cost item result in dissatisfied customers, but a low-cost high-frequency item can result in a poor-quality image. In any case, identifying the vital few problems should consider the impact on the customer of the problems.

Costs should be considered in identifying the vital few problems

9.6 Stratification by Two or More Factors

Previous sections discussed the value of making comparisons using different methods of stratification. It is often possible to stratify a performance indicator by many criteria. For example, a defect in an operation may be stratified by the following factors:

Day
Shift
Machine
Spindle, pallet, fixture, station, e.g.
Tool source
Operator
Source of incoming stock
Upstream processing paths

9.6 Stratification by Two or More Factors

Individual Pareto diagrams can be formed for each stratification factor, or it is possible to combine two factors on a single Pareto diagram using stacked bars.

An example involving two factors is an operation that has three machines and produces four types of defects a day.

Defect	Machine			Total
	A	B	C	
1	10	42	8	60
2	2	21	32	55
3	12	14	28	54
4	30	7	4	41
Total	54	84	72	210
N	200	350	400	950

A possible analysis is a Pareto diagram of the defects, as shown in Figure 9.14a. Typically, quality performance is presented in this format by using percentages of total production without stratification factors, such as machine. This Pareto diagram is not informative because the vital few defects are not apparent. An approach for improving the analysis is to re-evaluate the defect categories and use different defect classifications to, it is hoped, produce a more meaningful Pareto analysis. An alternative approach is to use additional stratification factors, such as the machine producing the defect. However, a routine analysis of the total defects for each machine in Figure 9.14b produces little additional information. It is necessary to use the percentage of defects on the left-hand scale since the number of parts N evaluated from each machine varies.

A more useful analysis can be obtained by considering two factors in the Pareto analysis—the type of defect and the machine. It is first necessary to convert the data to percentages.

Defect	Machine			% of Total inspected
	A	B	C	
1	5	12	2	6.3
2	1	6	8	5.8
3	6	4	7	5.7
4	15	2	1	4.3
Total	27	24	18	

A combined two-factor Pareto diagram can now be prepared in either of the two formats shown in Figure 9.14c and d. Notice that the order of the bars is different in Figure 9.14a

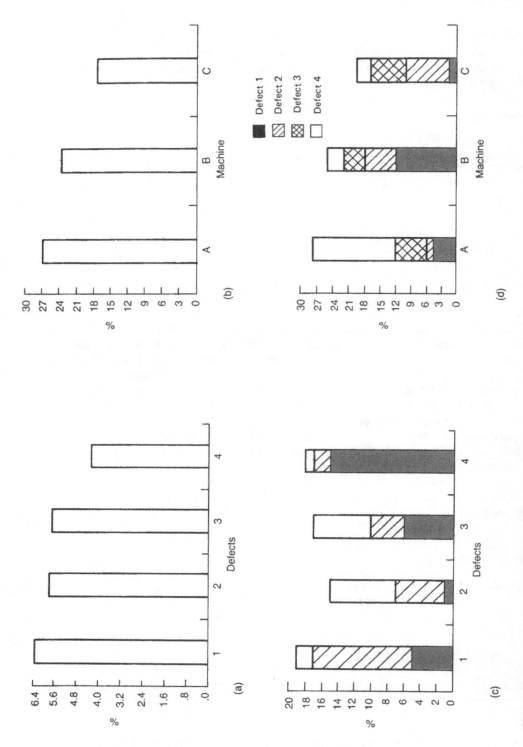

Figure 9.14 Four plots of the same machine defect data: (a) percentage of defects using total production; (b) percentage of defects using production by machine; (c) percentage of machine contribution to individual defects; (d) percentage of individual defects by

9.6 Stratification by Two or More Factors

and c. In Figures 9.14b, c, and d, percentages are computed using the production of an individual machine rather than total production. When production rates differ for a stratification factor, this adjustment is more useful for identifying problem areas. Clearly, machine A produces a significant portion of defect 4. Similarly, machine B for defect 1 and machines B and C for defect 2 are unusually large.

Suppose now we wish to study defect 4 in more detail to evaluate the source of the problem for machine A. Each machine at the operation has two spindles,

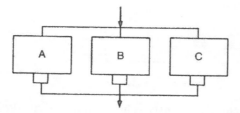

so a temporary color coding system is implemented to indicate which spindle produces a part. A check sheet is developed for the operation and produces the following summary results for the number of type 4 defects:

	Machine			
Spindle	A	B	C	Total
1	10	14	2	26
2	25	10	4	39
Total	35	24	6	65
N	250	600	300	1150

Converting to the percentage of each machine's production and plotting the Pareto diagram gives Figure 9.15:

	Machine			
Spindle	A	B	C	Total
1	4	2.3	.7	2.3
2	10	1.7	1.3	3.4
Total	14	4	2	5.7

It is apparent that spindle 2 on machine A is a significant part of the problem. Also, spindle 1 on machine A has a higher than expected percentage of defects when compared with

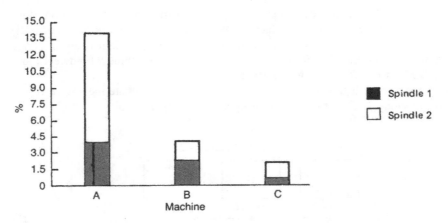

Figure 9.15 Two-factor Pareto analysis by machine and spindle.

machines B and C. Thus, there may be a special cause problem affecting machine A as well as a special cause problem affecting only spindle 2.

One of the difficulties in interpreting Pareto diagrams is knowing when additional stratification factors are useful. As illustrated earlier, one of the uses of additional factors is to change the interpretation from what at first appears to be a common cause system problem into a special cause problem. It is generally much easier to fix a special cause problem, so this practice is very useful for solving problems. In the previous example, it first appeared a common cause system problem existed with an equal frequency of defects. Classification by machine and later by spindle indicated special cause machine and spindle problems. Improving the operation involves making the machines perform in a similar manner. If all machines and spindles performed poorly, improvement is typically much more difficult.

Consider another example in which two machines with four spindles perform a roughing operation prior to finish machining on three machines:

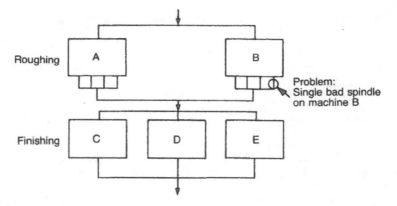

Assume a single bad spindle on machine B results in a defect at the finishing operation. Analyzing the defects at the finishing operation would show about an equal frequency of defects for machines C, D, and E. Thus, a common cause problem is assumed to exist. Some factor common to all finishing machines is causing the problem. It is then necessary to stratify rejects by the machine (A or B) producing the rough part. This analysis shows that machine B is the cause of the problem. Subsequent analysis by classifying defects by machine B spindle finally identifies the special cause spindle problem. The sequential use of additional stratification factors is a common problem-solving strategy.

> Additional classification variables can change interpretation of a common cause problem to a special cause problem

9.7 Warnings About Using Pareto Analysis

This chapter discusses Pareto diagrams used for the purpose of separating the vital few problem causes from the trivial many causes. However, there are several important considerations when applying Pareto analysis:

Customers. A Pareto analysis can produce a set of priorities that totally ignores customers, either the next user or the final user. Clearly, this is a poor way to establish improvement priorities. Customer concerns can be difficult to quantify but should not be ignored. The question must be answered, "Are we measuring what's important to our customers?" It is generally possible to use different methods of classification or different performance indicators to better represent customer concerns.

Stability. The data collected for a Pareto analysis may be derived from processes that are unstable. Interpretation of the resulting analysis can be difficult. The selective use of P charts is a good approach for evaluating stability.

Measurement. Operational definitions of defects are often unclear, so Pareto categories can be poorly defined. Also, there is a tendency to examine parts for only a single defect and to stop the inspection once a defect is found. This biases the Pareto analysis toward identifying the easy-to-find problems as the vital few.

Indicator. It is not always possible to ignore the trivial many problems because of the use of different performance indicators. For example, a low-frequency product problem can be associated with a high warranty cost and result in extreme customer dissatisfaction. Definition of the vital few and trivial many changes based on selection of the performance indicator.

Strategy. Groups select the top problem to solve when they are just starting to learn problem-solving methods. It is many times better to select an easier problem to learn and practice problem-solving skills. Also, the time and financial resources must be available to address a problem. Problem selection must consider the availability of both resources.

Time. Too short a time period can be used to establish a Pareto analysis. For example, Pareto analysis of a single day's problems is the same as using the "problem of the day" for establishing the vital few problems. It is necessary to evaluate the performance criteria over a representative time period.

> Those with data speak few words

9.8 Case Studies

The following examples illustrate a variety of applications of Pareto analysis.

9.8.1 Assembly Repair

The summary data for the assembly repair case study example appear in Table 9.4. The work sheet for the summary data appears in Table 9.5, along with the Pareto diagram in Figure 9.16.

366 9 Pareto Diagrams

Table 9.4 Summary of Chapter 8 Assembly Repair of a Car Air-Conditioning Evaporator

Defective item	Week 5/31, hour								Week 6/8, hour								Total
	1	2	3	4	5	6	7	8	1	2	3	4	5	6	7	8	
Sealer		1															1
Defective housing	4	1	10	4	3	4	3			2	2	1	5	1	3		43
Foreign material									1				1	1			3
Welding																	
Missing screws	13	7	11	3	16	8	15	12	24	25	20	15	35	30	25	15	274
Missing doors																	
Cams loose or missing		1	1						1								3
Missing spring																	
Cam screws loose or missing	2								6	4							12
Missing or misaligned gasket	1		1		1	2	4	3	5	6	3	4	6	4		1	41
Missing door arm									3								3
Missing or incomplete studs	1		1		3	3	4	1	1								14
Door shaft arm not connected	1		1														2
Missing recirculating duct	1					2	1	2			1				2		9
Scrap	4	1	9	1	2							1		1			19
Motor wires not connected	9	5	3		1												18
Missing motor clips									5	15	7	3	7	7	10	5	59
Incomplete parts										1	3		1	3	5		13
Total	36	16	37	8	26	19	27	18	46	53	36	24	55	47	45	21	514
Units repaired					105								161				266
Units produced					1395								1440				2835

9.8 Case Studies

Table 9.5 Assembly Repair Work Sheet

Defective item	Number of defectives	% Composition		Cumulative %		% of Total inspected
1. Missing screws	274	274/514 =	53.3		53.3	274/2835 = 9.7
2. Missing motor clips	59	59/514 =	11.5	53.3 + 11.5 =	64.8	59/2835 = 2.1
3. Defective housing	43	43/514 =	8.4	64.8 + 8.4 =	73.2	43/2835 = 1.5
4. Missing or misaligned gasket	41	41/514 =	8.0	73.2 + 8.0 =	81.2	41/2835 = 1.4
5. Scrap	19	19/514 =	3.7	81.2 + 3.7 =	84.9	19/2835 = 0.7
6. Motor wires not connected	18	18/514 =	3.5	84.9 + 3.5 =	88.4	18/2835 = 0.6
7. Missing or incomplete studs	14	14/514 =	2.7	88.4 + 2.7 =	91.1	14/2835 = 0.5
8. Incomplete parts	13	13/514 =	2.5	91.1 + 2.5 =	93.6	13/2835 = 0.5
9. Cam screws loose or missing	12	12/514 =	2.3	93.6 + 2.3 =	95.9	12/2835 = 0.4
10. Other	21	21/514 =	4.1	95.9 + 4.1 =	100.0	21/2835 = 0.7
Total	514		100.0		100.0	266/2835 = 9.4

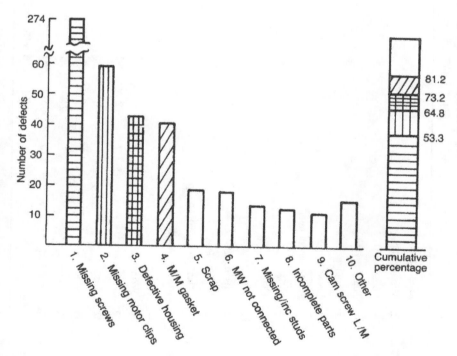

Figure 9.16 Assembly repair Pareto diagram.

Table 9.6 Door Panel Assembly Defects

Defect	6/3	6/4	6/7	6/8	6/10	6/11	6/14	6/15	6/16	6/17	6/18	Total
1. Misaligned clips												0
2. Missing clips		2		1		3			1			7
3. Loose vinyl	10	6	8	3	9	11	19	7	17	16	13	119
4. Foam covering slots												0
5. Misaligned carpet												0
6. Loose carpet	5	3	3	5	1	1	2	2	1	2	1	26
7. Loose anchor nuts					2							2
8. Dirty	3	2	3		1	4	3	5	3		1	25
9. Cuts	4	3	5	4	2	5	6	3	1	2	4	39
10. Missing bracket	3											3
11. Misaligned lens	6	2	1	2	1	3	3		2		1	21
12. Missing lens												0
13. Misaligned weatherstrip								1	2			3
14. Missing weatherstrip												0
15. Voids	3									1		4
16. Wrinkles	2	4	3	9	6	4	4	1		6	2	41
17. Scratches	2	2	2	4	1	5		1	1	2		20
18. Blisters												0
19. Other												0
Total defects	38	24	25	28	23	36	37	20	28	29	22	310
Pieces checked	570	505	367	429	423	402	419	349	402	429	457	4752
% Defective	6.7	4.8	6.8	6.5	5.4	9.0	8.8	5.7	7.0	6.8	4.8	6.5

9.8 Case Studies

It is apparent that the Pareto principle holds in this example since 4 (of 22) potential problems account for 81% of the defects. The data collection and Pareto analysis lead to several questions:

Why are missing screws such a problem?
Why were there so many more problems the week of 6/8 compared with 5/31 (327 versus 187)?
Why are problems present one week and not another?

Another way to view these results is to classify the defects by the source of the problem:

Supplier
Assembly process
Operator attention

This approach can suggest what group of problems should be addressed first.

9.8.2 Assembly Defects

The defects associated with a door panel assembly are summarized in Table 9.6 with the Pareto diagram given in Figure 9.17. In this example three defects account for about 65% of the total defects. Perhaps reclassifying the defects, as suggested in the previous exam-

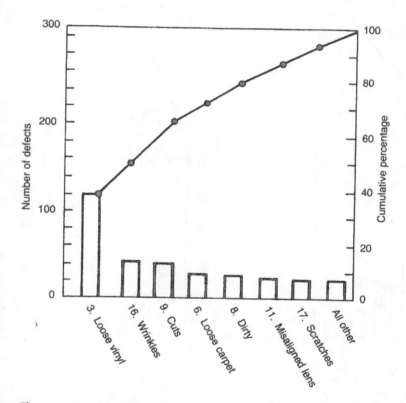

Figure 9.17 Pareto analysis for door panel assembly defects (summary for 6/3 through 6/18).

ple, would provide further insight. Notice that the overall level of defects is relatively stable for most categories. An analysis was performed for a longer period of time, and a cost Pareto diagram was added to the analysis that appears in Figure 9.18. It is apparent that cut vinyl increases in importance when cost is considered because of the need for complete replacement.

9.8.3 Leak Rejects

A week's Pareto analysis was kept on the air leak test rejects for a truck differential case. An employee group met weekly to review the prior week's Pareto diagram and the P chart for the overall reject rate. Problems were assigned a color so that the Pareto bars could easily be followed in the meetings. Different shadings are used to represent the problems in Figure 9.19. The weekly Pareto analyses and daily P charts were posted in the department. In the week of 9/5, corrective actions were taken to eliminate cases in which the manufacturing holes were not drilled through. The ongoing Pareto analysis serves as an effective verification tool. In the Pareto analysis for 9/12, it was found that loose-locking plate bolts was a new problem caused by the operator's early release of an automatic torquing gun. The problem was eliminated in subsequent weeks.

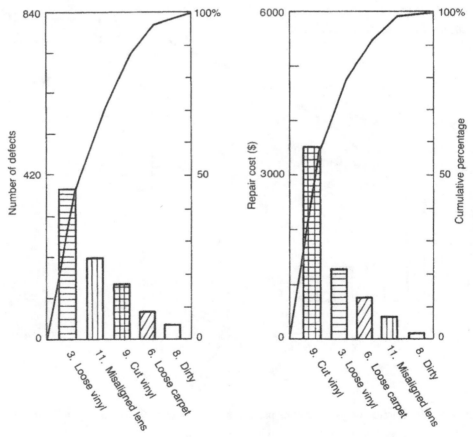

Figure 9.18 Defect and cost Pareto analysis for door panel assembly defects.

9.8 Case Studies

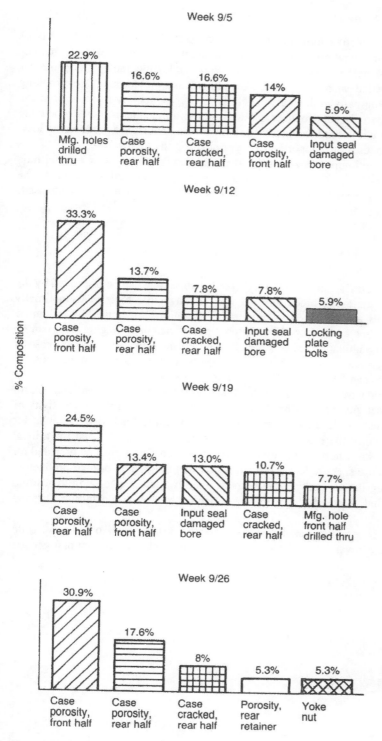

Figure 9.19 Pareto analysis of weekly air test rejects.

9.8.4 Productivity Analysis

Pareto analysis is an effective tool for analyzing productivity losses. The analysis of productive losses for an operation starts with determining the machine cycle time, including the loading and unloading of parts. From the machine cycle time, the optimum number of parts for any time period can be computed. The optimum assumes the machine is running continuously, producing parts at the stated cycle time. A productivity loss occurs if the optimum number of parts is not produced. Stated simply, a machine is either producing parts at a stated rate or it is not producing parts at a stated rate, incurring a productivity loss. The Pareto analysis focuses on categorizing and prioritizing the losses.

Using the machine optimum is a more effective approach than attempting to determine an acceptable, normal loss, which is a common industrial engineering practice. The optimum approach assumes that no loss is acceptable and that all losses can be addressed. Consider three common "normal" loss categories:

Tool changes
Operator breaks
Gaging

Each of these loss categories can be addressed. Tool change time can be improved by the use of any quick-change tools, block tool change, or longer life tools or by optimally scheduling when tools are changed. In production operations, operator breaks can be staggered so that the line control operation is continually manned. Gaging time can be reduced by the use of combination gages that make multiple measurements. In most cases it is possible to reduce productivity loss, but no loss must be assumed normal so that a useful prioritization of losses can be obtained from the Pareto analysis.

The analysis of productivity loss is performed for the critical few operations that control a department's productivity. Overall productivity can increase only if the productivity of these operations increases. If these operations are not known, stacked bar graphs for productivity, discussed in the next chapter, can be prepared. Assuming that the critical control operations are known, productivity loss data collection sheets must be prepared for these operations because specific losses are not adequately defined on normal production reports. An exploding Pareto analysis can be performed by first categorizing losses into generic categories and then analyzing these categories in detail in additional Pareto analyses.

It is useful to use generic loss categories first since the skills and personnel required to make improvements can vary in different categories. Consider several common generic categories:

No stock
Cycle time
Labor
Rejects
Mechanical downtime
Electrical downtime
Hydraulic downtime
Awaiting repair
Awaiting parts
Coolant
Tools

9.8 Case Studies

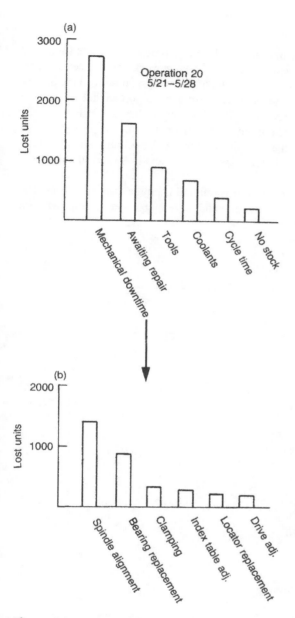

Figure 9.20 Exploding Pareto analysis for productivity losses: (a) generic productivity losses; (b) mechanical downtime.

The team required to address mechanical problems is probably different from the team required to address tooling problems. The generic losses categories can be used to prepare a standard Pareto diagram, as shown in Figure 9.20a. The significant loss categories can then be exploded into detailed losses, as shown in Figure 9.20b. Problem-solving teams should be used to address individual problem areas. The productivity losses Pareto diagrams use either lost units or lost hours (at a stated cycle time) on the left-hand scale.

9.8.5 Report Errors

A publication department started a system to record the errors present in reports submitted for printing. These errors resulted in extensive reprocessing work for the department, which increased overall publication costs. The results of the analysis produced the data in Table 9.7. The data collection procedure identified all errors in a report so that the number of errors (212) exceeds the number of reports (179). The Pareto diagram in Figure 9.21 shows that three categories account for about 60% of the errors and six categories account for 82% of the errors. In Figure 9.21, the line plot format for the cumulative percentage produces an awkward scaling so that the Pareto bars are quite small. A stacked bar format would be less cumbersome.

Unfortunately, a useful piece of information was omitted from the data collection. The number of reports with 0, 1, 2, 3, . . . , errors could have suggested an improvement strategy. Consider two hypothetical cases.

Case 1:   special cause problem

Number of report errors	0	1	2	3	4	5	6	7	8	Total
Number of reports	125	1	2	15	20	14	2	0	0	179

Table 9.7   Report Submission Errors

Category	Actual number of errors
Distribution list	70
Job work sheet	29
Pagination	26
Title page	20
General format	15
Sequence	13
Document approval form	5
B/W prints (quality)	5
Photo/drawing numbers	5
Table of contents	5
Reproduction card	4
Photomaster sheet	4
Caption	3
Others	8
	212
Number of reports $N = 179$	

9.8 Case Studies

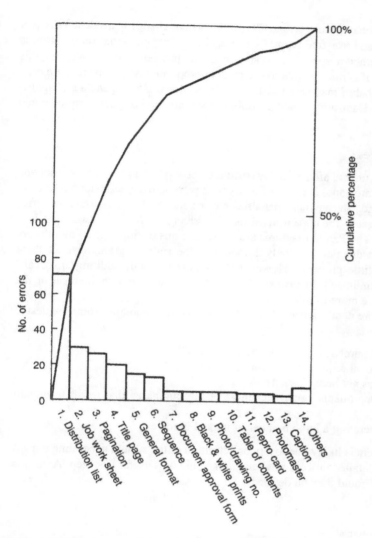

Figure 9.21 Pareto analysis of report submission errors.

Case 2: common cause problem

Number of report errors	0	1	2	3	4	5	6	7	8	Total
Number of reports	15	139	20	5	0	0	0	0	0	179

In the first case, the logical question arises of why 125/179 or 70% of the reports had no errors and the remaining 30% generated all the errors. Improvement might focus on comparing the two systems and making the desired changes. In the second case, 164/179 or

92% of the reports had one or more errors. Clearly, a system change is needed. Why is it not possible to submit an error-free report? Once again we see additional information is able to determine whether a special or common cause problem exists. Without this additional information, it is not possible to select an appropriate improvement strategy.

This example is typical of many administrative applications. Errors generate a significant amount of work. Data are needed to define problems and suggest improvement strategies.

9.8.6 Unpaid Invoices

The number of vendors unpaid after a 30 day period is a cost problem to both a vendor and the purchaser. Billings are handled many times by both parties, many telephone conversations are necessary, and late payment penalties can be incurred. An accounting office started to analyze the cause for unpaid invoices. The Pareto analyses for two periods is shown in Figure 9.22. The Pareto diagrams are somewhat misleading since 3 months are summarized in the left diagram and only 1 month in the right diagram. Overall there appears to have been little progress. However, the category "staff auditing approval" appears to have been eliminated. In any case, use of the percentage of total invoices on the left-hand scale would be more useful.

In this example, more detail is needed to further define the problem. Some suggested questions follow:

What elements in staff purchasing are causing the delays?
Why would local approval require more than 30 days?
Why were packing slips not being issued?
Are a few specific departments generating most of the purchases that result in unpaid invoices?
Are a few vendors generating a large number of unpaid invoices?

The lack of improvement is likely due to a poor definition of the system generating unpaid invoices. Preparing a detailed process flow diagram would be a useful first step. Areas for which data are needed could then be defined.

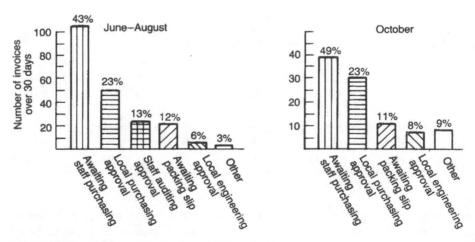

Figure 9.22 Pareto analysis of unpaid invoices.

Problems

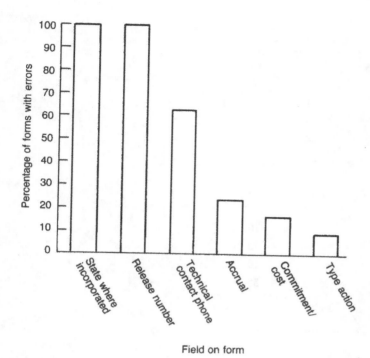

Figure 9.23 Pareto analysis of field containing errors on a data form.

9.8.7 Form Errors

A large part of the information generated in an organization is derived from a variety of standard preprinted forms. The errors generated by the completion or lack of completion of these forms is the cause of a great deal of "extra" work. One organization started to record the number of errors that occurred for each data field. The form was a relatively low-volume document but, as Figure 9.23 indicates, generated a high volume of work.

Many of the same questions arise as were asked in the report errors case study concerning the existence of a common or special cause. However, in this example *all* forms had errors in two fields—no wonder clerks spend excessive time processing this form! Extensive redevelopment of the form and the completion instructions reduced errors significantly. The data were able to suggest the system changes that were needed.

Problems

9.1 The defects in a car tailgate glass inspection appear below. Only one defect per part is assumed to occur. (a) Prepare a Pareto diagram for the number of defects using both the line and the stacked bar plot for the cumulative percentages. (b) Prepare a Pareto diagram using the percentage composition of the total defects (180). (c) Repeat b using the percentage of the total production (1200). (d) Interpret the Pareto analysis. (e) Prepare a Pareto diagram with a stacked bar plot using cost as the performance indicator. Compare the results to those for a. Suggest how improvement should be addressed.

Defect	Number	Cost ($)
Loose moldings	14	.25
Missing monogram	0	.50
Broken welds	1	.10
Loose ball studs	0	.10
Mislocated center molding	4	.15
Mislocated side molding	24	.10
Scratched molding	1	.10
Dented molding	44	.75
Scratched glass	7	5.25
Chipped glass	0	.10
Poor cleanup	79	.30
Mislocated handle holes	1	.10
Mislocated ball studs	5	.35
Total	180	
Number inspected $N = 1200$		

9.2 Prepare a P chart for the loose vinyl and total rejects in Table 9.6. How is this information useful for the Pareto analysis in Figure 9.17?

9.3 Prepare a Pareto analysis work sheet for the data in Table 9.7. Suggest a better measurement scale than actual number of errors for the report errors case study.

9.4 Analyze the data in Table 8.1 using appropriate Pareto diagrams. What do you conclude from the analysis?

9.5 A four spindle machine had both undersized (U/S) and oversized (O/S) inside diameter defects. The following was the resulting Pareto diagram. Give two possible explanations for the results. How could you investigate which explanation might be true?

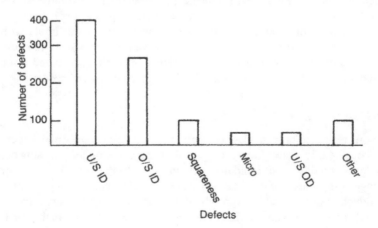

Problems

9.6 An operation with two machines each with two spindles was evaluated for a defect over a period of 3 months. (The number in parentheses indicates the number of parts evaluated.)

Machine, spindle	August	September	October	Total
A, 1	25 (500)	17 (400)	32 (600)	74 (1500)
A, 2	33 (450)	90 (550)	82 (350)	205 (1350)
B, 1	17 (300)	18 (220)	35 (400)	70 (920)
B, 2	67 (200)	93 (600)	18 (300)	178 (1100)

(a) Compare the defects for machines A and B over the 3 month period. Do common or special cause problems seem to exist? (b) Prepare a two-factor Pareto analysis using machine and month. Do common or special cause problems seem to exist? (c) Prepare a three-factor Pareto analysis using machine, spindle, and month. Do common or special cause problems seem to exist?

9.7 A sudden increase in defects on all machines at a finishing operation resulted in a team being formed to address the problem. To more clearly define the problem, the team decided to classify the defects by critical upstream stratification factors. Consider the following partial process flow diagram:

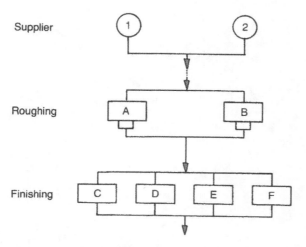

A color coding system was initiated for supplier 1 or 2 and machine A or B, and a check sheet was placed at each finishing operation. After a day, the following results were obtained:

Machine, rough	Supplier 1	Supplier 2
C, A	3 (50)	1 (46)
C, B	2 (45)	12 (38)
D, A	1 (47)	4 (55)
D, B	5 (36)	10 (53)
E, A	7 (42)	2 (46)
E, B	1 (49)	11 (39)
F, A	2 (64)	2 (53)
F, B	3 (41)	8 (46)

(a) Why did the team decide to evaluate upstream stratification factors? Is the problem necessarily with upstream processing? (b) Prepare a Pareto analysis that would assist the team in properly defining the problem.

10
Stratification and Graphs

10.1 Two Fundamental Tools

Previous chapters discussed the use of stratification and graphic displays of data. These topics are presented in more detail in this chapter. Both concepts are important individually for process control or problem analysis, but the use of the combined approaches in the form of stratified graphs is shown to be another powerful tool that can be used in many applications.

10.1.1 Stratification

As we drive down a highway in a mountainous area, it is not uncommon to see exposed layers of rock where part of a mountain or hill has been removed to make way for the highway. These layers or geologic strata form the mountain and can be quite different in appearance.

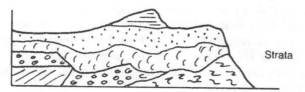

The term "stratification," derived from "stratum," is used in data analysis since many strata or groups of factors are involved in any collection of data. Some common stratification factors for manufacturing applications are:

Time (days, weeks, months)
Shift
Operators
Machines (e.g., spindles, fixtures, pallets, and stations)
Vendors
Lots

Why is stratification important? Suppose we attempt to find iron by drilling through a number of geologic strata. We do not take all the earth obtained from the drilled hole and dump it in a pile and evaluate the iron content of a mixed sample from the pile. The logical approach to finding iron is to collect a sample from each of the strata and evaluate the iron content. Similarly, finding a manufacturing problem often involves selecting the correct stratification factor that enables discovery of a problem. "Dumping" all factors together provides little insight. For example, suppose a problem is discovered in the output of a machine with four spindles. It is only natural to group or stratify the output for each spindle.

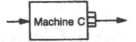

If one of the spindles is defective, the spindle stratification factor uncovers the problem. However, suppose machine C receives the output of machines A and B.

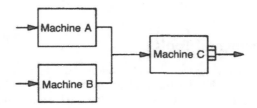

If the spindles on machine C are not the source of the problem, then the problem rate is about the same for each spindle. In this case, the source of the problem may be machine C (nonspindle components), A, or B. By collecting and analyzing the data utilizing the stratification factors, it is possible to evaluate whether machine A, B, or C is the root cause of the problem.

Unfortunately, the correct stratification factors for resolving a problem are generally not known prior to data collection. It is then necessary to record many different stratification factors. A safe strategy is to record factors that change during the period of data collection. Also, it is desirable to spread the data collection over a long enough time period to encompass several changing stratification factors that may be important. Preparation of a thorough check sheet (Procedure 8.1) noting stratification factors is a good practice to use prior to data collection. It is not possible to effectively determine all potentially important stratification factors without planning. Preparation of a process flow diagram as part of the planning activity is useful.

Record all stratification factors that change when collecting data

It is necessary to record stratification factors when using the comparison principle introduced in Chapters 5 and 9 and discussed in detail in Chapter 11. The logic is simple. In the machine C spindle example, the output of one spindle should be roughly similar to the output of any other spindle. Comparison of the separate output of each spindle is an effective method for evaluating potential spindle problems. As discussed in the previous chapter, use of the comparison principle can result in identification of a special cause problem. What at first appeared to be a common cause problem with machine C can become a special cause problem associated with a particular spindle. Figure 10.1 illustrates this process.

Stratification factors can separate special and common cause problems

10.1 Two Fundamental Tools

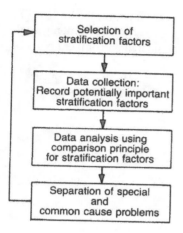

Figure 10.1 Relationship of stratification factors, the comparison principle, and special and common cause determination.

10.1.2 Graphs

Speak with data

A graph combines two useful concepts. First, a graph is based on data. The importance of data as a basis for action and communication has been repeatedly emphasized. A graph must start with data; opinion or emotion plays no role. Second, a graph is a type of picture. Generally, we think and recall information most effectively when we have a mental picture of the point being considered. Because of these two concepts, a graph is one of the most effective ways to both communicate information and analyze data. Essentially all the statistical analysis methods discussed in this book result in a graph or chart because of these desirable features. The resulting pictures serve as a primary method of analysis that directly communicates the results and serves as a basis for meaningful discussion.

A picture is worth a thousand words

Combining the comparison principle with graphic analysis results in a powerful analysis tool. The graph then becomes a visual comparison tool. In many cases a problem can be detected that would be difficult to detect by other methods. At the very least, it is always a good practice to graphically display the results of a comparison and other data analyses. In Figure 5.26, a comparison of histograms for different spindles (strata) indicated a spindle mean shift problem. In Figure 9.10, Pareto diagrams were used to compare defect rates for different machines (strata). These are examples of *stratified graphs* that uncover "stratification problems" utilizing the comparison principle. Finally, the most common stratification factor is time. If a machine has a defect rate of 20% in month 1 and 18% in month 2, we use a P chart to evaluate whether the change was significant. One of the problems with using only graphic comparisons is the possibility of overinterpreting differences. Often graphic displays can lead to more detailed analyses, or differences are so large that detailed methods are not necessary. Chapter 11 discusses several detailed comparison methods.

There are many different types of graphs. The intent here is to discuss the three basic types of graphs used in many industrial applications: (1) the bar graph, (2) the line (or XY) graph, and (3) the pie chart. Various combinations of these graphs are also commonly used. Most of the statistical graphs presented in this text can be considered one of these types of graphs:

Chapter	Graph	Type
2	Process flow diagram	Flowchart
3	Control charts	Line graphs
5	Histogram	Bar graph
9	Pareto diagram	Bar graph
11	Means bar plot	Bar graph
11	ANOM plot	Bar graph
12	Scatter plot	XY graph
13	Cause-and-effect diagram	Flowchart

This list adds a flowchart to the standard methods of graphical presentation.

10.1.3 Computer Plotting

Over the last several years, computer software packages have made it easy to produce high-quality graphs. It is no longer necessary to spend excessive amounts of time to prepare common graphic displays. To emphasize this point, most of the graphs in this chapter are prepared using a microcomputer with a standard plotter. It is now easy to use different shading, symbols, or line types to make stratified graphs. Perhaps the most powerful tool to show different strata is the use of different colors. Although this technique is not used here, the power of this approach should not be overlooked.

10.1.4 Typical Application

Most industrial performance indicators are related to quality, productivity, or cost. Improvement in these indicators is the objective of all organizations. The data in Tables 10.1 through 10.3 are typical of information available for many manufacturing operations. These data address only quality and productivity; costs are related to these indicators. As noted by Deming (1982), productivity is directly related to quality. Improving quality improves productivity. However, it is often useful to address both quality and productivity problems. The objective of the joint analysis is to eliminate variability in operation and performance of the manufacturing system. For example, the mechanical downtime noted in Table 10.2 can be expected to be related to several quality problems (whether we measure the affected characteristics or not). A machine that has minimal downtime due to effective PM, lubrication, and correct setup undoubtedly produces a higher quality part than would otherwise be the case. Conversely, a machine that produces few defects (e.g., Table 10.2) not only benefits productivity by producing good rather than scrap parts, but the reduced part-to-part variability improves production consistency. Downtime throughout a department is reduced because of consistent part locating, few jam-ups, even machining loads,

10.1 Two Fundamental Tools

Table 10.1 Department A Performance Indicators: Operation 20 Quality Rejects

Week	Production[a]	Oversize	Burrs	Porosity	Undersize	Other	Total
1	507	25	31	7	5	10	78
2	621	4	19	3	11	7	44
3	422	9	21	0	17	3	50
4	692	5	27	13	4	2	51
5	408	6	10	22	3	1	42
6	797	0	16	7	5	8	36
7	602	11	30	16	13	2	72
8	624	4	19	4	9	0	36
9	593	12	18	0	16	4	50
10	706	2	24	0	15	6	47
11	643	21	6	3	9	22	61
12	624	18	8	2	17	6	51
Total	7239	117	229	77	124	71	618
Prior quarter	7320	240	221	60	50	53	624

[a] Planned production, 620 units per week.

Table 10.2 Department A Performance Indicators: Operation 20 Nonoperating (N/O) Productivity Losses

Week	Mechanical downtime	Manning	Tool change	Gaging	No stock	Unexplained	Total
1	73	150	132	45	0	117	517
2	100	135	147	45	0	10	437
3	13	121	127	45	200	124	630
4	35	100	138	45	0	41	359
5	159	150	122	45	105	71	652
6	0	26	129	45	0	69	269
7	197	0	130	45	0	56	428
8	8	147	151	45	0	91	442
9	0	118	143	45	137	16	459
10	0	125	126	45	0	53	349
11	67	100	135	45	0	51	398
12	59	50	143	45	52	78	427
Total	711	1222	1623	540	494	777	5367
Prior quarter	875	1064	1611	540	607	583	5280

Table 10.3 Department A Performance Indicators: Productivity Analysis for Week 12

Operation	Optimum	Sold	Rejects	N/O	Cycle time
10	1087	654	15	418	0
20	1286	624	51	427	184
30	999	599	30	319	51
40	1575	593	21	650	311
50	1392	589	10	610	183

ªPlanned production, 620 units per week.

and predictable tool life. Quality and productivity are highly interrelated, so we use both performance indicators to address overall departmental performance.

The objective of improvement must be to eliminate variability from the manufacturing system

The data in Tables 10.1 through 10.3 are for a hypothetical department A assumed to have five operations with the following process flow diagram:

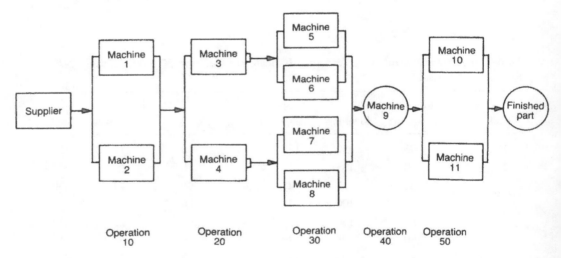

Table 10.1 is a standard listing of quality defects for a critical operation. Table 10.2 is a listing of productivity losses for the same operation. The productivity losses are listed by generic category. The important loss categories must be broken down further into individual items causing losses (exploding Pareto analysis), as discussed in a case study in Chapter 9. Focusing attention on a critical control operation maximizes the impact of any improvements. Also, performing a joint quality and productivity analysis helps to uncover overall operating problems that could influence both indicators. Tables 10.1 through 10.3 show changes over time, a key factor in most analyses. It is apparent that even in this simple, abbreviated example, graphic methods are necessary to effectively analyze the data.

10.2 Bar Graphs

The data in Table 10.3 show five categories related to weekly productivity:

1. Optimum: maximum number of parts assuming continous operation during all shifts
2. Sold: number of acceptable parts produced; may include repaired parts
3. Rejects: number of parts not acceptable; parts will be repaired or scrapped
4. N/O: number of parts lost because the machine is not operating (N/O)
5. Cycle time: number of pieces lost because the machine produces parts at a slower than optimum rate of production

These categories help address departmental productivity losses and identify critical control operations. This approach considers no loss acceptable; a complete accounting to machine optimum is performed.

10.2 Bar Graphs

A standard bar graph is a plot used to compare two or more groups on some measurement scale:

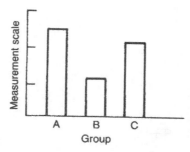

Some examples follow:

A	B	C	Measurement scale
Week 1	Week 2	Week 3	Number of defects in a department
Dept 1	Dept 2	Dept 3	% Defects
Tool 1	Tool 2	Tool 3	Number of parts
Chemical 1	Chemical 2	Chemical 3	Yield

If the groups are associated with time (e.g., day, week, or month), the earliest time is to the far left (group A). However, if the groups are not associated with time, the groups are typically ranked highest to lowest and are plotted as group A, B, C, and so on. The standard Pareto diagram is an example of this type of bar graph.

A bar graph is a useful way to display categories of performance indicators. The quality and nonoperating productivity losses for the 12 week period in Tables 10.1 and 10.2 are displayed in Figure 10.2. The purpose of these Pareto bar plots is to prioritize problem areas

Figure 10.3 is a bar graph of the lost units due to mechanical downtime. Notice that the

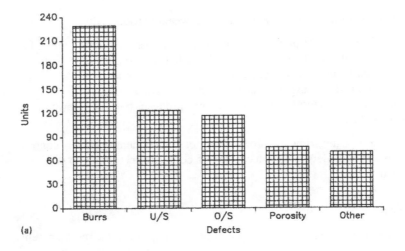

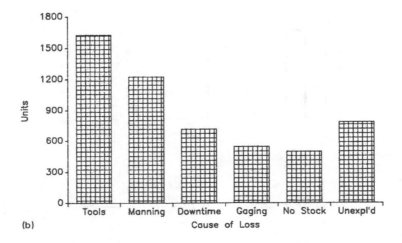

Figure 10.2 Pareto bar graphs: (a) quality rejects; (b) nonoperating productivity losses.

groups are weeks so there is a time ordering and a horizontal format is used for the bars. The purpose of this graph is simply to display numerical information. In the downtime example, the goal or objective is 0 units lost for downtime. In other cases, the goal may be different from 0 so a bar graph can be used to plot the performance deviation:

Performance = actual performance − goal deviation

Figure 10.4 is a plot of the production performance deviation from Table 10.1 to a schedule of 620 units. This plot can be used when the goal changes over time.

10.2.1 Multiple Factors

The standard bar graph uses two factors: a performance indicator (plotted on the measurement scale) and a factor that forms the grouping (or stratification) variable. It is often

10.2 Bar Graphs

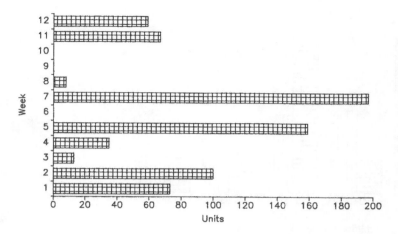

Figure 10.3 Bar graph of lost units due to mechanical downtime.

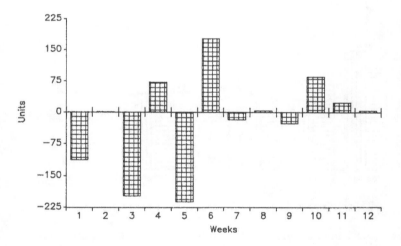

Figure 10.4 Bar graph of production performance deviation.

informative to select an additional factor to enable the bar plot to be used for comparison purposes. The most common factor is a prior time period. Figure 10.5 repeats Figure 10.2 with the prior quarter's performance. Because of the different production rates, it is necessary to use percentages for the reject analysis in Figure 10.5a. Assuming the same number of operating hours for the two quarters, the actual losses can be plotted as shown in Figure 10.5b.

Comparisons using factors other than time are also common. Figure 10.6a compares the rate of absenteeism for three departments. Again it is necessary to use percentages since the number of employees in each department is different. Placing the last quarter's results on the left side of the plot provides a convenient time reference. A three-dimensional version of the same plot appears in Figure 10.6b. This format conveys less information than the

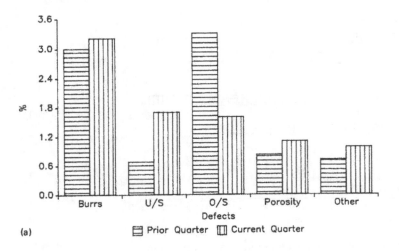

(a)

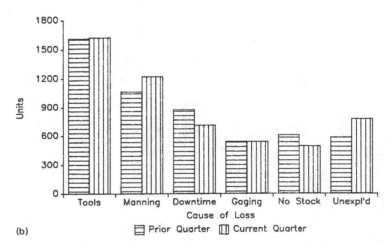

(b)

Figure 10.5 Two-factor Pareto bar graphs: (a) quality rejects; (b) nonoperating productivity losses.

two-dimensional version since it is difficult to read the scale (percentage absenteeism) for a bar height. These formats make it difficult to compare weekly performance for an individual department. Figure 10.6c is a rearrangement of the plot that enables direct within-department comparisons.

10.2.2 Stacked Bars

When multiple factors are considered, it is sometimes useful to stack the bars rather than place them side by side. The stacked configuration enables a visual evaluation of the total problem by the height of the group of bars. Figure 10.7a shows a standard two-factor stratification of burr rejects. The bottom spindle on machine 4 is the biggest problem. However, both machines 3 and 4 have a high rate of rejects from some common cause even without the bottom spindle problem. Figure 10.7b is the same plot in a stacked bar format;

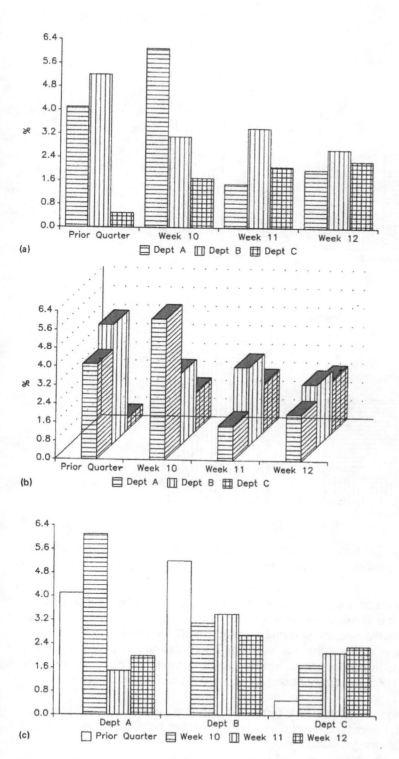

Figure 10.6 Bar graphs of absenteeism in three departments: (a) two-dimensional department comparison by week; (b) three-dimensional department comparison by week; (c) three-dimensional week-to-week comparison by department.

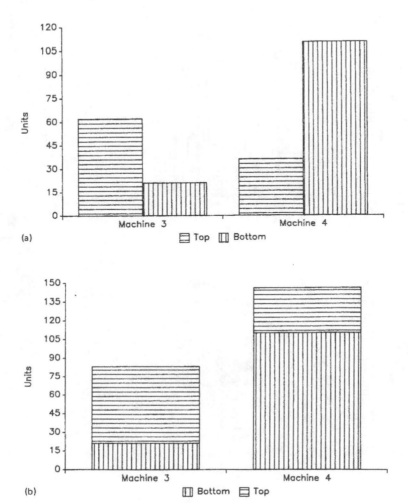

Figure 10.7 Bar graphs for operation 20, 12 week burr reject analysis: (a) multiple bars for top and bottom spindles; (b) stacked bars for top and bottom spindles.

it is easier to see that machine 4 has the larger number of rejects. Since the number of rejects is being compared, it is assumed that the number of units produced is the same for all four spindles. If this is not the case, the percentage of burr rejects from each spindle should be plotted. This analysis assumes that the operation reject data are collected noting the machine and spindle that generated each reject. Since this detail is often not routinely available, a special check sheet must be prepared and a study conducted to troubleshoot the reject problem.

It is convenient to use the stacked bar format to compare the contribution to burr rejects by machines 3 and 4 over time. Assuming equal production rates, this comparison can utilize any of the following measures:

Number of rejects
Percentage composition (by machine)
Percentage of units produced

10.2 Bar Graphs

Table 10.4 Comparison of Machine 3 and 4 Burr Rejects

Week	Machine	Units produced	Number rejects	% Composition		% Units produced	
1	3	263	11	11/31 =	35	11/263 =	4.2
	4	244	20	20/31 =	65	20/244 =	8.2
	Total	507	31				
2	3	285	5	5/19 =	26	5/285 =	1.8
	4	336	14	14/19 =	74	14/336 =	4.2
	Total	621	19				
3	3	201	10	10/21 =	48	10/201 =	5.0
	4	221	11	11/21 =	52	11/221 =	5.0
	Total	422	21				
4	3	435	21	21/27 =	78	21/435 =	4.8
	4	257	6	6/27 =	22	6/257 =	2.3
	Total	692	27				
5	3	133	2	2/10 =	20	2/133 =	1.5
	4	275	8	8/10 =	80	8/275 =	2.9
	Total	408	10				
6	3	375	3	3/16 =	19	3/375 =	.8
	4	422	13	13/16 =	81	13/422 =	3.1
	Total	797	16				
7	3	321	10	10/30 =	33	10/321 =	3.1
	4	281	20	20/30 =	67	20/281 =	7.1
	Total	602	30				
8	3	307	4	4/19 =	21	4/307 =	1.3
	4	317	15	15/19 =	79	15/317 =	4.7
	Total	624	19				
9	3	294	3	3/18 =	17	3/294 =	1.0
	4	299	15	15/18 =	83	15/299 =	5.0
	Total	593	18				
10	3	348	13	13/24 =	54	13/348 =	3.7
	4	358	11	11/24 =	46	11/358 =	3.1
	Total	706	24				
11	3	331	0	0/6 =	0	0/331 =	0
	4	312	6	6/6 =	100	6/312 =	1.9
	Total	643	6				
12	3	296	1	1/8 =	12	1/296 =	.3
	4	328	7	7/8 =	88	7/328 =	2.1
	Total	624	8				
Total	3	3589	83	83/229 =	36	83/3589 =	2.3
	4	3650	146	146/229 =	64	146/3650 =	4.0
	Total	7239	229				

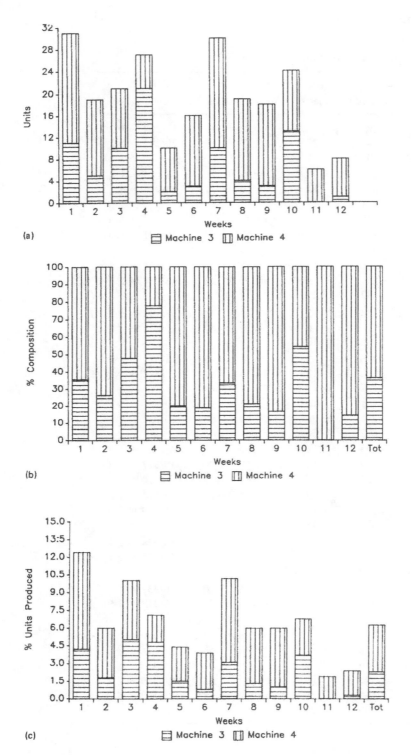

Figure 10.8 Stacked bar graph comparison of machine 3 and 4 burr rejects: (a) number of rejects; (b) percentage composition; (c) percentage of units produced.

10.3 Line Graphs

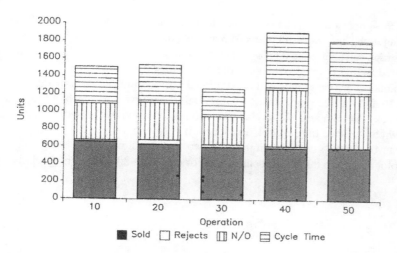

Figure 10.9 Stacked bar graph of productivity analysis for department A.

The computations can be obtained by following the example in Table 10.4. The results of the analysis in Table 10.4 are plotted in Figure 10.8. Since the production rates differ slightly, it is preferable to use Figure 10.8c. Also, the use of percentages enables the inclusion of a "total" category without lengthening the vertical plotting scale.

Stacked bar plots are used to break down a total into individual categories. In Table 10.3, the optimum machine production is broken down into individual components. Figure 10.9 is a stacked bar graph of the individual operations in department A. This graph is useful for analyzing a department's current performance as well as its potential performance. The direct impact of quality losses can also be assessed. However, it is not always possible to identify the critical control operations with this approach, since, in practice, other department operations balance their output to match the output of the control operation.

10.3 Line Graphs

A standard line graph is a plot of two factors. These factors represent two values or measurements for a unit that are plotted on the x and y axes of a two-dimensional plot.

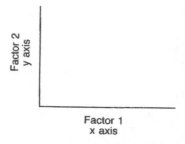

The case in which the x axis factor is time is considered here. The more specialized case in which the x axis is a general measurement scale is addressed in the Chapter 12 discussion of

XY scatter plots. The simplest line graph is the run chart introduced in Chapter 3. The line graph corresponding to the bar graph in Figure 10.3 appears in Figure 10.10. In this simple case, a bar or line graph can be used to display the data. Generally, if more than 10–15 time points are considered, a line graph provides a more usable display of the data.

One of the advantages of line graphs is that additional information can be provided on the graph. Figure 10.11 is a line graph of the average daily production for each of the 12 weeks. For each week, the high and low day's production is indicated by the vertical bars. A line is also shown to indicate the scheduled production (124 units).

A line graph can be used to consider more than two factors, similar to a bar graph. Figure 10.12 is a plot of the top three reject categories for operation 20. Different symbols

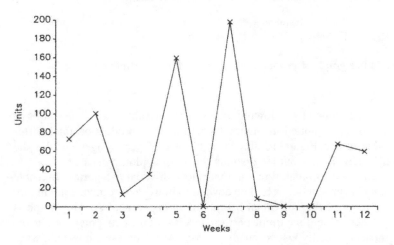

Figure 10.10 Line graph of lost units due to machine downtime.

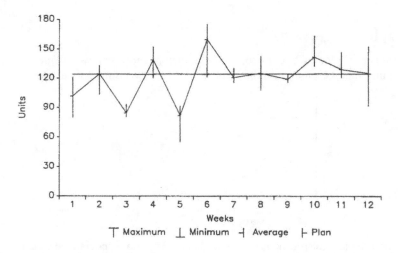

Figure 10.11 Line graph of average daily production for department A.

10.3 Line Graphs

or line styles can be used to indicate the various groups (i.e., burrs, undersized, and oversized). Percentages are used since the number of units produced varies by week. Figure 10.13 shows how absenteeism in different departments can be compared with a line graph. It is possible to display more information in a line graph (Fig. 10.13) than in a bar graph (Fig. 10.6). Unfortunately, there is a tendency to react to changes in performance measures that are due to random variability. P or C charts provide a useful alternative to standard line graphs in some cases.

Another type of line graph has two y axes. Two factors measured on different scales are plotted. The pattern of the up-and-down movement of each factor's line allows evaluation of the association of the two factors. Figure 10.14 is a plot of quality and productivity

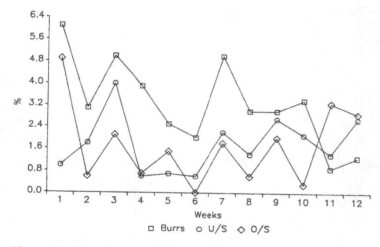

Figure 10.12 Line graph of percentage rejects.

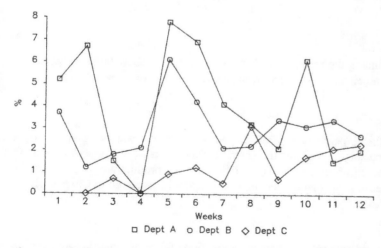

Figure 10.13 Line graph of absenteeism in three departments.

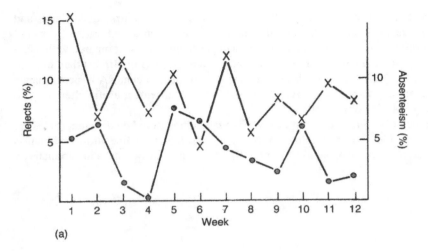

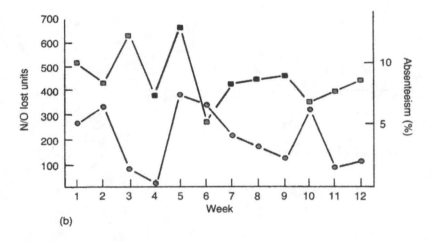

Figure 10.14 Evaluating the relationship between two factors using line graphs: (a) rejects (×) versus absenteeism (●); (b) nonoperating productivity loss (■) versus absenteeism (●).

losses compared with the percentage of absenteeism. (This plot was not prepared by a computer graphics package.) The declining trend for rejects seems to correspond to generally lower levels of absenteeism. The relationship for productivity is not clear. A better method of assessing correlation between two factors is discussed in Chapter 12.

10.4 Pie Chart

A pie chart (circle graph) provides a visual representation of the contribution each part has toward the total. If there are three defects to consider, the two pie charts represent different visual displays of a problem.

10.4 Pie Chart

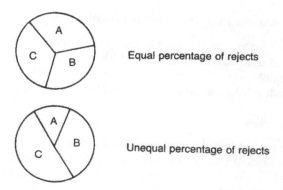

Equal percentage of rejects

Unequal percentage of rejects

The pie chart provides a natural way to communicate the composition of a total.

Drawing a pie chart requires the use of a protractor to draw the angle of a circle attributable to each category. Recall that a complete circle has 360°, regardless of its size. Table 10.5 illustrates the procedure used to obtain the angle associated with each category using the data in Table 10.1. For each reject, the percentage of total rejects is multiplied by 3.6 to obtain the angle to be used in the circle:

Circle angle = % total rejects × 3.6

Figure 10.15 is the resulting pie chart.

Table 10.5 Pie Chart Calculations

Reject	Number rejects	% Total rejects	Pie chart angle (°)
Burrs	229	229/618 = 37.0	37.0 (3.6) = 133
Undersize	124	124/618 = 20.1	20.1 (3.6) = 72
Oversize	117	117/618 = 18.9	18.9 (3.6) = 68
Porosity	77	77/618 = 12.5	12.5 (3.6) = 45
Other	71	71/618 = 11.5	11.5 (3.6) = 42
Total	618	100	360

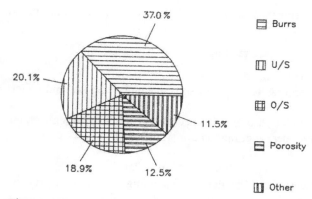

Figure 10.15 Pie chart of quarterly rejects.

10.5 Interpretation Problems with Graphs

Since a graph provides a picture, there is a tendency to immediately interpret the graph and suggest a course of action. The ease with which results are conveyed with a graph contributes to possible misuse of the information. The problem is not with graphic presentation but with a failure to understand and address the problems discussed here.

10.5.1 Sampling Plan

The measurements represented on a graph are the result of a data collection process. Failure to identify and understand this process leads to a potential misinterpretation of graphic results. For example, suppose histograms were prepared for a capability assessment of the two machines at operation 20.

The immediate conclusion is that machine 4 is worse than machine 3. We tend to assume the following:

The upstream process flow is the same for both machines.
Incoming material has not changed.
Parts were measured similarly (same person and gage).
The time over which samples were collected from each machine is the same.

Each of these sampling plan characteristics needs to be understood to clearly interpret the histograms.

An effective way to convey the sampling plan in a graph is to use an effective legend to summarize the sampling plan. For proper interpretation of a graph, the following questions should be answered. A good graph legend should contain this information.

1. Who? The individual(s) who collected parts, made the measurements, and prepared the graph should be listed.
2. What? What characteristic is being measured? If the data are coded, the coding should be made clear by stating

   Actual value = offset + scale × gage reading

   All stratification factors should be stated.
3. Where? The location in the process (e.g., operation, machine, spindle, and pallet) should be identified.
4. When? The time period over which parts are obtained should be stated, along with the sampling method (consecutive, periodic, random, or stratified) as discussed in Chapter 4.
5. Why? The purpose of the data collection should be briefly stated.
6. How? The measurement process should be defined, including the gaging instrument and any operational definitions.

10.5 Interpretation Problems with Graphs

7. How many? The number of measurements made should be stated.

These questions were addressed in Procedure 8.1 for preparing a check sheet. A good check sheet properly defines the sampling plan and thus enables proper interpretation of any resulting graphs.

10.5.2 Measurement System

Interpretation of any data (graphic or otherwise) requires an understanding of the measurement system. Operational definitions should be understood. Again, proper preparation of a check sheet encourages clear definitions. Appendix I discusses methods to evaluate gages. Evaluation of the measurement system should occur prior to data collection and preparation of graphs. Unfortunately, graphic representation of data makes it easy to overlook the measurement system.

10.5.3 Stability

In the ideal case we use graphic methods only to display data from stable processes. In practice, graphs are used for troubleshooting to help identify special causes that contribute to a lack of process stability. Consider the case in which we wish to compare the dimensions of a part produced on two machines. In the ideal case, process stability should be established for each machine and a comparison of the output made. However, a typical approach involves collecting 50 samples from both machines and making histograms. If we make a formal comparison test (Chap. 11) or simple graphic comparison, we do not know what a comparison will produce next week or next month, for example, since stability has not been established. Such factors as raw material, operator, and previous operations could cause a difference in the mean output from the machines. *These factors are totally unrelated to any possible machine differences.* We can search for machine differences when there is no difference. A method of minimizing the opportunity for misinterpretation is to use a balanced sampling plan as discussed in Chapter 11. Any lack of process stability makes consideration of the time over which parts are collected quite important. As the time between measured parts increases, the variability of the process increases. Short-term and long-term graphic results may be quite different.

10.5.4 Insignificant Differences

Graphs are often used to display performance indicators that receive close scrutiny in industry. Changes in these indicators can result in a variety of management reactions. An important question, often not addressed, is whether the change could be the result of random variation. A simple example involves a plot of the percentage of defective parts:

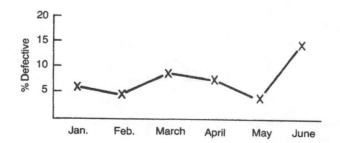

Is 15% defectives in June the result of special cause problems or is it the result of random variability? Certainly the change from 5 to 15% defectives seems alarming. However, only by using the *P* chart control limits can a proper evaluation be made. Similar problems exist with changes in other performance indicators. Caution must be exercised in using the comparison principle. Random variability must be separated from true differences. Control charts are one method of evaluating changes. Chapter 11 discusses a different approach.

10.5.5 Stratification

As discussed in the first section of this chapter, stratification is a powerful tool that should be used with graphic presentation. The failure to use stratified graphs can result in delaying the problem-solving process. Consider an operation that has three machines with the following defects:

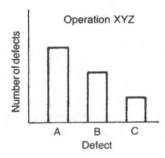 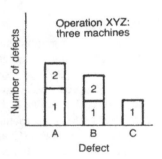

The Pareto analysis on the right clearly suggests that machine 1 is a problem; also, special cause problems must exist with both machines 1 and 2 since machine 3 has no defects. Teams interpreting graphic analyses often conclude that it is necessary to further stratify the results to make an effective interpretation. Unfortunately, the required stratification factors are not routinely collected.

10.5.6 Scaling

The visual impact of a graph can lead to a false impression because of the scale used on either the *x* or *y* axis. It is always good practice to use the same scales on graphs that are to be compared. However, scaling can cause other problems. Consider the three graphs of the same data:

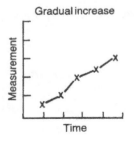 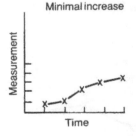 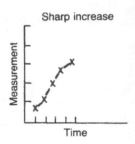

The visual interpretation changes with different plotting scales. This impression carries over to control charts,

Problems

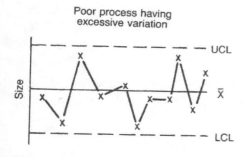

Poor process having excessive variation

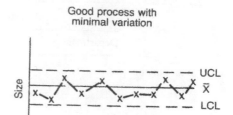

Good process with minimal variation

and bar graphs.

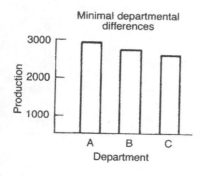

Minimal departmental differences

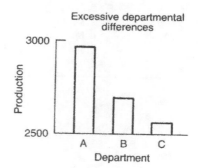

Excessive departmental differences

A false impression is always possible using graphs since some scale must be selected. Care must be used in the interpretation of the graphic results.

Problems

Use a computer graphics package and duplicate the figures given in this chapter.

10.1 The data for Figures 10.2 through 10.5, 10.9 through 10.12, and 10.15 are Tables 10.1 through 10.3.

10.2 The data for Figures 10.7 and 10.8 are from Table 10.4.

10.3 The data for Figures 10.6 and 10.13 are as follows:

	Absenteeism (%)		
	Department A	Department B	Department C
Prior quarter	4.1	5.2	.5
Week 1	5.2	3.7	1.1
2	6.7	1.2	0
3	1.5	1.8	.7
4	0	2.1	0
5	7.8	6.1	.9
6	6.9	4.2	1.2
7	4.1	2.1	.5
8	3.2	2.2	3.1
9	2.1	3.4	.7
10	6.1	3.1	1.7
11	1.5	3.4	2.1
12	2.0	2.7	2.3

11
Comparison Methods

11.1 Improvement and Troubleshooting Using Comparisons

A standard method used in scientific experiments for making improvements is to design a study comparing two or more groups. The data collected from the groups are then used to determine the "best" group. For example, suppose a drug company wants to evaluate whether a new drug is better than an existing drug. A study is designed in which both drugs are given to two groups of subjects. The rate of cure for a disease or time required for a cure is evaluated for each drug and the most effective drug determined. In agriculture, new types of fertilizers are compared with existing brands using a controlled test to make comparison of crop yields. A new type of cola or beer is evaluated using consumer preference tests comparing results with current brands. The list is endless: improvements are made in scientific studies using controlled tests that result in a data-based comparison of different groups.

How can industrial applications make improvements utilizing what is common practice in scientific studies? Consider two identical machines at an operation, each with two spindles:

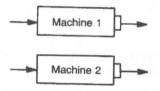

Comparison with new machines is possible, but we are seeking improvements utilizing what is available in current operations. Two key elements of the scientific studies can be used:

1. Form matched groups for comparison.
2. Collect controlled data from these groups.

The matched groups in this example are the four machining spindles. Simply stated, output for each spindle should be similar. If it is not, an improvement opportunity exists. Another approach for improvement involves comparing the current process output to the output after a key process parameter is altered. For example, the spindle speed or feed rate can be

405

changed, a new tool developed, or a new part material composition tried. This logic is the basis of the comparison principle first introduced in Chapter 5.

There are two comparison strategies for identifying process improvement opportunities. Each strategy involves comparing means and standard deviations obtained from controlled studies.

1. Compare all units producing a common characteristic. The process can be improved to at least the level of the best performing unit.
2. Compare current process output with output after selected process parameters (men, methods, machines, material, measurement, and environment) have been temporarily changed. The process can be improved by a better selection of parameters.

In many manufacturing applications, a variety of different matched groups can be formed from the multiple streams of output or stratification factors in a process. Examples of matched groups that occur in industrial settings are machines performing the same operations, heads or spindles on a machine, pallets or fixtures holding parts, filling stations for bottles or cans, and machine operators. In many applications, the improvement opportunity identified by the comparison principle has proven to be large. Also, in most cases minimal expense is required to keep like units producing comparable quality output.

This discussion has focused on making an improvement to a process using comparison methods. These same methods can also be used for troubleshooting process problems. Recall from Figure 5.26 that spindle differences increased the overall process spread. This factor alone can account for a lack of process capability. Thus, troubleshooting the source of a defect or identifying the cause of poor process capability can utilize the comparison principle. Teams involved in troubleshooting can frequently utilize comparisons to identify the causes of a problem.

It is sometimes useful to perform the comparison studies discussed in this chapter prior to implementation of control charts. A comparison study is designed to quickly detect differences between groups. It can be difficult to quickly detect differences in the mean levels of different units using control charts (Prob. 3.11). Differences in variability are even more difficult to uncover. Ideally, baseline control charts (Procedure 4.1) would be established for each unit, but this approach typically requires a significant study period. The comparison procedures discussed here enables quick detection of meaningful differences.

Comparison is not restricted to analyzing only the current process. The second part of the comparison principle states that improvement can be obtained by comparing the output of the existing process to the output of the process with selected process elements modified. Some examples of factors that can be altered in machining operations include manufacturing procedures, tooling design, coolant composition, clamping or part-holding design, material hardness, material composition, machining speeds and feed rates, cycle time, and temperature. The discussion in this chapter focuses on comparisons obtained by changing only a few factors. A more powerful approach is to change a number of factors using designed experiments, as discussed in Box et al. (1978). Taguchi methods have also proven useful in some applications (see Kackar, 1985).

11.1.1 Balanced Sampling

We have seen in Chapter 4 that the sampling plan used to obtain parts that form a subgroup on a control chart can greatly influence the ability of the chart to detect process problems. As subgroup samples were spread out over time, the sources of variability increased and the

11.1 Improvement and Troubleshooting Using Comparisons

control chart became less sensitive to process changes. The method of collecting the parts for making comparisons is equally important. It is necessary to minimize changes to any of the process parameters (men, methods, machines, material, measurements, and environment) other than the parameters being studied (e.g., different machines). Unfortunately, many process parameters are difficult to control and can be expected to change. Examples of parameters that are difficult to control are wearing of tools, temperature of machines, and incoming material. Even in a completely stable process, manufacturing conditions are not identical—variability is simply controlled and predictable. Thus, to make any comparison it is necessary to *balance* the influence of process parameters.

Minimize changing process parameters that are not being studied

There are two approaches to balancing the influence of process parameters. First, control the process parameters by selecting identical manufacturing conditions. For example, use common upstream operations if two machines are being compared. Use common sources of incoming material, operators, gaging, or tools, for example. Many process parameters can be easily selected. Unfortunately, some parameters are difficult to control or their importance to the process being studied is unknown. The second approach for minimizing the influence of process parameters is to use balanced sampling for the units being studied. The term "balanced sampling" simply means that the order in which parts are obtained from the process are balanced across all units being studied. For example, consider a machine that has three fixtures that hold parts. Each fixture rotates across a single common machining station:

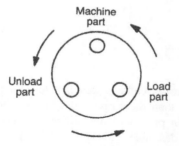

A balanced sampling plan to compare the fixtures is

Fixture 1	Fixture 2	Fixture 3
Time 1	Time 2	Time 3
Time 4	Time 5	Time 6
Time 7	Time 8	Time 9
.	.	.
.	.	.
.	.	.

where "time" refers to the production order in which parts are machined. In this example, as is true in many cases, the balanced sampling plan corresponds to the natural production order. The balanced sampling approach reduces the chance of detecting fixture differences resulting from changes in another important process parameter.

In some applications it is not realistic to use balanced sampling, particularly when process parameters are changed. Consider the case in which a new tool design is compared with the current design. A balanced sampling plan requires that the tool be changed after every part, which is not realistic. A compromise approach is to alternately sample parts throughout a tool's life, alternating the current and new tool designs. Assuming a 500 part tool life, the plan might be as follows:

Group	Part	Tool
I	1	Current tool 1
(Periodically sample *n* parts)		
I	500	Current tool 1
II	501	New tool 1
(Periodically sample *n* parts)		
II	1000	New tool 1
III	1001	Current tool 2
(Periodically sample *n* parts)		
III	1500	Current tool 2
IV	1501	New tool 2
(Periodically sample *n* parts)		
IV	2000	New tool 2

The groups to be compared are current tool 1 and new tool 1 and current tool 2 and new tool 2. If the new tool is better than the current tool, a consistent pattern should emerge. Although this data collection plan does not utilize balanced sampling, the groups are balanced over time, which is desirable. This approach is much more desirable than a single one-time comparison (e.g., compare groups I and II) since changes in process parameters are more likely to be spread across several groups. Selection of the better tool design is based on comparing groups I and III with groups II and IV to get an overall assessment.

In both these sampling plans, we attempt to spread out any time-related process changes over the groups or units being compared. No production process is static: all process parameters change to some degree over time. We see that the run chart is an integral part of the analysis in comparison studies. It is necessary to assess how any changes with time impact on the conclusions of the study.

11.1.2 Statistical Significance*

Examination of the means or standard deviations of two or more groups indicates that the values are not identical. Which differences are greater than would be expected from natural random variability? These differences are called statistically significant. When two or more groups exhibit statistically significant differences in their means or standard deviations, it is concluded that random variation is probably not the cause of the difference. Conversely, if

11.1 Improvement and Troubleshooting Using Comparisons

differences in group means or standard deviations are not statistically significant, it is concluded that the differences may be the result of random variability. Thus, statistical significance provides a benchmark for evaluating group differences.

A lack of statistically significant differences in means or standard deviations does not necessarily imply the groups are the same. The observed differences may be due to random variability from identical groups, or the sample size n may be too small, making it unlikely that differences are detected. Consider the extreme two-group example: in case 1 there are $n = 2$ measurements per group and in case 2 there are $n = 50$ measurements per group. Assume there are identical means for both cases:

	Group I		Group II	
Case	n	$\bar{X}$	n	$\bar{X}$
1	2	3.0	2	5.0
2	50	3.0	50	5.0

A plot of the case 1 points is shown in Figure 11.1a. An individuals histogram for case 2 is shown in Figure 11.1b. In case 1 there is no apparent difference between the groups, but an

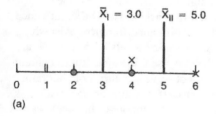

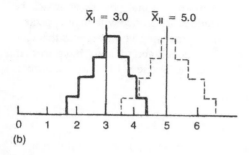

Figure 11.1 Influence of sample size in detecting differences between groups: (a) case 1, $n = 2$ measurements per group; unknown difference between group 1 (●) and group 2 (×); (b) case 2, $n = 50$ measurements per group; obvious difference between group 1 (—) and group 2 (---).

obvious difference exists for case 2. Increasing the sample size increases the likelihood of detecting true differences.

When a statistically significant difference is detected, it can be concluded that the observed difference in means or standard deviations are probably not the result of random variability. However, the differences may or may not be practically important. It is not uncommon to perform a comparison study and obtain a statistically significant difference between two groups that is of no practical consequence. For example, it may be possible to align a machining spindle to within about .001 inch of the desired location. If the specification limits are ±.010, a statistically significant difference between two spindles of, say, .0011 inch is of no practical consequence. In fact, an attempt to make the spindles the "same" may produce a worse result. Increasing the sample size makes detection of smaller differences possible, but some of these differences may not be practically important.

Statistical significance does not imply practical significance

Selection of the sample size required to detect differences can be obtained by a statistical procedure (e.g., Dixon and Massey, 1983, p. 308), assuming the standard deviation and size of the difference to be detected are known. Larger sample sizes are required to detect differences in standard deviations. In many manufacturing applications, the following guidelines have proven useful in detecting practically significant differences:

Mean differences: $\qquad n = 15$ or more per group.
Standard deviation differences: $\quad n = 30$ or more per group.

A concept associated with determining statistically significant differences is the significance level of the test used to assess whether significant differences exist. The significance level is the chance of making an error by concluding that significant differences exist when, in fact, differences are due to random variability. Significance levels of 5 and 1% are commonly used. In the text procedures only the 5% level is used; the appendixes contain tables for both levels. These tabled values for testing do not serve as a threshold but rather as a guide for evaluating whether differences may exist. If a value greater than z is statistically significant, values slightly less than z cannot be ignored. In many cases additional data collection may be warranted.

The comparison procedures in this chapter do not emphasize the use of formal statistical testing by necessarily optimal testing procedures. A more informal graphic approach is used in which separate group means and standard deviations are plotted with their expected variability. These plots have proven useful in identifying meaningful differences that can lead to significant quality improvement. The reader is encouraged to study more formal statistical comparison procedures in such texts as Dixon and Massey (1983) and Snedecor and Cochran (1980).

11.2 Variability Comparisons

Excess process variability is a common problem in manufacturing. This variability results in a lack of process capability (Chap. 7), which is also displayed as excessive variation on a range control chart. Chapter 6 discussed why some changes in variability are difficult to detect (e.g., Fig. 6.7) using control charts. Detection of even a moderate change in spindle variability on a three-spindle machine can be difficult (Fig. 6.18). A useful approach is to perform a comparison study of the variation of the groups or units. The typical measure of

11.2 Variability Comparisons

variability used is the standard deviation of each group. If differences are detected, it should be possible to reduce the variability of groups having the higher standard deviations to a level comparable to groups having the lower standard deviation.

Comparison of individual standard deviations for a group can be conveniently performed using the Hartley (1950) F_{max} test. Assuming there are n measurements made for each group (preferably using balanced sampling), the standard deviations can then be calculated for each group:

Group	1	2	3	$\cdots$	k
Standard deviation	s_1	s_2	s_3	$\cdots$	s_k

The F_{max} ratio can then be obtained by determining the smallest standard deviation s_{min}, and the largest standard deviation s_{max}:

$$F_{max} \text{ ratio} = \left(\frac{s_{max}}{s_{min}}\right)^2$$

If the F_{max} ratio is larger than the tabled value (see Nelson, 1987b) in Procedure 11.1 or Appendix III, a significant difference in variability probably exists. As with all the comparison procedures in this chapter, it is important to first plot a run chart and histogram for the groups being compared. This practice provides a visual assessment of any potential process changes and evaluation of potential group differences.

Example 11.1 Consider the $n = 30$ consecutive measurements in Table 11.1 obtained from five machines. The run chart for each machine appears in Figure 11.2. Machine 5 appears more variable than the other machines, and machine 2 appears to have a higher mean. However, there do not appear to be trends in the run chart. Figure 11.3 contains histograms for each machine. There are no obvious extreme measurements that might indicate a problem with the data. Again, problems with machines 2 and 5 are suggested by the histograms. The variability of the machines can be evaluated using the standard deviations:

Machine	1	2	3	4	5
SD (s)	1.66	1.91	1.63	2.30	3.99

The smallest s is 1.63 and the largest 3.99. Thus, the F_{max} ratio is

$$F_{max} \text{ ratio} = \left(\frac{3.99}{1.63}\right)^2 = 5.99$$

For $k = 5$ units and $n = 30$, the value in the table from Procedure 11.1 is about 2.8, which is less than 5.99. Thus, a significant difference in variability exists.

Once the F_{max} ratio is used to evaluate whether significant differences exist, how can the units with different variability be determined? A natural procedure is to construct an

Procedure 11.1 Variability Comparison

It is possible to evaluate differences in the variability of selected groups using an F_{max} test and a variation comparison (VC) chart. Let a constant number of measurements n be made on each of k groups, where the values are arranged as follows:

Measurement	Group			
	1	2	$\cdots$	k
1	X_1	X_1	$\cdots$	X_1
2	X_2	X_2	$\cdots$	X_2
.	.	.		.
.	.	.		.
.	.	.		.
n	X_n	X_n	$\cdots$	X_n
Mean	$\bar{X}_1$	$\bar{X}_2$	$\cdots$	$\bar{X}_n$
SD	s_1	s_2	$\cdots$	s_k

The number of measurements obtained from each group should be $n = 30$ or more. Smaller sample sizes have an increased risk of not detecting meaningful differences in variability. The time interval over which the measurements are collected should not be so large that between-group differences are masked by other process differences. Balanced sampling should be used if possible. The following steps are used for a variability comparison:

1. A run chart and histogram (Procedure 5.1) should be prepared for each group. These plots should be examined for process irregularities or very unusual measurements that might adversely influence the comparison. The histogram should represent an approximately normal, bell-shaped distribution, or transformations (Chap. 5) of the data should be considered.

2. Compute the mean and standard deviation for each group:

$$\bar{X} = \frac{X_1 + X_2 \cdots X_n}{n}$$

$$s = \sqrt{\frac{(X_1 - \bar{X})^2 + (X_2 - \bar{X})^2 + \cdots + (X_n - \bar{X})^2}{n - 1}}$$

Procedure 3.1 gives steps to use for computing s.

3. Find the smallest standard deviation s_{min} and the largest standard deviation s_{max}. Compute the ratio

$$F_{max} \text{ ratio} = \left(\frac{s_{max}}{s_{min}}\right)^2$$

If this ratio is larger than the value in the following table, a significant difference in variability between groups probably exists. A VC chart (step 4) can be used to identify the unusual groups. If the ratio is smaller than the tabled value, no difference in variability can be detected (go to step 5).

n	\multicolumn{9}{c}{k}
	2	3	4	5	6	7	8	9	10
5	9.6	15.5	20.6	25.2	29.5	33.6	37.5	41.2	44.8
10	4.0	5.3	6.3	7.1	7.8	8.4	8.9	9.4	9.9
16	2.9	3.5	4.0	4.4	4.7	4.9	5.2	5.4	5.6
21	2.5	3.0	3.3	3.5	3.7	3.9	4.1	4.2	4.4
31	2.1	2.4	2.6	2.8	2.9	3.0	3.1	3.2	3.3
61	1.7	1.8	2.0	2.0	2.1	2.2	2.2	2.3	2.3
Max	1.0	1.0	1.0	1.0	1.0	1.0	1.0	1.0	1.0

4. If difference in group variability exist, prepare a VC chart to assess which groups differ. For each group, prepare an upper and lower variation limit line.

Upper variation limit:

$$UVL = C_2 s$$

Lower variation limit:

$$LVL = C_1 s$$

Plot the interval (LVL, UVL) for each group along with the standard deviation. Groups with nonoverlapping intervals are likely to have different variability.

n	C_1	C_2
5	.60	2.87
10	.69	1.83
15	.73	1.58
20	.76	1.46
25	.78	1.39
30	.80	1.34
35	.81	1.31
40	.82	1.28
45	.83	1.26
50	.83	1.25
60	.85	1.22
70	.86	1.20
80	.86	1.18
90	.87	1.17
100	.88	1.16

5. If no group differences in variability exist, compute the pooled standard deviation:

$$s_p = \sqrt{\frac{s_1^2 + s_2^2 + \cdots + s_k^2}{k}}$$

which has statistical degrees of freedom,

$$f = (n - 1)k$$

Table 11.1 Consecutive Measurements from Five Machines

No.	Machine				
	1	2	3	4	5
1	3	5	−3	1	−4
2	1	−1	−3	1	−8
3	−1	4	0	0	−1
4	0	0	1	1	−3
5	0	2	0	−1	3
6	−1	1	−2	3	8
7	0	4	−1	1	2
8	−1	3	−4	2	−5
9	0	5	1	−1	2
10	0	4	0	2	−5
11	−5	3	−2	−2	9
12	−3	2	−1	−3	2
13	1	1	1	2	−2
14	−1	3	2	1	−1
15	0	5	−1	3	−5
16	1	1	2	−4	0
17	−1	7	0	2	−3
18	−2	6	0	0	1
19	−1	4	0	−1	−5
20	1	4	3	0	−7
21	0	3	2	1	−2
22	1	6	0	8	1
23	1	5	−1	4	4
24	−2	5	1	2	−4
25	2	4	1	3	2
26	1	2	0	2	2
27	0	5	0	0	0
28	−2	6	0	4	1
29	0	3	2	0	−3
30	3	3	1	0	−2
Mean	−.17	3.50	−.03	1.03	−.77
SD	1.66	1.91	1.63	2.30	3.99

11.2 Variability Comparisons

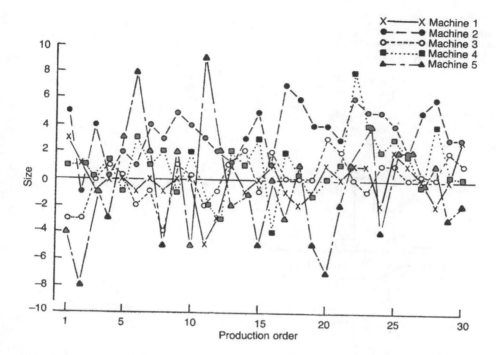

Figure 11.2 Run chart for the five machines in Example. 11.1.

interval for each group in which the standard deviation may be expected to occur. This interval accounts for the expected random variability that could occur in a stable process. The limits of the interval are given in Procedure 11.1 and are shown here.

Upper variation limit:

$$\text{UVL} = C_2 s$$

Lower variation limit:

$$\text{LVL} = C_1 s$$

Plotting the upper and lower limits for each group in what we shall call a VC chart (variation comparison) enables a visual assessment of possible group differences using the criteria shown in Figure 11.4. This comparison is valid whether or not the group means differ.

Example 11.1 (continued) The upper and lower variation limits can be computed for each group using Procedure 11.1 with the constants

$$n = 30 \quad C_1 = 0.80 \quad C_2 = 1.34$$

Multiplying each group s by these two factors enables computation of LVL and UVL:

Machine	LVL	s	UVL
1	1.33	1.66	2.22
2	1.53	1.91	2.56
3	1.30	1.63	2.18
4	1.84	2.30	3.08
5	3.19	3.99	5.35

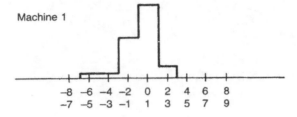

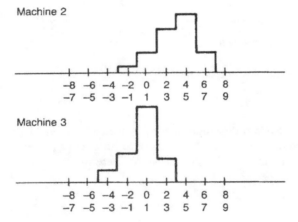

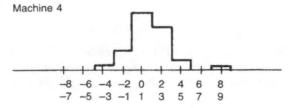

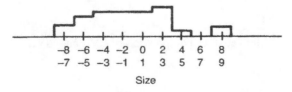

Figure 11.3 Histograms for the five machines in Example 11.1.

11.2 Variability Comparisons

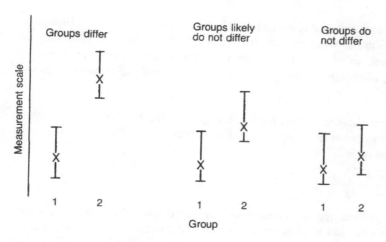

Figure 11.4 Interpretation of group interval differences. The likelihood of group differences decreases as the overlap of the group intervals increases.

These values are plotted on a VC chart in Figure 11.5. It is apparent that machine 5 has greater variability than the other machines.

It is interesting to note that the data in Table 11.1 were generated using the simulation procedure described in Chapter 6. The machines in this example correspond to

Machine	Population	Mean	σ
1	A	0	1.74
2	B	2	1.74
3	A	0	1.74
4	G	0	2.35
5	F	0	3.74

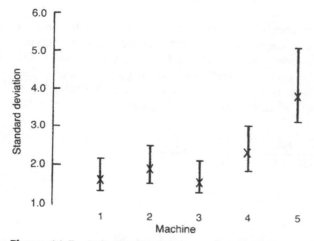

Figure 11.5 VC chart comparison of machine standard deviations for Example 11.1.

All population values are contained in the intervals in Figure 11.5, as expected. However, the comparison procedure was not able to clearly identify the increase in variability of machine 4. A different set of $n = 30$ measurements from each group might show a difference. However, increasing the sample size to $n = 50$ would make detection of the difference more likely.

A pooled or combined standard deviation is often useful if the F_{max} ratio suggests that there are no differences between group standard deviations. From Procedure 11.1 the pooled standard deviation is

$$s_p = \sqrt{\frac{s_1^2 + s_2^2 + s_3^2 + \cdots + s_k^2}{k}}$$

This value does not assume that the group means are necessarily the same. When the process is stable, s_p can be used as an estimate of the overall process standard deviation. A statistical value associated with s_p is the "degrees of freedom," which is

$$f = (n - 1)k$$

The use of the f value is discussed later in this chapter.

Example 11.1 (continued) Suppose from the VC chart in Figure 11.5 that we are willing to assume that machines 1, 2, and 3 have the same variability. A pooled estimate of variability is

$$s_p = \sqrt{\frac{(1.66)^2 + (1.91)^2 + (1.63)^2}{3}} = 1.738$$

which is close to the true value (1.74) used in simulating the data from Chapter 6. The degrees of freedom for s_p are $f = 29 \times 3 = 87$.

It is important to note that the F_{max} comparison procedure described in this section is not the optimal procedure from a statistical viewpoint. A test such as Levene's test (Snedecor and Cochran, 1980, p. 253) is generally a better testing procedure. Other procedures are discussed and compared by Gartside (1972) and Conover et al. (1981). The F_{max} test is a simple, easy-to-use procedure that provides useful information in many manufacturing situations. A formal test for determining which of the k variances differ is given by Nelson (1987a).

This section discusses variability comparisons when the sample size n is the same for each group. If the sample sizes differ and n_i measurements are obtained for group i, it is not possible to use the F_{max} ratio. However, Levene's test or Bartlett's test (Snedecor and Cochran, 1980, p. 252) can be used to evaluate overall differences in group variability. The VC chart can be prepared in the same manner as step 5 of Procedure 11.1 using the C constants appropriate for each group's n_i. If the variability is assumed the same, the pooled standard deviation becomes

$$s_p = \sqrt{\frac{(n_1 - 1)s_1^2 + (n_2 - 1)s_2^2 + \cdots + (n_k - 1)s_k^2}{(N - k)}}$$

where

$$N = (n_1 + n_2 + \cdots + n_k)$$

The degrees of freedom for s_p are

$$f = N - k$$

11.3 Location Comparisons

Evaluation of differences in location between groups is an important part of improving the capability of a process. Chapter 5 contained several examples (e.g., Fig. 5.26 or Prob. 5.7) in which significant process improvement could be realized by making machine adjustments so that different fixtures, spindles, or stations, for example, have a common mean for the characteristic being evaluated. Of course, separate control charts can be used for each group and the location and variability compared in baseline studies (Procedure 4.1). However, it is generally not possible to maintain separate control charts on the many different stratification factors in a process. The location comparison procedures discussed in this section provide a useful supplement to standard $\bar{X}$ charts. These procedures can detect group differences that can be difficult to identify using control charts in which the out-of-control signals are associated with "mixture" problems (Chap. 6).

Selection of the appropriate method to evaluate differences in location depends on whether it is reasonable to assume that each group has about the same variability. Figure

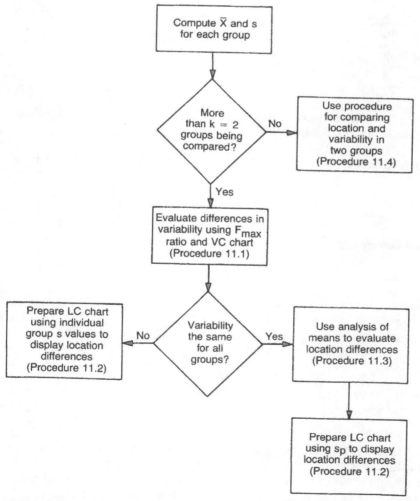

Figure 11.6 Overview of group comparison procedures.

11.6 gives an overview of the procedures discussed in this chapter. The LC chart (location comparison) is used to display group differences in location similarly to the use of the VC chart to display variability differences. The computations used to prepare an LC chart vary slightly depending on whether a common variability for all groups is assumed. The analysis of means procedure developed by Ellis Ott is used to evaluate location differences if a common group variability can be assumed. A special comparison method for $k = 2$ groups is given in Procedure 11.4. Throughout this section it is assumed that we have computed the individual k group means and standard deviations as noted in step 2 of Procedure 11.1.

11.3.1 LC Charts

An $\bar{X}$ chart displays the variability of subgroup means over time. If a stable process exists, these $\bar{X}$ values should vary between the control limits UCL and LCL. In a similar manner the individual group means obtained in a comparison study can be expected to vary between the upper and lower mean limits.

Upper location limit:

$$\text{ULL} = \bar{\bar{X}} + \frac{st}{\sqrt{n}}$$

Lower location limit:

$$\text{LLL} = \bar{\bar{X}} - \frac{st}{\sqrt{n}}$$

The value of t for the n measurements used to compute $\bar{X}$ is given in Procedure 11.2 along with the complete LC chart procedure.

Example 11.1 (continued) From Table 11.1 the means and standard deviations can be used with

$$n = 30 \qquad t = 2.04 \qquad \frac{t}{\sqrt{n}} = 0.37$$

to prepare the following table:

Machine	$\bar{X}$	s	$st/\sqrt{n}$	ULL	LLL
1	−.17	1.66	.61	.44	−.78
2	3.50	1.91	.71	4.21	2.79
3	−.03	1.63	.60	.57	−.63
4	1.03	2.30	.85	1.88	.18
5	−.77	3.99	1.48	.71	−2.25

The LC chart in Figure 11.7 clearly shows that machine 2 has an unusually high mean. This confirms the indication given in the run chart and histograms in Figures 11.2 and 11.3. Notice that the length of the LLL, ULL line is directly increased by the individual group standard deviation s. The larger variability of group 5 is clearly indicated in Figure 11.7.

11.3 Location Comparisons

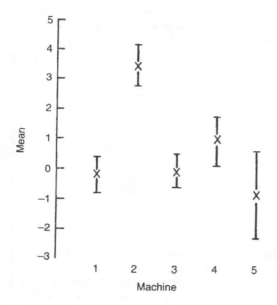

Figure 11.7 LC chart comparison of machine means for Example 11.1.

It is also possible to use LC charts when the group variabilities do not differ. In this case the pooled standard deviation s_p is used to compute ULL and LLL. The value of t must be obtained from Procedure 11.2 (step 5) using the degrees of freedom

$$f = (n - 1)k$$

This approach is the preferable LC chart since the t value is smaller, which results in a narrower distance between ULL and LLL. Thus, differences in location between groups can be more easily detected. For a common group variability, the upper location limit is

$$\text{ULL} = \bar{\bar{X}} + \frac{s_p t}{\sqrt{n}}$$

The lower location limit is

$$\text{LLL} = \bar{\bar{X}} - \frac{s_p t}{\sqrt{n}}$$

Example 11.1 (continued) Consider the measurements in Table 11.1, but exclude machine 5. In this case the $F_{\max}$ ratio does not indicate that variability differed between groups. The pooled standard deviation and degrees of freedom are

$$s_p = \sqrt{\frac{s_1^2 + s_2^2 + s_3^2 + s_4^2}{4}}$$

$$= \sqrt{\frac{(1.66)^2 + (1.91)^2 + (1.63)^2 + (2.30)^2}{4}}$$

$$= 1.89$$

$$f = (n - 1)k$$
$$= (30 - 1)4$$
$$= 116$$

Procedure 11.2 LC Charts

The purpose of a location comparison (LC) chart is to evaluate possible differences in location using means for selected groups. Let a constant number of measurements n be made on each of k groups where the values are arranged as follows:

	Group			
Measurement	1	2	$\cdots$	k
1	X_1	X_1	$\cdots$	X_1
2	X_2	X_2	$\cdots$	X_2
.	.	.		.
.	.	.		.
.	.	.		.
n	X_n	X_n	$\cdots$	X_n
Mean	$\bar{X}_1$	$\bar{X}_2$	$\cdots$	$\bar{X}_n$
SD	s_1	s_2	$\cdots$	s_k

The number of measurements obtained from each group should be $n = 15$ or more. Smaller sample sizes have an increased risk of not detecting meaningful differences in location. The time interval over which the measurements are collected should not be so large that any between-group differences are masked by other process differences. Balanced sampling should be used if possible. The following steps are used for a location comparison:

1. A run chart and histogram (Procedure 5.1) should be prepared for each group. These plots should be examined for process irregularities or very unusual measurements that could adversely influence the comparison. This histogram should represent an approximately normal, bell-shaped distribution, or transformations (Chap. 5) of the data should be considered.
2. Compute the mean and standard deviation for each group:

$$\bar{X} = \frac{X_1 + X_2 \cdots + X_n}{n}$$

$$s = \sqrt{\frac{(X_1 - \bar{X})^2 + (X_2 - \bar{X})^2 + \cdots + (X_n - \bar{X})^2}{n - 1}}$$

Procedure 3.1 gives steps to use for computing s.
3. Determine whether all k groups being compared have a common standard deviation using the F_{max} ratio and the VC chart (Procedure 11.1). Go to step 6 if the groups have different variability and to step 4 if there is common variability between groups.

Common Group Variability

4. If the k group standard deviations can be assumed not to differ, compute the pooled standard deviation s_p and its degree of freedom f:

$$s_p = \sqrt{\frac{s_1^2 + s_2^2 + \cdots + s_k^2}{k}}$$

$$f = (n - 1)k$$

5. Use the $\bar{X}$ for each group, and compute the upper and lower limit lines.

Upper location limit:

$$ULL = \bar{\bar{X}} + \frac{s_p t}{\sqrt{n}}$$

Lower location limit:

$$LLL = \bar{X} - \frac{s_p t}{\sqrt{n}}$$

Plot the interval (LLL, ULL) for each group along with the mean. Groups with nonoverlapping intervals are likely to have different means. The ANOM in Procedure 11.3 should be used for a more exact evaluation of possible group differences. If $k = 2$ groups, Procedure 11.4 should be used.

f	t
10	2.23
15	2.13
20	2.09
25	2.06
30	2.04
40	2.02
50	2.01
60	2.00
70	1.99
100	1.98
Max	1.96

Different Group Variability

6. If the k group standard deviations cannot be assumed the same, compute the upper and lower mean limit lines for each group. Use the $\bar{X}$ and s for each group.

   Upper location limit:

   $$ULL = \bar{X} + \frac{st}{\sqrt{n}}$$

   Lower location limit:

   $$LLL = \bar{X} - \frac{st}{\sqrt{n}}$$

   Plot the interval (LLL, ULL) for each group along with the mean. Groups with nonoverlapping intervals are likely to have different means. If $k = 2$, Procedure 11.4 should be used to evaluate group differences.

n	t
5	2.78
10	2.26
15	2.14
20	2.09
25	2.06
30	2.05
40	2.02
50	2.01
60	2.00
70	1.99
100	1.98
Max	1.96

From Procedure 11.2, the value of t corresponding to $f = 100$, which is close to 116, is 1.98. With $s_p = 1.89$, $n = 30$, and $t = 1.98$ ($s_p t/\sqrt{n} = 0.68$), the following table can be prepared:

Machine	$\bar{X}$	ULL	LLL
1	−.17	.51	−.85
2	3.50	4.18	2.82
3	−.03	.65	−.71
4	1.03	1.71	.35

The LC chart can be prepared from this table.

The LC chart can also be used when the sample size n differs between groups. If the variabilities differ between groups, ULL and LLL are computed using the s, t, and n values appropriate for each group. If the variabilities do not differ between groups, s_p is computed using the procedure described at the end of the previous section (for varying n_i). The f degrees of freedom value used in Procedure 11.2 are $f = N - k$, where N is the total number of measurements for all groups.

A team troubleshooting a chronic quality problem found that stratification at an upstream machining operation caused a variety of downstream processing problems. Differences in the alignment of the multiple spindles resulted in both location and variability differences. Fortunately, the machine's alignment characteristics did not change rapidly once proper adjustments were made. As an ongoing control method, the team developed check sheets from which location and variability comparison charts were prepared on a weekly basis. These charts were posted at the operation.

11.3.2 Analysis of Means

Ellis Ott (Ott and Snee, 1973; Ott, 1975) developed an analysis of means (ANOM) procedure that provides an overall evaluation of mean differences for k groups. This procedure assumes that all groups have common variability. The pooled standard deviation s_p with $f = (n - 1)k$ degrees of freedom is used as an estimate of the common standard deviation. The case for a constant n for all groups is presented here; Nelson (1983a) discusses the case for unequal n. The ANOM procedure provides a method to test for mean differences, but the LC chart can still be used to display the group means.

The ANOM procedure is based on computing an upper and lower decision line that is equally spaced above and below the overall mean $\bar{\bar{X}}$ (see, e.g., Schilling, 1973). Each mean is then plotted on an ANOM chart. If any mean is outside the decision lines, the mean is said to be different from the remaining means. Calculation of the decision lines is similar to that for the LC chart lines.

Upper decision line:

$$\text{UDL} = \bar{\bar{X}} + \frac{s_p h}{\sqrt{n}} \sqrt{\frac{k-1}{k}}$$

11.3 Location Comparisons

Lower decision line:

$$LDL = \bar{\bar{X}} - \frac{s_p h}{\sqrt{n}} \sqrt{\frac{k-1}{k}}$$

The value of the constant h depends on the number n of measurements used to compute the group $\bar{X}$ values and is given in Procedure 11.3 along with the complete ANOM procedure. The individual $\bar{X}$ values along with UDL and LDL are typically plotted to form an ANOM chart, which is used to evaluate mean differences.

Example 11.1 (continued) Assume machine 5 was not part of this example. The F_{max} test then concludes that there are no differences in the variability for machines 1 through 4. The overall mean from Table 11.1 is

$$\bar{\bar{X}} = \frac{\bar{X}_1 + \bar{X}_2 + \bar{X}_3 + \bar{X}_4}{4}$$

$$= \frac{(-.17) + 3.50 + (-.03) + 1.03}{4}$$

$$= 1.08$$

The pooled standard deviation was computed previously as $s_p = 1.89$ with $f = (30 - 1)4 = 116$ degrees of freedom. The value of h from Procedure 11.3 is $h = 2.5$. Thus, the upper and lower decision lines can be computed:

$$UDL = 1.08 + \frac{1.89 \times 2.5}{\sqrt{30}} \sqrt{\frac{4-1}{4}}$$

$$= 1.08 + 0.75$$

$$= 1.83$$

$$LDL = 1.08 - 0.75$$

$$= 0.33$$

The ANOM chart is given in Figure 11.8a. It is obvious that the machine means are different, with machine 2 having a higher mean. Excluding machine 2 from the mean calculations, the overall average is

$$\bar{\bar{X}} = \frac{(-.17) + (-.03) + 1.03}{3} = 0.28$$

The decision lines for $k = 3$ groups and $f = 116$ are

$$UDL = 0.28 + \frac{1.89 \times 2.37}{\sqrt{30}} \sqrt{\frac{3-1}{3}}$$

$$= .28 + .67$$

$$= .95$$

$$LDL = .28 - .67$$

$$= -.39$$

The ANOM chart is given in Figure 11.8b. We conclude that machine 4 may be different from machines 1 and 3.

Procedure 11.3 ANOM Charts

The purpose of an analysis of means (ANOM) chart is to evaluate possible differences in location using means for different groups. The ANOM procedure assumes the variability for the different groups is the same. Let a constant number of measurements n be made on each of k groups where the values are arranged as follows:

	Group			
Measurement	1	2	...	k
1	X_1	X_1	...	X_1
2	X_2	X_2	...	X_2
.	.	.		.
.	.	.		.
.	.	.		.
n	X_n	X_n	...	X_n
Mean	$\bar{X}_1$	$\bar{X}_2$	...	$\bar{X}_n$
SD	s_1	s_2	...	s_k

The number of measurements obtained from each group should be $n = 15$ or more. Smaller sample sizes have an increased risk of not detecting meaningful differences in location. The time interval over which the measurements are collected should not be so large that any between-group differences are masked by other process differences. Balanced sampling should be used if possible. The following steps are used for an ANOM:

1. A run chart and histogram (Procedure 5.1) should be prepared for each group. These plots should be examined for process irregularities or very unusual measurements that could adversely influence the comparison. This histogram should represent an approximately normal, bell-shaped distribution, or transformations (Chap. 5) of the data should be considered.
2. Compute the mean and standard deviation for each group:

$$\bar{X} = \frac{X_1 + X_2 \cdots + X_n}{n}$$

$$s = \sqrt{\frac{(X_1 - \bar{X})^2 + (X_2 - \bar{X})^2 + \cdots + (X_n - \bar{X})^2}{n - 1}}$$

Procedure 3.1 gives the steps for computing s.

3. Determine whether all k groups being compared have a common standard deviation using the F_{max} ratio and the VC chart (Procedure 11.1). If the variability between groups cannot be assumed the same, it is not appropriate to use this ANOM procedure. An LC chart should be prepared (Procedure 11.2).
4. Assuming the k groups have a common standard deviation, compute the pooled mean $\bar{\bar{X}}$ and standard deviation s_p with its degrees of freedom f:

$$\bar{\bar{X}} = \frac{\bar{X}_1 + \bar{X}_2 + \cdots + \bar{X}_k}{k}$$

$$s_p = \sqrt{\frac{s_1^2 + s_2^2 + \cdots + s_k^2}{k}}$$

$$f = (n - 1)k$$

5. Compute the upper and lower decision lines.

   Upper decision line:

   $$\text{UDL} = \bar{\bar{X}} + \frac{s_p h}{\sqrt{n}} \sqrt{\frac{k-1}{k}}$$

   Lower decision line:

   $$\text{LDL} = \bar{\bar{X}} - \frac{s_p h}{\sqrt{n}} \sqrt{\frac{k-1}{k}}$$

The h factor can be obtained from the following table or from the more complete table in Appendix III.

	Number of groups k								
f	2	3	4	5	6	7	8	9	10
3	3.18	4.18							
4	2.78	3.56	3.89						
5	2.57	3.25	3.53	3.72					
6	2.45	3.07	3.31	3.49	3.62				
7	2.37	2.94	3.17	3.33	3.45	3.56			
8	2.31	2.86	3.07	3.21	3.33	3.43	3.51		
9	2.26	2.79	2.99	3.13	3.24	3.33	3.41	3.48	
10	2.23	2.74	2.93	3.07	3.17	3.26	3.33	3.40	3.45
15	2.13	2.60	2.76	2.88	2.97	3.05	3.11	3.17	3.22
20	2.09	2.53	2.68	2.79	2.88	2.95	3.01	3.06	3.11
30	2.04	2.47	2.61	2.71	2.79	2.85	2.91	2.96	3.00
40	2.02	2.43	2.57	2.67	2.75	2.81	2.86	2.91	2.95
60	2.00	2.40	2.54	2.63	2.70	2.76	2.81	2.86	2.90
120	1.98	2.37	2.50	2.59	2.66	2.72	2.77	2.81	2.84
Max	1.96	2.34	2.47	2.56	2.62	2.68	2.72	2.76	2.80

6. The ANOM chart is prepared by plotting $\bar{\bar{X}}$ with a solid line and UDL, LDL with dashed lines. The group means are then plotted. If any group mean is beyond UDL or LDL, significant differences in group location probably exist.

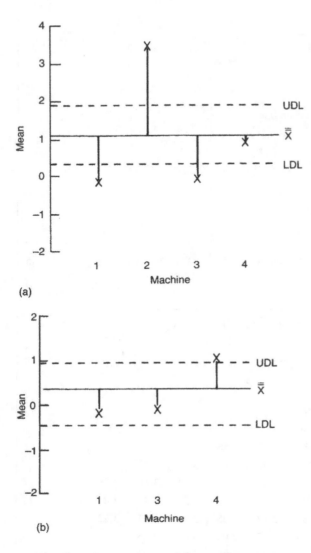

Figure 11.8 Comparison of Example 11.1 machine means using an ANOM chart: (a) machine 5 eliminated; (b) machines 2 and 5 eliminated.

11.4 Comparison of Two Groups (Different Parts)

The special case in which it is of interest to compare the location and variability for $k = 2$ groups occurs frequently. Some typical examples are listed here:

Before versus after an improvement: work is performed to improve part of a process. Comparison of measurement before and after the improvement provides an assessment of the improvement's impact.

Before versus after process events: the influence of process events can be determined by comparing units produced at different times. Examples of process events include machine start-up, tool change, use of new operator, and performance of preventive maintenance.

11.4 Comparison of Two Groups (Different Parts)

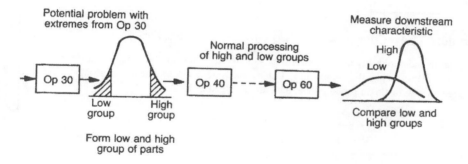

Figure 11.9 Comparison of upstream extremes after downstream processing.

Treatment versus no treatment: the effectiveness of a treatment can be determined by comparing units that have had a treatment with those that did not.

New versus old method: the effectiveness of a new method can be determined by comparing units produced by the new and the old methods.

Strata 1 with strata 2: two machines or two spindles on the same machine, for example, can be evaluated for differences.

Inexpensive versus expensive method: quantification of the benefit of an expensive method is obtained by comparison with an inexpensive method.

High versus low in-process extremes: a problem is encountered when it is suspected that high or low values of an upstream in-process characteristic are influencing downstream part quality. Groups of high and low upstream parts can be selected, as illustrated in Figure 11.9. Comparison of the downstream low versus high group means and standard deviations can detect process problems. It is not necessary to use the same characteristic for determining the upstream groups and making the downstream comparison. When the same parts are measured at two points in a process, Procedure 12.5 should be used.

Cause for attribute rejects: the cause for a reject may be related to a measurable part characteristic. Two groups can be formed by sorting good and reject parts. The cause of the reject can be an attribute (e.g., burr versus no burr) or variable (e.g., above USL versus below USL) characteristic. For each group a characteristic thought to be related to the reject condition can be measured and the means and standard deviations compared.

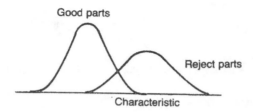

A difference in the means or standard deviations may be related to the cause of the reject.

The VC or LC charts can be prepared for the $k = 2$ groups. However, the comparison procedures given in Procedure 11.4 provide a better method of testing for group differences.

Procedure 11.4 Comparison of Two Groups (Different Parts)

The location and variability of independent measurements made on two groups of different parts can be compared using this procedure. Let n_1 measurements be made on group 1 and n_2 measurements be made on group 2. The data are arranged as follows:

	Group 1	Group 2
	X_1	X'_1
	X_2	X'_2
	.	.
	.	.
	.	.
	X_{n_1}	X'_{n_2}
No. measurements	n_1	n_2
Mean	$\bar{X}_1$	$\bar{X}_2$
SD	s_1	s_2

The number of measurements obtained from each group should be 15 or more for location comparisons and 30 or more for variability comparisons. Smaller sample sizes have an increased risk of not detecting meaningful differences. The time interval over which the measurements are collected should not be so large that any between-group differences are masked by other process differences. Balanced sampling should be used if possible. The following steps are used to compare variability and location:

1. A run chart and histogram (Procedure 5.1) should be prepared for each group. These plots should be examined for process irregularities or very unusual measurements that could adversely influence the comparison. The histogram should represent an approximately normal, bell-shaped distribution, or transformations (Chap. 5) of the data should be considered.
2. Compute the mean and standard deviation for each group. For group 1,

$$\bar{X}_1 = \frac{X_1 + X_2 + \cdots + X_{n_1}}{n_1}$$

$$s_1 = \sqrt{\frac{(X_1 - \bar{X}_1)^2 + (X_2 - \bar{X}_1)^2 + \cdots + (X_{n_1} - \bar{X}_1)^2}{n_1 - 1}}$$

Procedure 3.1 gives the steps for computing s. An analogous calculation is used for group 2.

Comparison of Variability

3. Compute the F_{max} ratio by dividing the largest group standard deviation by the smallest standard deviation:

$$F_{max} \text{ ratio} = \left(\frac{s_{max}}{s_{min}}\right)^2$$

If the ratio is larger than the value in the following table, a significant difference in variability between groups probably exists (go to step 6). If the ratio is smaller than the tabled value, no difference in variability can be detected (go to step 4). A VC chart (Procedure 11.1) can be used to display any group differences.

n for s_{min}	\multicolumn{12}{c}{n for s_{max}}											
	5	6	7	8	9	10	15	20	25	30	50	100	Max
5	6.4	6.3	6.2	6.1	6.0	6.0	5.9	5.8	5.8	5.7	5.7	5.7	5.6
6	5.2	5.1	5.0	4.9	4.8	4.8	4.6	4.6	4.5	4.5	4.4	4.4	4.4
7	4.5	4.4	4.3	4.2	4.2	4.1	4.0	3.9	3.8	3.8	3.8	3.7	3.7
8	4.1	4.0	3.9	3.8	3.7	3.7	3.6	3.4	3.4	3.4	3.3	3.3	3.2
9	3.8	3.7	3.6	3.5	3.4	3.4	3.2	3.2	3.1	3.1	3.0	3.0	2.9
10	3.6	3.5	3.4	3.3	3.2	3.2	3.0	2.9	2.9	2.9	2.8	2.8	2.7
15	3.1	3.0	2.9	2.8	2.7	2.7	2.5	2.4	2.4	2.3	2.2	2.2	2.1
20	2.9	2.7	2.6	2.5	2.5	2.4	2.3	2.2	2.1	2.1	2.0	1.9	1.9
25	2.8	2.6	2.5	2.4	2.4	2.3	2.1	2.0	2.0	1.9	1.9	1.8	1.7
30	2.7	2.5	2.4	2.4	2.3	2.2	2.1	1.9	1.9	1.8	1.8	1.7	1.6
50	2.6	2.4	2.3	2.2	2.1	2.1	1.9	1.8	1.7	1.7	1.6	1.5	1.4
100	2.5	2.3	2.2	2.1	2.0	2.0	1.8	1.7	1.6	1.6	1.5	1.4	1.3
Max	2.4	2.2	2.1	2.0	1.9	1.9	1.7	1.6	1.5	1.5	1.4	1.2	1.0

Comparison of Location: Common Group Variability

4. If no group differences in variability exist, compute the pooled standard deviation:

$$s_p = \sqrt{\frac{(n_1 - 1)s_1^2 + (n_2 - 1)s_2^2}{n_1 + n_2 - 2}}$$

which has degrees of freedom

$$f = n_1 + n_2 - 2$$

The value of s_p is the best estimate of the common group standard deviation.

5. Compute the quantities

$$a = \sqrt{\frac{n_1 + n_2}{n_1 n_2}}$$

$$T = \frac{\bar{X}_{max} - \bar{X}_{min}}{as_p}$$

where $\bar{X}_{max}$ is the larger mean and $\bar{X}_{min}$ the smaller mean. If T is larger than t in the following table for f degrees of freedom, the means are probably different. The LC chart (Procedure 11.2) can be used to display the differences using s_p for the common group variability.

f	t
10	2.23
15	2.13
20	2.09
25	2.06
30	2.04
40	2.02
50	2.01
60	2.00
70	1.99
100	1.98
Max	1.96

Comparison of Location: Different Group Variability

6. Compute the quantities

$$w_1 = \frac{s_1^2}{n_1}$$

$$w_2 = \frac{s_2^2}{n_2}$$

$$w = \sqrt{w_1 + w_2}$$

$$T = \frac{\bar{X}_{max} - \bar{X}_{min}}{w}$$

Compute the degrees of freedom for T:

$$u_1 = \frac{w_1^2}{n_1 - 1}$$

$$u_2 = \frac{w_2^2}{n_2 - 1}$$

$$f = \frac{(w_1 + w_2)^2}{u_1 + u_2}$$

Note that f is rounded down to the next closest integer. If the value of T is larger than t for f degrees of freedom in the table, then the group means are probably different. An LC chart can be used to display the group differences.

11.4 Comparison of Two Groups (Different Parts)

This section assumes there are n_1 measurements made from group 1 and n_2 measurements made from group 2. It is important to note that the n_1 measurements from group 1 and n_2 measurements from group 2 are made on different parts (i.e., the data are not paired). Procedure 12.5 discusses the case in which two paired measurements are made on the same parts and a comparison is desired. The following examples illustrate the Procedure 11.4 comparison procedure.

Example 11.2 An engineer was attempting to improve the surface micro finish of a drilled hole. Because micro finish is related to many process parameters, it was decided to make a short-term study comparing the current uncoated drill to a new coated drill. If results were favorable, a longer term study would be conducted. Each tool ran 25 parts, but 5 parts using the new tool were inadvertently not measured. Unfortunately, it was not reasonable in this situation to use balanced sampling by alternating use of the new and old tools. Consequently, the new tool was run after 25 parts were produced from a current tool. Both tools were new at the start of their runs. The data appear in Table 11.2.

1. A run chart indicated no unusual problems. The histogram is given in Figure 11.10 for both tools. The small sample size makes assessment of a pattern difficult.
2. The means and standard deviations appear in Table 11.2.
3. A difference in variability is evaluated using

$$F_{max} \text{ ratio} = \left(\frac{8.53}{8.43}\right)^2 = 1.02$$

   The tabled value for 20 and 25 is 2.0, so we conclude there is no difference in variability between the two tools.
4. The pooled standard deviation is

$$s_p = \sqrt{\frac{(24)(8.43)^2 + (19)(8.53)^2}{25 + 20 - 2}}$$

$$= 8.47$$

   which has $f = 43$ degrees of freedom.
5. The value of T is computed by the following

$$a = \sqrt{\frac{25 + 20}{25 \times 20}} = 0.3$$

$$T = \frac{39.36 - 34.45}{0.3 \times 8.47} = 1.93$$

The tabled value of t for $f = 43$ degrees of freedom is about 2.02. Thus, we conclude there is no significant difference in the two tools. The comparison charts are shown in Figure 11.11. The data suggest that there is not a large difference, if any, between the two tools. If the observed difference of about 5 μinches is important, a larger sample could be considered.

Example 11.3 A common problem that arises in manufacturing is evaluating the influence of incoming part characteristics on an operation. For example, consider a size dimension machined at an operation. The material hardness of incoming parts varies and potentially influences the machining of the part. It is possible to select both hard and soft

Table 11.2 Surface Micro Finish for Two Tools

No.	Current tool	Coated tool
1	38	28
2	35	45
3	53	41
4	25	40
5	39	47
6	47	34
7	39	31
8	41	43
9	46	23
10	54	36
11	36	44
12	40	45
13	47	29
14	31	32
15	38	25
16	48	21
17	40	31
18	21	43
19	39	22
20	32	29
21	39	
22	33	
23	28	
24	53	
25	42	
n	$n_1 = 25$	$n_2 = 20$
Mean	$\bar{X}_1 = 39.36$	$\bar{X}_2 = 34.45$
SD	$s_1 = 8.43$	$s_2 = 8.53$

parts and compare the final size characteristic of the two groups. There are two ways to conduct a study. If variable measurements are available for part hardness, a paired study can be performed. These studies are described in Chapter 12. Unfortunately, variable measurements are not always available or are unreliable, as with some hardness measurements (see Chap. 4). In these cases, parts can often be meaningfully grouped into extreme categories, such as "hard" and "soft" parts.

In this study parts are selected from the extremes of the hardness specification. If the hardness specification is correct, there is little difference between the machining of "hard"

11.4 Comparison of Two Groups (Different Parts)

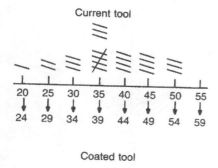

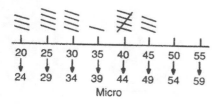

Figure 11.10 Micro finish histograms for Example 11.2.

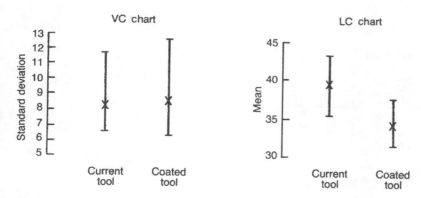

Figure 11.11 Tool evaluation comparison charts for Example 11.2.

and "soft" parts. Large differences in location or variability of final part size may imply that the hardness specification is too wide, as shown in Figure 11.12.

In this example, $n_1 = 20$ soft parts and $n_2 = 30$ soft parts were obtained for testing. Hard and soft parts were intermixed in the production run to minimize the influence of process changes. The final size characteristic was evaluated for all 50 parts. The data appear in Table 11.3.

1. A run chart indicated no unusual problems. The histogram in Figure 11.13 shows apparently more variability for soft parts.
2. The means and standard deviations appear in Table 11.3.

436   11 Comparison Methods

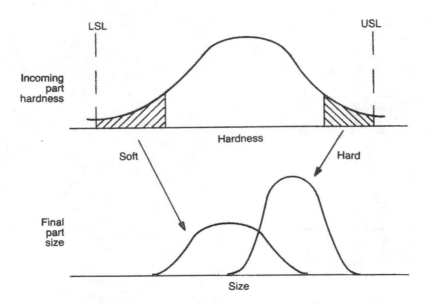

Figure 11.12   Forming groups based on incoming part hardness from Example 11.3.

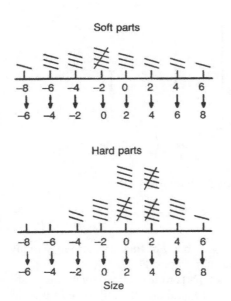

Figure 11.13   Size histograms for Example 11.3.

11.4 Comparison of Two Groups (Different Parts)

Table 11.3 Finished Size Obtained from Two Hardness Groups

No.	Soft parts	Hard parts
1	−1.2	6.6
2	−4.7	5.3
3	.1	1.3
4	4.8	4.4
5	−6.2	3.6
6	−4.3	1.9
7	.7	3.5
8	−.4	2.2
9	2.2	−.4
10	1.2	1.2
11	−2.8	−.6
12	6.4	3.7
13	−.3	2.5
14	−1.6	−1.7
15	−4.3	−3.0
16	2.4	.9
17	−1.4	0
18	5.3	3.6
19	−4.4	3.6
20	−2.2	2.4
21		4.6
22		2.7
23		4.0
24		2.8
25		−3.4
26		3.2
27		−.1
28		.1
29		1.4
30		.7
n	20	30
Mean	−.54	1.90
SD	3.51	2.36

3. Differences in variability are evaluated using

$$F_{max} \text{ ratio} = \left(\frac{3.51}{2.36}\right)^2 = 2.21$$

The tabled value for 20 and 30 is 1.9, so we conclude that there are differences in variability between hard and soft parts.

6. Evaluation of mean differences is performed using the following calculations:

$$w_1 = \frac{(3.51)^2}{20} = 0.616$$

$$w_2 = \frac{(2.36)^2}{30} = 0.186$$

$$w = \sqrt{0.616 + 0.186} = 0.90$$

$$T = \frac{1.90 - (-.54)}{.90} = 2.71$$

The degrees of freedom of T are computed as

$$u_1 = \frac{(.616)^2}{19} = .020$$

$$u_2 = \frac{(.186)^2}{29} = .0012$$

$$f = \frac{(.616 + .186)^2}{.020 + .0012} = 30$$

The tabled value of t is 2.04, so we conclude that the means are significantly different.

This study concludes that hard and soft parts differ in both location and variability. The comparison charts showing these differences appear in Figure 11.14. If these differences are large enough to influence the quality of the final part, the hardness specification should be narrowed.

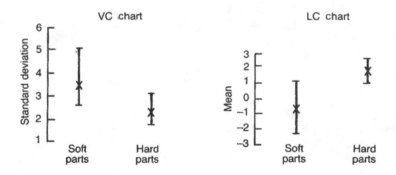

Figure 11.14 Size evaluation comparison charts for Example 11.3.

11.5 Case Studies

11.5.1 Fixture Differences

A milling operation was performed on a 30-fixture machine. Two separate stations performed the machining, as shown here:

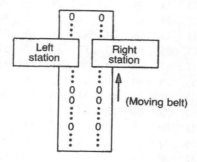

The 15 fixtures on each side of the operation were on a moving belt. The fixtures located the parts in the machining station. This operation was a roughing operation that had an in-process specification of ±50 for the dimension being machined. The final finishing operation showed unacceptable variability, so it was decided to start the troubleshooting process by studying the roughing operation. At 15 different times over the course of 2 days, parts were collected from the right and left stations for all 15 fixtures. The data are listed in Table 11.4.

The means and standard deviations are given in Table 11.5. The F_{max} ratios to evaluate variability differences follow.

Left side:

$$F_{max} = \left(\frac{14.0}{5.0}\right)^2 = 7.8$$

Right side:

$$F_{max} = \left(\frac{16.5}{5.4}\right)^2 = 9.3$$

The table in Procedure 11.1 does not extend to 15 groups. However, the value for 10 groups is 5.6, which is significantly below these ratios so that there is probably a variability difference between fixtures. Also, from the practical viewpoint the difference between the low- and high-variability fixtures is appreciable. If any doubt exists, additional parts can be collected from the low and high fixtures and another F_{max} test performed.

These fixtures were examined and it was discovered that the low-variability fixtures had a new locator pin and the high-variability fixtures had worn locator pins. The importance of this particular pin had not previously been appreciated. The limits for the VC chart appear in Table 11.5. The VC chart in Figure 11.15 shows that there are large differences in variability between fixtures. However, the right and left sides do not appear to differ. These variability differences in a roughing operation mean that the finishing operation receives parts that exhibit excessive variability that would impact on the finished quality as well as other factors, such as tool life.

Table 11.4 Milling Masline Data

Fixture	L	R	L	R	L	R	L	R	L	R
1	35	50	51	55	21	50	35	51	56	60
2	37	52	41	51	39	49	42	53	34	44
3	42	55	28	36	48	57	48	58	38	51
4	28	40	41	60	44	48	29	41	45	64
5	38	55	34	44	39	60	39	49	36	57
6	52	55	29	49	36	55	35	52	31	48
7	46	68	45	58	40	60	41	63	44	66
8	30	47	35	55	43	57	36	57	37	58
9	65	72	47	48	37	38	31	32	31	32
10	37	53	33	49	49	69	32	48	36	52
11	54	63	21	43	66	74	18	40	50	59
12	38	62	38	70	73	67	50	44	30	62
13	37	41	50	60	30	47	31	35	33	50
14	46	56	44	54	25	30	24	35	21	32
15	41	52	40	49	43	69	42	66	42	53
1	20	48	18	47	25	53	38	53	31	44
2	41	52	36	47	31	46	42	53	35	46
3	35	48	35	48	32	40	48	57	33	45
4	26	38	25	37	41	45	33	44	27	40
5	31	42	31	49	35	46	37	58	31	42
6	38	57	25	45	30	47	31	50	30	47
7	41	63	33	52	38	60	30	49	44	67
8	35	56	33	54	35	56	30	47	30	51
9	66	73	25	26	51	58	46	47	36	38
10	37	53	35	47	23	39	43	60	32	47
11	50	58	51	60	55	63	25	38	53	62
12	40	34	30	62	41	35	33	62	41	37
13	31	48	22	39	23	40	33	45	27	42
14	41	51	27	38	40	51	28	39	38	49
15	30	40	47	58	36	47	31	47	30	45
1	30	45	43	60	38	57	31	44	38	71
2	37	72	48	57	61	82	46	59	44	61
3	40	25	37	54	46	46	49	43	43	22
4	46	22	35	65	45	61	42	26	31	69
5	53	40	20	28	51	62	25	23	56	57
6	40	54	39	53	45	60	44	61	43	62
7	33	55	40	47	44	50	35	53	35	53
8	60	48	44	64	45	55	41	67	41	67
9	28	75	44	44	46	67	50	63	40	61
10	55	59	38	50	36	44	50	64	49	63
11	39	43	35	50	32	55	34	51	41	45
12	39	61	40	68	28	49	30	47	37	65
13	48	51	40	51	53	80	45	54	48	75
14	50	74	41	37	51	61	45	33	46	38
15	43	54	31	38	50	89	45	94	41	52

11.5 Case Studies

Table 11.5 Comparison Chart Values for Milling Data

Fixture	Variability comparison						Location comparison					
	Left side			Right side			Left side			Right side		
	s	LVL	UVL	s	LVL	UVL	$\bar{X}$	LLL	ULL	$\bar{X}$	LLL	ULL
1	10.8	7.9	17.1	7.3	5.3	11.5	34.0	28.1	39.9	52.5	48.5	56.5
2	7.2	5.3	11.4	10.3	7.5	16.3	40.9	36.9	44.9	54.9	49.2	60.6
3	6.8	5.0	10.7	11.1	8.1	17.5	40.1	36.4	43.8	45.7	39.6	51.8
4	7.9	5.8	12.5	14.3	10.4	22.6	35.9	31.6	40.2	46.7	38.8	54.6
5	9.9	7.2	15.6	11.4	8.3	18.0	37.1	31.7	42.5	47.5	41.2	53.8
6	7.4	5.4	11.7	5.4	3.9	8.5	36.5	32.4	40.6	53.0	50.0	56.0
7	5.0	3.7	7.9	6.9	5.0	10.9	39.3	36.5	42.1	57.6	53.8	61.4
8	7.8	5.7	12.3	6.4	4.7	10.1	38.3	34.0	42.6	55.9	52.4	59.4
9	12.2	8.9	19.3	16.5	12.0	26.1	42.9	36.2	49.6	51.6	42.5	60.7
10	8.6	6.3	13.6	8.3	6.1	13.1	39.0	34.3	43.7	53.1	48.5	57.7
11	14.0	10.2	22.1	10.4	7.6	16.4	41.6	33.9	49.3	53.6	47.9	59.3
12	11.0	8.0	17.4	12.7	9.3	20.1	39.2	33.1	45.3	55.0	48.0	62.0
13	10.0	7.3	15.8	12.7	9.3	20.1	36.7	31.1	42.3	50.5	43.5	57.5
14	10.1	7.4	16.0	12.6	9.2	19.9	37.8	32.2	43.4	45.2	38.3	52.1
15	6.4	4.7	10.1	16.4	12.0	25.9	39.5	36.0	43.0	56.9	47.9	65.9

Since variability differs between fixtures, the ANOM procedure should not be used. However, the location difference can be evaluated on the LC chart. Table 11.5 contains the limits for the LC chart that are plotted in Figure 11.16. The most obvious problem is that the right side ($\bar{\bar{X}} = 52.0$) has a higher mean than the left side ($\bar{\bar{X}} = 38.6$). Since the specification limits are ±50, it is apparent that the process is being targeted, perhaps to leave more material for the finishing operation. However, the right-side parts are clearly too high. From these data it is not possible to determine whether the right station was adjusted incorrectly or if all the right fixtures were a problem. The most likely cause would seem to be a station adjustment problem, but it was found that the right side of the belt carrying the fixtures was adjusted improperly.

Improvement in the operation must consist of replacing the worn locators to reduce variability and adjusting the right side of the belt carrying the fixtures. Because of the excessive variability, it is not possible to effectively evaluate smaller location difference. Notice that the width between the bars on the low-variability fixtures (e.g., 6 left or 7 right) in Figure 11.12 is narrow in comparison with the width on the high-variability fixtures (e.g., 9 right, 15 right). Since detecting location differences depends on the overlap of the bars, it is necessary to reduce variability before other location differences can be detected. Thus, the comparison procedures should be rerun after the corrections are made. It is generally good practice to reduce any variability differences prior to "fine-tuning" location.

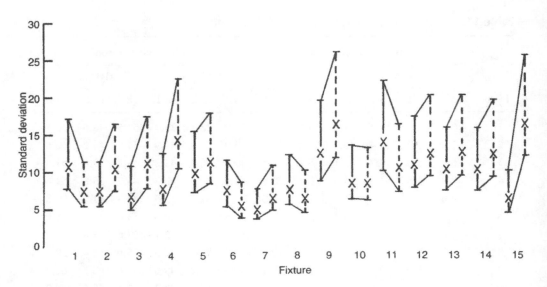

Figure 11.15 VC chart for milling data: left side (—); right side (---).

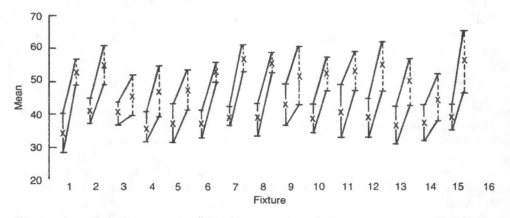

Figure 11.16 LC chart for milling data: left side (—); right side (---).

This example illustrates the value of comparison studies prior to the use of standard control charts. Although it would have been possible to study the process using control charts, the time required to detect and eliminate the large problems would probably have been longer. However, it is natural to use control charts to monitor the process once the problems have been corrected. The control limits for the improved process are narrower owing to the reduced process variability. Thus, a tighter control of the process is obtained.

11.5 Case Studies

11.5.2 Process Parameter Comparisons

A reamer was used to drill a valve bore in a transmission valve body. There were three different diameters in the bore. The surface finish of the bore (microinches) is a critical characteristic related to the feed rate (inches per minute) of the machining spindle. Since the machining spindle operates at a single speed and feed rate, it is difficult to optimize the surface finish for all three different diameters. A test was run at five different feed rates. For each feed rate, nine parts were machined. After each group of nine parts were run, the feed rate was changed. A balanced sampling plan would have required changing the feed rate after every part, which was not practical in this example. The same tool was used throughout the test. The tool life is generally several thousand parts, so tool wear was thought to be of little consequence.

The data from the tests appear in Table 11.6. The F_{max} ratios follow.

Diameter 1:
$$F_{max} = \left(\frac{16.9}{8.1}\right)^2 = 4.4$$

Diameter 2:
$$F_{max} = \left(\frac{21.7}{8.6}\right)^2 = 6.4$$

Diameter 3:
$$F_{max} = \left(\frac{19.4}{3.8}\right)^2 = 26.1$$

The tabled value from Procedure 11.1 is about 7.1, so only for diameter 3 (the smallest diameter) does the variability appear to differ between feed rates with the lower feed rates having lower variability. The VC chart in Figure 11.17 shows generally increasing variability with feed rate. Only a feed rate of 20 inches/minute appears to break the increasing pattern. For diameters 1 and 2 the pooled standard deviations are as follows.

Diameter 1:
$$s_p = \sqrt{\frac{(13.9)^2 + (14.0)^2 + (16.9)^2 + (12.5)^2 + (8.1)^2}{5}}$$
$$= 13.4$$

Diameter 2:
$$s_p = \sqrt{\frac{(8.6)^2 + (17.2)^2 + (21.7)^2 + (17.7)^2 + (20.1)^2}{5}}$$
$$= 17.6$$

with $f = 40$ degrees of freedom.

The LC charts for the different feed rates appear in Figure 11.18. Notice that it is convenient to plot the feed rates on a scaled x axis. The chart indicates that for all three diameters using the slower feed rate improves or does not deteriorate surface finish. When the groups are ordered, as with differing feed rates, it is informative to use a scaled x axis to

Table 11.6 Tool Test Data

	\multicolumn{5}{c}{Feed rate}				
	16.5	20	30	35	40
	\multicolumn{5}{c}{Diameter 1}				
	57	25	55	49	54
	27	26	40	27	54
	24	33	34	38	45
	25	29	88	41	44
	34	36	37	58	42
	48	55	56	62	57
	44	48	44	62	54
	47	57	65	49	64
	61	60	58	62	64
$\bar{X}$	40.8	41.0	53.0	49.8	53.1
s	13.9	14.0	16.9	12.5	8.1
LLL	31.8	32.0	44.0	40.8	44.1
ULL	49.8	50.0	62.0	58.8	62.1
	\multicolumn{5}{c}{Diameter 2}				
	29	54	51	57	67
	61	47	58	53	40
	43	43	42	54	60
	41	64	53	59	53
	48	39	52	56	57
	46	76	38	57	55
	41	69	48	55	62
	48	44	86	107	106
	39	89	104	48	89
$\bar{X}$	44.0	58.3	59.1	60.7	65.4
s	8.6	17.2	21.7	17.7	20.1
LLL	32.1	46.4	47.2	48.8	53.5
ULL	55.9	70.2	71.0	72.6	77.3
	\multicolumn{5}{c}{Diameter 3}				
	22	39	44	54	56
	28	29	56	42	50
	27	63	36	32	102
	28	42	51	46	93
	28	38	43	57	46
	28	41	50	80	71
	29	62	43	50	71
	21	38	49	50	52
	34	37	42	50	65
$\bar{X}$	27.2	43.2	46.0	51.2	67.3
s	3.8	11.5	6.0	13.0	19.4
LLL	24.3	34.5	41.5	41.4	52.7
ULL	30.1	51.9	50.5	61.0	81.9

11.5 Case Studies

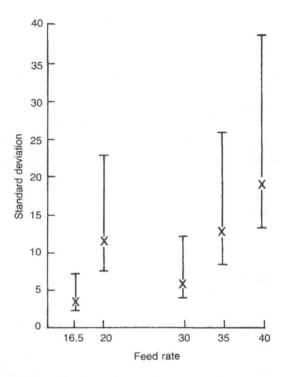

Figure 11.17 VC chart for surface finish.

evaluate a rate of change. Unfortunately, the conclusions in this study are not as clear as we would have liked. This is the consequence of using a small $n = 9$ sample size.

11.5.3 Machine Tryout

Recall from Chapter 7 that one of the essential parts of a machine tryout should be comparison of major sources of variation, such as fixture, pallets, and spindles. The machine tryout case study in Chapter 7 gave results from a tryout of an eight-fixture grinding machine. The data for run 5 on 6–29 appear in Table 11.7. Before it is meaningful to align the machine fixtures to obtain a common mean, it is desirable to have a common variability between fixtures. A logical question to ask from the data in Table 11.7 is why fixture 8 variability ($s = 1.0$) is lower than fixture 5 variability ($s = 4.1$). The F_{max} ratio,

$$F_{max} = \left(\frac{4.1}{1.0}\right)^2 = 16.8$$

is greater than the tabled value of 8.9 in Procedure 11.1, so the difference in variability is significant. It was discovered that the part was held too loosely in several of the fixtures. An attempt was made to make all fixtures like fixture 8.

A number of small changes were made on the machine, and in each case run charts, histograms, and comparison charts were made after each run. The data from run 8 on 7–23

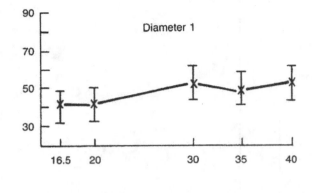
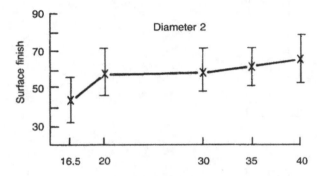
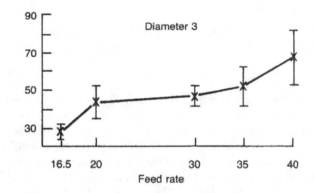

Figure 11.18 LC chart for surface finish.

11.5 Case Studies

Table 11.7 Flatness Data from Machine Tryout

	Fixture							
	1	2	3	4	5	6	7	8
				Run 5				
	13	12	13	11	12	13	11	11
	12	10	12	14	12	11	13	13
	10	8	15	11	11	10	12	11
	11	11	12	12	9	10	11	11
	12	10	10	11	11	12	11	12
	12	12	11	12	14	10	10	11
	13	10	10	11	10	11	15	13
	21	20	12	14	18	15	14	12
	14	14	10	14	20	15	18	10
	16	16	12	10	20	13	14	11
$\bar{X}$	13.4	12.3	11.7	12.0	13.7	12.0	12.9	11.5
s	3.1	3.5	1.6	1.5	4.1	1.9	2.4	1.0
				Run 8				
	21	16	20	18	20	16	18	16
	16	16	19	18	15	20	14	15
	18	15	20	18	18	15	19	14
	22	14	20	20	20	17	14	15
	22	16	16	20	16	14	20	14
	15	13	18	15	19	15	17	19
	22	14	24	15	17	18	20	14
	20	14	22	16	14	16	19	15
	16	22	21	15	22	20	14	14
	17	15	16	18	20	21	22	20
$\bar{X}$	18.9	15.5	19.6	17.3	18.1	17.2	17.7	15.6
s	2.8	2.5	2.5	1.9	2.6	2.4	2.9	2.2
LLL	17.3	13.9	18.0	15.7	16.5	15.6	16.1	14.0
ULL	20.5	17.1	21.2	18.9	19.7	18.8	19.3	17.2

made at the manufacturer's plant appear in Table 11.7. The F_{max} ratio to compare variability is

$$F_{max} = \left(\frac{2.9}{1.9}\right)^2 = 2.3$$

which is lower than the 8.9 tabled value in Procedure 11.1. Thus, there appears to be common variation between fixtures. The common pooled standard deviation is

$$s_p = \sqrt{\frac{(2.8)^2 + (2.5)^2 + (2.5)^2 + (1.9)^2 + (2.6)^2 + (2.4)^2 + (2.9)^2 + (2.2)^2}{8}}$$
$$= 2.5$$

with $f = 72$ degrees of freedom. Although it is good that the fixtures have common variability, there has been an apparent increase in variability from the machine builder's site to the manufacturer's floor. This change was discussed in conjunction with Table 7.4 concerning a decrease in capability.

Since the fixtures have a common variability, the ANOM procedure can be used to evaluate location differences. The grand mean is 17.49, and the ANOM decision lines are

$$UDL = 17.49 + \frac{2.5 \times 2.80}{\sqrt{10}}\sqrt{\frac{7}{8}}$$
$$= 17.49 + 2.07$$
$$= 19.6$$
$$LDL = 17.49 - 2.07$$
$$= 15.4$$

The ANOM plot in Figure 11.19 indicates that we would not conclude that there are fixture differences. However, the means vary between both extremes, so additional data collection may be warranted.

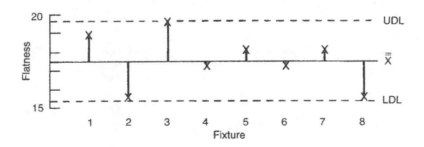

Figure 11.19 ANOM chart for flatness.

Problems

Problems

11.1 Verify the calculations of LLL and ULL in Table 11.5.

11.2 Verify the calculation of LLL and ULL in Table 11.7.

11.3 Prepare an ANOM chart for diameters 1 and 2 for the tool comparison case study.

11.4 A spot check of valve body flatness from the old grinding machine described in the Chapter 7 case study produced the results shown here. (a) Do the eight fixtures have different variability? (b) Do the eight fixtures have a common mean flatness? (c) Prepare the appropriate comparison charts.

	\multicolumn{8}{c}{Fixture}							
	1	2	3	4	5	6	7	8
	16	18	16	15	26	22	20	25
	22	23	21	23	15	17	13	23
	15	10	22	22	22	13	15	17
	21	18	22	25	19	30	23	21
	13	15	25	20	15	20	17	20
	11	17	15	20	15	18	16	22
	12	22	19	15	23	17	31	31
	11	11	15	25	19	16	15	22
	17	25	20	18	15	20	17	21
	25	21	23	17	11	21	28	30
$\bar{X}$	16.3	18.0	19.8	20.0	18.0	19.4	19.5	23.2
s	4.9	5.0	3.5	3.7	4.6	4.6	6.0	4.4

11.5 In Problem 4.1 a stratified control chart was used to monitor the concentricity of a gear blank outer diameter (OD) produced from an eight-spindle screw machine. The control chart was found to be reasonably stable. (a) Evaluate any location or variability differences between the spindles. (b) Prepare appropriate comparison charts.

11.6 The length measurements that follow were collected from the same parts as were studied for concentricity in Problem 4.1 and analyzed in Problem 11.5. A stratified control chart was prepared using two subgroups per 8 hour shift. The specification limits were 0–20 (with 1 = .001 inch). (a) Prepare a control chart for the length data and interpret the results. (b) Analyze the eight spindles for differences in variability and location.

| | | | | Spindle | | | | | | |
No.	1	2	3	4	5	6	7	8	$\bar{X}$	R
1	11	12	14	13	12	14	12	10	12.3	4
2	9	10	12	8	10	10	11	11	10.1	4
3	10	13	10	10	11	10	11	10	10.6	3
4	5	11	10	12	10	5	7	9	8.6	7
5	10	10	11	10	9	9	9	10	9.8	2
6	7	8	9	8	8	8	7	6	7.6	3
7	13	11	11	12	10	7	12	8	10.5	6
8	13	12	11	16	11	10	9	10	11.5	7
9	10	13	14	11	11	9	13	10	11.4	5
10	11	11	13	11	10	10	10	8	10.5	5
11	13	13	15	14	14	14	13	12	13.5	3
12	5	7	10	9	9	7	7	7	7.6	5
13	13	12	14	13	14	13	13	12	13.0	2
14	13	15	15	16	14	15	15	14	14.6	3
15	14	13	14	13	15	14	11	10	13.0	5
16	10	10	11	9	13	9	10	9	10.1	4
17	11	11	14	13	12	13	12	11	12.1	3
18	13	13	14	15	14	15	14	13	13.9	2
19	11	12	13	13	14	12	12	10	12.1	4
20	13	13	14	15	13	13	12	11	13.0	4
21	8	10	11	11	12	9	9	10	10.0	4
22	6	10	13	10	10	10	11	6	9.5	7
23	11	10	13	14	11	11	11	11	11.5	4
24	9	9	11	7	11	10	8	8	9.1	4

11.7 The 50 time periods that follow are size readings from the OD of a journal (± 10 specification) collected from a four-spindle machine over a 2 week period. (a) Prepare a stratified control chart, and interpret the results. (b) Compute the mean and standard deviation for each spindle. (c) Draw a histogram for each spindle and all combined spindles. Use the Procedure 5.2 work sheet format to compute the mean and standard deviation of each spindle. (d) Prepare and interpret the appropriate comparison charts. (e) What can you conclude about the value of control charts, histograms, and comparison charts?

Day	Time	Spindle 1	2	3	4	Day	Time	Spindle 1	2	3	4
11/7	8:10	0	1	−1	−1	11/14	7:31	2	2	−1	−2
	9:15	3	−3	−3	2		8:40	3	−2	−2	2
	10:05	5	−4	−1	−3		10:07	2	−2	−5	6
	11:16	3	−4	−4	−6		11:38	−2	1	−2	−2
	2:04	1	0	−4	1		12:33	1	3	−5	−4
11/8	7:00	5	−2	−4	0	11/15	9:07	0	−1	−4	−6
	9:15	1	0	−5	1		11:08	−1	−1	0	−4
	12:02	0	−2	−2	0		12:47	−2	0	−4	−5
	1:45	3	−1	−6	9		1:42	2	−1	−2	−4
	2:30	0	−2	−3	−2		2:30	1	−1	−4	−5
11/9	6:04	3	−2	−1	4	11/16	7:40	2	0	−3	9
	7:02	1	−1	−3	3		9:50	2	1	−4	−2
	9:25	7	−1	−3	0		10:32	0	0	−4	−1
	11:12	4	−3	−5	−6		11:22	−1	2	−4	4
	1:30	0	1	−1	−3		1:05	2	0	−3	4
11/10	10:04	0	−3	−4	2	11/17	7:08	0	−1	−3	−2
	10:30	4	−1	−2	−3		9:02	−1	5	−4	2
	11:45	−1	−1	−3	−4		10:12	4	2	−4	1
	1:15	8	0	−5	−2		12:08	8	0	−2	0
	2:30	−2	−2	−2	−2		1:22	1	−3	−3	−1
11/11	8:04	1	−1	−2	−5	11/18	6:15	0	−1	−3	−2
	9:15	1	1	−4	−2		7:05	−1	0	−3	−1
	11:05	1	−1	−1	0		9:20	0	−1	−4	7
	12:45	2	−1	−2	0		11:30	−2	2	−2	5
	1:30	4	5	−1	−3		12:47	−1	0	−3	1

11.8 It was known that heat treat increased the out-of-roundness of the inner diameter (ID) of a bore. This condition resulted in poor tool life and excessive scrap from an operation that hones the ID after heat treat. A team decided to investigate the possibility that an alternative method of loading the parts for heat-treat processing would improve roundness. A heat-treat fixture was modified so that 15 parts could be loaded horizontally (current method, parts 1–15) and 15 parts loaded vertically (test method, parts 16–30). This fixture could be run in the normal production process through the heat-treat operation. The roundness (USL = 1.5) of ID was measured (1 = .0001 inch) in two positions on the part. (a) What was the advantage of using a single fixture for processing? (b) Compare the positions. (c) Compare the two loading methods. Which loading method is preferable?

	Horizontal Load			Vertical Load	
Part	Position 1	Position 2	Part	Position 1	Position 2
1	3	2	16	1	3
2	0	6	17	5	4
3	1	6	18	2	3
4	1	7	19	6	4
5	1	7	20	0	3
6	1	7	21	1	1
7	2	2	22	2	5
8	5	9	23	2	3
9	1	6	24	5	0
10	0	9	25	2	5
11	2	5	26	3	4
12	1	3	27	2	3
13	2	8	28	2	4
14	3	7	29	5	5
15	0	5	30	2	3

11.9 A new procedure was developed to improve the yield (measured in grams) of a process. The new process was more expensive than the current process, so careful evaluation was necessary. The data for the current and new process follow.

Current process:

3.3, 5.5, 7.6, 3.3, 5.4, 3.8, 7.1, 5.8, 3.9, 3.5, 4.1, 8.3, 2.3, 4.0, 3.6, 6.0, 4.1, 2.8, 4.2, 3.1, 4.7, 3.6, 4.9, 2.1, 5.2

New process:

4.8, 5.6, 4.9, 6.1, 3.9, 4.6, 4.8, 3.6, 5.3, 6.1, 6.9, 5.6, 7.0, 5.9, 5.4

(a) Does the new process have significantly different yield or variability than the current process? (b) Prepare the comparison charts.

11.10 A special premachining heat treatment of transmission gear blanks was being evaluated. It was hoped that the heat treatment would stabilize the gear blank metal structure and enable machining a more uniform size characteristic. From the same lot of material, 50 blanks were selected for heat treatment and 25 blanks were selected as a control group. The data for the two groups follow.

Non–heat-treated blanks:

2.6   2.0   3.8   −1.9   −.3   −1.0   3.3   4.2   −5.4   2.1   −1.5
−2.7   .8   3.6   5.5   −.1   −.8   −1.8   2.4   4.4   1.2   3.4
2.0   −2.9   8.5

Problems

Heat-treated blanks:

−.5	.7	1.0	2.5	3.4	3.3	−.2	2.5	−5.5	−.8	5.2
−5.1	5.0	.6	2.3	2.4	1.6	.3	4.5	−1.7	.6	1.8
−3.2	−.4	−2.2	2.9	−.3	2.0	1.4	−2.0	0.3	.2	.9
0	1.3	.4	−.1	.5	1.4	.9	.8	−.5	−.5	1.5
−1.7	.9	−2.4	2.8	.9	−.2					

(a) Does the heat treatment cause any change in location or variability? (b) Prepare the comparison charts.

12
Scatter Plots

12.1 General Concepts

Understanding the relationship between characteristics increases our ability to control a process and detect problems. A scatter plot provides a method to evaluate these interrelationships and is thus an important tool for studying a process. Fortunately, preparation and interpretation of scatter plots is easy. Once personnel become accustomed to use of this tool, it is one of the most widely used statistical tools.

The phrase "relationship between characteristics" means simply to what extent changes in one characteristic are able to predict changes in another characteristic. Consider conducting a survey in a typical factory or office. A total of 50 individuals are randomly selected and asked their height, weight, and age. The resulting data can be displayed as follows:

Person	Height (inches)	Weight (pounds)	Age (years)
1	62	110	37
2	73	191	52
.	.	.	.
.	.	.	.
.	.	.	.
50	68	153	25

The characteristics of interest are height, weight, and age. The relationship between the characteristics involves identifying the following:

How does weight change with height?
How does weight change with age?
How does height change with age?

A scatter plot provides a graph for evaluating this relationship. The plot displaying the height and weight relationship can be prepared by drawing the scales for height and weight and plotting the individual height and weight pairs for each person. The pair for person 1 is 62, 110, which is plotted in Figure 12.1. The completed plot of all 50 individuals might

12.1 General Concepts

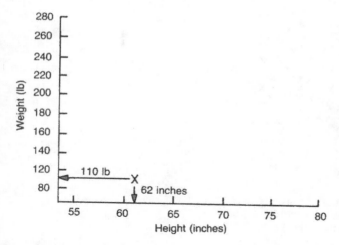

Figure 12.1 Example of plotting a measurement pair on a scatter plot. Person 1: height = 62 inches; weight = 110 pounds.

appear as shown in Figure 12.2a. The plot shows that as height increases, weight increases, but there is appreciable variability in this relationship. For example, a person 65 inches in height may vary in weight from 120 to 250 pounds. There does not appear to be any relationship between age and weight in Figure 12.2b. An individual 25 years old could have roughly the same weight as a person 40 years old. Thus, the lack of any trend in the scatter plot implies for the data being evaluated that there is no relationship between the characteristics.

In manufacturing applications, the measured units are parts or subassemblies in which two or more characteristics are measured for each unit. Consider the production of a washer,

where the inner and outer diameter (ID and OD) are of interest. For each part, measurement of the ID and OD could be made:

Part	ID	OD
1	1.217	2.015
2	1.215	2.014
.	.	.
.	.	.
.	.	.
50	1.219	2.017

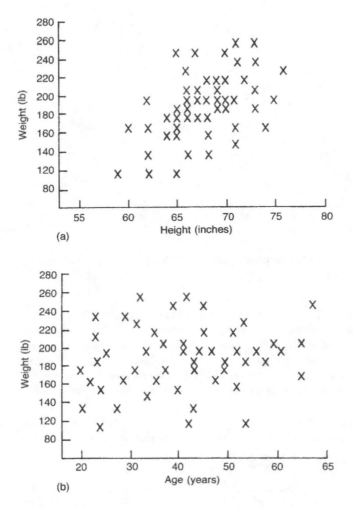

Figure 12.2 Scatter plots of height, weight, and age relationships: (a) height versus weight; (b) age versus weight.

A scatter plot of the ID versus the OD would enable visual evaluation of the relationship between the two characteristics. If the relationship is as shown in Figure 12.3, the quality of the two characteristics is closely related. Other characteristics, such as flatness, may be related to the ID or OD so that it may be useful to consider them to more completely understand the washer's production interrelationships. It is possible to measure the ID, OD, and flatness of a group of washers and prepare the scatter plots. Since multiple measurements are made on the same part, it is necessary to tag or number the parts so that measurements can be assigned to the correct parts.

12.1 General Concepts

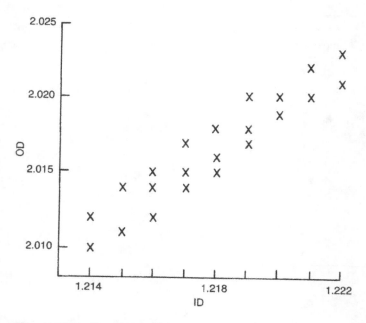

Figure 12.3 Example of related characteristics.

12.1.1 Types of Patterns

Graphs, such as control charts and histograms, provide a visual display of measurement patterns that convey information on process behavior. Similarly, the scatter plot pattern conveys the extent of the relationship between two characteristics. Figure 12.4 shows the five general patterns. The relationship between characteristics is referred to as the degree of "correlation" between the characteristics.

A high positive or negative correlation implies a linear trend between the characteristics. Conversely, no correlation between the characteristics implies that the value of one characteristic is not related to the value of the other characteristic. If a mild positive or negative correlation exists, the relationship between the characteristics exhibits significant variability. A linear trend line is much less clear. Interpretation of these patterns is generally not difficult, but two mistakes are commonly made.

First, it is assumed that if correlation exists between two characteristics, one characteristic in some way causes the value of the other characteristic. This is *not* true! Consider evaluating the relationship between a plant's heating oil use and the percentage of employees who are sick. If monthly values are evaluated over several years, the data can be arranged as follows:

Month	Heating oil use (bbl)	Average % employees sick
1	10,500	7.0
2	11,600	6.5
.	.	.
.	.	.
.	.	.
12	9,846	4.7
1	12,367	6.1
2	13,502	7.3
.	.	.
.	.	.
.	.	.

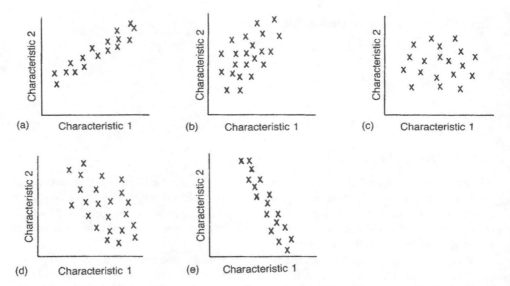

Figure 12.4 Typical patterns for scatter plots. (a) High positive correlation (r between 1.0 and .8); higher values of one characteristic are directly related to higher values of the second characteristic. (b) Mild positive correlation: higher values of one characteristic tend to be associated with higher values of the second characteristic, but predictability is variable. (c) No correlation ($r = 0$): there is no apparent relationship between the characteristics (over the range of values measured). (d) Mild negative correlation: higher values of one characteristic tend to be associated with lower values of the second characteristic, but predictability is variable. (e) High negative correlation (r between -1.0 and $-.8$): higher values of one characteristic are directly associated with lower values of the second characteristic.

12.1 General Concepts

The scatter plot might have the pattern in Figure 12.5. Concluding that heating oil use is correlated with employee sickness is correct. However, concluding that heating oil use causes employee sickness is not necessarily correct. Oil use increases in the winter, which may be related to increased employee sickness. Other possibilities may also exist. When correlation between two characteristics exists, it may be due to the existence of a common causal factor (e.g., temperature).

Correlation does not imply causation

It is necessary to use knowledge of the subject area to establish meaningful causation relationships. In manufacturing applications, it is generally possible to use knowledge of the process or product to conclude that the existence of correlation implies causation. For example, in the washer example in which the ID and OD were related, suppose the OD was punched on a die by first locating on the ID. Variability in the ID may logically result in variability in the OD. The judgment of causation is based on external process knowledge, not the scatter plot.

Suppose two characteristics are not correlated on a scatter plot. It is usually reasonable to assume that a problem with one of the characteristics is not related to the other characteristic provided the second mistake discussed next is not made. In many cases identifying the absence of correlation is valuable in resolving a problem. If two characteristics are not related, a problem with one characteristic could not have been caused by the other characteristic.

The second mistake commonly made in using scatter plots is to consider too narrow a range of the characteristics being evaluated. In the height, weight, and age example, we concluded from Figure 12.2b that age was not correlated with weight using the plant data collected. However, we know this result is not true! A person's weight increases with age when the age span 0 to about 20 years is considered, as shown in Figure 12.6. The data we evaluated from the adults in a plant, of course, did not detect this relationship since only adults were sampled.

In a manufacturing application, it is always advisable to spread out the parts being evaluated beyond the range of the specification limits. We see later that this practice

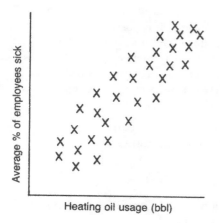

Figure 12.5 Example of correlation without causation.

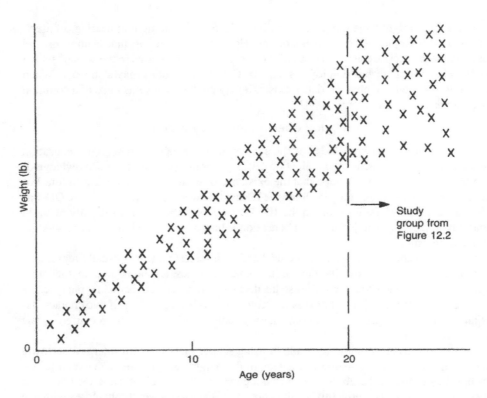

Figure 12.6 Scatter plot illustrating that correlation depends on the selection of sampling units.

enables evaluation of whether in-process specification limits are correct. Figure 12.7 illustrates a poor and good correlation study. Unfortunately, it is possible to control only one of the characteristics being evaluated and sometimes even this is not possible. The selection of extreme parts for one of the characteristics may require special production runs. The interpretation of the presence or absence of correlation between two characteristics is valid only for the range of data values considered. Thus, a good scatter plot analysis considers values slightly beyond the normal range of operations.

> For a correlation study, select parts that span
> slightly beyond the normal operating range

12.2 Construction and Analysis of a Scatter Plot

Preparation of a scatter plot is easy, as the following two cases illustrate. Consider $N = 10$ measurements for the values of two characteristics on a part:

12.2 Construction and Analysis

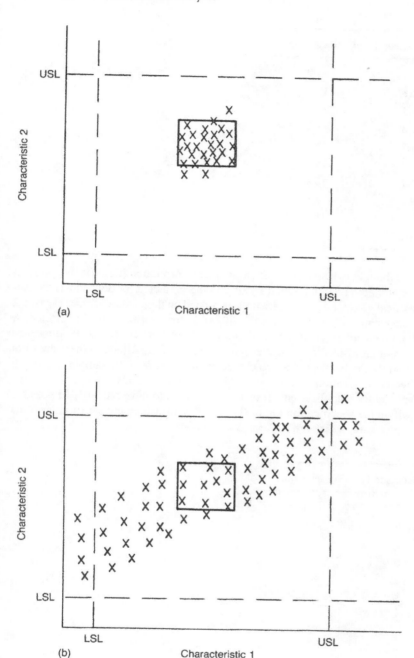

Figure 12.7 Two correlation studies that could conclude no correlation exists if only boxed area evaluated. (a) poor design with a narrow range considered, (b) good design which detects high positive correlation.

Part	Characteristic 1	Characteristic 2
1	9.4	13.2
2	2.8	19.4
3	3.2	14.7
4	1.8	11.4
5	4.7	13.9
6	2.9	14.7
7	3.1	16.4
8	9.2	17.6
9	5.4	16.7
10	3.2	14.1

The characteristic thought to be the possible cause of any correlation (i.e., the independent characteristic) is plotted on the horizontal x axis and the predicted (or dependent) characteristic on the vertical y axis. Let characteristic 1 be plotted on the x axis and characteristic 2 on the y axis. Using standard graph paper, a plotting scale for each axis must be selected. A scale of 0–10 is convenient for characteristic 1 and 10–20 for characteristic 2. Plotting each pair of values at a point results in the scatter plot in Figure 12.8. Although there does not appear to be a correlation between the two characteristics, the $N = 10$ sample size is too small for an effective evaluation. Sample sizes of $N = 30$–50 are advisable.

In many manufacturing applications, it is not uncommon to have only 10–15 possible gage readings within the specification limits. This limited measurement sensitivity results in coded scatter plot data, which are illustrated as follows:

Part	Characteristic 1	Characteristic 2
1	3	−1
2	5	2
3	4	3
4	0	−5
5	−1	−2
6	5	2
7	3	−1
8	−1	−2
9	5	2
10	4	0

A correlation table listing the frequency of the measurement pairs can be prepared from these data, as shown in Table 12.1. This table makes preparation of the scatter plot easier. Notice that several parts have the same pairs of values. Placing an "x" three times on the plot would be confusing, so the following plotting convention can be used:

12.2 Construction and Analysis

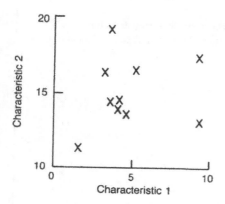

Figure 12.8 Sample scatter plot: unique points.

Table 12.1 Correlation Table

Total	2	1	0	0	2	2	3	Total
3						/		1
2							///	3
1								0
0					/			1
−1				//				2
−2	//							2
−3								0
−4								0
−5		/						1
	−1	0	1	2	3	4	5	

Characteristic 1

x   single value
    two values
    three values
    four values

The scatter plot in Figure 12.9 can be prepared directly from the correlation table. Procedure 12.1 is used to prepare a standard scatter plot.

Procedure 12.1 Scatter Plot Construction

A scatter plot is a graph used to display the relationship between characteristics. It is necessary to use pairs of measurements made on the *same* sampling units (e.g., parts) so the data appear as

Sampling unit	Characteristic 1	Characteristic 2
1	X_1	Y_1
2	X_2	Y_2
.	.	.
.	.	.
.	.	.
N	X_N	Y_N

It is a good practice to use a sample size of at least $N = 30$–50 sampling units. The usefulness of the scatter plot is greatly enhanced by attempting to spread out the range of values for one of the characteristics (labeled X) so the entire specification range is represented. Also, when the data are collected, all meaningful stratification factors should be recorded (e.g., supplier, operator, machine, spindle, and fixture). The following steps are used to construct a scatter plot:

1. Select the characteristic (labeled 1), if any, that for some reason is thought to predict the value of the other characteristic (labeled 2). This characteristic should be plotted on the horizontal x axis. The second characteristic is plotted on the vertical y axis. Find the minimum and maximum for the characteristics. Choose a convenient plotting scale so that the lowest scale value is smaller than the minimum and the highest scale value is larger than the maximum. Draw the scales on standard graph paper. The length of the two axes should be about the same.
2. When there are less than about 20 unique measurement readings for each characteristic, it is convenient to prepare a correlation table. List the range of different possible values for each characteristic in a two-dimensional table:

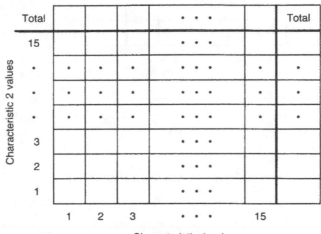

464

12.2 Construction and Analysis

3. Plot the values for each sampling unit. If a correlation table is used, the scatter plot can be plotted directly from the table. If repeated values occur in a particular location, use the following plotting convention:

   x single value
   ⊗ two values
   ⊗ three values
   ⊗ four values
   etc.

4. Use the patterns in Figure 12.4 to evaluate the relationship between the characteristics. Procedures 12.2 and 12.3 can be used to test for the presence of correlation.
5. A good practice is to prepare the histogram for each characteristic (Procedure 5.1). Histograms can be prepared directly from the "totals" column and row in the correlation table.

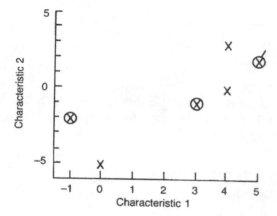

Figure 12.9 Sample scatter plot: duplicate points.

12.2.1 Sign Test for Correlation

The general patterns in Figure 12.4 are somewhat subjective, but it is possible to explicitly test for correlation between two characteristics. The sign test for correlation described in this section is based on a simple concept. Suppose we find the median value for characteristic 1 (median 1) and characteristic 2 (median 2). These values are then indicated by dashed lines on a scatter plot, as shown in Figure 12.10. The numbers, 1, 2, 3, and 4 are used to indicate the quadrant on the plot. If a positive correlation exists, quadrants 1 and 3 should contain significantly more points than quadrants 2 and 4, as shown in Figure 12.11a. Conversely, if the number of points in quadrants 2 and 4 is significantly greater than that in quadrants 1 and 3, a negative correlation exists, as shown in Figure 12.11b. Procedure 12.2 provides a method to compare the number of points in quadrants 1 and 3 (denoted A) with the number of points in quadrants 2 and 4 (denoted B). The sign test given in Procedure 12.2 is presented in Ishikawa (1982). Dixon and Massey (1983, pp. 405–406) and Nelson (1983b) present other sign tests for correlation. The critical values in Procedure 12.2 are from standard binomial percentiles.

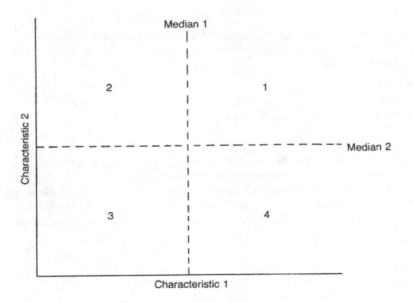

Figure 12.10

To illustrate the calculations of the sign test, consider the data in Table 12.2. The correlation table for these data is shown in Table 12.3. The scatter plot in Figure 12.12 can be drawn directly from this table. Following the steps in Procedure 12.2, the sign test for correlation can be performed.

1. Rank X and Y values. Order X values for characteristic 1:

   1, 2, 2, 2, 3, 3, 3, 4, 4, 4, 4, 5, 5, 5, (5, 5,) 5, 6, 6,
   6, 6, 6, 7, 7, 8, 8, 8, 9, 9, 10

   Order Y values for characteristic 2:

   -3, -3, -3, -2, -2, -2, -1, -1, -1, -1, -1, -1, -1,
   0, (0, 0,) 0, 0, 0, 0, 0, 1, 1, 1, 1, 2,
   2, 2, 2, 3

2. Form the listing in step 1, median 1 = 5 and median 2 = 0. Note that the median values could be obtained directly from the correlation table.
3. The labeled scattered plot is shown in Figure 12.12.
4. Count the points in each quadrant:

Quadrant	No. points
1	8
2	1
3	9
4	3
Total	21

$A = 8 + 9 = 17$
$B = 1 + 3 = 4$
$C = \min(A, B) = 4$
$Q = 9$
$SS = N - Q = 30 - 9 = 21$

12.2 Construction and Analysis

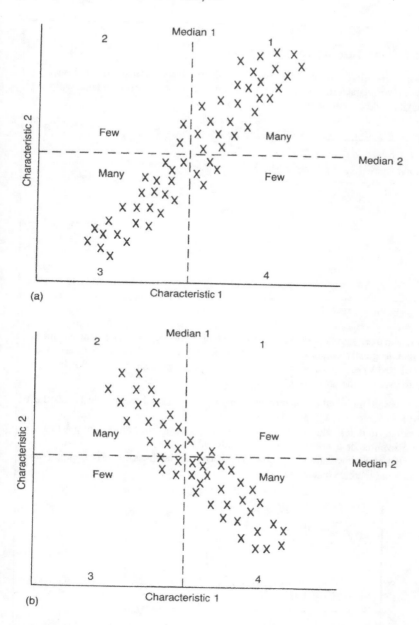

Figure 12.11 Comparing the number of points in quadrants using median lines: (a) positive correlation; (b) negative correlation.

Procedure 12.2 Sign Test for Correlation

The sign test for correlation provides a method to quantitatively evaluate the presence of correlation between two characteristics. Let pairs of measurements be made on the same sampling units (e.g., parts) so the data appear as

Sampling unit	Characteristic 1	Characteristic 2
1	X_1	Y_1
2	X_2	Y_2
.	.	.
.	.	.
.	.	.
N	X_N	Y_N

It is a good practice to use a sample size of at least $N = 30\text{--}50$ sampling units. Also, the usefulness of the scatter plot is greatly enhanced by spreading out the range of values for one of the characteristics (labeled X) so that the entire specification range is represented. The following steps should be used to evaluate the presence of correlation:

1. Rank from lowest to highest the values of each characteristic. For characteristic 1, rank the X values, and for characteristic 2 rank the Y values.
2. Find the median (middle) value for each characteristic. Recall from Procedure 3.1 that the median for N even is the average of the two middle values.
3. Prepare a scatter plot (Procedure 12.1). Draw the dashed median lines on the scatter plot, and label the lines and the four quadrants.

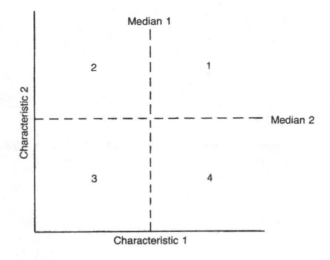

4. Count the number of points in each quadrant, and compute the following quantities:

   A = number of points in quadrants 1 and 3
   B = number of points in quadrants 2 and 4
   Q = number of points on either median line
   Total $N = A + B + Q$

5. Compute the test value C, where

   C = smaller of A and B = $\min(A, B)$
   Sample size for the test SS = $N - Q$

   If C is no greater than the table value c for the given value of SS, the characteristics are correlated.

   The maximum value c of computed C needed to conclude that a correlation exists:

SS	c	SS	c	SS	c
20–22	5	54–55	19	86–87	33
23–24	6	56–57	20	88–89	34
25–27	7	58–60	21	90–91	35
28–29	8	61–62	22	92–93	36
30–32	9	63–64	23	94–96	37
33–34	10	65–66	24	97–98	38
35–36	11	67–69	25	99–101	39
37–39	12	70–71	26	110	43
40–41	13	72–73	27	120	48
42–43	14	74–76	28	130	52
44–46	15	77–78	29	140	57
47–48	16	79–80	30	150	61
49–50	17	81–82	31	200	84
51–53	18	83–85	32		

The absence of correlation may be due to the presence of too narrow a range for a characteristic, inadequate measurement sensitivity, or poor gage repeatability.

Table 12.2 Example of Two Measured Characteristics on a Part

Part	Characteristic 1	Characteristic 2	Part	Characteristic 1	Characteristic 2
1	5	0	16	10	2
2	2	−1	17	7	0
3	7	−1	18	3	−2
4	6	2	19	6	1
5	9	2	20	6	−1
6	3	0	21	5	0
7	9	1	22	8	1
8	5	0	23	2	−2
9	8	−3	24	5	0
10	4	−1	25	3	−1
11	4	3	26	4	−3
12	6	0	27	5	−1
13	1	−2	28	6	1
14	5	0	29	4	−1
15	2	−3	30	8	2

Table 12.3 Correlation Table for the Data in Table 12.2

	Total	1	3	3	4	6	5	2	3	2	1	Total
	3				/							1
	2						/		/	/	/	4
	1						//		/	/		4
Characteristic 2	0			/		𝑯𝑯𝑯	/	/				8
	−1		/	/	//	/	/	/				7
	−2	/	/	/								3
	−3		/		/				/			3
		1	2	3	4	5	6	7	8	9	10	

Characteristic 1

12.2 Construction and Analysis

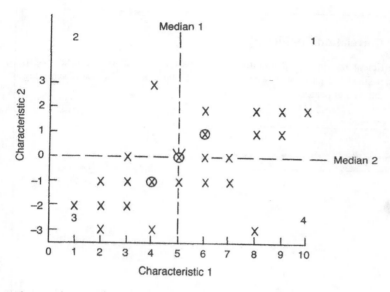

Figure 12.12 Scatter plot of the data in Table 12.2.

5. The sample size maximum value in the table in Procedure 12.1 is $c = 5$, so we conclude that there is correlation between the characteristics.

12.2.2 Correlation Coefficient*

The extent of the relationship between two characteristics is quantified by the correlation coefficient r. The value of r ranges between -1 and 1. A perfect correlation of -1 or 1 implies that there is a direct linear relationship between the two characteristics, as shown in Figure 12.13. For most practical applications, $r = -1$ or 1 is never observed, but these values serve as benchmarks to evaluate other correlations. Figure 12.4a and e shows more typical patterns for high positive or negative correlations. Another benchmark is the absence of any correlation, where $r = 0$ as shown in Figure 12.4c.

Procedure 12.3 gives the method used to compute r. Generally, r can be obtained using a calculator or computer routine. Rarely are the calculations in steps 2, 3, and 4 performed directly. The correlation testing procedure in step 5 must be used to determine whether r is significantly different from 0; that is, is there a relationship between the two characteristics? When the histograms for the two characteristics represent a standard bell-shaped normal distribution, the correlation testing method in Procedure 12.3 is more sensitive than the sign test method in Procedure 12.2. When nonnormal histograms are encountered, the sign test procedure is the appropriate evaluation method.

Using the previous $N = 30$ observations, the calculations in Procedure 12.3 are illustrated as follows:

1. Compute the means and standard deviations:

$\bar{X} = 5.27 \quad s_X = 2.30$
$\bar{Y} = -0.23 \quad s_Y = 1.59$

Procedure 12.3 Correlation Coefficient

The correlation coefficient provides a quantitative measure of the extent of the relationship between two characteristics. Let pairs of measurements (X, Y) be made on the same sampling units (e.g., parts) so the data appear as

Sampling unit	Characteristic 1	Characteristic 2
1	X_1	Y_1
2	X_2	Y_2
.	.	.
.	.	.
.	.	.
N	X_N	Y_N
Mean	$\bar{X}$	$\bar{Y}$
SD	s_X	s_Y

It is good practice to use a sample size of at least $N = 30$–50 sampling units. Also, the usefulness of the correlation testing procedure is greatly enhanced by spreading out the range of values for one of the characteristics (labeled X) so that the entire specification range is represented. The following steps should be used to evaluate the presence of correlation:

1. Compute the mean and standard deviation of the two characteristics (Procedure 3.1):

Characteristic	Mean	SD
1	$\bar{X}$	s_X
2	$\bar{Y}$	s_Y

2. Compute the quantity

   $$F = X_1Y_1 + X_2Y_2 + \cdots + X_NY_N$$

3. Compute the quantities

   $$G = N\bar{X}\bar{Y}$$
   $$H = (N - 1)s_X s_Y$$

4. Compute the correlation coefficient

   $$r = \frac{F - G}{H}$$

   The value of r is between -1 and 1.

5. Ignoring the plus or minus sign of r, if the value of r is greater than the following tabled value, a significant correlation probably exists between the characteristics. The procedure in this step assumes that the histogram represents an approximately normal, bell-shaped distribution for the Y characteristic.

The minimum value of computed r needed to conclude that a correlation exists:

N	Minimum value	N	Minimum value
5	.88	25	.40
6	.81	30	.36
7	.76	35	.33
8	.71	40	.31
9	.67	45	.29
10	.63	50	.28
11	.60	60	.25
12	.58	70	.24
13	.55	80	.22
14	.53	90	.21
15	.51	100	.20
16	.50	150	.16
17	.48	200	.14
18	.47	300	.11
19	.46	400	.10
20	.44	500	.09

The absence of correlation may be due to the presence of too narrow a range for a characteristic, inadequate measurement sensitivity, or poor gage repeatability.

A few extreme X or Y values can significantly influence the value of r. When this occurs, the usefulness of r diminishes since it is not desirable to have a few measurements overly influence a computed value. Procedure 12.2 should be used when extreme values are present.

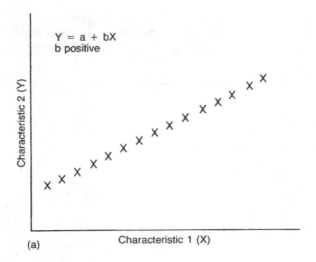

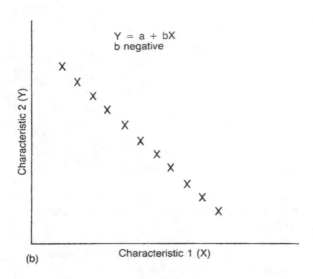

Figure 12.13 Example of perfect correlation: (a) perfect positive correlation ($r = 1$); (b) perfect negative correlation ($r = -1$).

2. Compute the quantity

$$F = (5 \times 0) + (2 \times -1) + (7 \times -1) + \cdots + (8 \times 2) = 22$$

3. Compute the quantities

$$G = 30(5.27)(-0.23) = -36.36$$
$$H = 29(2.30)(1.59) = 106.05$$

12.2 Construction and Analysis

4. Compute the correlation coefficient

$$r = \frac{22 - (-36.36)}{106.05} = 0.55$$

5. The tabled value for $N = 30$ is 0.36, so we conclude that there is correlation between the two characteristics.

12.2.3 Equality: No Difference Benchmark Line

In several types of studies, it is useful to prepare scatter plots with the same horizontal and vertical axes. These cases occur when a characteristic is measured a first time (usually plotted on the horizontal axis) and is then measured later a second time (usually plotted on the vertical axis). The relationship between the first and second measurements is of interest. Using the same scale on both scatter plot axes makes the 45° line a benchmark in which the first measurement equals the second, indicating that no change has occurred, as shown in Figure 12.14. A simple application of this technique is when we measure a group of parts twice with the same gage. Ideally, all values fall on the 45° line. However, measurement error causes differences between the first and second gage readings. Quantification of gage repeatability error is discussed in Appendix I, but a simple scatter plot can be used to display the gaging error, as shown in Figure 12.15a and b. Another application of the same concept is when the second measurement is by a much more accurate gage than the first measurement. In this case gage bias is effectively displayed, as shown in Figure 12.15c and d. Alternatively, the second gage reading may be by a different operator who used the same gage as was used to make the first measurement. In all cases a visual display of the scatter plot with the 45° benchmark line enables direct interpretation of the data.

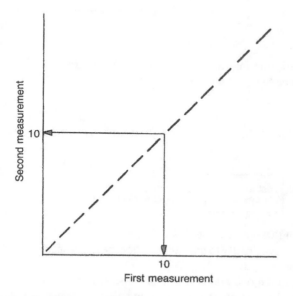

Figure 12.14 Sample benchmark line. This is a no-difference line where the first measurement = the second measurement.

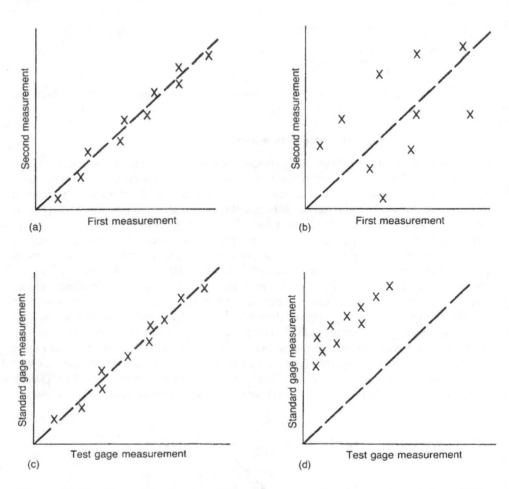

Figure 12.15 Use of the benchmark line in gage studies: (a) good gage repeatability; (b) poor gage repeatability; (c) no gage bias; (d) test gage biased low.

Another application of the benchmark line is when a test treatment or operation is evaluated.

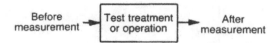

Again the scatter plot conveys the influence of the test, as shown in Figure 12.16. The scatter plot benchmark line $X = Y$ equates directly to the 0 point when the differences $Y - X$ are plotted on a histogram, as shown in Figure 12.16.

The influence of sequential production operations is often of concern. A part characteristic can be unintentionally altered at various stages of production. Suppose Op 30 produced a final part characteristic and it is suspected that downstream operations 40 and 50 change the characteristic:

12.2 Construction and Analysis

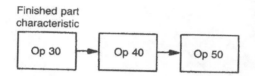

Again a simple scatter plot can be used to address this important problem, as shown in Figure 12.17. For this analysis to be effective, a wide range of parts from Op 30 should be selected, including parts beyond the specification limits.

12.2.4 Stratified Plots

The importance of stratification factors has been repeatedly emphasized throughout this text. In many cases the key to solving a problem or making an improvement is collecting and analyzing data using a key stratification factor. Stratification factors are also important for assessing the relationship between characteristics. Consider the height versus weight example in Figure 12.2 in which an obvious stratification factor is sex. The plot probably appears as shown in Figure 12.18, where males are taller and heavier than females. Use of the coding M = male and F = female helped indicate this important relationship in the plot.

In some cases coding of a key stratification factor enables discovery of the underlying cause of a problem. Consider attempting to discover why the heat treatment of parts introduces unexpected variability. By measuring a dimensional characteristic of parts before and after the heat-treat operation, it is possible to code the furnace where the heat

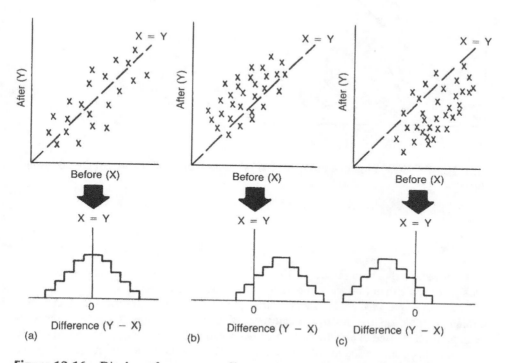

Figure 12.16 Display of treatment effects using scatter plots or histograms: (a) no treatment effect; (b) after greater than before; (c) after less than before.

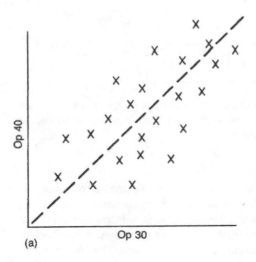

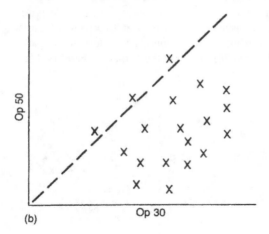

Figure 12.17 Evaluating operation effects: (a) no influence of Op 40; (b) Op 50 lowers value.

treat occurred, as shown in Figure 12.19. The scatter plot makes it apparent that furnace 2 is less predictable than furnace 1 and may be the cause of a problem.

In each example in the next section, it is possible to use a stratification factor and examine potential differences in the categories forming the factor. A single or separate scatter plot can be used to plot the categories. As noted in Procedure 8.1, the critical step is to plan the check sheet used to collect data so that all potentially meaningful stratification factors are recorded. Many studies are inconclusive owing to the lack of recording and analyzing potential differences between stratification factors.

12.3 Applications

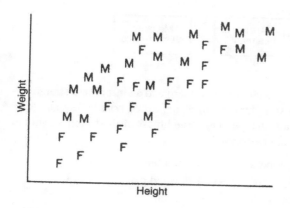

Figure 12.18 Scatter plot stratified by sex: F, female; M, male.

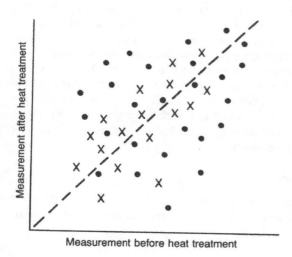

Figure 12.19 Scatter plot stratified by heat-treat furnace: furnace 1 (×); furnace 2 (●).

12.3 Scatter Plot Applications

Why is the presence or absence of a relationship between characteristics important? No universal concept dictates that the presence or absence of correlation is good or bad. Interpretation of the results depends on the situation. The following general applications are presented to illustrate the variety of uses of scatter plots.

12.3.1 Treatment Effects

Improvements are often made by studying the influence of an added operation or treatment in a process. The typical treatment effect study measures a characteristic before the treatment and again after the treatment is given.

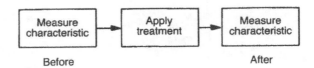

For example, suppose the straightness of a shaft is a problem and an in-process straightening operation is being evaluated. Part straightness can be measured both before and after the straightening operation to assess any change. The straightness of the final part should also be measured, if other operations are performed.

$$\begin{array}{ccc} \text{Straightness} & \text{Straightness} & \text{Straightness} \\ \text{before new operation} \rightarrow & \text{after new operation} \rightarrow & \text{of finished part} \\ \text{(in-process)} & \text{(in-process)} & \text{(final)} \end{array}$$

Improvement must address in-process results, but final results are related to customer satisfaction. Scatter plots can be used to assess how final straightness depends on the two in-process results.

This type of study results in a classic paired data set in which the before measurement on the x axis is compared to the after measurements on the y axis. The scatter plot can be evaluated as shown in Figure 12.20. Since the same characteristic is measured, the 45° line provides a useful benchmark for evaluating the influence of the treatment. A formal testing procedure to compare location and variability is given later in Procedure 12.5.

In this type of study, a strong correlation is usually expected. The treatment is expected to influence the characteristic in a predictable way. It is often useful to compute the prediction line discussed later in the chapter. A lack of correlation implies that there is no predictable influence of the treatment (over the range of the before measurements). Lack of predictability may be an indication of a problem.

12.3.2 Process Influence: Robustness

The complexity of manufacturing processes can make troubleshooting difficult when problems occur at a downstream operation. Consider a problem encountered at operation

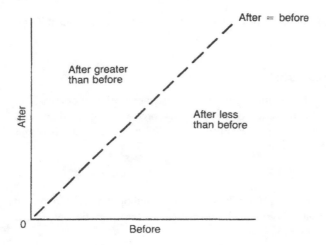

Figure 12.20 Evaluating treatment effects.

12.3 Applications

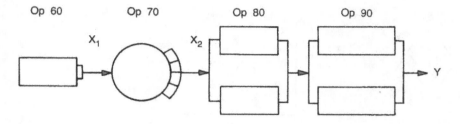

Figure 12.21 Process flow diagram for process influence study.

90 shown in Figure 12.21. Possible causes are thought to be associated with operations 60 and 80. Measuring (say, $N = 50$) the same parts after each operation results in a paired data arrangement.

Part	X_1, Op 60	X_2, Op 80	Y, Op 90
1	X_1	X'_1	Y_1
2	X_2	X'_2	Y_2
.	.	.	.
.	.	.	.
N	X_N	X'_N	Y_N

The scatter plots may appear as any of the patterns shown in Figure 12.22. Establishing the relationship between an upstream operation (either Op 60 or Op 80) and a downstream operation (Op 90) can be important in troubleshooting. If a high correlation exists, a problem at an upstream operation can cause a problem in downstream processing. Conversely, the absence of correlation over a representative range of production for a characteristic suggests troubleshooting efforts should be focused elsewhere. The downstream process is tolerant of the normal range of upstream output. Either type of information is valuable for troubleshooting.

This application of process influence considers only three operations. Of course, a larger number of operations or characteristics at an operation can be considered. The only drawback is that the spread of the measurements may become too narrow. In this example, there may be a large spread of measurements after Op 60 but a narrow spread after Op 80. Again, the narrow spread decreases the likelihood of a significant correlation with Op 90. A separate group of parts may need to be collected after Op 80 to obtain a wide range of measurements.

When a treatment effect was evaluated, the same part characteristic was measured before and after the treatment. In process influence robustness applications, the same or a different characteristic may be used. The straightness of a machined bar may be examined at different processing stages (same characteristic). Alternatively, the influence of concentricity at different processing stages on final straightness may be of interest (different

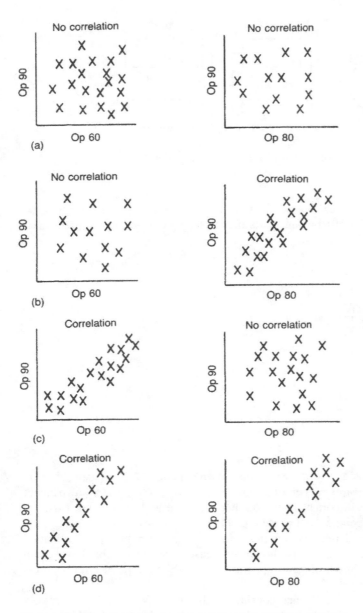

Figure 12.22 Analyzing process influence with scatter plots. (a) Neither Op 60 nor Op 80 influences the output of Op 90 (process robustness); (b) Op 80 influences the output of Op 90; Op 60 does not. (c) Op 60 influences the output of Op 90; Op 80 does not. (d) Both Op 60 and Op 80 influence the output of Op 90.

12.3 Applications

characteristics). When the same characteristic is evaluated at different stages, Procedure 12.5 can be used to formally test for changes in mean level or variability.

Understanding the relationship of a characteristic produced in an operation to incoming part characteristics must consider two cases. First, it is generally desirable to have a high correlation between a part feature at different stages of processing (e.g., rough, semi-finished, and finished). This enables effective targeting and in-process, upstream control. Quality is "built up" in the process. The absence of correlation in this case implies that other outside influences introduce variability that may control final part quality, making process control more difficult. Second, it is desirable that a characteristic produced in an operation not be correlated with other incoming part characteristics. Processing is then insensitive to upstream variability. This is the essence of process robustness. Typically, an operation is robust for only a limited range of other incoming characteristic values.

12.3.3 In-process Specification Limits

When in-process and final characteristics have a high correlation, scatter plots can be used to determine the required in-process specification to attain a stated final specification requirement. Typically, product requirements determine final specification limits. In-process limits are determined by the manufacturing requirements necessary to meet final limits. However, historical limits or trial and error is typically used to determine in-process limits. A standard scatter plot correlation study provides a quantitative alternative to these traditional approaches.

A standard correlation study can be performed considering in-process and final characteristics that are thought to be related. If the resulting correlation is low, other factors control the final characteristic, as explained in the previous discussion. If the correlation is high, Figure 12.23 illustrates how the two sets of limits can be related. The elliptical area containing the points must be such that all in-process points produce a final characteristic within the required limits. Draper and Smith (1981, Sec. 1.4) give the statistical method that can be used to quantitatively account for the variability around the prediction line:

Final value = $a + b$ in-process value

In these studies, it is essential to select samples well beyond the normal range of the in-process characteristic.

12.3.4 Cause-and-Effect Relationships

In Chapter 13 a cause-and-effect (C & E) diagram is discussed as a method of identifying potential causes of a problem. After identifying potential causes, it is necessary to use the statistical tools presented in this text to evaluate whether a potential cause is a root cause of a problem. Consider the surface finish of a machined part that exhibits poor capability with some parts beyond the specification limits. A C & E diagram is developed for this problem and indicates that material hardness is a potential cause, as shown in Figure 12.24.

By selecting a wide range of hardness values, it is possible to evaluate whether hardness can be a contributor to poor surface finish. Figure 12.25 presents two relationships. The test for correlation between hardness and surface finish is an evaluation of whether hardness variation is one of the factors causing poor surface finish. The absence of correlation suggests that hardness is unrelated to poor surface finish and that other potential causes should be evaluated. A low but significant correlation may imply that the potential cause does contribute to the effect, but other causes may also exist. Even if correlation is

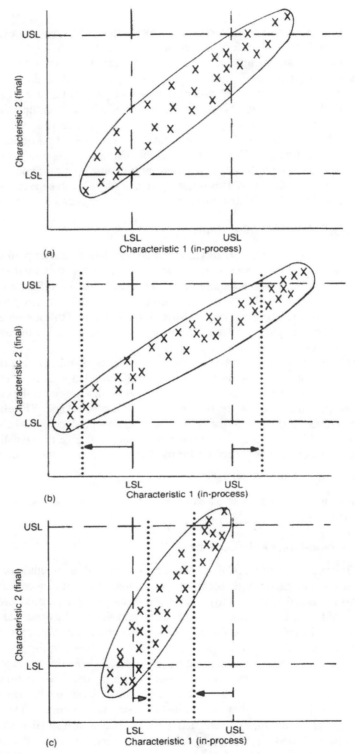

Figure 12.23 Using scatter plots to determine in-process specification limits: (a) ideal specifications; (b) expansion of specifications possible; (c) contraction of specifications required.

12.3 Applications

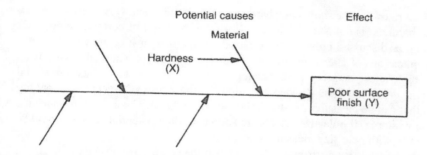

Figure 12.24 Investigating potential causes from a C & E diagram using scatter plots.

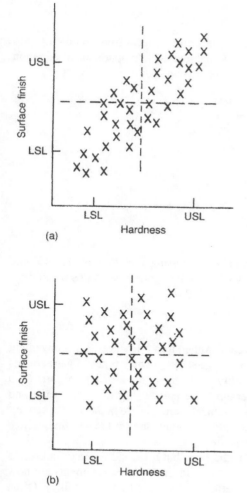

Figure 12.25 Using correlation to assist in determining problem causes: (a) correlation exists (causation possible); (b) no correlation (causation not possible).

shown to exist, it must then be shown that hardness exhibits sufficient variability to cause a problem (i.e., hardness is not a latent cause). Also, the existence of correlation does not imply causation, as discussed previously. A causal mechanism must be shown to exist.

In many applications of C & E diagrams, the problem or effect being addressed is an unacceptably high reject or scrap rate. Scatter plots can be used to evaluate potential causes, but two cases must be considered. First, if the cause of rejects is a quality characteristic that can be measured, it should be measured directly instead of measuring a reject rate. For example, if hole size rejects are addressed then potential causes should be correlated directly with hole size measurements, not a reject rate.

The second case involves an attribute effect, such as the presence or absence of burrs. In this case the procedures discussed in Chapter 11 should be used. If the hardness of a part is thought to be related to burrs, parts with and without burrs can be sorted into two groups and the hardness evaluated. By plotting histograms and using the comparison methods in Procedure 11.4, the influence of hardness can be evaluated.

12.3.5 Symmetrical Parts

Recall from Chapter 11 that one method of making improvements or troubleshooting is to make meaningful comparisons. Symmetrical parts enable useful comparisons of common dimensions. Consider a simple case of an engine connecting rod:

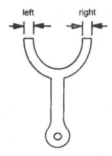

Comparison of right and left side dimensions can be made using a scatter plot. The 45° line provides a convenient benchmark for evaluating whether the right and left sides are similar. A formal comparison method is given in Procedure 12.5.

12.3.6 Gage Analysis

A formal approach for evaluating gages is given in Appendix I. However, it is convenient to use scatter plots to display the data obtained in gage studies. Three types of gage studies are considered. First, a typical gage repeatability study measures a group of parts on a single gage and then in random order remeasures the parts using the same gage and operator. Differences between the two measurements are due to a lack of gage repeatability, as illustrated in Figure 12.26a. Ideally, all points are on the 45° line, which occurs when the first and second measurements are equal.

A second type of gage study is the same as the first, but a second operator is used to make the second group of measurements. Differences in the two measurements are now due to both a lack of gage repeatability and operator differences (reproducibility). Often operator bias is of concern, as illustrated in Figure 12.26b.

12.3 Applications

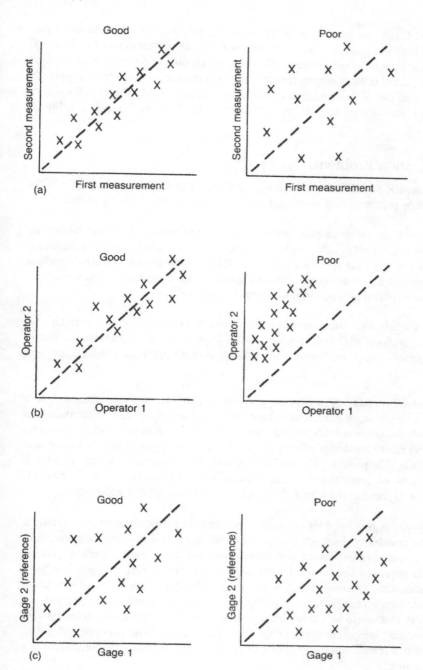

Figure 12.26 Displaying measurement variability with scatter plots: (a) gage repeatability; (b) gage repeatability and reproducibility; (c) gage accuracy.

A third type of gage analysis is to compare two gages that measure the same part characteristic. If both gages obtain the same measurements, all plotted points are on the 45° line. However, each gage has some repeatability error so a scatter of points centered around the 45° line is expected if there is no difference between the gages. Gage bias is typically evaluated by making one of the measurements using a reference standard gage. Differences predominantly on one side of the 45° line are due to gage bias, as illustrated in Figure 12.26c.

12.4 Application Problems

A number of potential pitfalls arise when a scatter plot analysis is used. Awareness of the following potential problems helps ensure that a useful scatter plot analysis is obtained.

Limited Range. If the range of both characteristics is too narrow, low correlations are typically observed. Generally, it is possible to select one characteristic over a wide range, preferably values beyond the specification limits. There is no need to randomly sample parts; a better approach is to select parts spanning the specification range. It is sometimes necessary to make special production runs to obtain extreme parts.

Measurement Variability. Inadequate gage repeatability can result in low correlations. It is important to perform gage repeatability studies (App. I) prior to conducting a scatter plot study. It is sometimes possible to use nonproduction, high-precision gaging equipment for the study.

Poorly Defined Objective. Establishing correlation between two characteristics does not establish causation or predictability—knowledge of the subject area must make that inference. A well-thought-out study should attempt to define relationships between characteristics based on prior knowledge of the process. The logical approach for data collection then becomes clear. The presence of correlation in a study then confirms or refutes what is hypothesized based on process knowledge. Use of the data summaries in Chapter 13, which utilize a well-thought-out check sheet, requires this disciplined approach.

Process Stability. A process does not need to be stable to perform a scatter plot study if the cause of the instability does not change the relationship between the characteristics being studied. In many cases, scatter plot studies are used to discover a cause of process instability. This approach can be used quite successfully. However, it is possible for a cause of process instability to result in changing the relationship between characteristics. It is difficult to interpret scatter plot results in this case. Consider unstable temperatures in a chemical process that cause unstable yields. Attempting to study yields versus process parameters other than temperature can produce unusual results. When unexpected results are encountered in a scatter plot study, a lack of process stability should be entertained as a possible explanation.

Dependence on Additional Characteristics. It is not unusual for many process parameters or part characteristics to be interrelated. Selecting only two characteristics to study on a scatter plot may produce unexpectedly low correlations due to the dependence on other characteristics. It is possible to collect any number of characteristics and examine combined "multiple regression" relationships (Draper and Smith, 1981).

12.4 Application Problems

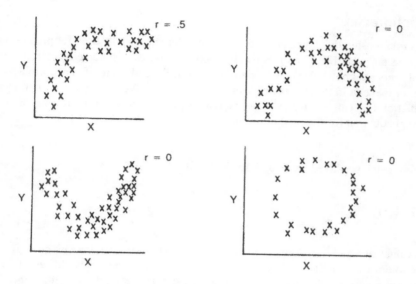

Figure 12.27 Nonlinear patterns.

Nonlinear Patterns. The scatter plot patterns in Figure 12.4 can be easily interpreted, but nonstandard patterns are possible. These patterns do not cause a problem unless the user fails to plot the data on a scatter plot and relies on a reported correlation coefficient, often from a computer program. As the plots in Figure 12.27 illustrate, plotting the data can easily uncover unusual patterns or erroneous results that a single correlation coefficient cannot convey.

Extreme Points. When an extreme measurement is obtained, it is not obvious how to interpret the data. Extreme points can provide an indication of the existence of a problem, which may help uncover an unidentified special cause. The existence of an extreme point is valuable information about a process. If one extreme value is produced, others are possible. It may be necessary to repeat the study to determine whether the extreme point is truly unusual. However, a low-frequency special cause is not easily repeated. Caution must be used in testing for correlation when an extreme point is present. The sign test in Procedure 12.2 is not influenced by an extreme value. However, the presence of one or more extreme points can cause the correlation coefficient test in Procedure 12.3 to indicate the existence of correlation when none is present.

Plot the data

Tagging Parts. The use of scatter plots requires that pairs of measurements be obtained. Parts are often tagged or numbered to enable correct identification and pairing of measurements. The study must ensure that special handling or tagging does not influence the processing or measurement of the parts.

12.5 Prediction Line*

The scatter plot relationship displayed between two characteristics can be used for prediction if two criteria are met. First, the relationship between the two characteristics must make sense: only process knowledge can assess this criteria. Second, a significant correlation from either of the tests in Procedures 12.2 or 12.3 must exist. Assuming these criteria are met, a prediction line can be displayed on the scatter plot, as shown in Figure 12.28. The mathematical equation for the prediction line is

$$Y = a + bX$$

If $a = 1$ and $b = 2$, the predicted value of Y for $X = 5$ is

$$Y = a + bX$$
$$= 1 + (2 \times 5)$$
$$= 11$$

Given that the values of a and b are known, the value of Y can be predicted for a given X. It is not desirable to draw the prediction line by eye since an accurate line can be difficult to obtain, as shown in Figure 12.29. Procedure 12.4 gives a simple method for obtaining the best "least-squares" prediction line.

Calculation of the prediction line can be illustrated using the $N = 30$ measurements from Table 12.2. Following the steps in Procedure 12.4:

1. Compute the means and standard deviations:

   $\bar{X} = 5.27 \quad s_X = 2.30$
   $\bar{Y} = -.23 \quad s_Y = 1.59$

2. Compute the quantity

   $F = (5 \times 0) + (2 \times -1) + (7 \times -1) + \cdots + (8 \times 2) = 22$

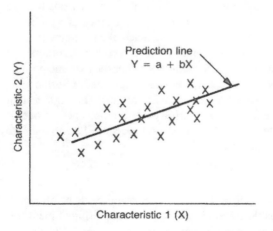

Figure 12.28 Scatter plot with a prediction line.

12.5 Prediction Line

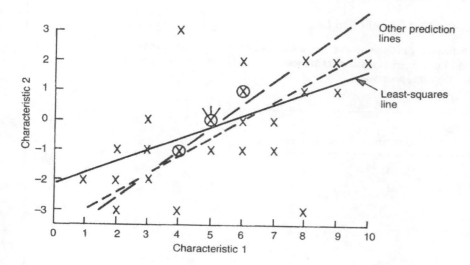

Figure 12.29 Selecting the best prediction line.

3. Compute the quantities

   $G = 30(5.27)(-.23) = -36.36$
   $I = 29(2.30)(2.30) = 153.41$

4. Compute the slope term

   $$b = \frac{22 - (-36.36)}{153.41} = 0.38$$

5. Compute the y axis intercept term:

   $a = -.23 - (.38 \times 5.27) = -2.23$

   The prediction line is

   $Y = -2.23 + 0.38X$

   Selecting values of $X_1 = 1$ and $X_2 = 10$, the corresponding values of Y are

   $Y_1 = -2.23 + (.38 \times 1) = -1.85$
   $Y_2 = -2.23 + (.38 \times 10) = 1.57$

The pairs of points (1, −1.85) and (10, 1.57) can be plotted on the scatter plot and connected to indicate the prediction line, as shown in Figure 12.29.

Notice that there is significant variability around the prediction line. A value of $X = 6$ produces a predicted value of

$Y = -2.23 + (0.38 \times 6) = .05$

However, the observed variability in Figure 12.29 at $X = 6$ ranges between −1 and +2. Variability around a predicted value is expected. It is possible to quantify the degree of

Procedure 12.4 Prediction Line

A prediction line can be computed from paired data when it is found there is a correlation between the X and Y characteristics. Let pairs of measurements be made on the same sampling units (e.g., parts) so the data appear as

Sampling unit	Characteristic 1	Characteristic 2
1	X_1	Y_1
2	X_2	Y_2
.	.	.
.	.	.
.	.	.
N	X_N	Y_N

The usefulness of the prediction line increases if the range of values is spread out for one of the characteristics (labeled X). The prediction line should not be used outside the range of values used to compute the line. The following steps should be used to compute the prediction line $Y = a + bX$:

1. Prepare a scatter plot of the data. Compute the mean and standard deviation of the X and Y characteristics (Procedure 3.1):

Characteristic	Mean	SD
1 (X values)	$\bar{X}$	s_X
2 (Y values)	$\bar{Y}$	s_Y

2. Compute the quantity
$$F = X_1Y_1 + X_2Y_2 + \cdots + X_NY_N$$

3. Compute the quantities
$$G = N\bar{X}\bar{Y} \qquad I = (N - 1)s_X^2$$

4. Compute the slope term
$$b = \frac{F - G}{I}$$

Note that if the correlation coefficient r is available (Procedure 12.3), then b can be computed directly:

$$b = r\frac{s_Y}{s_X}$$

12.5 Prediction Line*

5. Compute the y axis intercept term

   $a = \bar{Y} - b\bar{X}$

6. The prediction line is

   $Y = a + bX$

   which can be drawn on the scatter plot. Select any two convenient values of X denoted by X_1 and X_2, and predict Y values:

   $Y_1 = a + bX_1$
   $Y_2 = a + bX_2$

   The pairs (X_1, Y_1) and (X_2, Y_2) can be plotted on the scatter plot and connected to obtain the prediction line.

   A few extreme X or Y values can significantly influence the prediction line a and b values. When this occurs, the usefulness of the line diminishes since it is not desirable to have a few measurements overly influence a computed value.

expected variability using more advanced statistical methods (e.g., Draper and Smith, 1981, Sec. 1.4).

The prediction line can be used in many of the applications discussed in the previous section. Several examples illustrate the use of the prediction line.

Example 12.1 Consider hot machined parts measured to control a process. However, the part measurements decreased as the part cooled to room temperature. Since the specifications were stated at room temperature, the relationship of hot to cold measurements was of interest. Over the range of the specification limits, hot parts were measured and allowed to cool before a remeasurement. A possible scatter plot and prediction line are given in Figure 12.30a. The values of a and b can be computed for the prediction line:

$\text{Cold} = a + b \text{ hot}$

Thus, for any hot measurement it is possible to predict the cold measurement.

As often occurs, the specification limits are stated for a final product that depends on other characteristics. Thus, the specification limits are often associated with the plotted Y characteristic. It is easy to compute the corresponding X characteristic value:

$\text{USL} = a + bX \qquad X = \dfrac{\text{USL} - a}{b}$

$\text{LSL} = a + bX \qquad X = \dfrac{\text{LSL} - a}{b}$

Thus, the cold specification limits can be converted to estimated hot part specification limits:

$\text{USL}_{hot} = \dfrac{\text{USL}_{cold} - a}{b}$

$\text{LSL}_{hot} = \dfrac{\text{LSL}_{cold} - a}{b}$

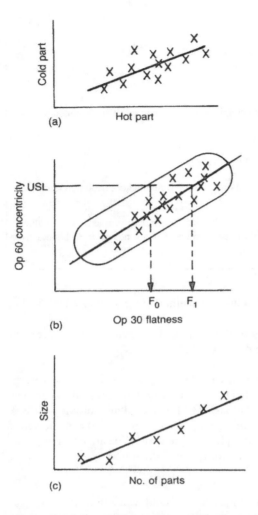

Figure 12.30 Prediction line applications: (a) adjusting for part temperature; (b) determining in-process specification; (c) evaluating tool wear.

In practice the hot limits should be narrower than this result suggests since it is necessary to account for variability about the prediction line (Draper and Smith, 1981, Sec. 1.7).

Example 12.2 Consider the flatness of a locating surface at Op 30 that influences the concentricity of a dimension at Op 60. The scatter plot and prediction line are shown in Figure 12.30b. The predicted flatness associated with USL for concentricity is

$$F_1 = \frac{\text{USL} - a}{b}$$

12.6 Comparison of Two Groups

However, we wish to ensure that USL is not likely to be exceeded so we cannot use F_1 owing to variability about the prediction line. The scatter plot can be used to select a new value flatness F_0, shown in Figure 12.30b, which reasonably ensures that no concentricity values are above the USL.

Example 12.3 Consider the typical tool wear problem from Chapter 4. The standard prediction line can be obtained:

Average size = $a + b$ no. of parts

so the change in size over any period can be predicted. Plotting the actual size values as shown in Figure 12.30c on the plot enables visual evaluation of the variability associated with the prediction line.

Example 12.4 Relating process parameters to part characteristics is one of the most useful applications of scatter plots. For example, we can predict hardened case depth for a part from the furnace temperature using the prediction line

Case depth = $a + b$ temperature

Another example is predicting the surface finish for a part from the machining speed and feed rate. When process parameters are adjustable, the prediction line provides a useful method of targeting the process.

12.6 Comparison of Two Groups (Same Parts)

Scatter plots provide a graphic comparison of two groups of measurements made on the same parts. Use of the 45° benchmark line for a visual comparison is often all that is required to evaluate a difference between the two groups. However, it is sometimes useful to supplement this method with a quantitative comparison test. The method described in this section is particularly useful for comparing variability in the two groups that is not reliably assessed using visual comparisons. The procedures discussed here parallel Procedure 11.4, in which two groups of measurements made on different parts are compared. In this section comparisons are made on two groups of measurements made on the *same* parts. This is referred to as having paired measurements (Chap. 11 was for independent measurements). In many practical applications, paired measurements are preferred since it is possible to more consistently detect differences between two groups. It should be emphasized that the comparison tests are meant to supplement, not replace, graphic displays, such as histograms, comparison plots, and scatter plots.

The sampling method directly influences which comparison methods are meaningful. If parts are obtained to specifically span the specification range, scatter plots and prediction lines are more useful. The location comparison method in Procedure 12.5 can be used if the differences between paired measurements has an approximately normal, bell-shaped distribution. However, because of the nonrandom sampling, process capabilities or any comparisons of variability using Procedure 12.5 are not meaningful. Conversely, when random sampling is used, both location and variability can be meaningfully tested using Procedure 12.5 provided that the individual group measurements have an approximately normal distribution. It is generally desirable to collect parts over a representative time period to effectively characterize the process performance. However, random sampling may not provide sufficient spread in the data to make scatter plots or prediction lines useful.

Procedure 12.5 Comparison of Two Groups (Same Parts)

The location and variability of two groups of paired measurements made on the same parts can be compared using this procedure. Let pairs of measurements (X, Y) be made on N sampling units (e.g., parts). The X and Y measurements form the two groups. The data and differences between the groups can be arranged as follows:

Sampling unit	Group 1	Group 2	Difference
1	X_1	Y_1	$D_1 = X_1 - Y_1$
2	X_2	Y_2	$D_2 = X_2 - Y_2$
.	.	.	.
.	.	.	.
.	.	.	.
N	X_N	Y_N	$D_N = X_N - Y_N$
Mean	$\bar{X}$	$\bar{Y}$	$\bar{D}$
SD	s_X	s_Y	s_D

The number of measurements obtained from each group should be $N = 15$ or more for location comparisons and $N = 30$ or more for variability comparisons. Smaller sample sizes have an increased risk of not detecting meaningful differences. If only location shifts are of interest, it is good practice to spread out the range of values for one of the characteristics by selecting some extreme parts. If both location and variability differences are of interest, random sampling over a representative time period should be used. The following steps should be used to compare variability and location:

1. Run charts and histograms (Procedure 5.1) should be prepared for the X, Y, and D data. A scatter plot (Procedure 12.1) for the X and Y data should be prepared. These plots should be examined for process irregularities or unusual measurements that could adversely influence the comparison procedures.
2. Compute the mean and standard deviation for each group. For group 1,

$$\bar{X} = \frac{X_1 + X_2 + \cdots + X_N}{N}$$

$$s_X = \sqrt{\frac{(X_1 - \bar{X})^2 + (X_2 - \bar{X})^2 + \cdots + (X_N - \bar{X})^2}{N - 1}}$$

Procedure 3.1 gives the steps for computing s_X. An analogous calculation is performed for group 2.

Comparison of Variability

3. Compute the F_{max} ratio by dividing the largest group standard deviation by the smallest standard deviation

$$F_{max} \text{ ratio} = \left(\frac{s_{max}}{s_{min}}\right)^2$$

This test procedure is only valid if the histograms of both X values and Y values represent an approximately normal, bell-shaped distribution.

4. Compute the correlation coefficient r using Procedure 12.3.
5. Perform the following series of calculations:

$$J = 4r^2 F_{max}$$
$$K = (F_{max} + 1)^2$$
$$L = F_{max} - 1$$
$$M = \frac{L}{\sqrt{K - J}}$$

If M is greater than the following tabled value, significant differences in variability probably exist.

The minimum value of computed M needed to conclude that a variability difference exists:

N	Minimum value	N	Minimum value
5	.88	25	.40
6	.81	30	.36
7	.76	35	.33
8	.71	40	.31
9	.67	45	.29
10	.63	50	.28
11	.60	60	.25
12	.58	70	.23
13	.55	80	.24
14	.53	90	.21
15	.51	100	.20
16	.50	150	.16
17	.48	200	.14
18	.47	300	.11
19	.46	400	.10
20	.44	500	.09

Comparison of Location

6. Compute the difference between the group 1 (X) and group 2 (Y) measurements, where $D = X - Y$ as shown earlier. This test requires only that the histogram of the differences represent an approximately normal, bell-shaped distribution.
7. Compute the mean and standard deviation of the difference values:

$$\bar{D} = \bar{X} - \bar{Y}$$
$$s_D = \sqrt{\frac{(D_1 - \bar{D})^2 + (D_2 - \bar{D})^2 + \cdots + (D_N - \bar{D})^2}{N - 1}}$$

8. Compute the test value:

$$T = \sqrt{N}\frac{\bar{D}}{s_D}$$

Ignore any sign of T. If T is larger than the following tabled value of t, a significant difference in location probably exists between the two groups.

N	t	N	t
5	2.78	18	2.11
6	2.57	19	2.10
7	2.45	20	2.09
8	2.36	25	2.06
9	2.31	30	2.05
10	2.26	35	2.03
11	2.23	40	2.02
12	2.20	45	2.02
13	2.18	50	2.01
14	2.16	60	2.00
15	2.14	70	1.99
16	2.13	100	1.98
17	2.12	Max	1.96

Some common applications of paired measurements on the same parts are discussed here.

Operation Effect. An operation or treatment is performed on parts to change some characteristic of the part:

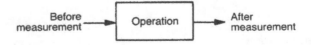

A measurement is made on each part before the operation and again after the operation on the same characteristic. It is necessary to number the parts to enable pairing the before and after measurements. The change or difference between the measurements should be computed:

Difference = after − before

Positive differences then indicate an increase due to the operation.

12.6 Comparison of Two Groups

Common Measurements on a Part. A common measurement, such as metal hardness, may be obtained at two (or more) locations on a part. The within- and between-part variability may be of interest and can be evaluated using comparison of means and standard deviations. If the means of the measurements taken at two locations on a part are not significantly different, then the within-part variability is minimal. The variability between parts at the two measurement locations can be evaluated using the standard deviation at the two locations.

Different Measurements on a Part. Two different characteristics, such as bore size and concentricity, may be related. It is possible to evaluate whether the two characteristics are correlated using the sign test (Procedure 12.2) or the correlation coefficient r (Procedure 12.3). In some applications, such as two size characteristics, it is meaningful to evaluate whether the variability is the same for the two characteristics. It is often necessary to use coded data as a deviation from nominal to have common scaling.

The method used to make these comparisons appears in Procedure 12.5. The variability comparison procedure is given in Snedecor and Cochran (1980, p. 190) and Nelson (1982a). The location comparison is performed by a standard paired t-test, also given in Snedecor and Cochran (1980, p. 84). The following examples illustrate the calculations in Procedure 12.5.

Example 12.5 A group wanted to study the effectiveness of a straightening operation performed on a steel shaft. If parts could be produced within the .015 inch upper specification limit, the operation could be eliminated. A study was conducted by first collecting $N = 30$ parts over a period of several days. These parts were numbered and randomly straightened over a period of several days to obtain a meaningful representation of the process. Parts were measured before and after the straightening operation.

The data along with the differences appear in Table 12.4. The data are coded so that

Actual value (inches) = .001 coded value

The data are analyzed by following the steps in Procedure 12.5.

1. A run chart indicated no unusual problems. Histograms in Figure 12.31 show a decrease in straightness variability after the straightening operation. The scatter plot in Figure 12.32 shows the generally lower after straightness measurements. Points above the benchmark line indicate a decrease in straightness from the straightening operation. The influence of measurement variability is not known.
2. The means and standard deviations are as follows:

	Before	After
$\bar{X}$	12.1	11.0
s	2.1	1.3

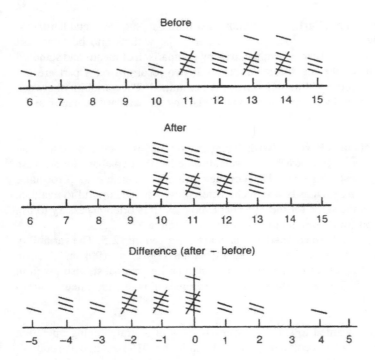

Figure 12.31 Histograms of straightness.

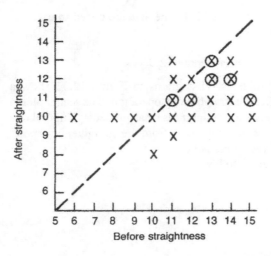

Figure 12.32 Scatter plot of straightness.

12.6 Comparison of Two Groups

3. The F_{max} ratio is

$$F_{max} \text{ ratio} = \left(\frac{2.1}{1.3}\right)^2 = 2.61$$

4. The correlation coefficient using Procedure 12.3 is $r = .40$.
5. The following calculations are required to test for differences in variability:

$$J = [4 \times (.40)^2] \times 2.61 = 1.67$$
$$K = (2.61 + 1)^2 = 13.03$$
$$L = 2.61 - 1 = 1.61$$
$$M = \frac{1.61}{\sqrt{13.03 - 1.67}} = .48$$

Since M is greater than the tabled value of .36, we conclude that there is a significant decrease in variability due to the straightening operation.

6. The differences between the before and after measurements appear in Table 12.4.
7. The mean and standard deviation of the differences are

$$\bar{D} = -1.10 \qquad s_D = 2.0$$

8. The test value is

$$T = \frac{\sqrt{30} \times 1.1}{2.0} = 3.01$$

which is larger than the tabled value of 2.05. Hence we conclude that there is a significant increase in the straightness (i.e., a lower mean) due to the straightening operation.

If the process before and after straightening can be assumed stable, the process spreads and capabilities are as follows:

	Before straightening	After straightening
Upper process limit (UPL = $\bar{X} + 3s$)	18.4	14.9
Process capability (CPU)	.46	1.0

Thus, the straightening operation is necessary to maintain process capability. This study characterizes the benefit of the straightening operation. The mean is lowered slightly, but more important for capability is the appreciable reduction in variability.

This study used random sampling over a period of time to represent the process performance. This approach allowed assessment of process capability (assuming stability), but it has two drawbacks. First, stability is not assessed so it is not possible to reliably predict process performance. However, because the difference in performance is large, it is reasonable to conclude that elimination of the straightening operation is not justified.

Table 12.4 Shaft Straightness Before and After Straightening Operation

No.	Before	After	Difference
1	9	10	1
2	12	11	−1
3	14	12	−2
4	13	13	0
5	14	13	−1
6	13	10	−3
7	11	12	1
8	13	13	0
9	12	11	−1
10	11	11	0
11	14	10	−4
12	15	10	−5
13	10	10	0
14	15	11	−4
15	10	8	−2
16	13	12	−1
17	15	11	−4
18	11	9	−2
19	11	10	−1
20	13	12	−1
21	6	10	4
22	8	10	2
23	14	12	−2
24	14	11	−3
25	12	10	−2
26	12	12	0
27	13	11	−2
28	11	13	2
29	11	11	0
30	14	12	−2
Mean	12.1	11.0	−1.1
SD	2.1	1.3	2.0

12.6 Comparison of Two Groups

Second, there is a risk that a reasonable spread of the data was not obtained so that the relationship between the before and after measurements is not clear.

Example 12.6 A location on a casting was difficult to machine so a study was performed comparing a normal location (location 1) with the difficult-to-machine location (location 2). Brinell hardness readings were taken in the two locations. The $N = 25$ parts were collected over several lots of parts to represent normal production hardness variability. The data along with the differences appear in Table 12.5. The analysis follows the steps in Procedure 12.5:

Table 12.5 Part Hardness in Two Locations

No.	Location 1	Location 2	Difference
1	229	247	−18
2	219	220	−1
3	219	227	−8
4	219	219	0
5	216	230	−14
6	213	217	−4
7	217	224	−7
8	217	219	−2
9	216	229	−13
10	223	235	−12
11	214	222	−8
12	223	221	2
13	221	223	−2
14	227	236	−9
15	215	224	−9
16	221	232	−11
17	215	203	12
18	210	207	3
19	229	219	10
20	217	235	−18
21	218	232	−14
22	213	208	5
23	225	231	−6
24	222	233	−11
25	220	222	−2
Mean	219.1	224.6	−5.5
SD	5.0	10.0	8.0

504   12 Scatter Plots

1. A run chart showed no unusual problems. The histograms in Figure 12.33 show the more variable hardness values in location 2. The scatter plot in Figure 12.34 indicates a possible relationship between the two locations, as expected. The lack of a stronger relationship may be due to hardness measurement variability or within-part variability.
2. The means and standard deviations are as follows:

	Location 1	Location 2
$\bar{X}$	219.1	224.6
s	5.0	10.0

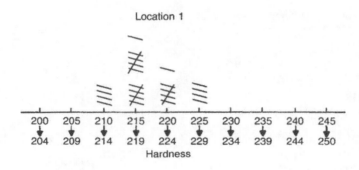

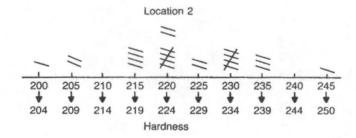

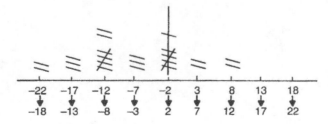

Figure 12.33   Histograms of hardness.

12.6 Comparison of Two Groups

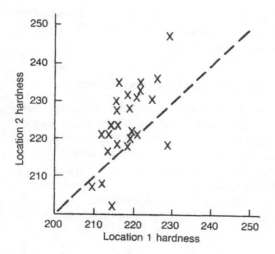

Figure 12.34 Scatter plot of hardness.

3. The F_{max} ratio is

$$F_{max} \text{ ratio} = \left(\frac{10.0}{5.0}\right)^2 = 4.0$$

4. The correlation coefficient using Procedure 12.3 is $r = 0.61$.
5. The following calculations are required to test for differences in variability:

$$J = [4 \times (.61)^2] \times 4.0 = 5.95$$
$$K = (4.0 + 1)^2 = 25.0$$
$$L = 4.0 - 1 = 3.0$$
$$M = \frac{3.0}{\sqrt{25.0 - 5.95}} = .69$$

Since M is greater than the tabled value of .40, there is a difference in variability in the two locations, with location 1 having about half the variability of location 2.

6. The differences in the hardness values at the two locations appear in Table 12.5.
7. The mean and standard deviation of the differences are

$$\hat{D} = -5.5 \qquad s_D = 8.0$$

8. The test value is

$$T = \frac{\sqrt{25} \times 5.5}{8.0} = 3.44$$

which is much larger than the tabled value of 2.06. Hence we conclude that location 2 has significantly higher mean hardness.

These results can be related to the difficulty in machining location 2. The shift in the mean is not desirable but is probably not the cause of the problem. The increased variability causes hardness values that are unusually high and low. These extremes are probably beyond the range that the machine tooling can accommodate. The casting vendor must make process improvements to reduce the variability of location 2.

12.7 Relationship Between Scatter Plots and Control Charts

Scatter plots provide a useful tool for guiding the selection of characteristics that should be monitored using control charts. Typically, a large number of potentially important part characteristics should be monitored on an ongoing basis. It is often not practical to monitor all these characteristics, so some selection is required. Chapter 14 discusses a list of criteria that should be used in the selection process. One of the criteria is the interrelationship between characteristics. Stated simply, it is not necessary to monitor two characteristics that are highly related. For example, if two dimensions are cut by the same tool, it is frequently not necessary to monitor both characteristics. Consider two size characteristics charted using a subgroup size of $n = 3$:

Subgroup	Part	Size 1	Size 2
1	1	10	7
	2	1	2
	3	5	4
2	4	3	1
	5	9	6
	6	4	2
	.	.	.
	.	.	.
	.	.	.

A control chart for this type of data plotted on different scales appears in Figure 12.35. Little is gained by charting both characteristics. It is apparent in Figure 12.35 that the two characteristics are related. However, the extent of the relationship is more easily assessed using a scatter plot.

The existence of a significant correlation using the sign or correlation coefficient r tests is not sufficient justification to monitor only one characteristic. The degree of correlation may be low, so there is insufficient predictability. It is possible to compute the prediction line and evaluate the degree of predictability. Also, it is possible to use the following subjective criteria for the correlation coefficient r that have proven useful in manufacturing applications:

Separate charts justified: r between -0.5 and 0.5
One chart may be justified: r between -0.8 and -0.5 or r between 0.8 and 0.5
One chart justified: r between -1.0 and -0.8 or r between 1.0 and 0.8

The question then arises of which characteristic should be measured. Using gage repeatability and ease and cost of measurement, it is possible to make a logical choice.

> Scatter plots can be used to help select characteristics for control charting

12.8 Target Charts

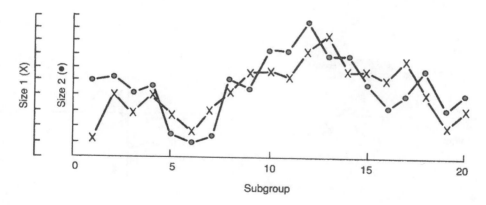

Figure 12.35 Control chart pattern for related characteristics.

12.8 Target Charts*

Engineering specifications need to address a variety of part characteristics to ensure proper product function. Size is probably the characteristic most frequently discussed in the quality control literature. However, position dimensions are equally important characteristics but have received less attention due, in part, to the difficulty in gaging. Position has an engineering blueprint symbol ⊕ and is also referred to as true position (App. II). Many machining part specifications are based on size and position part characteristics. Controlling size and position is a large part of the control of part quality.

The recent utilization of high-precision (±.00002 inch) coordinate measuring machines (CMM) has enabled a more disciplined approach to controlling part position characteristics. These machines provide a complete spatial representation of the position of a part feature. In the simplest case, consider measuring the position of a hole center at some fixed depth, which results in the target chart in Figure 12.36a. The "target" is the position the center of the hole should be relative to a part reference datum. The CMM coordinate data is typically analyzed using deviation from the target value:

Deviation = actual distance − target distance

The point $x = 0$, $y = 0$ on the plot becomes the target. Engineering specifications state a true position circle diameter centered at the target that should contain all hole centers, as shown in Figure 12.36b. A target chart is a type of scatter plot in which the axes correspond to distances that establish a position measurement.

The x,y coordinates from a group of parts can be used to calculate the required alignment move for a machine to obtain the best possible centering around the target. The center or centroid of five measured part positions is simply the mean of the individual x and y coordinates:

$$x_0 = \frac{x_1 + x_2 + x_3 + x_4 + x_5}{5}$$

= mean of individual x coordinates

$$y_0 = \frac{y_1 + y_2 + y_3 + y_4 + y_5}{5}$$

= mean of individual y coordinates

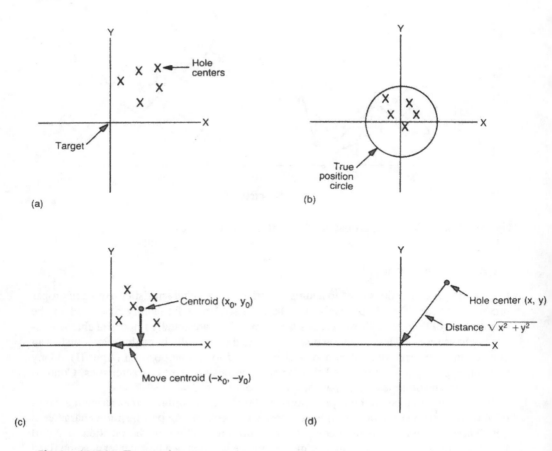

Figure 12.36 Target plot concepts.

The alignment move is then $-x_0, -y_0$, as shown in Figure 12.36c. Again the scatter plot "picture" graphically conveys the needed adjustment.

If the distance of a hole center to the target point is less than the radius of the true position circle, the hole position meets the engineering specification. The distance of a hole center (x, y) to any other point (a, b) is

Distance from (x, y) to $(a, b) = \sqrt{(x - a)^2 + (y - b)^2}$

Thus, the distance to the target point $(0, 0)$ is $\sqrt{x^2 + y^2}$, as shown in Figure 12.36d. If we let $\bar{X}_t$ and s_t denote the mean and standard deviation of the distances from the hole centers to the target point, then the position capability of the process is

$$\text{CPU}_t = \frac{\text{USL} - \bar{X}_t}{3s_t}$$

as shown in Figure 12.37. Of course, the stability of the process needs to be established. If $\bar{X}_t$ is beyond USL, a 0 capability is assumed.

Another useful distance is the distance from hole centers to their centroid (x_0, y_0). Let the mean and standard deviation of these distances be denoted $\bar{X}_c$ and s_c, respectively. The

12.8 Target Charts

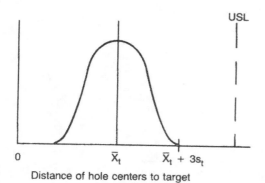

Distance of hole centers to target

Figure 12.37 Capability of meeting true position specification.

capability computed using these quantities evaluates whether the position is capable after the prescribed move of $(-x_0, -y_0)$, assuming the move is 100% successful. The distance $\bar{X}_c + 3s_c$ gives the radius of the circle that contains most (99.73%) hole centers. The potential capability of the process assuming a move is 100% successful is

$$\text{CPU}_c = \frac{\text{USL} - \bar{X}_c}{3s_c}$$

Example 12.7 Consider the following positional measurements ($1 = .0001$ inch) in which the radius specification of the true position circle is 30:

Part	x	y	Distance to target	Distance to centroid
1	30	−13	32.7	9.5
2	2	−26	26.1	24.3
3	29	−32	43.2	22.9
4	22	5	22.6	15.6
5	21	13	24.7	23.6
Mean	20.8	−10.6	29.8	19.2
SD	11.3	19.4	8.4	6.4

A target chart appears in Figure 12.38. An alignment move of −20.8 along the x axis and 10.6 along the y axis is indicated. The current capability is $\text{CPU}_t = 0$, and a successful alignment move would produce $\text{CPU}_t = \text{CPU}_c = 0.56$. The radius of the actual circle is $19.2 + (3 \times 6.4) = 38.4$, which is greater than 30. It will be necessary to reduce the variability of the hole center positions, as well as make an alignment move, to attain a capable process. More than five points are generally used so that the required move can be determined more accurately.

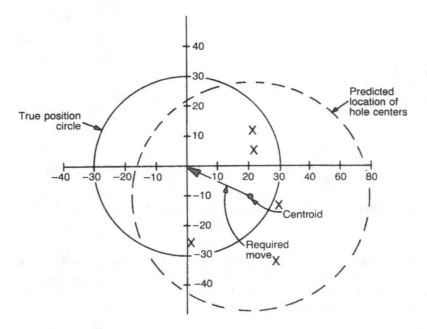

Figure 12.38 Target plot for Example 12.7.

The preceding discussion addresses only the two-dimensional position of a hole center. A third dimension is typically obtained by measuring the coordinates of a hole at the top and bottom of the hole, as shown in Figure 12.39. The coordinates of the top and bottom hole centers can be plotted together on a target chart to obtain a three-dimensional perspective, as seen in Figure 12.40a.

Another application of target charts involves the alignment of roughing and finishing machining stations. If a hole is machined using two (or more) machining stations, the initial hole is the "rough" hole and the final hole is the "finish" hole. It is essential for roughing and finishing stations to be properly aligned. Improper alignment of roughing stations can cause oval finished holes and poor tool life. It is possible to obtain the position of hole

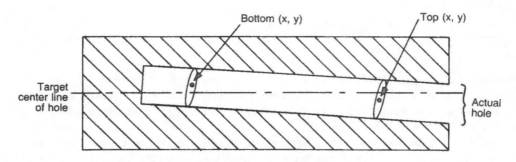

Figure 12.39 Three-dimensional hole position.

12.9 Case Studies

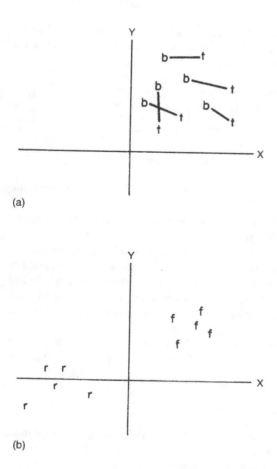

Figure 12.40 Special target plots. (a) Three-dimensional target plot: bottom hole center, b; top hole center, t. (b) Rough and finish hole alignment: finished hole center, f; rough hole center, r.

centers from rough and finished part. Plotting a target chart using both sets of positions enables evaluation of potential aligment problems, as indicated in Figure 12.40b.

Target charts are a useful tool for displaying position measurements and should be utilized in any defect prevention system with true position specifications. One of the benefits of target charts over other analysis tools is that a specific corrective action is indicated. It is obvious to any user when a problem exists. The value of routine monitoring using three charts should not be overlooked.

12.9 Case Studies

12.9.1 Targeting a Process

A small aluminum valve used in a transmission was processed on a grinding operation. After grinding, the valves were tumbled in deburring media to remove any burrs. The valves were then anodized, which consisted of a continuous chemical process that placed a thin coating of aluminum oxide on the valve. The purpose of the coating was to increase a valve's wear resistance. The process flow follows:

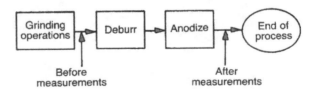

The deburring process removed some material and anodizing added material. Overall the planned net change was zero.

A number of process parameters could be changed in both the deburring and anodizing operations to attain the desired zero net change. The targeting problem is difficult, since, as in most manufacturing applications, it is necessary to balance several part characteristics. There may be from one to six different diameters on a valve. Also, there are a large number of different valves to be processed, and it is undesirable that the deburr and anodize parameters be changed for every valve. Thus, it is necessary to balance the many process parameters to accommodate different valve characteristics as well as completely different valves.

Measurements before deburring and after anodizing were made on a valve that had two diameters. Parts were collected randomly over a representative time period. The data in Table 12.6 are coded so that

Large diameter: actual value (mm) = 12.99 + .0001 coded value.
Small diameter: actual value (mm) = 10.36 + .0001 coded value.

The specification is 0–100. The 50 valves were selected over several production runs in an attempt to spread out the range of the data.

The summary values for the two diameters are as follows:

	Large diameter			Small diameter		
	Before deburr	After anodize	Difference	Before deburr	After anodize	Difference
Mean	28.7	34.0	5.3	31.6	59.5	27.9
SD	20.2	20.0	10.9	9.7	16.2	17.9

It is apparent that the large-diameter variability is the same before deburring and after anodizing. The T value from Procedure 12.5 for comparing the means is $T = (\sqrt{50} \times 5.3)/10.9 = 3.4$, which is greater than the tabled value in Procedure 12.5, so we conclude that the process mean does increase slightly after anodizing. Figure 12.41a confirms this conclusion since there are more points above the no-difference benchmark line. Also, Figure 12.41a shows that the before measurements are highly correlated with the after measurements ($r = 0.85$). This implies that it is possible to target the finished large-diameter valves by targeting the diameters in the grinding process.

12.9 Case Studies

Table 12.6 Transmission Valve Measurements[a]

Large diameter				Small diameter			
Before deburr	After anodize	Before deburr	After anodize	Before deburr	After anodize	Before deburr	After anodize
−5	5	25	25	35	50	25	35
15	10	20	20	35	65	30	25
15	20	25	25	50	45	30	25
15	20	35	40	40	75	45	55
0	20	25	25	30	70	25	75
0	5	10	15	30	45	40	75
20	20	20	20	15	25	40	70
15	25	65	60	30	40	30	55
10	5	10	20	45	75	30	45
30	25	35	30	40	90	15	45
65	50	65	70	15	75	45	70
20	15	15	30	20	90	30	50
40	25	20	15	30	85	35	60
40	30	60	40	25	80	35	45
50	35	55	50	30	80	20	65
40	40	15	25	55	50	35	75
25	25	10	30	45	60	20	65
35	50	50	70	25	65	25	55
25	40	40	60	40	50	30	65
35	55	−5	10	35	55	30	45
20	40	95	110	40	55	35	75
20	40	25	40	30	45	30	55
45	55	10	30	20	50	25	60
25	35	40	55	15	45	40	70
20	35	50	60	45	75	15	75

[a] Note that columns 1, 2, 5, and 6 contain data for parts 1–25; the remaining columns contain data for parts 26–50.

The variability of the small-diameter measurements after anodizing is significantly greater than before deburring using Procedure 12.5 ($M = 0.47$). This is not desirable since it implies that deburring and anodizing add variability. However, notice that the after anodizing variability for the small diameter ($s = 16.2$) is less than the after anodizing variability for the large diameter ($s = 20.0$). The T value for comparing the means is $T = (\sqrt{50} \times 27.9)/17.9 = 11.0$, which is significant, so the deburring and anodizing operations increase the mean diameter. Figure 12.41b again confirms this conclusion, with most points above the benchmark line. Unfortunately, Figure 12.41b also shows the lack of

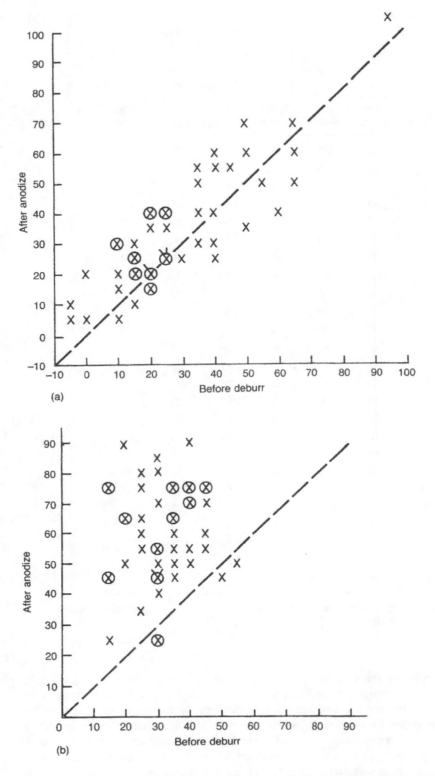

Figure 12.41 Scatter plots of large (a) and small (b) valve diameters before deburring and after anodizing.

12.9 Case Studies

correlation ($r = 0.11$) between the before and after measurements. For example, a valve of 30 before deburring produces a valve after anodizing ranging between 25 and 85. This lack of correlation along with increased variability makes control of the final after anodizing small diameters difficult.

Combining the analysis of the two diameters, we conclude that the process parameters should be adjusted so that the mean amount of material added is reduced. Potential causes for the lack of correlation for the small diameter should be explored so that controlling the process is easier.

12.9.2 Setting an In-process Specification Limit

A cold heading operation produced a rough blank that was finished machined into a transmission gear. The overall length (OAL) of the blank was critical in the machining line but could only be indirectly controlled at the cold heading operation. The cold heading operation used a series of progressive dies, which resulted in material being extruded on the B end of the gear in Figure 12.42. The thrust face dimension addressed in Problem 12.3 was also critical.

The main control of OAL was thought to be initial weight of the gear blank. A study was planned in which a range of initial weights were selected and the resulting OAL values measured. The data appear in Table 12.7 and are plotted in Figure 12.43.

The correlation ($r = 0.95$) between weight and OAL is apparent in Figure 12.43. This strong correlation implies that OAL can be controlled by selecting an appropriate range of weights. The prediction line relating the two characteristics is

$$\text{OAL} = -2.54 + 0.0082 \text{ weight}$$

This equation implies that increasing the weight by 1 increases the OAL by about .008 inch. A weight of 950 g is predicted to have an OAL of 5.25:

$$\text{OAL} = -2.54 + (.0082 \times 950) = 5.25 \text{ inch}$$

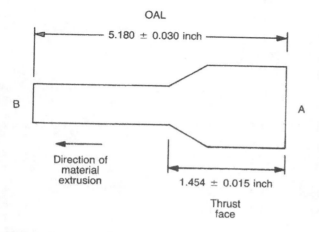

Figure 12.42 Rough gear blank.

Table 12.7 Gear Blank Measurements

Part	Weight (g)	OAL (inches)
1	955.0	5.315
2	958.0	5.328
3	954.0	5.307
4	952.0	5.279
5	950.0	5.243
6	952.5	5.293
7	955.0	5.278
8	956.0	5.313
9	950.0	5.257
10	950.5	5.264
11	951.5	5.260
12	950.5	5.250
13	955.0	5.311
14	958.0	5.316
15	957.0	5.289
16	949.5	5.277
17	951.5	5.258
18	960.5	5.307
19	956.5	5.311
20	956.5	5.301
21	950.5	5.245
22	950.5	5.273
23	948.5	5.259
24	942.5	5.192
25	958.0	5.337
26	949.5	5.247
27	944.0	5.196
28	956.5	5.308
29	942.0	5.176
30	949.0	5.247
31	948.0	5.243
32	950.5	5.247
33	942.5	5.192
34	944.5	5.202
35	957.5	5.311
Mean	951.8	5.270
SD	4.89	0.0423

12.9 Case Studies

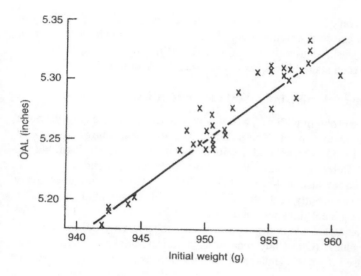

Figure 12.43 Scatter plot of gear blank measurements.

The prediction line is drawn in Figure 12.43. Rearranging the prediction equation so that we can determine the required weight range gives

$$\text{Weight} = \frac{\text{OAL} + 2.54}{.0082}$$

The OAL in-process specification is USL = 5.21 and LSL = 5.18, so

$$\text{Weight}_{upper} = \frac{5.21 + 2.54}{.0082} = 945.1$$

$$\text{Weight}_{lower} = \frac{5.15 + 2.54}{.0082} = 937.8$$

The interval (937.8, 945.1) should be narrowed for expected variability in the prediction process (see Draper and Smith, Sec. 1.7, 1981). It is apparent in Figure 12.43 that there is variability around the prediction line.

Using the interval (937.8, 945.1) presents a problem since the process spread for the blank weight was found to be about 7 g. Two alternatives exist. First, the process that cuts the gear blanks should be improved to narrow the process spread. Second, the OAL specification range of (5.15, 5.21) should be evaluated. A correlation study between the rough and finished machining operation should be performed.

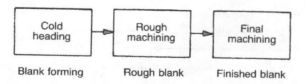

Since (5.15, 5.21) is an in-process specification, it may be possible to widen the specification limits without changing the finished part specification. A study could be performed by selecting blanks across the range of weights within the process spread. Ideally, a number of blanks with extreme high and low weights should be selected. Critical characteristics of these blanks can then be measured after the roughing and finishing operations. The correct range of OAL values for each stage of the process can then be determined.

12.9.3 Troubleshooting: Selecting Important Characteristics

A team was investigating a problem with a flywheel in which the concentricity of the outer diameter (OD) was difficult to control. The standard $\bar{X}$ and R chart showed a lack of stability. Few parts were found to be beyond the upper specification of .010 inch, but the lack of process stability offered an opportunity to improve the process and reduce variability. The team prepared a cause-and-effect diagram (Chap. 13) that contained many potential causes for the concentricity problem. The two most likely potential causes were thought to be the incoming steel thickness and hardness, shown in Figure 12.44. Two separate studies were conducted obtaining 115 hardness-concentricity values and 130 thickness-concentricity values. (One study measuring hardness, thickness, and concentricity could have been used if a sufficient spread of the data was obtained.) The data are coded so that

Concentricity: actual value (inches) = .001 coded value.
Steel hardness: actual value (Rockwell B) used.
Steel thickness: actual value (inches) = .01 coded value.

The scatter plot in Figure 12.45 displays the study results. The sign test for the hardness-concentricity correlation shows 45 of 115 points in quadrants 1 and 3, which from Procedure 12.2 is marginally significant. The sign test for the thickness-concentricity correlation shows 43 of 130 points in quadrants 1 and 3, which is significant. Thus, we conclude that thickness and to a lesser extent hardness are related to concentricity. This does not directly establish a causal relationship, but process knowledge dictated that thickness and hardness could cause concentricity problems. Since lower concentricities are desirable, the negative correlations imply that higher hardness and thickness levels are desirable. A follow-up study evaluating concentricity could be performed using steel with 75–80 hardness and 86–92 thickness. This would provide verification of the scatter plot results.

12.9.4 Troubleshooting: Finding the Source of Process Changes

A gear blank went through three machining steps prior to a heat treat to harden the gear. After heat treat, the ID of the gear was honed to attain its finished ID size. Ovality in the ID of the gear shown in Figure 12.46 made the honing operation subject to high scrap and

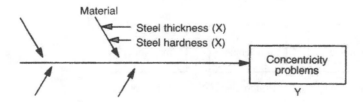

Figure 12.44   Partial C & E diagram for a concentricity problem.

12.9 Case Studies

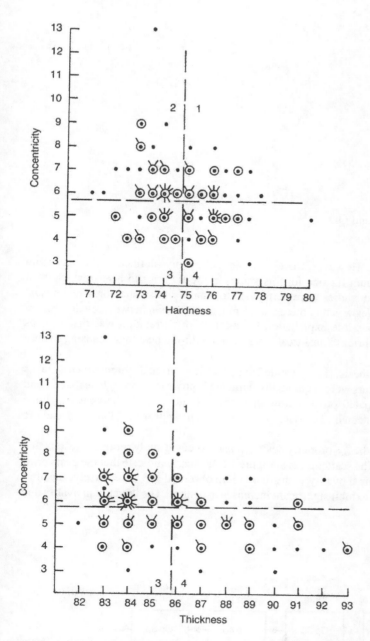

Figure 12.45 Scatter plot of steel hardness and thickness versus concentricity.

520   12   Scatter Plots

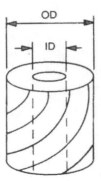

Figure 12.46   Gear ID and OD.

rework and poor tool life. However, there was minimal ovality (defined as the maximum ID size minus the minimum ID size for this study) in the ID prior to heat treat. Several before and after heat treat studies were performed to quantify this behavior. The unusual conclusion from the studies was that there was minimal correlation between ovality before and after heat treat. A possible explanation for the lack of correlation was that stresses introduced in machining prior to heat treat were released during heat treat, causing ovality changes.

To test the stress hypothesis, it was decided to collect 15 gears at different process stages prior to heat treat. The gears were sequentially numbered, processed through heat treat, and remeasured. The abbreviated process flow diagram in Figure 12.47 shows where three groups of gears were collected. The data (1 = .0001 inch) appear in Table 12.8 and are plotted in Figure 12.48.

As is true in many studies, a properly designed and executed study produces results that need minimal analysis. The scatter plots in Figure 12.48 clearly indicate that the gears prior to heat treat exhibit minimal ovality with a maximum observed at 3. Conversely, the after heat-treat measurements exhibit significant increases in ovality. The increase in ovality due to Op 10 is

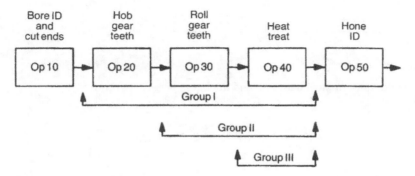

Figure 12.47   Partial process flow diagram for gear ID study.

12.9 Case Studies

Op 10 effect = mean after Op 40 − mean after Op 10
= 3.5 − 1.0
= 2.5

Assuming the stress effects are cumulative in Op 10 and Op 20, the increase in ovality due to Op 20 is

Op 20 effect = mean after Op 40 − mean after Op 20 − Op 10 effect
= 8.2 − 1.3 − 2.5
= 4.4

Similarly, the increase in ovality due to Op 30 is

Op 30 effect = mean after Op 40 − mean after Op 30 − Op 10 effect − Op 20 effect
= 8.3 − 1.1 − 2.5 − 4.4
= 0.3

Thus, we conclude that the boring (Op 10) and honing (Op 20) operations are responsible for changes that occur during heat treat. Subsequent studies showed that refixturing the

Table 12.8 Gear Blank Ovality at Different Processing Stages

	Group I		Group II		Group III	
	Op 10	Op 40	Op 20	Op 40	Op 30	Op 40
	1	3	0	5	1	9
	1	1	1	10	0	12
	2	6	1	7	2	5
	0	5	2	4	1	8
	1	3	3	10	2	5
	1	6	2	9	1	13
	2	1	1	13	2	11
	1	4	1	7	0	5
	0	4	2	7	1	6
	1	3	1	6	2	7
	1	2	2	8	1	7
	1	4	1	10	0	8
	0	3	1	15	2	8
	1	3	0	8	1	10
	2	4	2	4	1	10
Mean	1.0	3.5	1.3	8.2	1.1	8.3
SD	.66	1.5	.82	3.1	.74	2.6

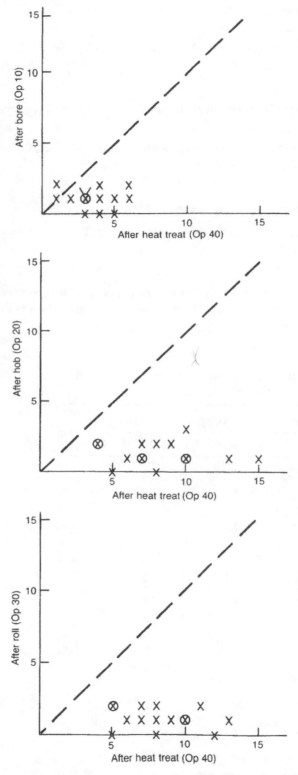

Figure 12.48 Scatter plots of gear blank ID size changes.

Problems

parts in the heat-treat process significantly reduced ovality changes. The new fixtures were designed so that stresses would be relieved more uniformly, causing less distortion.

This case study illustrates the iterative nature of problem solving. The results of one study lead to additional questions. Root causes of problems rarely surface after a single study. The importance of appropriate data collection and analysis is presented as part of a comprehensive problem analysis system in Chapter 13.

12.9.5 Correlation Between Measurements

A new in-process gage was to be purchased to measure the flatness of a large aluminum part. The rought part flatness was determined in the initial stage of the process; the final finished flatness was determined at the end of the machining line. Ideally, the same type of gage is used in both places, but this was not possible since the intermediate machining operations altered the rough part locating positions.

The gages could be temporarily altered to enable measurement of the same parts to assess the degree of correlation between the two gages. The flatness (1 = .001 inch) was measured on 30 parts using each gage for the initial evaluation of the new in-process gage design. The higher, more variable in-process gage readings are apparent in Figure 12.49a. The gage builder made a number of small refinements to stabilize the positioning of the part in the gage. After several subsequent studies, the in-process and final gages agreed reasonably well, as seen in Figure 12.49b. For the refined design, the 45° line approximately splits the data, implying there is no mean shift between the measurement procedures.

The parts span too narrow a range of the 0–30 specification in both studies. Since parts were collected consecutively from the process, the narrow range is expected. It would have been better to collect parts over a period of time or specially generate high- and low-flatness parts. This sampling plan would have enabled comparison of the gages over the whole range of intended use.

Problems

12.1 In the first case study, a large and small valve diameter was measured at two stages in the production process. (a) Prepare a scatter plot of the large (X) versus the small (Y) diameters at the two stages. Interpret the benchmark line. (b) Are the two characteristics correlated using both the sign and correlation coefficient r tests? (c) Using the case study results and parts a and b, determine which of the following strategies is best for placement of control charts:

1. Chart both diameters before deburring and after anodizing (four charts).
2. Chart only large or small diameter before deburring and after anodizing (two charts).
3. Chart both diameters before deburring (two charts).
4. Chart both diameters after anodizing (two charts).
5. Chart only large or small diameter before deburring (one chart).
6. Chart only large or small diameter after anodizing (one chart).

12.2 Product specifications were written for static back-out torque (newton-meters) of a bolt. The dynamic installation torque (pound-inches) was used to set up and monitor the bolt installation station. A study of 60 random bolts was conducted by measuring first the dynamic torque at installation and then the static back-out torque required for loosening. (Note that 0.113 × pound-inches = N−m.) (a) Prepare a scatter plot of dynamic (X)

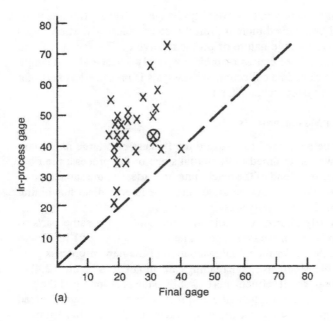

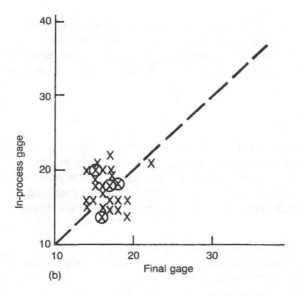

Figure 12.49 Gage correlation study: initial (a) and refined (b) in-process gage design.

versus static torque (Y). Draw a benchmark line where dynamic = static torque. Interpret the plot. (b) Would it be necessary for dynamic to be equal to static torque for effective monitoring? Are the two measurements correlated using both the sign test and the correlation coefficient r test? (c) What are the process implications of these results? (d) Prepare histograms of the two torque measurements. Interpret the results. What influence would measurement variability have on the correlation?

Bolt	Static	Dynamic	Bolt	Static	Dynamic	Bolt	Static	Dynamic
1	10	90.8	21	14	92.9	41	15	90.4
2	10	91.9	22	12	92.9	42	14	92.1
3	9	91.2	23	12	92.1	43	12	91.5
4	10	89.6	24	13	92.3	44	12	90.3
5	10	92.9	25	12	91.1	45	13	93.8
6	10	94.8	26	12	91.6	46	13	92.3
7	9	91.1	27	12	95.7	47	14	92.6
8	10	91.4	28	13	91.7	48	12	91.4
9	10	90.5	29	14	94.9	49	12	94.1
10	10	92.3	30	13	92.7	50	15	93.0
11	10	91.4	31	12	93.1	51	13	92.9
12	10	89.8	32	12	93.6	52	12	91.4
13	10	91.8	33	13	91.2	53	13	93.3
14	10	89.4	34	15	91.3	54	15	94.3
15	10	92.7	35	12	91.2	55	12	95.4
16	10	93.2	36	12	91.6	56	13	94.9
17	10	90.3	37	14	92.9	57	13	91.9
18	10	89.9	38	12	90.4	58	14	92.3
19	10	89.9	39	12	92.0	59	13	91.9
20	10	94.2	40	12	92.9	60	15	91.7

12.3 In the second case study, the OAL of a gear blank was shown to be related to the weight of the blank before a cold heading operation. After retargeting the gram weight used in the case study, a second study was performed to examine the thrust face (TF) dimension. It was hoped that controlling the initial gram weight would control OAL, which would then control the thrust face dimension. (a) Prepare a scatter plot of the OAL (X) versus the thrust face dimension (Y). Draw the specification limits on the plot. Interpret the plot. (b) Are the two characteristics related? Justify your conclusions? (c) What is the prediction line? (d) Is the OAL targeted properly? (e) The die pattern determines the relationship between OAL and thrust face size. Why would it be desirable to make die alterations to obtain the following relationship?

Thrust face = $4.044 - 0.5$ OAL

Using the results from the case study, what gram weight should be used for the new process?

Part	TF	OAL	Part	TF	OAL	Part	TF	OAL
1	1.439	5.209	13	1.455	5.193	25	1.467	5.190
2	1.438	5.221	14	1.454	5.183	26	1.465	5.190
3	1.448	5.192	15	1.454	5.207	27	1.462	5.191
4	1.449	5.190	16	1.458	5.195	28	1.467	5.180
5	1.442	5.205	17	1.454	5.209	29	1.470	5.182
6	1.443	5.230	18	1.455	5.186	30	1.465	5.186
7	1.451	5.190	19	1.462	5.188	31	1.468	5.168
8	1.446	5.209	20	1.456	5.186	32	1.470	5.168
9	1.446	5.219	21	1.460	5.211	33	1.467	5.187
10	1.452	5.230	22	1.460	5.181	34	1.475	5.161
11	1.449	5.223	23	1.463	5.194	35	1.469	5.188
12	1.447	5.231	24	1.461	5.201			

Mean 1.4568   5.1964   SD .0097   .0178 $r = -.75$

12.4   The surface micro (microinches) finish of a steel part was improved by a burnishing operation. A study of 30 randomly selected parts was made to quantify the improvement. (a) Prepare a scatter plot for before and after burnishing. Interpret the plot. (b) It is apparent that burnishing improves the surface finish. Does this imply that the before and after measurements are correlated? Determine whether the before and after measurements are correlated using the sign test and the correlation coefficient test. (c) Does burnishing change the variability of the surface finish? (d) What is the prediction line? If the final specification is 35 μinches why is this line important?

Part	Before	After	Part	Before	After
1	20	5	16	36	22
2	20	6	17	39	12
3	18	8	18	40	14
4	19	11	19	41	12
5	23	7	20	41	27
6	26	6	21	43	26
7	26	6	22	44	26
8	26	6	23	45	31
9	22	10	24	46	14
10	31	10	25	46	15
11	32	10	26	46	20
12	36	9	27	48	29
13	36	10	28	48	31
14	36	12	29	49	20
15	36	19	30	49	30

12.5 A runout measurement for a hole position exhibited a lack of capability with USL = 50. The process flow appeared as

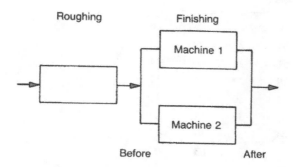

After the roughing operation, 50 parts were collected by specially selecting parts that spanned beyond the 0–70 in-process roughing specification. These parts were randomly assigned to machine 1 or 2 for finish machining. (a) Prepare a histogram of the before and after measurements. Interpret the results. (b) Is it necessary to compare the two groups of before measurements? How would the study be influenced if the two groups of before measurements differed appreciably? (c) Prepare a single stratified scatter plot. Interpret the plot. (d) Does each machine improve the runout? Does the variability change before and after finish machining? (e) Can Procedure 11.4 be used to compare the output of machines 1 and 2? How can the difference in mean output be determined? (f) Compute a prediction line for each machine individually and for the combined machines. Interpret the results. What might the study have shown if it were run using typical before measurements in the 10–40 range? (g) What might be the capability problem at the operation? Could this conclusion have been obtained by comparing the normal production output from machines 1 and 2?

	Machine 1		Part	Machine 2	
Part	Before	After		Before	After
1	30	16	26	45	16
2	79	61	27	76	23
3	71	51	28	13	14
4	39	29	29	22	13
5	22	11	30	16	12
6	76	53	31	74	27
7	64	41	32	40	14
8	71	56	33	63	33
9	66	49	34	71	22
10	77	50	35	32	18
11	53	30	36	46	19
12	15	6	37	10	2
13	20	13	38	8	1
14	26	18	39	12	8
15	54	37	40	30	18
16	12	21	41	39	22
17	54	39	42	20	6
18	75	53	43	61	24
19	11	9	44	76	22
20	14	5	45	7	7
21	77	57	46	69	29
22	35	26	47	61	10
23	19	19	48	66	28
24	73	47	49	71	31
25	38	20	50	71	30
Mean	46.8	32.7		44.0	18.0
SD	24.9	18.2		25.2	9.2

Combined before: 45.4, 24.8; after: 25.3, 16.1.

Machine	Before and after correlation r
1	.97
2	.84
1 + 2	.80

Problems

12.6 The poor capability for final flatness of a large aluminum component was thought to be due to the casting flatness or the rough in-process flatness. A cause-and-effect diagram was used by a team to help identify potential causes:

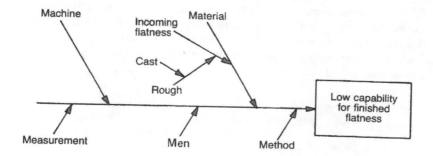

Random castings were selected and measured at the three stages of processing. (a) Prepare a scatter plot of casting versus finished flatness and rough versus finished flatness. Interpret the plots. (b) Are the casting and rough flatness measurements related to finished flatness using both the sign and correlation coefficient tests? (c) Is the casting flatness related to the finished flatness? (d) Interpret the flatness behavior through the process. Should process improvements address other C & E potential causes?

Part	Cast	Rough	Finish	Part	Cast	Rough	Finish	Part	Cast	Rough	Finish
1	15	13	8	18	20	21	15	35	21	18	10
2	13	16	7	19	5	9	6	36	29	25	11
3	22	23	11	20	21	19	6	37	10	10	2
4	2	10	10	21	14	11	9	38	30	25	7
5	16	18	11	22	17	17	12	39	25	24	13
6	23	17	13	23	23	20	11	40	8	11	5
7	30	21	9	24	11	11	11	41	23	17	6
8	8	13	3	25	22	22	10	42	6	10	8
9	26	20	13	26	19	15	10	43	27	23	9
10	10	13	9	27	23	15	13	44	22	17	9
11	25	21	13	28	13	14	6	45	19	18	11
12	11	12	4	29	29	23	14	46	26	23	7
13	13	15	8	30	17	15	6	47	26	25	13
14	18	19	11	31	25	17	15	48	13	13	11
15	17	14	6	32	29	24	11	49	29	22	11
16	18	16	10	33	8	13	12	50	26	22	14
17	26	18	14	34	15	14	5				

12.7 A critical dimension was found difficult to control owing to suspected heat-treat distortion. A special run of parts was made that spanned about 70% of the ±7.5 (1 = .0001 inch) specification width. (a) Prepare a scatter plot of the before and after heat-treat measurements. Interpret the plot. (b) Are the before and after measurements related? (c) Is it possible to determine whether the heat-treat operation changes variability from the study? What can be said about differences in means before and after heat treat? (d) What is the prediction line? (e) Use the prediction line to determine how the process should be targeted. Is this method of determining a process target better than evaluating a mean shift? Why? (f) Where should control charts be placed?

Part	Before	After	Difference	Part	Before	After	Difference
1	−3.7	−5.0	−1.3	26	4.6	3.1	−1.5
2	−2.7	−4.0	−1.3	27	4.9	3.3	−1.6
3	−3.6	−5.0	−1.4	28	4.6	3.4	−1.2
4	−3.7	−5.0	−1.3	29	4.9	3.3	−1.6
5	−3.9	−5.0	−1.1	30	4.9	4.0	−.9
6	−2.5	−4.8	−2.3	31	.6	−.8	−1.4
7	−2.4	−4.2	−1.8	32	1.1	−.5	−1.6
8	2.2	.2	−2.0	33	.8	−.3	−1.1
9	−3.4	−4.5	−1.1	34	.2	−1.4	−1.6
10	−3.0	−4.6	−1.6	35	.5	−1.0	−1.5
11	−2.0	−4.0	−2.0	36	1.7	.4	−1.3
12	−1.3	−2.9	−1.6	37	1.7	.2	−1.5
13	−1.1	−2.9	−1.8	38	1.0	−.2	−1.2
14	−1.1	−2.8	−1.7	39	1.4	−.5	−1.9
15	−2.0	−3.7	−1.7	40	1.6	.0	−1.6
16	−1.1	−2.0	−.9	41	2.5	.7	−1.8
17	−.8	−2.2	−1.4	42	2.5	.4	−2.1
18	−1.0	−3.1	−2.1	43	2.2	1.1	−1.1
19	−1.0	−2.4	−1.4	44	2.4	1.0	−1.4
20	−.9	−2.8	−1.9	45	2.1	.1	−2.0
21	5.6	4.8	−.8	46	2.8	1.4	−1.4
22	5.6	4.5	−1.1	47	3.5	1.9	−1.6
23	5.6	4.8	−.8	48	3.2	2.0	−1.2
24	6.5	4.8	−1.7	49	2.8	1.8	−1.0
25	5.5	4.8	−.7	50	3.3	1.6	−1.7

12.8 It is known that gear lead changes during heat treat. The targeting of the gear hobing operation was of interest. The −5 to 0 before heat-treat specification was used to obtain a −6 to +6 after heat-treat specification (1 = .0001 inch). Two different machines used for hobing were studied by randomly collecting 47 gears from each machine. (a) Prepare a scatter plot by machine for gear lead before and after heat treat. Interpret the plots. (b) Are the before and after lead measurements related? Would a prediction line be useful? (c) Compare the two machines for the variability and amount of shift (i.e., the difference) due to heat treat. (d) Is the -5 to 0 in-process specification appropriate? (e) Does the heat-treat operation change the lead variability? Do both machines behave similarly? (f) How can we improve the process?

	Machine 1				Machine 2		
Part	Before	After	Difference	Part	Before	After	Difference
1	−3.6	2.7	6.3	48	−3.4	1.2	4.6
2	−2.5	3.5	6.0	49	−4.4	3.9	8.3
3	−2.5	3.6	6.1	50	−4.2	4.3	8.5
4	−1.9	3.0	4.9	51	−3.9	4.0	7.9
5	−3.2	3.6	6.8	52	−3.8	3.5	7.3
6	−3.2	2.3	5.5	53	−3.5	2.3	5.8
7	−3.2	4.5	7.7	54	−3.2	.8	4.0
8	−3.2	4.0	7.2	55	−3.8	1.4	5.2
9	−2.9	4.0	6.9	56	−4.3	3.4	7.7
10	−3.1	4.7	7.8	57	−4.1	3.4	7.5
11	−2.7	5.5	8.2	58	−4.1	1.4	5.5
12	−3.1	5.2	8.3	59	−3.7	.3	4.0
13	−3.3	4.5	7.8	60	−3.5	1.9	5.4
14	−3.8	3.6	7.4	61	−4.3	3.1	7.4
15	−2.4	5.2	7.6	62	−4.1	2.6	6.7
16	−2.6	5.5	8.1	63	−3.8	1.2	5.0
17	−3.2	4.0	7.2	64	−3.0	2.6	5.6
18	−2.9	3.0	5.9	65	−4.3	4.0	8.3
19	−2.6	2.5	5.1	66	−3.1	5.2	8.3
20	−3.1	3.5	6.6	67	−3.0	3.1	6.1
21	−2.4	3.9	6.3	68	−4.5	3.5	8.0
22	−3.4	3.1	6.5	69	−4.8	4.5	9.3
23	−2.9	3.9	6.8	70	−4.1	2.7	6.8
24	−3.4	3.5	6.9	71	−4.6	5.0	9.6
25	−2.7	3.4	6.1	72	−3.6	4.2	7.8
26	−2.9	5.2	8.1	73	−3.6	2.3	5.9
27	−3.4	4.3	7.7	74	−3.6	2.6	6.2
28	−3.4	2.9	6.3	75	−3.8	2.8	6.6
29	−2.9	3.3	6.2	76	−3.4	5.1	8.5
30	−2.7	5.1	7.8	77	−3.8	2.5	6.3

	Machine 1				Machine 2		
Part	Before	After	Difference	Part	Before	After	Difference
31	−3.4	4.4	7.8	78	−4.2	3.7	7.9
32	−3.1	4.3	7.4	79	−4.1	4.4	8.5
33	−2.2	4.9	7.1	80	−4.3	3.5	7.8
34	−3.8	3.4	7.2	81	−4.1	3.8	7.9
35	−2.6	4.0	6.6	82	−5.0	3.2	8.2
36	−2.8	2.3	5.1	83	−3.3	4.0	7.3
37	−2.9	3.3	6.2	84	−4.3	2.2	6.5
38	−1.8	3.2	5.0	85	−4.6	2.5	7.1
39	−2.6	3.9	6.5	86	−3.4	6.0	9.4
40	−2.3	2.6	4.9	87	−4.4	5.0	9.4
41	−3.0	2.6	5.6	88	−4.6	3.4	8.0
42	−3.0	5.1	8.1	89	−4.2	4.6	8.8
43	−1.9	3.4	5.3	90	−3.1	3.4	6.5
44	−2.1	3.3	5.4	91	−4.6	1.5	6.1
45	−3.3	4.0	7.3	92	−4.3	4.8	9.1
46	−2.5	3.8	6.3	93	−4.0	2.2	6.2
47	−2.7	3.0	5.7	94	−3.5	4.0	7.5
Mean	−2.87	3.80	6.67		−3.94	3.21	7.16
SD	.48	.87	1.00		.50	1.28	1.44

12.9 A transmission gear blank was machined using the following process flow:

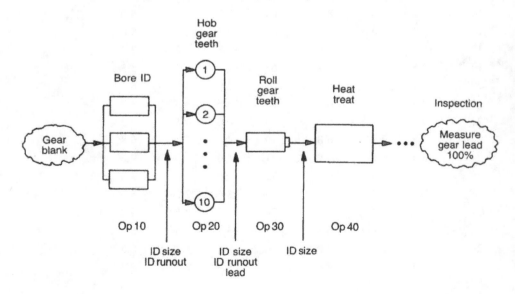

Problems

Excessive scrap for ID size, ID runout, and gear lead was observed at the final automatic 100% gear checking gage. It was thought that the lack of capability was possibly owing to the gear ID size or runout since several operations clamped or located in the ID. A study was conducted in which 50 random parts were measured for ID size (± 5 specification) and runout (USL $= 10$) after Op 10 and Op 20. Only size was measured after Op 30. All 10 hobs were used along with two roller spindles. (a) Is the size or runout coming into Op 20 related to the gear lead produced in Op 20? Is the absence of a relationship desirable? (b) Does the size or runout change after hobing? Does the size change after rolling? (c) Is there any lead stratification? Assume a common pooled standard deviation $s_p = 1.4$. (d) Do all machines change runout the same in Op 20? Assume a common pooled standard deviation $s_p = 0.87$?

	Op 10			Op 20			Op 30	
Part	ID	R/O	Machine	Lead	ID	R/O	Spindle	ID
1	2.5	1.0	1	3.5	-1.0	2.0	1	-.5
2	3.0	.5	1	1.7	-1.0	1.5	2	-.5
3	3.0	1.0	1	5.4	-1.0	2.0	1	-1.5
4	6.0	1.5	1	-2.4	1.5	2.5	2	1.5
5	-2.0	1.5	1	4.2	-7.0	1.0	1	-2.5
6	2.0	1.0	2	5.1	.5	2.0	2	-.5
7	2.0	1.5	2	3.9	-1.0	2.0	1	-1.5
8	6.0	3.0	2	5.0	4.0	3.0	2	3.5
9	2.0	1.0	2	2.9	-1.0	2.0	1	-1.5
10	1.0	2.0	2	2.6	.5	2.5	2	-1.0
11	2.0	1.5	3	.8	-1.0	1.5	1	-1.0
12	5.0	2.5	3	3.1	3.0	2.0	2	3.5
13	2.0	1.5	3	3.9	-2.0	2.0	1	-.5
14	1.5	1.0	3	3.3	-2.0	2.0	2	-2.0
15	.5	2.0	3	3.3	-3.0	1.5	1	-3.0
16	.5	1.0	4	5.8	-2.0	2.0	2	-2.0
17	1.0	1.0	4	5.3	-2.0	1.5	1	-2.5
18	1.0	5.5	4	6.2	-3.0	3.0	2	-3.0
19	2.5	2.5	4	1.3	-1.0	3.0	1	-1.0
20	2.5	1.0	4	5.5	-1.0	1.5	2	-1.0
21	1.0	1.5	5	1.7	-3.0	2.0	1	-3.0
22	3.5	1.0	5	1.8	1.0	1.5	2	1.5
23	4.0	1.5	5	2.8	1.0	2.0	1	1.0
24	1.5	2.5	5	.2	-2.0	2.0	2	-2.5
25	2.5	1.5	5	2.5	-.5	2.5	1	.0
26	.0	1.5	6	2.2	-3.0	1.0	2	-3.5
27	2.5	4.5	6	1.9	1.0	4.0	1	1.0
28	3.0	1.5	6	1.8	.0	2.5	2	1.0

	Op 10		Op 20				Op 30	
Part	ID	R/O	Machine	Lead	ID	R/O	Spindle	ID
29	1.0	1.5	6	1.9	−2.0	2.0	1	−2.0
30	3.0	2.5	6	2.0	−1.0	1.0	2	.0
31	2.0	1.0	7	4.2	.0	3.0	1	1.0
32	3.0	1.0	7	3.3	.0	2.0	2	−.5
33	3.0	2.0	7	1.2	−1.0	3.0	1	−1.5
34	2.0	1.5	7	3.3	−1.0	1.5	2	−1.0
35	1.0	2.5	7	3.8	−3.0	1.0	1	−3.0
36	3.0	1.0	8	4.2	1.0	2.0	2	2.0
37	3.0	.5	8	5.1	.5	2.5	1	.5
38	2.0	2.0	8	4.6	−1.0	1.5	2	−1.5
39	2.5	1.0	8	4.2	−.5	2.0	1	−.5
40	2.0	1.0	8	3.3	−1.0	1.5	2	−1.5
41	4.0	3.0	9	−.1	.0	2.5	1	1.0
42	1.0	1.5	9	.3	−2.0	2.5	2	−2.0
43	3.0	1.5	9	.0	−1.5	1.0	1	−.5
44	2.0	2.0	9	−.2	−2.0	2.5	2	−1.5
45	2.0	1.0	9	−.4	−2.0	1.5	1	−2.0
46	1.0	2.0	10	−3.3	.0	5.0	2	−1.0
47	2.5	2.0	10	−2.8	1.5	5.0	1	1.0
48	2.0	1.0	10	−2.4	1.0	4.0	2	.5
49	2.5	1.5	10	−4.1	2.0	5.0	1	1.5
50	.0	1.5	10	−4.3	−2.5	3.5	2	−3.0
Mean	2.19	1.67		2.18	−.79	2.27		−.71
SD	1.41	.91		2.63	1.81	.99		1.67

12.10 An end-cutting finishing reamer is a tool used to machine a hole. The end-cutting feature, in addition to the normal side cutting, is added to enable the reamer to determine a hole's position independent of the position of the rough starting hole. In this case, position is measured as a deviation (specification ±50) from a target location along a single axis established by a part datum. Four holes were drilled in a part. A two-station two-spindle roughing operation and a single, high-precision finishing spindle with an end-cutting reamer were used to machine the part:

Problems

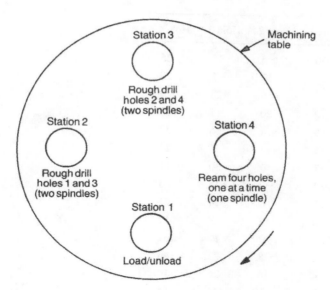

The hole position of the finished parts was unacceptable, and it was necessary to troubleshoot the operation. The final true position of the holes for 32 random parts follows. (a) Determine which of the six possible hole pairs are related. (b) Prepare scatter plots for the pairs of holes with high correlations. Interpret the plots. (c) Can you determine the cause of the problem? Does the end-cutting reamer control the hole position quality?

Part	Hole 1	Hole 2	Hole 3	Hole 4
1	41	7	2	19
2	41	−2	−8	30
3	35	2	7	32
4	28	21	12	2
5	37	45	6	−23
6	25	15	5	17
7	31	10	−5	24
8	37	15	8	12
9	28	9	0	23
10	38	5	−1	14
11	37	1	2	22
12	36	−4	−3	26
13	34	−2	8	30
14	34	−9	7	25
15	32	4	3	23
16	40	3	−3	23
17	31	−10	6	37

Part	Hole 1	Hole 2	Hole 3	Hole 4
18	30	−5	−15	25
19	35	−10	−5	20
20	5	6	23	24
21	2	20	17	22
22	8	6	36	32
23	15	14	25	27
24	−7	6	30	31
25	−4	−2	35	33
26	2	8	14	21
27	6	18	17	12
28	12	−2	17	32
29	2	−8	23	45
30	4	16	26	14
31	17	−16	15	44
32	1	−10	20	31

12.11 A capability assessment of a two-machine grinding operation revealed a .001 inch runout specification was not being met. Each machine had 10 fixtures that positioned the parts for grinding. Comparison plots revealed significant stratification between fixtures. The runout in a fixture was thought to be related to the runout in the part machined on the fixture. Runout data for a fixture and a part machined on the fixture follow. (a) Prepare a scatter plot. Is runout in a fixture related to part runout? (b) What is the correlation between fixture and part runout? What is the prediction line? (c) What value of fixture runout should be maintained to meet the specifications? Should we use the value from the prediction line? (d) Suggest a better way to study the relationship by reducing the influence of part-to-part variability.

No.	Fixture runout	Part runout	No.	Fixture runout	Part runout
1	1.5	1.34	11	.2	.28
2	1.0	.48	12	.4	.22
3	1.0	.68	13	.1	.30
4	1.5	1.22	14	.0	.26
5	1.0	.90	15	.2	.22
6	.4	.34	16	.4	.54
7	.5	.42	17	.5	.54
8	.8	.70	18	.2	.18
9	1.1	.88	19	.1	.32
10	.1	.30	20	.1	.28

Problems

12.12 The first case study in Chapter 11 examined fixture differences for a milling operation. (a) In Figure 11.16 the LC chart indicates that the right side is higher than the left side. Does this imply anything about correlation between the values from the two sides? (b) Prepare a stratified scatter plot of the left (X) versus the right side (Y) for fixtures 11 and 12. Interpret the plot. (c) What is the relationship between the right and left side for each of the fixtures? (d) Interpret the result. What process factors could explain the results in Figure 11.16 and the scatter plot?

12.13 A subassembly that controlled hydraulic pressure was difficult to assemble because of a spring interference condition. A new spring was designed for easier assembly, but it was important that no shift in the hydraulic pressure reading occur due to the new spring. To evaluate the new spring, 24 subassemblies were first tested with the old spring and the pressure readings recorded. Using the same 24 subassemblies, the old spring was replaced by the new spring and the pressure remeasured. The spring load (1.10 ± 0.5 pounds) directly impacts on the hydraulic pressure so the loads for the two groups of springs were determined. (a) Prepare scatter plots of spring load versus pressure. How strong is the relationship? What might explain any lack of correlation? (b) Do the spring loads for the two groups differ? Why is this important for comparing the pressure measurements? (c) Prepare a scatter plot of the two pressure comparisons. Does the new spring change the pressure?

Part	Old design Pressure	Old design Spring load	New design Pressure	New design Spring load
1	29.7	1.09	28.5	1.10
2	28.8	1.09	30.1	1.09
3	28.8	1.11	27.5	1.13
4	30.3	1.09	28.7	1.15
5	29.7	1.12	29.7	1.09
6	29.9	1.13	27.8	1.11
7	29.3	1.06	28.9	1.08
8	29.5	1.10	28.2	1.10
9	27.0	1.10	27.5	1.11
10	29.8	1.12	30.0	1.08
11	31.1	1.14	29.2	1.14
12	28.8	1.10	27.6	1.13
13	29.4	1.10	28.5	1.11
14	27.8	1.09	28.5	1.11
15	31.3	1.11	31.0	1.10
16	30.5	1.10	29.9	1.08
17	31.5	1.09	30.0	1.09
18	32.3	1.11	31.6	1.08
19	29.7	1.11	29.6	1.10
20	29.1	1.07	28.7	1.13

	Old design		New design	
Part	Pressure	Spring load	Pressure	Spring load
21	29.5	1.09	28.3	1.12
22	30.2	1.09	29.0	1.11
23	29.7	1.10	28.9	1.12
24	31.0	1.12	29.9	1.13

12.14 Problem 4.6 presented a standard tool wear situation. The means for 25 subgroups along with the hours of use follow. (It is assumed there was 1 hour of machine run time between subgroups.) (a) Compute the prediction line. (b) How long would it take for the tool to wear so parts span the (± 10) specification range. (c) Why should the tool not be allowed to wear from -10 to $+10$? What is a good tool change strategy?

Hour	Size	Hour	Size
1	−4.4	14	.8
2	−5.2	15	1.2
3	−5.0	16	1.6
4	−4.8	17	2.4
5	−3.8	18	2.2
6	−3.2	19	2.8
7	−2.6	20	3.2
8	−2.6	21	3.4
9	−1.8	22	4.0
10	−1.2	23	3.6
11	−1.2	24	4.6
12	−.2	25	5.0
13	.2		
Mean		13.0	−.04
SD		7.36	3.28
$r =$	.99		

12.15 An electronic gage at an operation was used to measure a part diameter where (1 = .001 inch). A total of 10 parts were measured by the gage and a reference standard CMM gage to evaluate possible gage bias. (a) Prepare a scatter plot of the CMM (X) versus the electronic gage (Y) measurements. (b) Perform a test to determine whether any gage bias exists. Is any bias significant relative to the $\pm .7$ specification?

Problems

Part	Electronic gage	CMM gage	Difference
1	−.15	.2	−.35
2	.20	.3	−.10
3	2.00	2.5	−.50
4	−.75	.5	−1.25
5	−.75	.0	−.75
6	−.15	.5	−.65
7	−.60	.5	−1.10
8	−.50	.0	−.50
9	.40	.6	−.20
10	.40	.5	−.10
Mean	.01	.56	−.55
SD	.826	.715	.397

13
Cause-and-Effect Diagrams and Problem Solving

13.1 General Concepts

The tools discussed in earlier chapters are essential for building a defect prevention system (DPS) and can be utilized for either process control or problem analysis. However, the existence of "tools" does not ensure implementation will be effective. What is needed is a system to organize and direct the application of the tools toward controlling processes and solving problems. This chapter presents a system to analyze industrial problems, and Chapter 14 addresses the entire defect prevention system. A number of systems can be used to attain the desired end result—a continuous improvement strategy. The important point is that training in statistical tools must be combined with an implementation and follow-up system. Finally, as emphasized in Chapter 1, the most important component of the strategy is the joint participation of management and employees in the implementation process.

> Management and employee participation is an essential component
> of any improvement strategy

In any manufacturing facility, many problems are "solved" on a daily basis. Solutions can involve the replacement of machine parts, use of new tooling, or training employees, for example. However, a number of problems are never completely solved and others are simply never addressed. The often cited explanation for the continued existence of these problems is that sufficient time does not exist for personnel to solve these difficult, lingering problems. However, it is the absence of an organized approach to analyzing problems, rather than a lack of time, that makes problem resolution difficult in many cases. The standard reaction-oriented problem solving discussed in Chapter 1 fails when a team approach is required, the collection and analysis of information or data is necessary, and a period of time is needed to complete the iterative problem analysis process. A systematic, data-oriented approach with a follow-up system is not only beneficial but essential to resolving these difficult problems.

A flowchart of 10 problem-solving steps found useful in many industrial applications appears in Figure 13.1. This series of steps provides a uniform, disciplined approach for analyzing quality, productivity, and engineering problems. Each step is discussed in later chapter sections. An integral part of this problem-solving system are the following four elements:

1. Problem analysis system steps summary
2. Cause-and-effect diagrams

13.1 General Concepts

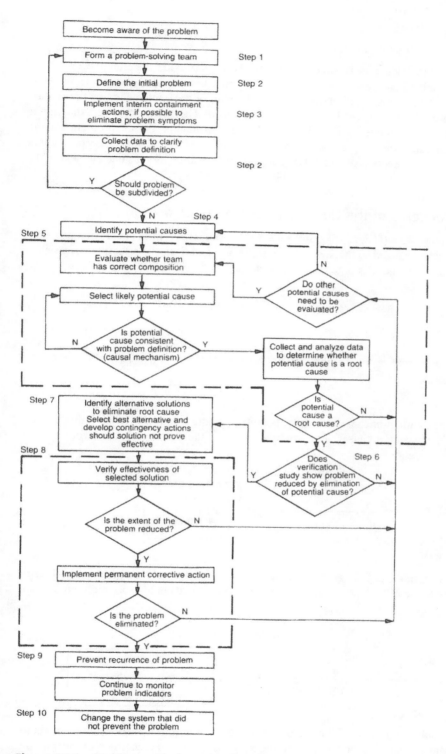

Figure 13.1 Problem analysis system flowchart.

3. Action plan
4. Data summaries

Each of these elements is simple; collectively they provide an effective system to organize, monitor, and report on the problem-solving process in Figure 13.1. With the exception of the cause-and-effect diagram, the elements of the implementation process are discussed after the problem-solving steps are presented. The cause-and-effect diagram is discussed first since it has traditionally been used as a critical part of the problem-solving process. Although the diagram is useful, experience has found that this tool cannot substitute for a complete problem-solving system: it should be used as one element in the overall system.

13.2 Concepts of the Cause-and-Effect Diagram

The cause-and-effect (C & E) diagram is a simple listing of *potential* causes of a problem. The diagram is arranged so that a statement of the problem (i.e., the effect) appears on the right, and the left side lists potential causes of the stated problem.

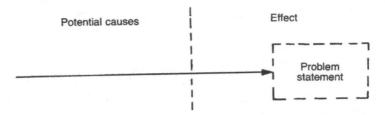

Each of the main branches has many potential subbranches or "twigs" that further subdivide the potential causes. For example, suppose the stated problem involves setup scrap at an operation:

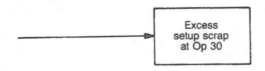

If the 5M elements C & E diagram were used, a main branch would be "men." This branch would seek to identify all potential causes related to men that could cause high setup scrap:

In this example the training of the personnel is a consideration. Training factors thought to be important are the adequacy of the training and testing the comprehension of the training material:

13.2 Concepts of the C & E Diagram

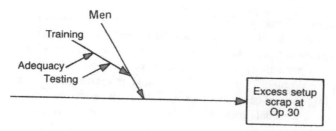

Other typical considerations that would start a new subbranch of men might include knowledge, absenteeism, or capability, for example.

The C & E diagram provides a convenient format to organize diverse thoughts concerning the potential causes of a problem. This diagram was developed by Professor Kaoru Ishikawa for use in the Japanese quality circles. The diagram is also called an Ishikawa diagram in honor of its originator or a fishbone diagram because of its shape.

13.2.1 Brainstorming

Preparation of a C & E diagram should be performed by a team of 5–10 individuals formed to solve the initially stated problem. The diagram visually displays the ideas of the group members and serves to stimulate new ideas. Thus, it is essential that a group be used to prepare the diagram. A freely flowing brainstorming process should be used to generate the list of potential causes. The meeting of the group members should be conducted using the following guidelines.

Leader. A leader should direct the activities of the team. During the sessions in which the diagram is prepared, the leader should act as a facilitator to encourage participation and new ideas. Generally, the leader should be a nonparticipant in cause generation, with the duty of facilitating the brainstorming process. There are differing opinions on how the leader should be selected. If supervisory personnel are part of the team, it is natural to have one of them serve as a leader. This practice facilitates implementation and builds the leadership skills of supervisors. The leader may, however, be any team member; the situation dictates the selection process.

Participation. One of the duties of the leader is to attempt having everyone participate in generating potential causes. It is possible to have group members take turns, but this practice is often too formal and restrictive. A better practice is for the leader to attempt to bring out ideas from reluctant participants. For example, the leader can ask, "Sam, what do you think about . . . ?"

Idea Building. One of the main reasons that groups are used to prepare C & E diagrams is to enable group members to build on each other's ideas. One person's initial idea often generates many new ideas; the result is always greater than each individual's initial list of ideas. The leader can facilitate this process: "Sam, can you expand on that idea?"

Criticism. It is imperative that there be no evaluation of the quality of any idea. A potential cause may be only remotely possible or, in fact, impossible. However, group members must not in any way criticize an individual with the idea. Criticism will lead to nonparticipation by some group members. Often after the C & E diagram has been completed, a review can eliminate any causes that are not feasible.

I'm the Problem. When a team assembles to discuss a problem, there is a natural defensiveness on the part of every person. The brainstorming process cannot function if every individual assumes his or her activity is not the cause of the problem. Each individual must assume his or her activity *is* the cause of the problem and ask, "How could what I do possibly cause the problem?" Since each person is an expert on his or her own activity, many meaningful potential causes will arise. If the leader starts the brainstorming process by getting a few individuals to assume "I'm the problem," other team members will generally be willing to follow.

Individuals. When problems are discussed, there is a temptation to blame individuals or other organizations. Again this is bad practice since it does not focus on specifics. Rather than saying Mr. X or Department X is not doing their jobs, focus on what specifically Mr. X or Department X is doing that might potentially cause the problem. This practice changes a destructive process to a constructive one.

13.2.2 The Effect

The lack of clear definition of the problem (i.e., the effect) makes arriving at the solution more difficult than is necessary. Consider the setup scrap example. Logical questions to more clearly define the problem would be

What is the percentage of scrap?
What is the time trend of the percentage of scrap?
Is the process stable?
What are the defects that make up the scrap (Pareto chart)?
Are the operational definitions of the defects clear?
Is the scrap about the same between machines, shifts, or operators, for example?

In many examples, the answers to these types of questions provide clear problem definition that can lead to a problem solution.

A clear problem definition is essential

Unfortunately, it is common practice to attempt solving a problem prior to a clear problem definition. The reaction-oriented problem-solving approach has little tolerance for "wasting time" on clearly defining a problem. This thought process is used when a C & E diagram is attempted prior to a clear problem definition. The first step in preparing a diagram should be a precise statement of the problem. Data should be used and the problem stratified into logical components. It is often possible to subdivide problems into natural groups. In most cases, this practice will lead to quicker problem resolution.

A consultant was asked to facilitate a team problem-solving meeting at which a C & E diagram was to be prepared. At the meeting his obvious first question was "What's the problem?" In the ensuing discussion, it became clear to the eight team members that each of them had a slightly different view of the problem. Over the following 2 week period, data were collected to clearly define the problem. Rather than one problem, the team agreed that there were three problems. The resulting C & E diagrams were relatively simple and helped direct the problem-solving process to meaningful data collection and root cause identification.

13.2 Concepts of the C & E Diagram

13.2.3 Potential Causes: Twig Building

An important part of constructing a C & E diagram is building branches or twigs that serve to expand from a general potential cause to more specific potential causes. Thus, the least useful type of C & E diagram appears as

When few twigs are present, there is a lack of sufficient definition of potential causes. The potential causes listed on a C & E diagram serve to direct the problem-solving process. A listing of general causes serves little use in directing the problem-solving effort.

It is possible to obtained a detailed breakdown of potential causes that form the C & E diagram twigs by using the simple construction question

What variability (in the next higher cause level) could result in the stated problem?

Consider the previous setup example in which "men" was a general potential causes category. The first subcause levels could be generated by asking the question

What variability in the way "men" set up Op 30 could cause excessive scrap?

The construction logic is simple. If there were no variability in the way personnel set up Op 30, then they could not be a cause of the scrap since all parts would be good (or all would be scrap). In this example, the question was answered by identifying the following subcauses:

Training
Knowledge
Absenteeism
Capability

Thus, there was possible variability in each of these areas that could cause excessive setup scrap. The subdivision process continues for each category. Consider the "Knowledge" subcause; the construction question is

What variability in the Knowledge of "men" could cause excessive setup scrap at Op 30?

The sources of variability representing the sub-subcauses were identified as

Experience (are employees with the correct experience being used?)
Evaluation (How was knowledge evaluated?)

The subdivision process should continue until reasonably detailed potential causes are identified so that information or data can be collected.

Thus, the process of generating the twigs involves repeated use of the construction question. The diagram twigs can be represented as shown in Figure 13.2. The exact format of the diagram is not critical, and it is frequently redrawn after the initial brainstorming session.

In some applications, the previous construction question is not appropriate. For exam-

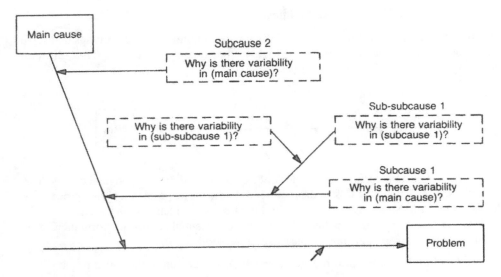

Figure 13.2 Generation of potential causes for a cause-and-effect diagram.

ple, if all parts are defective and redesign is necessary, focusing on variability is not meaningful. An alternative construction question is

What factor in (the next higher cause level) could result in the stated problem?

The combination of the two construction questions covers most applications. When possible, the first question should be used since focusing on variability is often useful in resolving problems.

13.3 Types of Cause-and-Effect Diagrams

13.3.1 Process Flow

The first step in understanding a manufacturing or administrative work process is to prepare a process flow diagram. It is useful to take the process flow diagram and use the construction question

What variability in (stage in the process) could result in the stated problem?

This approach has two advantages. First, the team is forced to prepare a process flow diagram. When this is done, all the alternative paths in which work is performed should be identified. Detailing the alternative ways in which work is accomplished can lead to a solution of the problem. Comments such as the following are frequently encountered:

"You mean we do . . . ! The procedure says that. . . ."
"Last year we didn't do it that way! We did. . . ."
"Our shift doesn't use that method! We do. . . ."
"That's what the book says, but you can't do it that way! You really need to. . . ."

Exploring the alternative ways in which work is accomplished can resolve the "hidden" problems that may contribute to or be the root cause of the problem.

13.3 Types of C & E Diagrams

The second major advantage of a process flow C & E diagram is that all individuals become familiar with the process. Since all group members may not be directly involved with the work process, defining how the work is accomplished is essential to solving a problem. The preceding advantages make it useful to start the potential cause-generating process by preparation of a process flow C & E diagram. A 5M or stratification diagram can then be prepared if necessary.

Some of the advantages and disadvantages of the process flow C & E diagram follow:

Advantages
  Forces preparation of the process flow diagram
  Forces consideration of the entire process as potential causes of a problem
  Identifies alternative work procedures
  Educational for members not familiar with the entire process
  Easy to use since most group members will be familiar with the process
  Can be used to anticipate process problems by focusing on sources of variability

Disadvantages
  Easier to overlook potential causes (e.g., material or measurements) since people may be too familiar with the process
  Hard to use on long, complicated processes
  Same potential causes may appear many times

Figure 13.3 is a process flow C & E diagram for oil level variability problems in a transmission.

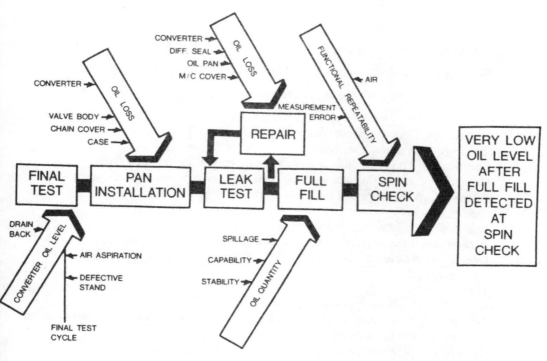

Figure 13.3 Example of a process flow C & E diagram.

13.3.2 The 5M Diagram

The 5M diagram is the most commonly used of the three general types of C & E diagram. Since problems are associated with a work process, the 5M diagram is used to consider the major elements of a work process:

Men
Methods
Machines
Materials
Measurements
Environment (optional)

These six elements define a work process, so it is natural to consider the potential causes of a problem being defined by these general categories. Some texts do not consider measurement separately, but experience has shown that it is always useful to assess measurement potential causes since they often contribute to a problem and in some cases are a root cause.

The basic diagram appears as

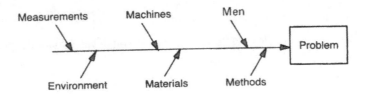

The standard construction question is

Why is there variability in (e.g., men) that could result in the stated problem?

Some of the issues that arise in each category are listed below.

Men
  Knowledge: Does an operator know what the job is?
  Training: Is an operator trained?
  Certification: Have operators demonstrated mastery of the required skills?
  Capability: Can any operator reasonably be expected to accomplish the task?
  Alternates: What happens if the usual operator is absent?
Method
  Standardization: Are job procedures defined clearly?
  Exceptions: Is the recovery procedure clear if the standard procedure cannot be used?
  Operational definition: Are operational definitions (e.g., target values) defined in the procedures?
Machine
  Capability: Do all machines demonstrate process capability?
  Differences: Have comparisons between machines, spindles, stations, or fixtures, for example, identified meaningful differences?
  Tooling: Are tool change intervals well defined and adequate?

13.3 Types of C & E Diagrams

    Adjustments: Are the criteria for machine adjustments clear?
    Maintenance: Is a preventive maintenance program in place, and is it adequate?
Material
    Variability: Is the variability of critical characteristics known?
    Changes: Have any material process changes occurred?
    Suppliers: What is the influence of any multiple suppliers?
Measurement
    Availability: Is the required gaging available?
    Operational definition: Are the characteristics to be measured operationally defined?
    Sample size: Are a sufficient number of parts being measured?
    Repeatability: Is gage repeatability sufficiently high?
    Bias: Does any gage bias exist?
Environment
    Cycles: Do problem cycles or patterns exist?
    Temperature: Does temperature influence operations?
    Testing: Has operational performance been evaluated over a realistic range of conditions?

Some of the advantages and disadvantages of the 5M C & E diagram follow:

Advantages
    Forces consideration of major elements of the process associated with a problem
    Can be used when detailed process knowledge is not known
    Focuses on the process not the product
Disadvantages
    Too many potential causes can be identified in a single branch (e.g., machine)
    Tendency to overlook small process details
    Not educational for group members not familiar with process

Figure 13.4 is a 5M C & E diagram for a machine downtime problem. It is sometimes useful to subdivide major potential cause categories. In Figure 13.4 the "tools" and "maintenance" categories were separated from the machine category.

13.3.3 Stratification

One of the disadvantages of the 5M C & E diagram is that many potential causes may be listed on a single branch. Consider a reliability problem encountered during the development phase of a product. There are probably few issues with men, machines, or measurement but many material or method (design) issues. When this occurs, it is often possible to stratify the potential causes into major areas of concern. If this approach is used, the risk is that the selected major categories of potential causes will not contain the root cause of the problem. However, the stratification C & E diagram is a natural choice in many applications in which the potential cause categories can be easily subdivided. For example, a finished product can be divided into its subassemblies. It is also possible to use the stratification diagram following the 5M diagram. For example, to address a machine reliability problem, the 5M diagram could be prepared by addressing men, method, materials, measurements, and environment as potential causes. The machine could then be divided into natural subsystems and a stratification C & E diagram prepared.

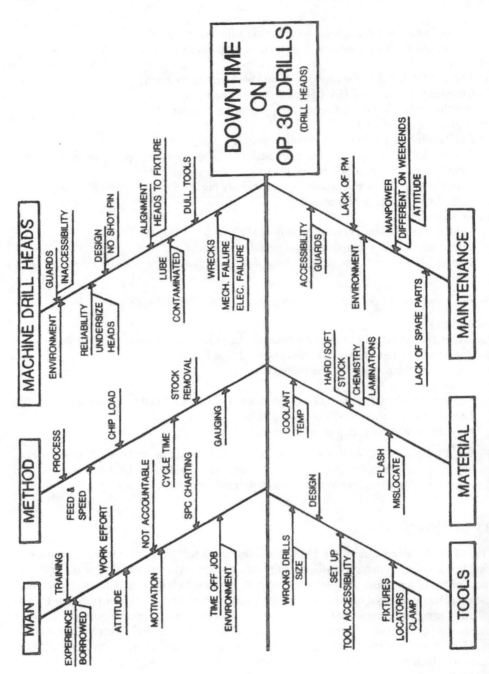

Figure 13.4 Example of a major elements C & E diagram.

13.4 Investigation of Potential Causes

Some of the advantages and disadvantages of the stratification C & E diagram follow:

Advantages
  Provides a clear grouping of potential causes that can be addressed directly in the problem analysis.
  Diagram is generally less complex than would be obtained using other approaches.
Disadvantages
  Important potential causes can be easily overlooked
  Defining major subdivisions can be difficult.
  Greatest knowledge of potential causes is required.
  Greatest knowledge of the product or process is required.

Figure 13.5 contains a stratification C & E diagram for a transmission gear noise problem. The natural strata (i.e., subdivisions) of potential causes are the gear subassemblies within the transmission. Notice that measurement and method concerns are listed for most subassemblies. Thus, the 5M concern areas can be carried over into the preparation of the stratification C & E diagram.

13.4 Investigation of Potential Causes

When first introduced to C & E diagrams, people sometimes mistakenly believe the diagram will in some way solve a problem. Experience quickly shows that the diagram is a simple, convenient tool for identifying and organizing the potential causes of a problem, but root cause determination requires further analysis. The C & E diagram is not intended to identify which potential causes are root causes. This evaluation requires the collection and analysis of data. Figure 13.6 diagrams the situation in which it is clear that determining whether a potential cause is a root cause can be a time-consuming process. The reaction-oriented "problem-solving" process from Chapter 1 is abandoned in lieu of data collection and analysis of potential causes.

> C & E diagrams provide a method of identifying and organizing potential causes of a problem

How is it possible to determine which potential causes to investigate first? The most common approach is for the team to prioritize the potential causes using their experience. Each of the selected potential causes should be evaluated using two criteria. First, a detailed description of the problem (5W2H) should be used to evaluate whether the potential cause is consistent with the problem description. For example, "tools" may be a potential cause of a machining defect, but the presence of the defect on only one machine, when several machines use the same tool, may eliminate tools from further consideration.

It should then be determined if a plausible causal mechanism exists in which the potential cause being considered could result in the problem. A statement should be made about how the potential cause could have resulted in the problem. For example, if it cannot be stated how "inadequate coolant" could have caused burrs on a part, it may be possible to eliminate the coolant from consideration. Often data will be needed to test the causal mechanism. The selected potential causes, which are consistent with a detailed description of the problem, and have a plausible causal mechanism, will then be evaluated quantitatively using data collected by the team. The drawback in this process over reaction-oriented problem solving is that time and effort are expended by evaluating potential causes that are eventually determined not to be a root cause. This loss is an expected part of a root cause-oriented problem-solving process.

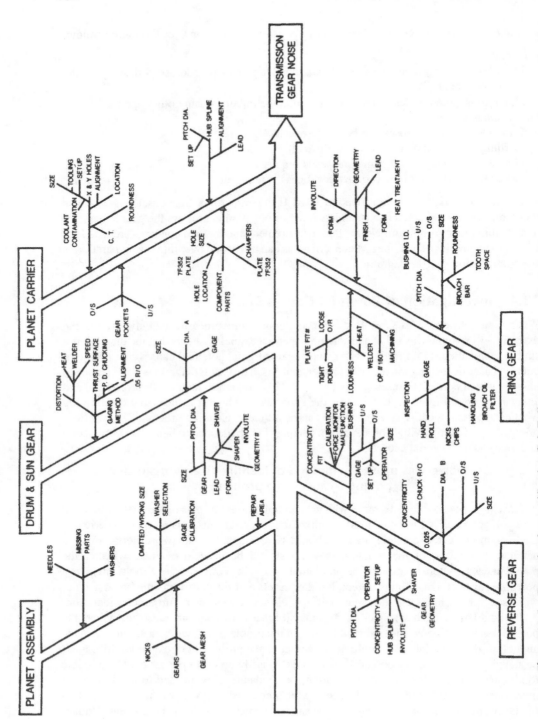

Figure 13.5 Example of a stratification C & E diagram.

13.4 Investigation of Potential Causes

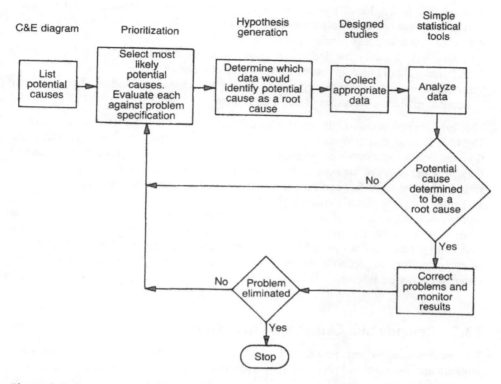

Figure 13.6 Process for identifying whether a potential cause is a root cause of a problem.

For each potential cause being investigated, the team must decide what data or information can be used to decide whether a potential cause is a root cause. This process can lead to stating a hypothesis that will be either proved or disproved by the data (see later discussion of the data summaries). The process of determining what data to collect for evaluating whether a potential cause is a root cause is one of the most important parts of the team problem-solving process. The collective experience of the group can provide the simplest, most expedient evaluation method. Once the appropriate data are available, the simple statistical tools discussed in earlier chapters can be used to evaluate a potential cause.

The number of actions required to address a potential cause can be large. Also, it is highly advisable to concurrently evaluate several potential causes since we wish to solve the problem as fast as possible. In many applications, several potential causes can be identified as root causes, so we must simultaneously evaluate more than one potential cause of a problem.

Many teams fail to solve a problem in a timely manner because of their inability to manage the required actions to investigate multiple potential causes. The best way to manage the problem-solving process is to use a standard action plan to identify *who* will do *what* by *when*?

The major elements of the action plan follow:

Potential cause or action: For each potential cause to be evaluated, list the series of actions required to conduct the study to determine whether a potential cause is a root cause.

Planned completion: Identify when each action is planned to be completed. If the planned date is delayed, the new planned date should be listed below the original date.

Actual completion: Identify when the action was completed.

Responsibility: Identify the individual responsible for the action. Generally, a single individual should be responsible for the completion of an action.

Status and conclusions: Briefly give the status of any actions in progress. For completed actions, give the conclusion of any data analysis.

This system provides a simple, effective method for managing a diverse set of activities. The C & E diagram action plan approach provides an effective potential cause identification and follow-up system to investigate potential causes. Team meetings can start by reviewing the status of the actions on the action plan. Future actions, timing, and responsibilities will be identified and placed on the action plan during the meeting.

Table 13.1 gives part of the action plan for the gear noise example. Notice the number of problem-solving actions that are occurring concurrently, all being monitored by the action plan. A team member is likely to work on several actions involving different potential causes. In general, actions should be sufficiently detailed so that some specific actions can be evaluated for completion at the next team meeting. If actions are defined too broadly, progress cannot be assessed.

13.5 Benefits of a Cause-and-Effect Diagram

The process used to prepare a C & E diagram and investigate the potential causes of a problem are discussed in Procedure 13.1. Some of the benefits of this process follow:

Root cause orientation: Identifying potential causes of a problem and then collecting information or data to determine if a potential cause is a root cause clearly focuses on surfacing root causes. Once root causes are addressed, the problem should be eliminated.

Potential cause definition: The C & E diagram is used to identify as many potential causes as possible at an early stage in the problem-solving process. This practice encourages a comprehensive analysis of possible problem causes. Since many potential causes are identified early in the problem-solving process, several causes can be simultaneously analyzed.

Team analysis required: Preparation of the C & E diagram is performed by a team that benefits from the diverse experience of the team members. Also, a synergy is often established since one person's ideas can often be expanded and built upon by other team members.

Variability focus: The twig-building process focuses on identifying sources of variability in the process that may be the cause of the problem. Since variability is the cause of many problems, the C & E diagram uses a natural approach.

Management tool: The C & E diagram combined with an action plan provides a natural management tool to evaluate the effectiveness of the problem-solving effort and monitor its progress. Since these tools are easily understood, they can be used at the lowest organizational level.

Anticipation of problems: It is not necessary to actually experience a problem to prepare a C & E diagram. Before any problems occur, it is possible to ask, "What could cause (e.g., a quality problem) at this stage of the process?" The C & E diagram can thus help

13.5 Benefits of a C & E Diagram

Table 13.1 Action Plan Format for Gear Noise Example

Potential cause/actions	Planned completion	Actual completion	Responsibility	Status/conclusions
1. Involute form				
Capability study	7/1	7/3	Jones	Process stable but interference condition possible ($C_{pk} = .7$)
Order six new broach shells	7/5		Peters	Received five of six; last shell due 8/1
Gage study	7/11		Jones	
	8/5			
2. Nicks				
Pareto study on nick location	7/2	7/5	Smith	Pareto analysis indicates 65% of nicks on long gear end
Process study on where nicks occur	7/8	7/8	Williams	80% of nicks on long gear end occur after Op 60
Revise roller conveyor system	8/12		Peters	
3. Roundness				
Perform gage study	7/3	7/3	Jones	Acceptable gage error (15%)
Initiate SPC charts	7/4	7/4	Rust	Process appears initially stable
Evaluate capability	7/20			
4. Welder characteristics				
Evaluate different hardness			Berger	
Obtain different hardness gears	7/10	7/9	Jones	Five sets of different hardness gears obtained
Test in vehicle	7/11		Tower	
Analyze data				
Determine characteristics in different transmissions	7/30		Tower	
Compare penetration in different welders	7/15	7/15	Jones	Comparison study performed; no differences detected
5. Op 160				
Evaluate stock removal capability	7/10		Smith	
Correlate stock removal with noise performance	7/20		Rust	

Prepared by: _____ Problem: Gear noise Updated: 7/15

Procedure 13.1 Preparation of a Cause-and-Effect Diagram

A cause-and-effect (C & E) diagram is a listing of potential causes of a problem. The steps used to prepare a C & E diagram follow:

1. Define the initial problem. Collect enough data so that the problem can be adequately defined by a clear statement. The severity of the problem should be quantified.
2. Select team members. Using the initial statement of the problem, select team members with the appropriate process or product knowledge, work experience, and training. Team members should agree that representation on the team is adequate. Generally, a team size of 5–10 people is appropriate. As new information becomes available, the team composition should be evaluated to ensure that the needed skills are available.
3. Select a leader. A team leader should be designated by either management selection or group preference.
4. Prepare a process flow. Prepare a standard process flow diagram. Identify all inputs, outputs, and sources of variation in the work process. Focus on having the team identify and evaluate alternative paths used to accomplish work.
5. State the problems. The initial problem definition should be used to direct team members to collect the needed data to quantify the problem using who, what, where, when, why, how, and how many (5W2H). Subdivisions of the initial problem into more manageable sub-problems should be considered. A clear statement of each problem should be made. The team composition should be analyzed to ensure that the appropriate areas are represented.
6. Select the type of C & E diagram. The team should select the C & E diagram appropriate to address the problem: process flow, 5M, or stratification. It is sometimes useful to prepare two or three C & E diagrams for a problem in the preceding order.
7. Prepare the C & E diagram. Use one of the two construction questions to obtain the twigs on the diagram. (1) What variability (in the next higher cause level) could result in the stated problem? (2) What factor in (the next higher cause level) could result in the stated problem? Utilize a brainstorming process in which ideas freely flow and no criticism is offered. Each team member should assume that his or her activity is responsible for the problem and ask, "How could what I do possibly generate the problem?"
8. Assign priorities. Have the team select the top several potential causes to investigate. These causes should be examined to evaluate whether they are consistent with a detailed problem description. Also, determine whether a possible causal mechanism exists so that the potential cause could result in the problem.
9. Prepare an action plan. Have the team decide what data should be collected to determine whether a potential cause is a root cause. For each potential cause to be investigated, decide who will do what by when. The leader should prepare and distribute an action plan. Subsequent meetings will start by obtaining a status on the actions to be completed and then the team will decide future actions.
10. Allocate resources. The resources available for the team to solve the problem should be evaluated by management. In particular, time should be made available for group members to accomplish their assigned tasks.

13.6 Problem Analysis System

to anticipate problems so that appropriate controls can be implemented. With all the experience that often exists, why not use a C & E diagram to prevent problems?

13.6 Problem Analysis System

Chapter 1 discussed the problem reaction wheel that serves as the basis for addressing many manufacturing problems. Too often, problems are not solved since the root causes of a problem are not identified. Problem symptoms are addressed, and Band-Aid fixes are used. The problem analysis system presented here is a data- and fact-driven approach to solving problems. However, we mut realize the demands of the manufacturing operations cannot wait for the analysis approach to be completed before any action is taken. Interim "fix-it" actions are necessary to continue operations. The difference in the two approaches is a realization that analysis is required (and will be performed) to obtain permanent problem solutions.

13.6.1 Analysis Process

The C & E diagram provides a list of potential causes that may be root causes for the problem being analyzed. How can we decide whether a potential cause is a root cause? Clearly, an analysis process is required. Data are needed to decide which potential causes are important. Problem-solving teams often flounder after a C & E diagram is prepared unless a transition is made from the initial brainstorming approach to a more disciplined analysis process focusing on data rather than ideas and opinions. Proper training of team leaders and members is required to enable a smooth transition. Failure to make this transition may not only lead to an inability to solve the problem, but generate hostility within the team. When data are not used to help direct the team's efforts, dominant individuals will impose their opinions on the group, which negates the potential benefit of teamwork.

Many individuals are under the mistaken belief that data-oriented problem solving can be accomplished by collecting relevant data on a problem, analyzing the results, and determining the correct solution. This belief is incorrect for two reasons. First, the analysis process is iterative. Once data are collected and analyzed, new questions often arise so that another data collection and analysis iteration is necessary. Second, many problems can have more than one root cause. Data collected while investigating one potential cause may not address other important potential causes. Thus, several potential causes should be studied using the data collection and analysis process.

Figure 13.7 depicts the iterative process typical in the analysis of a problem. The process starts with selecting a potential cause from the C & E diagram. Each iteration consists of six steps:

1. Hypothesis generation. State how the potential cause could result in the problem. Identify the causal mechanism that enabled the potential cause to result in the problem.
2. Design. What type of data can most easily prove or disprove the stated hypothesis? Develop a plan of how the study will be conducted. Identify the actions on an action plan.
3. Preparation. Obtain the required materials to conduct the study. Develop the data collection check sheet.
4. Data collection. Collect the required data.
5. Analysis. Using simple statistical tools, prepare graphic displays of the data.

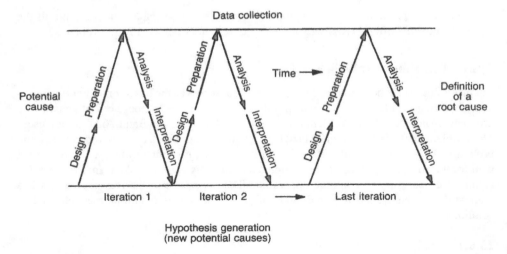

Figure 13.7 Iterative problem analysis.

6. Interpretation. Does the data support the hypothesis? Are other studies necessary to evaluate the hypothesis being considered? Identify other factors that may be related to the problem (i.e., other potential causes).

After each set of data is analyzed and interpreted, it is not unusual to add new potential causes to the C & E diagram. Often these potential causes are pursued before other C & E diagram causes since they are suggested by data rather than opinion. The C & E diagram is used to start the problem analysis process, but subsequent direction is largely based on observed data.

The three types of C & E diagrams have several main branches. These branches are often somewhat independent and can be pursued separately. Recall from the previous section that it is beneficial to explore several causes at the same time. Figure 13.8 shows the iterative nature of the analysis process and how simultaneous investigation can lead to quicker definition of all root causes. Also, since the analysis process is iterative, simultaneous investigation reduces the overall time required to define root causes. Too often, teams link the investigation of potential causes and become discouraged when the first few investigations are not fruitful.

The iterative nature of the analysis process combined with the simultaneous investigation of several potential causes can become a complex process to manage. The use of an action plan is essential for controlling the process. Maintaining an action plan for the team's activities accomplishes the following:

Provides team leader with easy to maintain follow-up system
Identifies expected tasks for each team member
Serves as an agenda for meetings
Enables management to monitor progress
Provides subtle pressure on team members to completely agree upon tasks

The most difficult part of the problem analysis process for many managers is to accept the iterative nature of the process. That data collection and analysis leads to more questions

13.6 Problem Analysis System

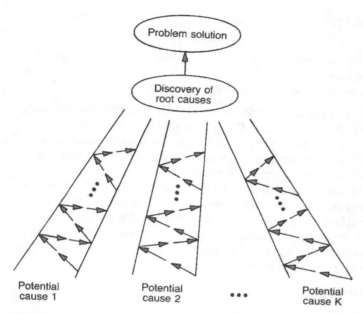

Figure 13.8 Addressing multiple potential causes.

and the need for more data rather than an immediate problem solution is difficult to accept. The cause of this lack of acceptance is that each level of management successively asks the lower levels *when* a problem will be solved. A response that data are being collected and the problem studied (with no resolution date known) is too often unacceptable. Generally, there is little concern for the process of solving problems—only planned actions and hopeful results are required. The problem analysis system provides a clear alternative to the traditional approach. However, it is necessary for management to focus on evaluating and monitoring the problem-solving process rather than defining end results with required timing. Also, it is useful for managers to expedite the analysis process by allocating resources to accomplish the action plan tasks.

13.6.2 Ten Problem-Solving Steps

The process of solving a problem is diagramed in the flowchart in Figure 13.1. The 10 steps provide an organized approach that can be used to address a wide variety of industrial problems. Experience has shown that problem-solving efforts often fail because a systematic approach is not used. To consistently solve problems, teams require a disciplined procedure to guide their activity. It should be emphasized not all problems use all 10 steps. Many "simple" problems can be solved utilizing the knowledge of key individuals who implicitly use some of the problem-solving steps. However, there exists many other difficult, often recurring problems that cannot be solved by experience alone. The following steps can be used to address these problems.

Step 1. Form a Team. When a problem is not solved quickly using the individual experience of key personnel, it is necessary to form a team. Deciding on the composition of the team is a critical first step that often predetermines the team's success. Questions to be addressed in selecting team members follow:

Does the team have the necessary process or product knowledge and work experience?
Does the team have the necessary technical skills?
Have sufficient time and resources been allocated for team members?

Since special skills are often required to investigate the cause of a problem, obtaining selected advisors is beneficial in some cases. The team's advisors are specialists in the required technical areas. Typically, they have limited time and help only to direct the work effort. However, other team members must be able to dedicate appropriate amounts of time to accomplish the tasks detailed in the team's action plan.

Generally, management will select both team members and advisors. However, too often the group's skills and available time of members are not adequately evaluated. A team leader is often selected by management. In employee participation groups, the leader is sometimes selected by the group members. Caution must be exercised so that the selection process does not cause division of group members. At the first meeting of the team, the responsible manager should make the following items clear:

Problem being addressed and its importance
Team members
Team advisors
Team leader
Time allocated for members to participate in team activities
Type of management reporting expected

Input from team members on these items is important to keeping communications open.

The size of teams should be kept reasonably small, between 5 and 10 members (with several advisors possible). As the size of the team increases, effectiveness decreases. Experience has shown that a committee's ability to take action is limited; a similar comment pertains to large problem-solving teams. It is preferable to have several individuals heavily committed to solving the problem rather than a large number of individuals with partial responsibility for "participating" in the problem-solving effort. The selected team should be considered the champions of the problem, with a significant part of their time assigned to the problem analysis. Unfortunately, organizational practice often results in team members minimally committed to a problem-solving team. All employees have assigned responsibilities, and participating on a problem-solving team is simply an added task. This practice contributes to the lack of effectiveness of many teams.

As part of the management reporting and review system, the composition of the team should be periodically evaluated. As data are collected, new facts become available for directing the team's effort. New skills and advisors may be required to investigate a potential cause. The team members and advisors must be continually evaluated to ensure that the correct skills are present.

Step 2. Define the Problem.   In the discussion of the effect part of the C & E diagram, it was noted that frequently too little attention is given to adequately defining the problem. However, proper definition of a problem requires time for collecting the required data. For many industrial problems time is initially not available; some action is necessary, often immediately, to salvage a bad situation. Thus, it is necessary to define the problem in two phases. The first phase identifies symptoms and quantifies the severity; the second phase

13.6 Problem Analysis System

focuses on a clear specification of the problem by conducting three different types of analysis.

When a problem is initially encountered in phase 1, the general symptom forms the basis of the problem definition.

Machine x is down owing to electrical problems. No backup machine or alternate process is available.

The scrap rate increased on x date from 3 to 25%.

Customer warranty claims on x engine component are 10%.

Failure of durability tests of a transmission component at 50,000 miles will delay product launch.

The extent of the problem (25%, 10%, and 50,000 miles) and its consequences are usually known. After the initial phase of symptom definition, an interim corrective action (step 3) may be required. This phase of problem definition is the same as used in the problem reaction wheel. However, there is a clear recognition in the problem analysis system that analysis of the problem will be conducted.

In the second phase, an in-depth analysis is required to clearly define a problem. There are many examples in which the analyses suggested here for a complete problem definition result in the solution being identified. The analysis starts with preparation of a process flow diagram to clearly define the work process and alternative paths. Also, team preparation of the diagram ensures that all individuals are familiar with the process. Following preparation of the diagram, three analysis methods are used to properly define the problem: (1) 5W2H, (2) stratification, and (3) comparison.

The analysis starts with quantifying the 5W2H elements:

Who. Identify individuals associated with the problem. Characterize customers that are complaining. Which operators are having difficulty?

What. Describe the problem adequately. Does the severity of the problem vary? Are operational definitions clear (e.g., defects)? Is the measurement system repeatable and accurate?

Where. Locate the problem. If a defect occurs on a part, where is the defect located? (Use location check sheet.) What is the geographic distribution of customer complaints?

When. Identify the time the problem started and its prevalence in earlier time periods. Do all production shifts experience the same frequencies of the problem? What time of the year does the problem occur?

Why. Any known explanation contributing to the problem should be stated.

How. In what mode of operation did the problem occur? What procedures were used?

How many. What is the extent of the problem? Is the process in statistical control? (e.g., P chart)

The 5W2H questions help to characterize the problem for further analysis in the next two parts of the problem definition.

Some problems arise from customer complaints. An internal customer's complaint could involve one department complaining that they cannot effectively use the output of another department. An external customer complaint could involve a customer complaining to a dealer that his car transmission "shifts funny." Too frequently, the wrong problem is solved and the customer complaint is not addressed. It is imperative that the customer complaint be clearly understood. The only method to ensure this is to have direct customer contact. For internal customers, it is advisable to have representatives from the complaining organization as part of the problem-solving team. In many cases, this approach is the only way a problem can be truly solved. External customer complaints typically require direct interviews to understand why the customer is not satisfied. It is not unusual for a customer complaint to be misrepresented by a company reporting system that classifies problems in prearranged standard categories. Part of the 5W2H problem definition is to clearly state the customer complaint.

The customer complaint must be clearly defined

An assembly department had an ongoing problem with selectively fitting a gear into a machined pocket. Size ranges of the gear were classified as A (smallest), B, C, and D (largest). Not only was the scrap rate of the machined parts high since no gear would fit in the pocket, but a larger number of parts were continually returned to the machining department to be reworked. The machining department would often find nothing wrong with the returned parts. Each department had numerous reasons why the other department was responsible for the problem. A team was formed with members from both departments. After clearly identifying both processes, they discovered that the gaging position for the pocket was different in the two departments.

The stratification analysis determines the extent of the problem for all the relevant stratification factors.

Is the problem the same for all shifts?
Do all machines, spindles, or fixtures have the same problems?
Do customers in various age groups or parts of the country have similar problems?

The important stratification factors will vary with the problem being considered, but most problems will have several factors. Check sheets can be used to collect relevant data. Essentially, this analysis seeks to develop a Pareto diagram for the important factors. The hope is that the extent of the problem will not be the same across all factors. The reason for any differences can lead to identifying a root cause.

When the 5W2H and stratification analyses are performed, it is important to consider a number of different indicators. For example, a customer problem identified by warranty claims may also be reflected by various in-plant indicators. Customer survey results should be related to internal indicators. In some cases, analysis of a problem can be expedited by correlating different problem indicators to clearly identify the problem to be addressed.

It has been said that there are no new problems, only different manifestations of old problems. In step 4, where we identify potential causes, prior similar problems will be investigated. In the problem definition step, it is often useful to quantify the existence of the problem in similar situations—a comparison analysis. The criteria used to match similar situations will vary with the type of problem. Identifying effective matches and evaluating the presence of the problem provides useful information to generate potential causes and possible problem solutions. If the comparison analysis identifies a comparable situation in which the problem does not exist, the analysis can focus on differences in where the problem is occurring and where it is not occurring.

Once the three analyses have been completed, it is sometimes possible to divide the problem into separate problems. It is easier to address smaller problems because fewer root causes will be involved. In the ideal case, a single root cause would be responsible for each problem to be addressed. If the problem is separated, different teams may be required to address each problem.

All three methods for problem definition are not used for every problem. However,

13.6 Problem Analysis System

collectively the different analyses provide a comprehensive description of a problem. Examples for the four types of problems presented earlier follow. A complete specification of the problem involves all the main statements.

Example 13.1 Machine Downtime

Machine x is down because of electrical problems. No backup machine or alternate process is available.

Machine x experienced 40% downtime in the prior 6 months. Downtime failures have been 70% electrical. Electrical downtime has been 50% switch related (Pareto diagram for 5W2H analysis).

Switch failures have been due to shorting caused by decomposition of wire shielding (5W2H analysis).

The same switches in other machines have exhibited no failures (comparison analysis).

The machines that exhibit no failures use different machining coolant (stratification analysis).

The machines that use the coolant used by the problem machine have an alternative switch design (comparison analysis).

Example 13.2 Excessive Scrap

The scrap rate at operation 20 increased on May 15 from 3 to 25%.

In the prior 6 months, the scrap varied between 0 and 5% and was stable (P chart for 5W2H analysis).

The two machines performing the operation have scrap rates of 23 and 27% (stratification analysis).

Tool usage at the operation has increased 30% on both machines (5W2H analysis).

No changes in the upstream process have been noted on control charts (5W2H analysis).

The vendor says no process changes have occurred, but control charts were not used (5W2H analysis).

Other similar steel parts supplied by same vendor have experienced 5% increased scrap (comparison analysis).

Example 13.3 Warranty Claims

Customer claims on the engine component are 10%.

Monthly claims varied between 5 and 15% and are not stable (5W2H analysis).

Most claims (60%) are from hot weather states, and claims increase by 50% during the summer months (Pareto diagram for stratification analysis).

All claims result from a scored y value that occurs in customer usage. All bushings are scored around the entire circumference (5W2H analysis).

Other engines use a differently designed bushing for the same application (comparison analysis).

Other bushings of the same design do not score in the engine (comparison analysis).

Example 13.4 Test Failures

Failure of durability tests of a transmission around 50,000 miles will delay component product launch (5W2H analysis).

A test conducted on 10 components. Failures occur at 45,000, 47,000, 52,000, 53,000, and 54,000 miles. The remainder of the components completed the 100,000 mile test (5W2H analysis).

Analysis of failed components showed heat fractures and burned components in area y (5W2H analysis).

Other transmissions with similar designs show no component failures (comparison analysis).

The lubrication system for the component is a new design in the current transmission (comparison analysis).

The prototype components are manufactured by two vendors. Four of the five failed components were from vendor A (stratification analysis).

Step 3. Implement Interim Containment Actions. Often it is possible to take containment actions to reduce or eliminate the problem symptoms. These actions should not be confused with actions to eliminate the root cause of a problem. Actions to address problem symptoms clearly have a different focus than those to eliminate problem causes. The intent of this step is to highlight the Band-Aid actions typical of the problem reaction wheel approach. Explicitly stating the interim actions helps to focus attention on an undesirable situation. It should be emphasized that there is no intent to sacrifice the quality of products in order to maintain production schedules. Generally, the tradeoff is adding cost to the product. Several examples follow:

Off-standard labor used to deburr the part
100% sorting implemented with control charts
Parts double-washed to eliminate residue
Machine cycle time reduced to eliminate problem
Tools changed every 50 parts instead of every 200 parts
Parts purchased from the vendor rather than manufactured in-house

Clearly, all these containment actions add cost to the product. A real danger in interim actions is the use of 100% sorting. Sorting does not guarantee that parts are within the specification limits. Sorting, in general, is cited as only 80% effective. If sorting is used, check sheets or control charts should be used for the affected characteristic to quantify the extent of the problem.

The effectiveness of interim actions should be evaluated when the actions are implemented on an ongoing basis. Too often, minimal verification of interim action is performed since there is an anticipation that the actions are only temporary. Since interim actions involve a change in the normal process, an opportunity exists to significantly degrade quality. Extreme care should be taken to ensure that quality standards are maintained.

Step 4. Identify Potential Causes. The C & E diagram discussed earlier is the tool used to identify potential causes of the problem. The diagram is prepared after phase 2 of the problem definition is completed in step 2. The information gained in step 2 enables the group to clearly state the "effect" part of the diagram. Too frequently, teams take an initial problem statement and define the effect in general, unclear terms. A poor C & E diagram will likely result from a poorly defined effect. Additionally, the team's analysis directions, typically gained from a good C & E diagram, cannot be planned well.

One of the three types of C & E diagrams (process flow, 5M, and stratification) should be selected. Generally, clear definition of the problem and a good process flow diagram gained from step 2 enable an appropriate selection of one or more of the C & E diagrams.

13.6 Problem Analysis System

The team should prepare the diagram using the guidelines in Procedure 13.1. In identifying potential causes, each team member should assume that his or her activity is responsible for the problem and ask, "How can what I do possibly generate the problem?"

The C & E diagram is one method of identifying potential causes; another is by performing a time line or comparison analysis. These two common analysis tools can be a very useful part of identifying potential causes. If the frequency of a problem changed dramatically, it is apparent that something has changed. Determining what changed, when may identify potential causes of a problem. Preparing a time line of the frequency of the problem and all process changes is a common problem analysis tool. Unfortunately, in a manufacturing environment many changes can occur and it is often difficult to identify the critical changes that resulted in the problem. In many cases, the event that caused the problem is identified only after the cause is identified by other methods.

A comparison analysis is performed by making comparisons with similar products or processes identified in the analysis in step 2 with the intent of identifying critical differences that suggest potential causes of the problem. It has been said that there are few new problems. To the extent that this is true, a comparison analysis will help identify potential causes. In some cases an immediate solution (or at least an interim solution) can be obtained. Unfortunately, the exact situation and problem may not exist, but enough similarity may exist to suggest potential causes to the problem being analyzed. These causes can be added to the C & E diagram prepared by the team.

An obvious solution to the problem should not be overlooked when performing a comparison analysis. Selecting the right comparison can often lead to a quick problem solution. This approach has the added benefit that an existing solution has been field tested and may only need adaptation to the existing situation. Too often the NIH (not invented here) mentality exists. The NIH practice is never stated as a policy since it is obviously illogical but is implicitly practiced when a comparison analysis is not performed.

Do not reinvent the wheel!

The objective of preparing the C & E diagram is to identify several potential causes to be investigated. These causes are often circled on the diagram. Each of the selected potential causes should be evaluated relative to the complete problem specification statement in step 2. If the potential cause is inconsistent with the problem specification, it should be eliminated from further consideration. Also, if a plausible causal mechanism does not exist, further evaluation of the potential cause should be reconsidered (see Sec. 13.4).

Example 13.1 (continued)   Machine Downtime

A current switch cannot function over prolonged exposure to existing machining coolant. The current coolant may cause shorting.
The switch materials may be defective, enabling reaction with coolant causing shorting.
Loose or improperly installed connectors resulting in intermittent electrical signals cause electrical faults that result in machine downtime.
Mechanical clamping pressure is set improperly, resulting in part mislocation and machine wrecks.

Example 13.2 (continued)   Excessive Scrap

A new inspector has been assigned to the department and may be using a different evaluation procedure.

Steel hardness or microstructural changes may make old tool change frequencies inadequate.

The tool vendor has changed, and tool quality has not been established.

Gage repeatability and bias have not been evaluated. The gage may be generating false rejects.

Machine preventive maintenance has not been performed for 1 month. Machines may not be capable.

Control charts have not been examined, and the process may not be stable, resulting in excessive scrap.

Example 13.3 (continued)   Warranty Claims

Material hardness may not be adequate so that normal wear results in scoring.

Inadequate lubrication distribution may result in higher than expected friction, resulting in scoring.

Extreme engine temperatures may result in low oil viscosity, causing poor lubrication and the resulting scoring.

Customers may use improper viscosity oil, which may provide poor lubrication and result in scoring.

Example 13.4 (continued)   Test Failures

The prototype vendor may have used a lower grade of steel, causing early failure.

Material specifications may not have sufficient strength for the current high rpm application, causing early failure.

The lubrication system may be inadequate, causing overheating and early failures.

Dimensional clearances may be inadequate for lubrication flow, causing early failures.

At least the top four or five potential causes should be listed in the 10-step summary. Identifying several potential causes forces the team to address multiple causes rather than a single cause. As noted earlier, many problems are the result of several root causes. An implicit part of the problem analysis systems discussed here is investigating potential causes by using parallel rather than series actions. Too often the "try one, try another" approach is used. The series approach requires excessive time and can make the team discouraged when the first few investigations are not fruitful.

Step 5. Analyze Potential Causes.   The analysis of the potential causes listed in step 4 is the time-consuming part of the problem analysis process. However, the analysis steps distinguish this approach from the problem reaction wheel. At this stage, the iterative problem analysis begins. Iterations for each potential cause consists of the following six steps: (1) hypothesis generation, (2) design, (3) preparation, (4) data collection, (5) analysis, and (6) interpretation. These problem analysis process steps were discussed earlier in the chapter. For each potential cause being investigated, the action plan helps the team step through the process.

Investigate potential causes using parallel actions

The action plan should address how the team plans to investigate each potential cause. Use of the action plan is critical to resolving the problem. The action plan is the single tool used to enable planning, monitoring, and follow-up on the team's efforts.

The reporting system used to summarize the progress on each of the 10 steps is presented

13.6 Problem Analysis System

in a later section. The main reporting performed in step 5 is a summary statement of each of the past investigations and a brief list of future plans. Reference to appropriate data summaries (see Procedure 13.4) should be given. Detailed plans are contained in the action plan. For management reporting, the dates associated with major events are usually given.

Example 13.1 (continued)   Machine Downtime

6/18: Off-line tests indicate that the coolant decomposes switch plastic. Shorting results after 1000 hours of operation.
6/20: Ongoing monitoring for electrical overloads has indicated no overload problem.
6/21: Thirty percent of connectors are loose. A possibility of intermittent signal exists.
7/1: A new connector design is to be evaluated.
7/3: Machine clamping pressure is to be compared on other similar machines.

Example 13.2 (continued)   Excessive Scrap

6/5: The new and old inspector are evaluated using the same 100 parts. No major differences are detected.
6/10: Histogram (see attachment) of steel hardness shows that the material is on the upper part of the specification limit ($C_{pk} = .8$). Microstructural results will be obtained.
7/1: The supplier is to be visited to evaluate hardness controls.
6/22: A comparison plot (see attachment) of new and old tools indicated no significant differences.
6/20: A gage study showed .002 inch bias in dimension x and a 54% gage repeatability error. Rehab of gage to be completed by (7/15).
7/5: Preventive maintenance will be performed on two Op 20 machines.
7/1: Control charts have been updated. A follow-up system to be initiated.

Example 13.3 (continued)   Warranty Claims

7/20: A new test using harder material is being planned.
7/1: The new lubrication distribution is to be tested using larger oil hole.
6/20: Frictional wear was evaluated on a test engine every 10,000 miles. Progressive wear was detected.
7/3: A high-temperature–low-viscosity oil test is currently in progress.
7/30: Analysis of oil viscosity on failed engine will be performed.

Example 13.4 (continued)   Test Failures

6/20: Analysis of two types of steel indicates no significant differences in hardness.
7/30: The steel microstructure is to be evaluated and compared.
7/15: An accelerated test is planned using high-strength steel.
7/30: A high-pressure pump and increased lubrication flow to be tested.
6/5: Finite element analysis indicated that no clearance problem exists.
8/15: Clearance in critical interfaces is to be increased by 30% and tested.

Step 6. Validate Root Causes.   The studies conducted in step 5 will generate three types of causes depending on the following:

Does the potential cause result in the problem?
Is the potential cause present?

The three types of causes are root cause, latent cause, and null cause.

Cause		Can cause the problem	Is present
Potential cause → Analysis →	↗ Root cause	Yes	Yes
	→ Latent cause	Yes	No
	↘ Null cause	No	Yes or no

The analysis conducted in step 5 showed which potential causes were capable of causing the problem. However, a cause must be present to result in the problem. Step 6 seeks to validate the identification of root causes, so we must ensure that a cause is present. Consider the example in which it was determined that variations in hydraulic pressure (potential cause) could cause mislocation of a part in a machining station, which results in a machine wreck (problem). Studies could artificially induce hydraulic pressure variations that result in a wreck. However, do pressure variations actually exist during normal operation of the machine? Determination of whether a cause is present is an essential part of declaring a cause to be a root cause.

Latent causes result in the problem but are not present, so a problem cannot be solved by addressing latent causes. Too often, problems are thought to be solved by addressing latent causes. The challenge in some problem analyses is not to determine what factors could cause the problem (i.e., latent causes), but which causes are actually present. In many cases, it is important to address the latent causes of a problem to ensure that the problem will not recur. This is the essence of prevention. In our example, a pressure modulation valve could be installed to help ensure that pressure oscillations would not cause a future machine wreck.

One test used to validate that a potential cause is a root cause is to generate the effect (problem) when the cause is present and eliminate the effect when it is absent. A with/without study is a simple test to determine if the effect can be created *with* the cause and is eliminated *without* the cause present. Artificially generating the problem is a simple but effective approach to validate a root cause. In other cases, it may be necessary to specially monitor the process to evaluate whether the presence of the cause results in the problem. In the machine wreck example, a pressure-monitoring recorder could be used to correlate pressure oscillations with machine wrecks.

Example 13.1 (continued)   Machine Downtime

Decomposition of switch plastic by the machining coolant is a root cause of the switch-related electrical downtime.

All existing switches in production showed some degree of decomposition when coolant was used. The switch showed no decomposition in other coolants.

The chemical lab verified that the switch plastic chemically reacts with the coolant.

Loose connectors are the latent cause. Connectors are not sufficiently loose to cause a problem.

Example 13.2 (continued)   Excessive Scrap

A change in metal hardness and microstructure is the root cause of the excessive scrap and increased tool usage.

13.6 Problem Analysis System

Control charts indicate a lack of metal hardness stability. The microstructure of metal indicates increased perlite.

The supplier change to a new heat-treat process corresponds with the increase in scrap.

Comparison plot (see attachment) of hardness using current and old (obtained from parts warehouse) parts indicates an increased hardness of 20 Brinell points with a 30% increase in variability.

Example 13.3 (continued) Warranty Claims

Improper usage of low-viscosity oil in hot weather conditions is root cause of premature failure of the component.

Accelerated bench testing using improper oil reproduced the failure at a mileage equivalent to 50,000 customer miles.

Analysis of 10 field failures all indicated that customers were using improper low-viscosity oil.

Example 13.4 (continued) Test Failures

Insufficient clearance on a piston seal ring restricted lubrication flow to the direct clutch and is considered the root cause of the test failure.

A comparison of failed and passed transmissions showed that all failed transmissions had clearances below the passed transmissions.

The durability test was rerun using seal rings from two failed transmissions in two passed transmissions. Excessive wear was noted.

Material hardness of the failed component is a latent cause. Increased hardness is required to reduce the risk of field failures.

Step 7. Identify Alternative Solutions. After the root causes of a problem are identified, there is a natural tendency to want to implement a solution quickly. The actual method of addressing the root cause is too often not evaluated sufficiently. Any problem can be solved in a number of ways, and it is useful practice to evaluate several approaches. An integral part of identifying alternative solutions is determining how similar problems were solved by other groups. Again, the NIH outlook causes individuals to "reinvent the wheel." A thorough analysis of different approaches to eliminating a root cause is a critical part of the problem-solving process.

One approach to generate various alternative solutions is to use the brainstorming process. The same format used to prepare the C & E diagram can be used to generate alternative solutions. Also, the C & E diagram can be used with "the effect" now stated as "to eliminate the root cause x." The type of C & E diagram will vary with the root cause being addressed. The resulting alternative solution C & E diagram is a good method to summarize the different approaches to solving a problem and provide a useful future reference should the problem recur.

One alternative that should be considered is redesign of the part or process with the goal of eliminating the opportunity for the problem to occur. Too frequently, a tendency exists to seek Band-Aid fixes to a problem. The hope is for the permanent solution to be one more "fix" away. In some cases, the root cause of the problem is not being addressed. However, in other cases the part or process needs to be redesigned. No amount of fine-tuning will eliminate the problem. Defect prevention requires permanent foolproof solutions.

Once a solution is selected, there is a tendency to "wait and see if it works." To pursue work on several possible solutions at the same time is often not considered. However, it is frequently possible to significantly reduce the time required to permanently correct a problem if a practice is made of simultaneously pursuing several alternative solutions. When one alternative fails, it is not necessary to start a new approach from the beginning. Anticipating failure and developing contingency actions is not a natural human activity but is important for timely problem resolution. Timing of major actions should be stated, and action plans on details of the implementation should be kept.

Example 13.1 (continued)   Machine Downtime

9/12: Change of switch plastic to metal is requested.
9/30: Change of coolant to a new type is to be evaluated as a contingency action.
10/15: A design for shielding plastic from the coolant is to be evaluated.
9/10: Connectors used for other applications are suitable for this machine and will be installed.

Example 13.2 (continued)   Excessive Scrap

8/15: The supplier requested that the old heat-treat process be utilized, but insufficient capacity exists. New process parameters are to be evaluated.
8/17: A new coated tooling is to be evaluated.
8/30: New suppliers are being evaluated as a contingency action.

Example 13.3 (continued)   Warranty Claims

9/30: Material hardness is to be increased.
10/15: Oil distribution holes are to be enlarged to enable a robust cooling design.
12/10: Owners are to be notified not to use improper oil.
3/30: An increase of 10% in cooling capacity is to be evaluated as a contingency action.

Example 13.4 (continued)   Test Failures

10/15: The clearance on the piston seal ring will be increased.
10/30: Lubrication holes on the direct clutch are to be increased in diameter by 15%.
9/15: An increase in hardness is to be evaluated as a contingency action.

Step 8. Verify and Implement Permanent Corrective Actions. The typical method of evaluating whether a problem solution is effective is to wait for implementation of the "fix" and see if the problem is eliminated. Although on the surface this seems a logical approach, it is for two reasons inadequate for effective problem solving. First, it is not unusual for a significant time period to be required for full implementation of a problem solution. If new methods, materials, machines, or gages, for example, are required, several months may be necessary for delivery. Second, the customer feedback time may be significant, particularly if product durability issues need to be addressed. Thus, effective problem solving requires verification of solutions, utilizing pilot tests to simulate final implementation. A number of approaches can be utilized:

Prototype parts implementing the final solution
Special processing approximating the proposed process for eliminating the root cause
Controlled production lots for customer evaluations
Controlled field tests, such as the use of fleets

13.6 Problem Analysis System

Each problem will have different types of pilot tests. Rarely is it not possible to conduct a pilot test. A sufficient number of parts should be evaluated to thoroughly test the solution.

Verification cannot await implementation

An automatic bolt feeder continually jammed, causing assembly line downtime. An engineer led a team investigation into the problem. The team determined the root cause of the problem was that oil on the bolts used to prevent rust clogged the bolt feeding mechanism. At a meeting reporting to management, the team presented their solution, which was to change the bolt specification to add a special permanent rust preventive coating on the bolt. This solution was reported to be an industry standard for solving bolt feeding problems but increased the cost of the product. When asked how he could be sure the problem would be fixed, the engineer replied, "It's the accepted solution!" Asked if he could think of any way to test this proposal, before the new bolts were available, the engineer replied, "We could temporarily wash the bolts."

Whatever type of pilot test is performed, two questions should be considered:

1. Is the customer satisfied with the solution?
2. Are any new problems created?

The problem solution must evaluate whether the customer believes that the problem is resolved. This assessment cannot await full implementation of the solution. If the problem originated with customers external to the company, select pilot groups of customers must be allowed to evaluate the corrected product in an actual usage situation. It is easier to evaluate the satisfaction of internal customers. Assessing whether a problem solution creates another problem is difficult since there are an unlimited number of potential problems. Experience must be used to identify the most likely potential problems.

Many problem "fixes" implemented have minimal impact on the problem. Not only are significant implementation costs incurred, with no benefit, but customers remain dissatisfied. Inadequate verification is a major weakness of many problem-solving efforts.

Example 13.1 (continued)   Machine Downtime

11/10: Experimental switches using the new plastic are evaluated in production. No failures have been observed after 200 hours.

11/30: The new coolant is evaluated in a production machine. The switch deterioration is eliminated, but the machine operators find the coolant smell displeasing.

Ongoing: New connectors are being evaluated on other machines.

Example 13.2 (continued)   Excessive Scrap

8/20: A trial run of new heat-treat process parameters resulted in 1% scrap and tool life equal to prior usage. Process capability is established ($C_{pk} = 1.5$).

9/20: A coated tooling provides a 25% increase in tool life with 10% scrap (reduced from 25%). New tools used starting 8/30.

9/15: A trial run of alternative supplier parts is to be evaluated.

Example 13.3 (continued)   Warranty Claims

11/15: A special run of 100 hardened parts is to be evaluated in Texas using improper, low-viscosity oil.

12/15: A special run of 100 hardened parts with enlarged oil holes is to be evaluated in Texas using improper, low-viscosity oil.

2/28: Dealers and owners will be notified about improper oil use.

Example 13.4 (continued)   Test Failures

11/1: Finite element analysis verified that increased seal ring clearances will result in 10% reduced friction.

11/15: Stress curve analysis indicated no benefit from increased component hardness.

12/15: New prototypes will be available for bench testing.

Step 9. Implement Ongoing Controls. A key part of the problem analysis process is ensuring that problems will not recur. Unfortunately, the most common and least desirable ongoing control is inspection. As previously stated, inspection is not 100% effective. The frequency of a problem may be reduced, but it will not be eliminated. The use of inspection-based controls can take many forms. Consider the typical corrective action taken for an assembly error made by an operator: "operator instructed on proper procedure and reminded to check his work." A preferable approach is to eliminate the opportunity for an error. Defect prevention thinking often suggests that the part or process be redesigned to foolproof the system. The best ongoing "control" is to eliminate the opportunity for a problem to occur.

> A small semifinished valve was produced on a five-spindle screw machine. When the end of a steel bar feeding a machine was reached, a short valve resulted. Operators were instructed to choose the short valves from a large tub of valves when they reached the end of the steel bars. The tub was then 100% inspected. The customer continually received short valves, which caused a variety of process problems. After one bad incident, the customer's representatives visited the supplier. The supplier's owner claimed he knew the problem was now solved. He had reprimanded the responsible operator and withheld some of his pay. He was also now going to provide 200% inspection as an extra service—at no charge! The customer indicated that none of the actions were acceptable and that a foolproof solution was required. The owner was surprised that his actions were not acceptable. After some investigation, he found that the screw machine could be fitted with special limit switches to stop the machine when a short bar feed occurred.

The objective of ongoing controls is to ensure that a problem will not recur. When redesign is not possible, it is often justified to use control charts or other methods, such as check sheets, preventive maintenance, and training. The operation control plan (Chap. 14) should reflect the new control methods. The control objective must address the 5M sources of variation to ensure a stable process. Finally, the ongoing control system should be tested by attempting to artificially induce the problem.

Try to make the problem happen

Example 13.1 (continued)   Machine Downtime

12/13: The production version of the experimental switches is installed. The preventive maintenance check list is updated to test switches after 500 hours.

12/5: A new connector design is installed that minimizes the risk of a loose connector. No ongoing control is needed.

13.6 Problem Analysis System

Example 13.2 (continued)   Excessive Scrap

8/20: The supplier is to utilize new heat-treat process parameters on an ongoing basis, verified by a checklist system.

8/20: The supplier is to utilize X bar and R charts for material hardness of part. The microstructure of every lot is to be evaluated. The results will be submitted with every shipment.

8/25: A tool life check sheet is to be utilized to monitor Op 20 machining trends.

Example 13.3 (continued)   Warranty Claims

3/20: No field failures are obtained using increased hardness with larger oil holes. A more robust design minimizes the opportunity for a problem. Design changes are submitted immediately.

3/20: Hardness X bar and R chart controls are to be utilized by the supplier and included in all shipments.

Example 13.4 (continued)   Test Failures

The test results indicate that seal ring clearance eliminates the problem.

Manufacturing process controls will include control chart monitoring of seal ring outer diameter and hub inner diameter to ensure consistent clearance.

Step 10. Change the System That Did Not Prevent the Problem.   A problem would not have occurred if the management or operating system prevented it. The existence and solution to an individual problem is important, but the real benefit to the organization is preventing similar problems from occurring. Only then will the problem analysis system become an integral part of the DPS. Generalizing from the existence and solution of an individual problem to the management or operating system is essential. Managers must understand why their systems allowed a problem to develop—the same system will allow future problems to occur.

The first step in generalizing is to prepare a process flow diagram of the process that should have prevented the problem. Frequently, there is no clear responsibility for addressing a potential problem area or there is no follow-up system. Once the process is clearly understood, the solution is often obvious. Corrections to the system should address responsibilities for follow-up activities. Too often, the "system" relies on individual initiative rather than good management practices. Finally, an attempt should be made to standardize methods so that a unified, well-organized system exists.

Example 13.1 (continued)   Machine Downtime

2/5: The machine purchasing review procedure will be updated to include chemical testing for compatability of plastic and rubber components with machining coolants. A standard sign-off procedure will be used.

1/30: All machines in the plant will be analyzed for compatability.

3/20: All old-style connectors will be replaced with the new connector design.

1/5: Connector designs will be standardized.

Example 13.2 (continued)   Excessive Scrap

9/20: The system to establish prior approval of supplier process changes will be updated to include trial runs of heat-treat changes.

Procedure 13.2 Problem Analysis System Steps

1. Form a team
    Designate team members and leader; keep team size small (say, 5–10 members).
    Ensure the team has the required process knowledge and work experience.
    Evaluate the need for special skills, and designate advisors. Solicit feedback from team members.
    Identify the expected time commitment of team members.
    Designate frequency and type of reporting concerning team's progress.
    Continually re-evaluate the need for special skills and designated advisors.
2. Define the problem
    Phase 1. State symptom, extent, and consequence of the problem. (Go to step 3 if necessary.)
    Prepare a process flow diagram.
    Start an action plan to define the problem. Identify who will do what by when.
    Phase 2. Identify who, what, where, when, why, how, and how much (5W2H). Quantify the problem extent for relevant stratification factors. Perform a comparison analysis of similar situations in which the problem may be expected to occur. Use all available indicators.
    Contact the customer to ensure complaints are clearly defined.
    Subdivide the problem into natural problem groups.
3. Implement interim containment actions
    State all problem symptom containment actions and time when implemented.
    Perform tests to evaluate effectiveness. State results.
    State procedures for ongoing evaluation of effectiveness (e.g., control charts and check sheets).
    Action plan should be used to coordinate implementation of interim actions.
4. Identify potential causes
    Define "the effects" for a C & E diagram clearly using data from phase 2 of step 2.
    Prepare a 5M process flow, or stratification C & E diagram (or any combination) for each effect.
    Team members should each assume their activity causes the problem and ask themselves, "How could what I do possibly generate the problem?"
    Prepare a time line analysis if the problem was not always present. Identify what changed when.
    Perform comparison analyses to determine if the same or a similar problem existed in related products or processes.
    Identify past solutions and root causes that may be appropriate for the current problem.
    Identify the top few potential causes, develop a plan for investigating each cause, and update the action plan.
    Evaluate a potential cause against the total problem description. Does a plausible causal mechanism exist so that the potential cause could result in the problem?
5. Analyze potential causes
    Use the iterative process to analyze each potential cause:
        Hypothesis generation: how does potential cause result in the problem?
        Design: what type of data can most easily prove or disprove the hypothesis?
        Preparation: obtain materials, and prepare a check sheet.
        Data collection: collect the data.
        Analysis: use simple, often graphic, methods to evaluate data.
        Interpretation: is the hypothesis correct?
    Investigate several potential causes independently using parallel actions.
    Use the action plan to manage each state of the analysis process for each potential cause being studied.

13.6 Problem Analysis System

6. Validate root causes
   Clearly state root causes, and specify data used to determine the root cause.
   Verify whether the root cause factors are present in the product or process.
   Conduct a with or without study to verify the root cause. Can we generate the problem?
   State latent causes and null causes.
7. Identify alternative solutions
   Evaluate how other groups solved similar problems.
   Use a brainstorming process to generate an alternative solution C & E diagram.
   Consider redesign of the part and process to eliminate the possibility of the problem (foolproofing).
   Anticipate the failure of the primary solution. Develop contingency actions.
8. Verify and implement permanent corrective actions
   Artificially fabricate a solution to allow actual process or field evaluation.
   Field test using pilot customer groups.
   Carefully verify that another problem is not generated by the solution.
   Stress customer and user evaluation after implementation.
   If possible, test and verify contingency actions.
9. Implement ongoing controls
   Ensure the problem will not recur.
   Seek to eliminate inspection-based controls.
   Address the 5M sources of variation.
   Test the control system by simulating the problem.
10. Change the system that did not prevent the problem
    Prepare a process flow diagram of the management or operating system that should have prevented the problem and all similar problems.
    Make needed changes to the system, and address system follow-up responsibilities.
    Standardize practices.
    Use an action plan to coordinate required actions.

10/1: All heat-treated parts will require control charting of hardness and microstructural analysis of all lots.

9/25: Standardized tool life check sheets are to be developed for all leadoff operations.

Example 13.3 (continued)   Warranty Claims

3/30: The durability test procedure will be updated to include tests assuming improper customer maintenance (e.g., low oil and low water).

Example 13.4 (continued)   Test Failures

No changes needed since the test procedure prevented the field problem, which is the intent of the system.

> Every problem is an opportunity to change the system

Overview. A summary of the 10 problem-solving steps appears in Procedure 13.2. Within each step several actions are typically performed. However, every problem will not necessarily perform each action. The uniqueness of each problem will dictate the appropriateness of the actions. The important element is that the 10 steps provide a generic system to address a variety of industrial situations. If utilized on an ongoing basis, this approach will generate the continuous improvement necessary to move toward a DPS.

13.6.3 Problem Analysis Tools

The first part of this book presented the simple statistical tools needed to examine the stability of a process. These methods are generally passive in nature since process parameters are not being changed, we are simply quantifying the time-related behavior of the process. The tools discussed in Part II of this book are useful not only in quantifying current performance, but in comparing the process performance when process parameters are changed. This active experimentation with process parameters is performed using design of experiments methods (e.g., Box et al., 1978, or Kackar, 1985), of which the comparison methods discussed in Chapters 11 and 12 are the simplest case.

Problems can often be identified using the passive approach. If the current process or product is acceptable but the problem is caused by an undesirable change, passive methods can indicate when the change occurred. Once the time is identified, it is often possible to determine what event brought about the change. Causes for the changes can then be eliminated. A typical example of a process change is an out-of-control process or multiple populations in a histogram.

When the process or product has never been acceptable, the problem must be addressed by changing the process or product parameters using active experimentation. Comparison of the current and modified process or product will indicate if a solution has been found. It is also good practice to use active experimentation to make the process or product more insensitive (robust) to factors that could cause a future problem: active defect prevention. The ability to actively experiment with the process or product is essential for making improvements. Frequently, problem-solving approaches are too narrow and attempt to fine-tune the existing system. Consequently, problems reappear.

The 10 tools listed in Table 1.1 provide the foundation for the problem-solving process. It is the use of these simple, effective tools with a disciplined follow-up system to address problems that increases the likelihood of a team's success. It is therefore essential that team members receive adequate training in use of the tools prior to participating in a problem-solving team. Procedure 13.3 lists tools and practices that are commonly used in each of the 10 problem-solving steps.

13.6.4 Problem Analysis Reporting

One of management's responsibilities is to ensure that the teams addressing problems are utilizing the correct problem-solving *process*. Too frequently, managers attempt to dictate a solution to every problem. The result is that employees become discouraged. "He'll do it *his* way no matter what I say!" and gradually grow to lack confidence in their own abilities. Focusing on the process of solving a problem results in managers coaching employees and facilitating actions to minimize employees' frustrations with the company's bureaucracy. A concise reporting format is required for managers to effectively review a team's progress and problem-solving effectiveness.

As stated in the first part of this chapter, four elements are used for reporting progress in the 10-step problem analysis system:

1. Problem analysis system steps summary
2. Cause-and-effect diagrams
3. Action plan
4. Data summaries

Procedure 13.3   Problem Analysis Tools

1. Form a team
2. Define the problem
    Process flow diagram to define the process
    Pareto analysis to select priority problems and stratify performance measures
    Control charts to indicate special causes
    Check sheets to define 5W2H
    Action plan to coordinate problem definition actions
3. Implement interim containment actions
    Check sheets to evaluate effectiveness of actions
    Control charts and histograms with intensive sampling for process monitoring
    Action plan to coordinate interim fixes
4. Identify potential causes
    Brainstorming to develop the potential causes
    Cause-and-effect diagrams to identify and organize potential causes
    Failure mode effect analysis (FMEA) to identify potential causes from observed failure mode
5. Analyze potential causes
    Check sheet to collect data
    Comparison plots, histograms, and stratified graphs to evaluate stratification factors or different process or product parameters
    Scatter plots to evaluate relationships between characteristics
    Gage studies to evaluate the measurement system
    Action plan to manage analysis steps
6. Validate root causes
    Comparison plots, histograms, and stratified graphs to validate cause (e.g., with and without comparison)
    Stratified graphs to validate presence of root cause factors
    Action plan to manage validation actions
7. Identify alternative solutions
    Brainstorming to solicit ideas
    Alternative solution C & E diagram to address potential areas for solutions
8. Verify and implement permanent corrective actions
    Control charts and histograms to evaluate process stability, and capability
    Check sheets to collect process or product evaluation information
9. Implement ongoing controls
    Control charts and check sheets to monitor process performance
    Comparison plots to periodically ensure stratification factors are not influencing process output
    Process control plan (Chap. 14)
10. Change the system that did not prevent the problem
    Process flow diagram to define the management system that did not prevent the problem
    Action plan to coordinate needed changes

These four elements can be used to present a team's progress. Even for complex problems, fewer than 5–10 pages are required to adequately summarize the status of a problem-solving effort. A quick update for a problem can be obtained simply by examining the summary of the 10 steps. It is important for all participants to appreciate the iterative nature of the problem-solving process so that action for action's sake is not encouraged. Also, understanding the use and application of simple statistical tools is required. These tools provide a basis for the data analysis and thus guide the problem-solving efforts. Each of the four reporting elements is discussed here.

Element 1. Ten Problem Analysis System Steps Summary. The results from the 10 steps can be reported in a concise manner. When appropriate, the timing of major events should be given. However, the problem-reporting format is not meant to replace an action plan that contains the detailed tasks required to accomplish the problem analysis steps. One feature of the 10 steps is they are performed in a sequential manner for a particular root cause. If a problem has two root causes, the analysis process could be on step 5 for cause 1 and step 7 for cause 2. In the Table 13.2 example, the update occurred 6/25 when the analysis process was in step 5. The sequential progression through the 10 steps provides the discipline required to address difficult problems.

A reporting format of important elements is used, but it is not meant to give detailed explanations. Statements should contain conclusions resulting from data collection and analysis. Reference should be given to any attached graphic data summaries. Updating the 10-step summary is easy because of the concise sequential nature of the report.

Element 2. Cause-and-Effect Diagram. It is good practice to include the C & E diagrams that are prepared. A sparse diagram with few causes or many main branches and few "twigs" indicates a shallow potential cause analysis. This can be an indication that team members need more training or that additional advisors, having a broader perspective, are needed. Thus, evaluation of the C & E diagram gives a quick assessment of the group's potential for solving a problem. Generally, a poor C & E diagram indicates that the team's ability and/or commitment to solving the problem should be examined.

Element 3. Action Plan. An action plan should be maintained on an ongoing basis. This document simply identifies who will do what by when. The format provides both a follow-up system and an agenda for group meetings. Also, the status and conclusion section provides brief summary information.

Element 4. Data Summaries. Since the problem analysis steps are based on data, it is possible to accumulate a large amount of information during the investigation of a problem. This information is frequently re-examined during the analysis process since the interpretation of past results may change based on new information. Also, information from the analysis process may be examined later because of its relevance to another problem. Therefore, it is essential to properly collect and summarize the information. The background material from the data collection analysis process should be retained (e.g., notebook sections). The action plan can serve as a convenient index. The guidelines in Procedure 13.4 have proven useful. Most of the results discussed in the data summary will be displayed graphically.

Procedure 13.4 Data Summary Format

In the titling information, state the problem title, date, and who prepared the summary in the same format as the summary of the 10 problem analysis steps. A good practice is to have the appropriate action plan tasks refer to a data summary page. Using a bullet point format, briefly address the following areas.

Hypothesis

State the hypothesis of how the potential causes being analyzed could result in the problem. When an exploration study is being performed, the objective of the study rather than a hypothesis should be stated.

State briefly what past studies have shown relative to this potential cause.

Study

Sketch the part, and show where measurements are being made.

Sketch part of the process flow to identify where in the process the parts are being collected. (Alternatively, refer to the process flow diagram.)

State how the parts are being collected. Identify random, consecutive or periodic sampling plan, for example. State the reason for the selected data collection plan. (A check sheet should be properly prepared using Procedure 8.1.)

Identify major stratification factors. State how many parts are collected from each major stratification factor (e.g., machines, spindles, fixtures, or stations).

Data

Identify process parameters, vendors, or lot numbers, for example, so that study conditions can be re-created if necessary.

State the time period over which the parts were collected and who measured the parts.

State the gage identification number, percentage repeatability, and bias. State any coding of the measurements.

List any computer file reference. Retain a listing of the raw data and check sheets.

Conclusions

List each conclusion and the rationale for the conclusion.
Refer to the attached graphs or tables.
Suggest the logical next steps in the analysis.

Table 13.2 Problem Analysis System Steps Summary Report

	Problem:	Excessive scrap in Op 20 Department 733
Prepared by:	J. J. Smith,	Date started: June 1, 198x
	Team leader	Date updated: June 23, 198x

1. Form a team
    Members
    V. L. Brown, QC Engineer   J. J. Smith, Supervisor
    A. Z. Jones, Mfg. Engineer   A. A. Williams, Jobsetter
    B. W. Lamb, Product Engineer   A. P. Zorn, Operator Op 20
    S. W. Moore, Buyer
    Advisors:
    R. M. Anderson, QC Metallurgist
    O. M. Boone, Supplier Rep
2. Define the problem
    The scrap rate at Op 20 increased on May 15 from 3 to 25%
    In the prior 6 months, the scrap rate varied between 0 and 5% and was stable (P chart attached)
    The two machines performing the operation have scrap rates of 23 and 27%
    Tool usage at the operation has increased 30% on both machines
    No changes in upstream process have been noted on control charts
    Vendor says no process changes have occurred, but control charts are not used
    Other similar steel parts supplied by the same vendor have experienced 5% increased scrap
3. Implement interim containment actions
    Tools changed at twice the normal frequency
    Off-standard labor used to 100% check parts
    Control charts started on hardness (see attachment)
4. Identify potential causes
    New inspector has been assigned to the department and may be using a different evaluation procedure
    Steel hardness or microstructural changes may make old tool change frequencies inadequate
    Tool vendor has changed, and tool quality has not been established
    Gage repeatability and bias have not been evaluated; gage may be generating false rejects
    Machine preventative maintenance has not been performed for 1 month; machines may not be capable
    Control charts have not been maintained, and process may not be stable, resulting in excessive scrap
5. Analyze potential causes
    6/5:   New and old inspector evaluated using same 100 parts; no major differences detected
    6/10: Histogram (see attachment) of steel hardness shows material is on the upper part of the specification limit ($C_{pk} = .8$); microstructural results will be obtained
    7/1:   Supplier to be visited to evaluate hardness controls
    6/22: Comparison plot (see attachment) of new and old tools indicated no significant differences
    6/20: Gage study showed a .002 inch bias in dimension x and a 54% gage repeatability error; rehab of gage to be completed by (7/15)
    7/5:   Preventive maintenance will be performed on two Op 20 machines
    7/1:   Control charts have been updated; follow-up system to be initiated
6. Validate root causes
    A change in the metal hardness and microstructure is the root cause of the excessive scrap and increased tool usage
    Control charts indicate a lack of metal hardness stability; microstructure of metal indicates increased perlite

13.6 Problem Analysis System

Supplier change to a new heat-treat process corresponds to the increase in the scrap rate
Comparison plot (see attachment) of hardness using current and old (obtained from parts warehouse) parts indicates an increased hardness of 20 Brinell points with a 30% increase in variability

7. Identify alternative solutions
   - 8/15: Supplier requested to utilize old heat-treat process, but insufficient capacity exists; new process parameters to be evaluated
   - 8/17: New coated tooling to be evaluated
   - 8/30: New supplier being evaluated as a contingency action
8. Verify and implement permanent corrective actions
   - 8/20: Trial run of new heat-treat process parameters resulted in 1% scrap and tool life equal to prior usage; process capability established ($C_{pk} = 1.5$)
   - 9/20: Coated tooling provides 25% increases in tool life with 10% scrap (reduced from 25%); new tools used starting 8/30
   - 9/15: Trial run of alternative supplier parts to be evaluated
9. Implement ongoing controls
   - 8/20: Supplier to utilize new heat-treat process parameters on ongoing basis; verified by checklist system
   - 8/20: Supplier to utilize X bar and R charts for material hardness of part; microstructure of every lot to be evaluated; results will be submitted with every shipment
   - 8/25: Tool life check sheet to be utilized in monitor Op 20 machining trends
10. Change the system that did not prevent the problem
    - 9/20: System to establish prior approval of supplier process changes will be updated to include trial runs of heat-treat changes
    - 10/1: All heat-treated parts will require control charting of hardness and microstructural analysis of all lots
    - 9/25: Standardized tool life check sheets to be developed for all leadoff operations

13.6.5 Roadblocks to Solving Problems

The lack of effective problem solving is prevalent in many industrial settings. The causes are varied but often include some of the following elements:

Problems not addressed: Problems are allowed to linger year after year with little attention. A "that's the way it is!" attitude prevails.

No problem system: The lack of a disciplined system to prioritize, analyze, and follow up on problems is not implemented. Action plans are not used. A "fight the problem of the day attitude" exists. Management reviews consist of examining hastily prepared problem status reports rather than assessment of the organization's problem analysis process.

Management patience: A lack of management knowledge of the problem-solving process (Fig. 13.1) makes all levels of management demand to know when a problem will be solved. Each level of management demands and is comforted by a list of *actions* to solve a problem—the longer the list, the greater the comfort. Root cause analysis and verification are not part of the management system.

No data: Problems are "solved" using experience or "engineering know-how" with little or no emphasis on the collection of facts through the analysis of data.

Band-Aid fixes: Use of the problem reaction wheel encourages only interim containment actions. Prevention is not emphasized.

One more fix away: Vested interests lead to an attitude that a permanent solution is one more fix away. The inability to admit failure and completely foolproof a process results in lingering problems.

Symptom fixes: Lack of analysis results in an inability to separate problem symptoms from causes. Potential causes are not systematically analyzed to arrive at a root cause. A long list of "corrective actions" typically results.

Wrong team: Selection of team members is based on convenience rather than needed skills. Cross-functional teams are not used. The skills necessary for the team are not periodically re-evaluated.

Time commitment: Employees are assigned to a team, but time allocation is not discussed. All other employee assignments are maintained. Problem champions are not designated.

They are the problem: Management practice, poor team selection, and opinion- (rather than data-) oriented analysis result in a defensiveness on the part of all team members. An unwillingness to consider "I'm the problem" by all members dooms the problem-solving effort.

Part III
Defect Prevention System

14
Building a Defect Prevention System

Many people have been mistakenly led to believe that the use of SPC charts is the key element needed for improved quality. Experience shows that the implementation of SPC charts results in initial quality improvement due, in part, to the increased attention to quality. This concern for higher quality is often initially associated with an increased level of inspection, which takes a variety of different forms. However, a sustained, continuous improvement will not be obtained.

Deming (1982) and Juran and Gryna (1980) discuss the essential management changes necessary to attain meaningful breakthroughs in quality performance. These management changes can produce a company culture in which quality performance is the number one management concern. A natural consequence of this attitudinal change is the need to establish defect prevention systems. It is no longer acceptable to have customers experience problems with products or services the company provides. In fact, the organization's goal is to have customers boast that the company's products surpass what competitors provide. Only then will a true competitive edge be attained.

As stated in the first chapter, the 10 simple statistical tools (SST) discussed in this text are insufficient to establish a defect prevention system (DPS). Both management participation (MP) and employee participation (EP) must be combined with the SST, as indicated in Figure 1.2. The 10 steps discussed in this chapter provide an implementation strategy to work toward a DPS. The importance of a comprehensive implementation plan cannot be overemphasized. Many SPC-based efforts to improve quality have failed owing to the absence of a well-conceived plan. A central part of the DPS is the process control plan, which selectively utilizes SPC charts, along with a number of other tools. Another central part of the DPS is the problem analysis system discussed in the previous chapter. Process control techniques alone are not sufficient for continuous improvement of production and management systems. An overview of the 10 DPS implementation steps appears in Procedure 14.1.

14.1 Establish Open Communications

Every individual in an organization must believe that he or she has an outlet to vent concerns about work. The obvious, and most natural, line of communication is the individual's immediate supervisor. It is essential that supervisors be trained in how to foster this first line of communication. An employee attitude that "No one cares about me or what

Procedure 14.1 Defect Prevention System

1. Establish open communication.
   Train supervisors in personal interaction skills.
   Inform employees of quality problems.
   Ensure that employees accept responsibility for their quality performance.
   Encourage employee teams to address quality problems in their work areas using the problem analysis system.
2. Change operating systems. Address critical prevention systems: standardized operating procedures, housekeeping, preventive maintenance, training, production flow, tool control, and gage control.
3. Initiate customer feedback systems.
   Develop a system to ensure customer problems are well defined.
   Relate in-house indicators to customer problems.
   Develop feedback systems to monitor performance with internal customers.
   Use accelerated customer field testing of major product changes.
4. Develop key quality and productivity indicators.
   Implement system to monitor over time all quality defects, including scrap, rework, setup parts, and test stand rejects, from all sources within a department.
   Implement system to identify sources of productivity variability. Key indicators include preventive maintenance ratios, machine downtime, and causes of lost production.
   Prepare weekly graphic summaries of key indicators.
5. Utilize problem-solving teams.
   Develop Pareto analysis from key quality and productivity indicators.
   Form teams to address top problems.
   Use 10-step problem analysis system to solve problems.
6. Define process relationships.
   Prepare a process flow diagram that identifies the process parameters potentially contributing to variability in process output.
   Prepare process relationship table
     List part characteristics by operation, and identify type code.
     Prepare a characteristic matrix noting incoming and outgoing influences.
     Assign importance levels considering safety, function, and customer criteria.
     Assign importance to process control factors.
     Determine control level from the preceding analyses.
7. Develop and implement process control plans.
   Prepare an operation control plan for each process operation.
     List part characteristics and critical process parameters with the control levels, specifications, and gaging information.
     Select a control method considering both the control level required and the process capability.
     Select a sampling plan that defines when, how many, and the method in which measurements are to be made.
     Report the last three evaluations of process capability, and note any unstable processes.
     Report the last three evaluations of gage repeatability.
   Utilize a follow-up system to ensure the preceding plans are implemented.
     Develop action instructions for all control methods.
     Develop an alarm system for the department.
     Assign responsibilities, and monitor implementation using a timing chart.
8. Develop an incoming material defect prevention system.
   Seek to establish single-source, long-term relationships.

14.1   Establish Open Communications

> Work with suppliers to develop their defect prevention systems and problem-solving skills.
> Continually reduce inventories; move toward just-in-time material control.
> 9. Emphasize management evaluation of systems.
> Require management involvement in the improvement and problem-solving process.
> Focus management on evaluating the systems by which work is accomplished and problems prevented.
> Change systems that did not prevent problems.
> 10. Develop a continuous improvement mindset.
> Focus management on establishing prevention-oriented systems.
> Eliminate short-term performance targets.

I think'' is largely the result of a poor relationship with the supervisor. Regardless of other organizational programs and financial benefits, the employee's attitude toward the entire organization will be colored by this poor relationship. Conversely, a good employee-supervisor relationship can overcome many organizational deficiencies. Thus, training supervisors in personal interaction skills is an essential ingredient for open communication.

Part of the communication process involves sharing information. How can anyone improve if actual performance numbers are not shared? Quality performance facts should include not only in-house indicators but customer-reported problems. Weekly meetings in which supervisors review performance numbers can be an effective tool for sharing information and encouraging suggestions. Posting performance indicators within a department can also be effective, provided follow-up meetings are held.

Once quality information is shared, a logical question is, "Do employees accept responsibility for their quality performance?" If not, the "system" may be such that the employee cannot control his or her quality performance. It is management's responsibility to take whatever actions are required to ensure that individual responsibility is accepted. Defects cannot be prevented unless every individual can fairly accept responsibility for the quality of the output he or she produces. A visible symbol of changing the system to encourage individual responsibility is the use of the stop buttons for an assembly line so that operators can stop the line when problems are encountered.

Employee teams have been used with varying success in involving individual employees in improving the quality performance within their work area. If employees are involved in making improvements, the communication process has a good foundation. Establishing departmental teams, also called quality circles, can be an effective way to encourage open communication and channel any resulting suggestions. As has been previously discussed, it is good practice to encourage teams to make improvements by focusing on data analyzed with the 10 simple statistical tools rather than opinions. This practice reduces the risk of confrontation due to differences in opinion. It is unfortunate employee teams have not been as successful in the United States as many would have hoped. One reason is that teams can only prosper when a good employee-supervisor relationship exists.

There are many indicators for the state of open communication. The following questions focus on some of the indicators relating to quality improvement:

Are employees aware of how their quality performance impacts on customers? Are they kept informed about their quality performance?

Do employees offer many suggestions for improving quality? Is the response time for suggestions reasonable?

Do employees feel free to report quality-related problems? Is there evidence of not reporting measurements beyond the specification limits?

Do employees maintain and use control chart data? Are out-of-control signals addressed? Do supervisors monitor control charts daily?

Are employee teams used? Are important problems being addressed?

14.2 Change Operating Systems

A system that prevents quality defects does not exist in a vacuum—the entire production environment must be addressed. Significant improvements in quality performance will not be gained by a piecemeal approach. Most of the management operating systems that need to be addressed are well known. Management has no excuse for not implementing needed changes.

Standardized operating procedures. Standardization has three elements. First, procedures must be developed. Second, procedures must be communicated to affected employees. Consider the procedure to set up a machine. A checklist should be developed on all the elements required to perform a proper setup. This procedure should then be displayed near the machine so that this procedure is communicated. Employees should be trained to ensure that the checklist items are understood. Individuals who have not received the training are not allowed to set up the machine. This standardization process is appropriate for most daily operating procedures for both manufacturing and administrative processes. Finally, follow-up on the use of procedures must be enforced. Many problems are caused by not adhering to existing procedures.

Housekeeping. The condition of the workplace reflects the discipline present in the manufacturing operation. High-quality products are not produced in an unkept work environment. Management and employees collectively share the responsibility of maintaining good housekeeping practices.

Preventive maintenance. Nothing is more obvious in a manufacturing organization than the need for PM. Productivity is of course related to a machine's ability to operate. However, it is equally obvious that a machine's maintenance condition influence the quality of its output. Defect prevention requires effective PM systems.

Training. The need for training is evident; unfortunately, much of the current training is not targeted to meet immediate job needs. It is much easier to develop generalized training that, although useful in some cases, does not address current organizational needs. Rarely is there effective follow-up to training for evaluating whether individual performance has improved. Few organizations maintain a personal employee development plan that includes training received, as well as training needs.

Production flow. The flow of parts in processes should be arranged to reduce the sources of variability. The typical crossover flow

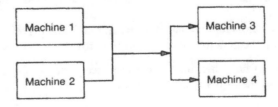

should be replaced by parallel flows:

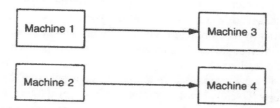

Targeting production flows, as discussed in Chapter 2, reduces the sources of variation. This change can often be implemented by simple color coding of parts.

Tool control. Control of the quality of tools used to machine parts is an important part of the prevention system. Tool quality directly impacts on part quality, yet tools are routinely controlled using ad hoc inspection-based methods. Internal and external tool suppliers must be required to implement a DPS.

Gage control. Monitoring the condition of production gages is critical to maintaining the measurement system. Periodic gage repeatability studies (App. I) quickly point to gaging problems prior to the occurrence of production problems.

In these cases, prevention is the cornerstone of all activities. Anticipating problems is the essence of prevention.

14.3 Initiate Customer Feedback Systems

Many quality improvement efforts focus on improving in-house quality indicators that cause scrap and rework with the associated increased cost. However, the relationship between in-house indicators and customer-experienced problems may not be high. Too often, quality performance is assessed by indicators that are easy to measure rather than focused on the customer. In most cases a high internal rate of quality problems is associated with customer problems. However, the converse is not necessarily true. Good internal quality performance does not assure satisfied customers. Customer problems must be clearly understood and defined, which requires a feedback system.

Proper definition of customer problems is essential to ensure that the correct problems are being addressed. In many cases a company's warranty system can be used to identify problem areas. However, direct discussion with customers is important since customer perceptions are not adequately defined by most warranty systems. In many cases, special audits, surveys, or customer test groups are required to obtain detailed, timely problem definition. Also, durability results beyond the warranty period need to be examined using special field test groups.

The results of customer quality concerns need to be related to in-house indicators. Additional in-house indicators may be needed to monitor potential customer problems. In the ideal case, internal indicators directly correlate with customer concerns so that problems can be prevented. It must be realized that customer feedback systems are essentially inspection-based methods to identify problem areas. The presence of a customer problem means that a company's prevention systems have failed. The problem analysis system (Chap. 13) should be used to address the problems to ensure that root causes are defined and prevention practices initiated. Verification of the solution to a problem requires monitoring appropriate in-house indicators and revisiting selected customers.

The term "customer" does not imply only the end user of a product. There are many intermediate internal "customers" in the production of any product. Feedback systems should be in place for internal customers as well as external customers. For example, a machining department sends a component to a subassembly operation. If the subassembly operations have difficulty assembling the component into the subassembly, the component may be defective, which could result in an end-user customer problem. Internal customers can provide valuable feedback information.

When significant changes are made in a product, an accelerated customer test is often warranted. These tests typically involve commercial users evaluating the changed product. For example, various automotive fleets provide important accelerated testing of major changes. This practice helps to prevent a widespread problem from developing from any product change.

14.4 Develop Key Quality and Productivity Indicators

In most accounting systems, scrap costs are a budget item that receives management attention because of its impact on overall cost performance. This system is inadequate for identifying and prioritizing quality improvement efforts. Some of the necessary features of a good quality indicator system follow:

Scrap. All out-of-specification conditions should be identified. Attribute defects, such as nicks and scratches, must also be defined.
Rework or reject. All out-of-specification conditions should be identified. To direct improvement efforts, scrap and rework are sometimes combined into a "reject" category since they are equally important for measuring quality performance. (Ideally, no parts should be reworked.) All parts are classified as either acceptable or rejected. The parts input to a department minus those sold after the first evaluation (without repair) are rejects. Thus, in-line rejects, setup scrap, and repairs, for example, are all included as rejects.
Sources. The operation generating each reject category must be identified. Outside suppliers as well as all in-house operations must be considered.
Time. Historical trends on a week-to-week basis should be summarized graphically. Recording rejects by shift is often useful for a more detailed analysis.
Capability. The process capability indices C_p and C_{pk} should be reported for variable characteristics when the process is stable. In most cases the process is not stable when excessive rejects are encountered.
Computerization. It is convenient to automate the summarization of the reject data to reduce the response time required for graphic summaries. The system should be designed to include the preceding features.

One of the problems with most systems is that a part is rejected for only one out-of-specification condition when several problems may exist. This practice does not present a problem if there is a common cause of the reject. In this case, one indicator of a problem is as good as another. Multiple out-of-specification conditions may result from several causes, however. Correcting the cause of one condition does not reduce the reject rate if another reject condition remains.

The indicator system should not only address quality problems. Most managers will acknowledge that there is a direct relationship between improved quality attained by process control and improved productivity:

Improved quality through process control → improved productivity

14.4 Develop Key Indicators

However, improving productivity through reducing productivity variability can lead to improved quality. Consider several examples:

- An operator had an ongoing problem in setting tools on a machine. Tool change times were excessive. A redesign of the part and a new type of tool eliminated excessive tool change time—and improved the operator's attitude.
- A central coolant system was continually being dumped since the coolant lost its tool lubrication characteristics owing to contamination. The lost production time due to tool breakage was eliminated by reducing the sources of coolant contamination and monitoring the coolant's chemistry on a daily basis.
- A machine exhibited chronic downtime due to the unstable alignment of the machining spindle that controlled the location of a hole in a part. The downtime was eliminated by redesigning the machine's aligment system.
- A machining department exhibited erratic productivity performance due to a variety of different causes of machine downtime. Operators were frustrated because of the crisis atmosphere that continually existed. Many problems were eliminated by implementing an intensive preventive maintenance program. Operator morale improved.

In many cases productivity performance directly impacts on quality. Eliminating the causes of productivity variability will improve quality. It is no accident that high quality is associated with minimal variability in productivity. It could be argued in each of the previous examples that an effective process control system would have eventually indicated a quality problem. Why use process control methods to indirectly discover the obvious? It is more efficient to use productivity indicators to directly address the causes of productivity variability:

Improved productivity through reducing productivity variability → improved quality

The fear, of course, in addressing productivity concerns is that earlier manufacturing practices will reappear when production quantities were stressed over quality. Focusing on sources of productivity variability is directed at improving the process rather than monitoring only the end result.

An effective DPS implements a system of key productivity indicators. Three areas should be addressed:

1. Preventive maintenance. A PM system should be developed so that every machine has a checklist of maintenance items that should be periodically performed. This system should be continually updated to address any causes of machine downtime. A management follow-up system should monitor the effectiveness of PM. The ratio of crisis to preventive maintenance is a good indicator.
2. Machine downtime. A system that identifies the frequency and duration of individual machine downtime is needed. Causes for the downtime as well as the corrective actions should be identified. It is useful to partition the total downtime into logical categories, such as awaiting repair, awaiting spare parts, and repair time.
3. Causes of productivity losses. The causes of lost productivity should be defined. The intent is to identify major problem areas, not to quantify minor problems. This system should use production personnel rather than standard industrial engineering logging. Balancing machine losses with machine optimum cycle times (Chap. 10) is preferable to using traditional "allowable losses." Any productivity loss can be addressed.

Each of these categories should be monitored over time using graphic displays. Pareto analysis can be used to select problems to be addressed by the problem analysis system.

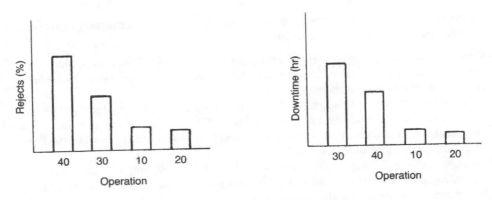

Figure 14.1 Relating quality (a) and productivity (b) losses.

14.5 Utilize Problem-Solving Teams

The solution of important existing problems must be an ongoing activity in a DPS. The quality and productivity problems identified in step 4 form the basis for a Pareto analysis, as discussed in Chapters 9 and 10. Clearly, it is not possible to address all problem areas at the same time. The Pareto principle says that 80% of the benefit will be gained by solving the top 20% of the problems. Because quality and productivity indicators are collected by manufacturing operations within a department, it is possible to prepare a Pareto analysis by operation. It is not unusual to have a similar ranking of problem operations when considering either a quality or productivity indicator, as shown in Figure 14.1. In some cases correcting the cause for machine downtime will eliminate a quality problem. Correspondingly, eliminating a quality problem will improve productivity—often more than would be expected from merely eliminating scrap.

Once the Pareto analysis is available, it is possible to form teams to address the top problems. Management must clearly allocate employee time so that participation on a team is not "one more task" added to a long list of tasks. As discussed in Chapter 1, managers

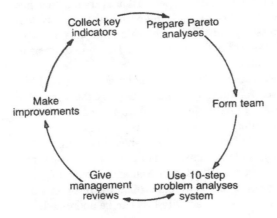

Figure 14.2 Improvement cycle.

14.6 Define Process Relationships

should selectively participate on the teams. The purpose for participating is not only to help solve a problem but to identify organizational systems that need to be changed.

The problem-solving teams should utilize the problem analysis system (PAS) summarized in Procedure 13.2. This system provides a step-by-step approach to address any problem. A key part of the system is maintaining an effective action plan that serves as a basis for management reviews.

The improvement cycle is a continuous process, not a one-time event, as indicated in Figure 14.2. The intent of the cycle is to develop a system that makes continuous improvement a natural consequence of team problem-solving efforts. Only then will the organization evolve toward a DPS.

14.6 Define Process Relationships

A critical part of developing a DPS is implementing a process control system. The standard approach utilizes SPC charts, such as the $\bar{X}$ and R chart. However, the complexity of most manufacturing processes combined with the availability of personnel makes it impractical to use control charts for a large number of the potentially important characteristics. Typically, a few critical characteristics are selected for SPC charts and other characteristics are monitored by inspection. In most manufacturing operations, no detailed control plan addresses the use of SPC charts as well as other process control tools. A procedure that can be used to develop a comprehensive control plan is given in steps 6 and 7. These two steps focus on completing the process relationship table and the operation control plan table, respectively.

Understanding the relationship between process characteristics is a necessary prerequisite to developing a process control plan. The start of understanding these relationships is to prepare a process flow diagram that identifies the major sources of variation (e.g., Fig. 2.9). Any process parameters contributing to variability should be identified. One method of identifying these parameters is to have a group prepare a process flow C & E diagram with the effect the variability of process output. This effort should focus on process parameters (e.g., coolant, tools, machine speeds, and temperature), not individual part characteristics. Also, the effect being addressed is the variability of process output, not the variability of process parameters. The variability of a process parameter may be irrelevant to the variability of a process output.

The remainder of step 6 involves completion of the process relationship table, which identifies the relationship between all part characteristics. The analysis process required to complete the table is illustrated here using a simple example in which a rough gear blank is processed into a semifinished part. The finished part is shown in Figure 12.46. A rough-cut hole is machined by the supplier along with a groove that distinguishes the long end from the short end. The process flow diagram appears in Figure 14.3. The completed process relationship table appears in Table 14.1. The table is prepared by addressing the following areas.

14.6.1 Descriptive Information

The information contained in Table 14.1 has been found to be appropriate for machining processes. Different situations may need to modify the contents to adapt it to the process being studied. The titling information contains simple descriptive information to define the process. The first two columns of the table are used to describe the characteristics. First, the

594   14 Building a Defect Prevention System

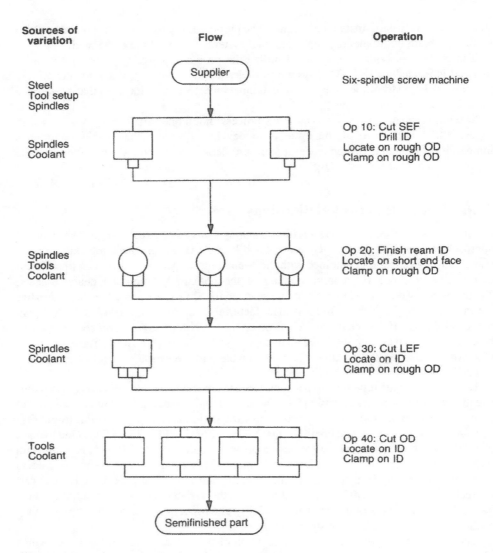

Figure 14.3 Process flow diagram for gear blanks.

14.6 Define Process Relationships

Table 14.1 Department Process Relationship Table

Operation and characteristics	Type	Controlling characteristics		Importance[a]			Control factors[a]					Control level
		Incoming	Outgoing	Safety	FCN	Customer	Men	Method	Machine	Tools	Material	
Incoming material												
IM-1 OD	IP	—	7		0		1	2	2	2	0	3
IM-2 ID	IBP	—	1		0		1	2	2	2	0	0
ID-3 hardness	IBP	—	2		2		1	0	0	0	3	1
Op 10												
10-1 ID size	IBP	3	1		0		1	3	2	2	3	1
10-2 squareness SEF	IBP	1	2		0		1	2	1	3	2	1
Op 20												
20-1 ID size	FBP	3	5		2		1	1	3	2	3	2
20-2 squareness SEF	FBP	2	1		1		1	1	2	2	2	1
Op 30												
30-1 length	FBP	3	1		2		3	2	1	3	1	3
30-2 squareness LEF	FBP	2	1		1		2	2	1	2	1	1
30-3 parallelism SEF to LEF	FIP	5	0		2		2	1	1	2	1	2
Op 40												
40-1 OD size	FIP	1	0		3		1	1	2	3	1	2
40-2 concentricity ID to OD	FIP	1	0		3		1	1	2	2	1	3

[a] Classification: 0 or blank, unimportant; 1, minor importance; 2, moderate importance; 3, critical.

Part name: Gear blank
Part no.: 1234

Date updated: 3/17/86
Name: Jones

characteristics generated at each operation are listed and numbered in order within the operation (e.g., 10-1, 10-2, 20-1, and 20-2). The critical incoming material dimensions are also listed.

For most applications, it is convenient to classify the characteristic. In machining applications, some useful classifications are as follows:

Code	Description
IP	In-process characteristic that will be modified later and is not a blueprint requirement
FIP	Final in-process characteristic that is not a blueprint requirement
IBP	In-process blueprint characteristic that will be modified later
FBP	Final blueprint characteristic
IS	In-process product safety-related characteristic
FS	Final product safety-related characteristic

The classification is for illustration only; the categories depend on the application.

14.6.2 Controlling Characteristics

The relationships between part characteristics can be defined in two ways. First, it is possible to conduct a study in which a scatter plot (Procedure 12.1) could be prepared. This study would use either the sign test (Procedure 12.2) or correlation coefficient (Procedure 12.3) to evaluate the relationship between the two characteristics. The second approach is more practical since it is not feasible to prepare scatter plots for the hundreds of characteristics that may be present. It is usually sufficient to use engineering analysis to identify expected relationships. The gear blank example is typical. The process flow diagram in Figure 14.3 identifies part-locating and clamping surfaces for each operation. Since these surfaces position a part in a machining station, it is obvious that the quality of the output relates to the variability of the two surfaces, particularly the locating surfaces. Another type of relationship involves upstream roughing operations. Generally, it is reasonable to assume that the quality of a finished characteristic is related in some way to the quality of any upstream processing of the semifinished characteristic.

A method of displaying the relationships in a process is to prepare a characteristic tree, as shown in Figure 14.4. This tree identifies the relationship between characteristics using arrows. The more arrows coming into a characteristic, the more the outgoing quality depends on other upstream characteristics. Conversely, the more arrows going out from a characteristic, the more the characteristic controls the quality of other downstream characteristics. When a characteristic has a large number of incoming or outgoing arrows, it is a good candidate for some type of process control.

The characteristic tree can become unwieldy when a large number of characteristics in a complex process are considered. It is generally easier to prepare a characteristic matrix, as shown in Table 14.2. This format enables a tabular display for any process. Listed down

14.6 Define Process Relationships

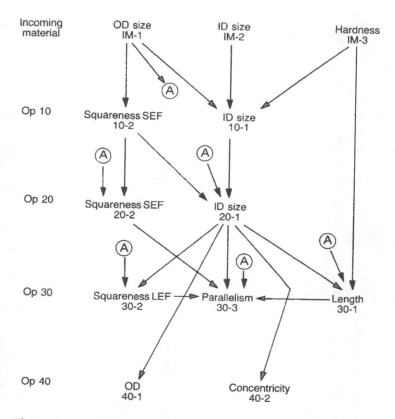

Figure 14.4 Characteristic tree for gear blank example. Note that all OD size (IM-1) clamping relationships are not indicated.

the rows are the different process operations in the same format as the process relationship table. The columns contain a corresponding list. The table is completed by going across each row and noting which column characteristics influence the row characteristic. Thus, it can be stated that

Row characteristic _____ is influenced by _____ column characteristic(s).

It is often convenient to use different symbols to denote the type of influencing relationship (e.g., C = clamp and L = locator).

The total of the entries in any row (excluding the I = identity) corresponds to the number of incoming arrows in the characteristic tree. The larger this number, the more upstream characteristics influence a characteristic, resulting in a situation that is difficult to control. After completing all rows, it is possible to compute the number of entries in a column that correspond to the number of outgoing branches. The larger this number, the more downstream characteristics are influenced by a characteristic. These values can then be placed in the process relationship table in the incoming (row totals) and outgoing (column totals) columns of the "controlling characteristics" section of the table. These numbers are used later to assign a priority ranking.

Table 14.2 Characteristic Matrix for Gear Blank Example[a]

Influenced characteristic[b]	IM-1 OD size	IM-2 ID size	IM-3 hardness	10-1 ID size	10-2 squareness SEF	20-1 ID size	20-2 squareness SEF	30-1 length	30-2 squareness LEF	30-3 parallelism	40-1 OD size	40-2 concentricity	Total[c]
10-1 ID size	C,L	X	X										3
10-2 squareness SEF	C,L			I									1
20-1 ID size	C			X	L								3
20-2 squareness SEF	C				L	I							2
30-1 length	C		X			L	I						3
30-2 squareness LEF	C					L	X	I					2
30-3 parallelism	C					L	X	I	X				5
40-1 OD size						C,L				I			1
40-2 concentricity						C,L			I				1
Total[d]	7	1	2	1	2	5	1	1	1	1	0	0	

[a] Matrix coding: C, clamping; L, locating; X, related characteristics; I, identity.
[b] Row characteristic ___ is influenced by ___ column characteristic(s).
[c] Equivalent to number of incoming arrows from characteristic tree.
[d] Equivalent to number of outgoing arrows from characteristic tree.

14.6 Define Process Relationships

14.6.3 Importance

There are several reasons a characteristic may be important; three criteria are considered. First, characteristics influencing the safety with which a product can be used are obviously critical. Any lack of product safety is unacceptable. Second, characteristics influencing product function, including durability, must be identified. Finally, there are characteristics that influence the way a customer feels about a product that influence neither the product's safety nor its function. For example, engine noise or the shift feel of a transmission are customer perceptions related to engine and transmission characteristics.

Each of the three "importance" criteria generates a separate evaluation of a characteristic (additional criteria may also need to be considered in other applications). Classifying the characteristics into categories, such as

0 = unimportant
1 = minor importance
2 = moderate importance
3 = critical

is useful. In some cases, such as the gear blank example, the safety and customer categories may not need to be considered.

14.6.4 Control Factors

In Chapter 2 the 5M process elements were defined as follows:

Men
Methods
Machines
Materials
Measurements
Environment

In the definition of the process relationships, it is useful to identify which of these elements are the controlling factors in each process operation. This classification influences the characteristics we wish to control, the sampling plan, and the method used to control the process. Juran and Gryna (1980) developed the control factor concept by identifying which factors are dominant within a process: setup, machine, operator, or component. In the process relationship table, the selection of subdivisions of control factors can be tailored to the application. In many machining applications, it is useful to use a separate "tools" category.

Some examples of activities that have a high degree of dominance in the major elements areas are listed in Table 14.3. Each of the major elements has varying importance in a process. The process relationship table lists the major elements and assigns an importance ranking from 0 to 3 as was done in the previous section.

14.6.5 Control Level

This section of the table is a summary of three types of classifications: controlling characteristics, importance, and control factors. Each characteristic is classified into the following control categories:

Table 14.3 Dominance of Major Process Elements in Selected Activities

Men	Methods	Machines	Materials	Measurements
Setup	Setup	Tool wear	Batch operations	Batch operations
Tool adjustment	Assembly	Tool compensation	Metal cutting	Monitors (e.g., torque)
Assembly	Die operations	Grinding		
Design		Drilling		Functional testing
Maintenance		Cutting		Weight filling
Soldering		Staking		Tool compensation
Filing		Honing		
Repairing				
Painting				
Welding				
Packing				

0 = no monitoring
1 = minimal monitoring
2 = moderate control
3 = intensive control

Assignment of these numbers is subjective but should be guided by the results of the three sets of classifications. It may be argued that this entire process is subjective and is therefore of minimal value. However, the important consideration is not the actual values in the table but the education process required to complete the entries. A reasonably detailed analysis of the process is required. Considering each characteristic from the three viewpoints ensures that a complete analysis has been conducted. Also, the traditional approach for selecting the control level involves implicit, sometimes arbitrary, selection criteria. The explicit system used to complete the table provides a desirable alternative.

Assignment of the control level should attempt to integrate all the row entries for a characteristic. Under process influence, the larger the number of outgoing branches, the greater the influence on downstream characteristics. Conversely, the larger the number of incoming branches, the greater is the influence of upstream characteristics. The larger the number in either case, the greater is the desired control level. Obviously, a large number of outgoing branches implies that control is required to ensure trouble-free downstream processing. Control over a characteristic with a large number of incoming branches provides a check on upstream processing. Unfortunately, an out-of-control condition may be due to any of the influencing characteristics and, thus, be difficult to troubleshoot.

The importance of a characteristic for safety, product function, or customer appeal directly influences the control level. The higher the importance level, the greater is the required control level. The 5M elements are primarily used to select a control method and sampling plan but may also influence assignment of the control level. Characteristics influenced by a large number of process elements are subject to more sources of variation and can be more difficult to control, increasing the required control level.

14.7 Develop and Implement a Plan

Two other factors govern the selection of control level. First, characteristics of an operation may be influenced by similar sources of variation. Controlling all characteristics at the same level is a waste of effort. These relationships should be indicated in the characteristic matrix. A typical example of this relationship is two characteristics machined by the same tool. Finally, the control level is typically high for final blueprint characteristics (FBP). If these characteristics are stable, it can be reasonably assumed that the upstream process is performing adequately. A problem arises when a lack of stability or process capability exists. The absence of effective upstream control will make troubleshooting very difficult.

14.7 Develop and Implement a Process Control Plan

The process relationships defined in Section 14.6 provide a basis for establishing an effective control plan. The control level summarizes the importance of all characteristics considering several criteria. However, an additional step is required to formulate an effective control plan. There are three additional elements required before a control plan can be specified:

1. Process parameters. Control of the 5M elements in addition to part characteristics is a desirable "preventive" approach to process control.
2. Capability. The control method and sampling intensity depend, in part, on the process capability for a characteristic.
3. Gaging. The ability of a process to be controlled and targeted partly depends on the effectiveness of the measurement system.

These three considerations will be integrated with the control level from the process relationship table for each process operation. The combined operation control plan tables provide a total department process control plan. An example is shown in Table 14.4 for operation 30 of the gear blank example. This tabular information provides a convenient method of reviewing both the intent and effectiveness of the control plan. The analysis and understanding of the process required for preparation of this table enables formulation of an effective control plan. Since various disciplines are involved, a team approach should be used for preparation of the table. The table is completed by addressing the following areas.

14.7.1 Descriptive Information

This part of the table starts by listing the part characteristics from the process relationship table. Any other important part characteristics not used in this table, such as appearance attributes, should also be listed. It is then necessary to determine which process parameters are important to the operation. A scatter plot analysis relating process parameters to part characteristics can be used to quantify any unknown relationships. Generally, experience is used to identify the critical parameters. Typical examples of important process parameters might include the following:

Temperature and atmospheric composition of a furnace
Tool size
Coolant concentration or composition
Hydraulic oil usage
Vibration characteristics of a machine

Table 14.4 Operation Control Plan Table

No.	Part characteristic, process parameters	Control level	Specification LSL/USL	Gaging No.	V/A	Method	Divisions	Control method	Sampling intensity	Capability C_p, C_{pk}	Capability Date	Gage repeatability %	Gage repeatability Date
1	Length	3	2.000/2.001	AB-102	V	Air column	11	Stratified $\bar{X}, R$ comparison plot	Four every 2 hours, consecutive weekly control chart data	1.3, 1.2 1.3, 1.1 NS-AP3	6/7 7/3 8/15	21 32 18	6/7 7/11 8/15
2	Squareness	1	0/.001	AB-506	V	Dial indicator	11	Measurement scale check sheet	One per hour	2.1, 2.0	4/5	15 20 18	5/3 7/18 8/10
3	Parallelism	2	0/.001	AB-506	V	Dial indicator	11	Run chart	One per hour	2.2, 2.2 2.0, 2.1	4/5 6/1	21 32 25	5/3 7/18 8/10
4	Appearance	1	—	—	A	Visual	—	Attribute check sheet P chart (nicks)	Five per hour, random One per day ($n = 40$)	—	—	—	—
5	Tool size	2	1.521/1.522	—	V	Micrometer	21	Run chart	At tool change	—	—	10 9 12	6/1 7/20 8/2
6	Spindle alignment	2	0/.0005	—	V	Dial indicator	6	Run chart	One per week	—	—	25 31 40	6/20 7/22 8/15
7	Coolant oil contamination	2	0/5%	ZP-102 —	V V	CMM Graduated cylinder	51 11	Target chart $\bar{X}, R$	One per month Three per week at 0900 hours	— 2.0, 1.5 2.1, 1.7 2.1, 2.0	— 4/20 6/12 8/15	10 25 31 32	7/22 6/15 7/21 8/12

Department: _____ Part name: _____ Date updated: _____
Operation: _____   Part no.: _____   Name: _____

14.7 Develop and Implement a Plan

The control factors section of the process relationship table can be used to identify which of the major elements influence the operation. Important elements can often be monitored by using key process parameters. For example, if a machine condition is an important control factor, it may be possible to monitor the machine's use of hydraulic oil or the vibration characteristics of a moving component. The rationale of this approach is that changes in these machine elements ultimately influence part characteristics. The prevention approach in DPS requires the control of process parameters that ultimately influence part characteristics. Some approaches to process control separate the control of part characteristics and process parameters. Control of the process is most effective when these elements are integrated. The process parameters listed in the operation control plan table may be part of other tracking systems, such as a PM checklist. Inclusion in the table implies that monitoring or analysis is required using tools, such as a run chart.

The control level is carried over from the process relationship table to the operation control plan table; a control level is also assigned to the process parameters. Specification limits are then placed in column 4 of the table. The importance of gaging to process control has been continually emphasized. The type of gaging used at the operation is identified in columns 5–8. The gage number is listed first in column 5 for quick reference. Column 6 identifies whether the gage provides a variable or attribute (V or A) measurement. Column 7 identifies the gaging method. Column 8 identifies the number of gage divisions (ignoring any interpolation) within the specification limits. For example, if the specification limits are 5–10 and the gage is measured in whole units

there are 6 divisions. If half-units were used, there are 11 divisions. The number of gaging divisions influences the measured capability of a process. Ideally, a minimum of 10 divisions should be present, and 15–20 divisions will provide a better evaluation of process capability. In some cases several gaging methods may be used. A gage at the operation may provide routine monitoring, and a high-precision gage located at a different site, such as a coordinate measuring machine (CMM), would provide verification of the control system.

14.7.2 Control Method

The remaining columns in the table define the control methods and the capability of the process. This part of the table will change over time as the control methods or capabilities change. Column 9 identifies the control methods selected to monitor the performance of the process. Some typical choices include the following:

Control charts: $\bar{X}$, R, or s charts; $\bar{X}$ and R charts; dot control charts; P charts; c charts
Measurement scale check sheets
Run charts
Attribute check sheets
Comparison plots
Target charts

Other choices tailored to unique applications are possible. In some cases more than one control method is appropriate. For example, in Table 14.4 a stratified $\bar{X}$ and R chart is used to monitor the gear length. Machine and spindle (two machines each with four spindles) differences are evaluated on a weekly basis using a comparison plot of the weekly control chart data.

Selection of the appropriate control method should consider control level and capability. The higher the control level, the more likely it is that control chart monitoring is required. The lower the capability, the more likely it is that control chart monitoring is required:

Control level	0	1	2	3
	No monitoring	⟶	Intensive control	
Capability	3 or greater	⟶	Less than 1	
Control method	None	⟶	Control chart	

Also, a characteristic that historically has exhibited a lack of stability is a good candidate for control charting. In industrial applications many characteristics may be associated with an operation and it is not possible, or necessary, to prepare control charts for most characteristics. The analysis required for the process relationship and operation control plan tables allows selection of the critical few characteristics.

Implicit in this discussion is the availability of capability data for every characteristic. For a new or existing process, it is good practice to evaluate the capability of most characteristics to establish an effective baseline. This information is a critical input in designing the ongoing control plan. It is also good practice to periodically re-evaluate the capability of characteristics that are not routinely monitored on a control chart. The tabular listing in the operation control plan table clearly indicates where capability data are available and their timeliness.

The typical approach for selecting characteristics for control charting relies on process experience. Generally, experience enables the selection of characteristics with a high control level, although all important characteristics are usually not identified. However, experience typically provides inadequate insight into which characteristics lack capability. One of the main benefits of the table is the attention that secondary and in-process characteristics receive. The control system often involves charting a few characteristics and spot-checking the rest. As the discussion of the control methods indicates, a variety of intermediate methods provide valuable monitoring for the control system. Thus, experience is a necessary input for the process relationship table, but it lacks the thoroughness required to develop the prevention-oriented control plan defined in the operation control plan table.

14.7.3 Sampling Intensity

An important part of the control plan is establishing an appropriate sampling plan. The following elements must be identified:

When: what time should samples (parts) be collected?
How many: what number of samples should be collected?
Method: what type of sampling method should be used (consecutive, periodic, random, or stratified)?

A number of factors must be considered to select an appropriate sampling plan:

1. Control level and capability
2. Process flow and control factors

14.7 Develop and Implement a Plan

3. Consequences of an out-of-control process
4. History of process stability

No explicit method is available for selecting the "best" sampling plan. Subjective evaluation of these factors enables a reasonable approach.

Just as control level and capability influence the selection of the control methods, they also influence the sampling intensity. A characteristic that has a high control level requirement should be considered for intensive sampling. As discussed in Chapter 7, the greater the capability, the lower is the chance that a process change will result in an out-of-specification condition. Thus, the higher the capability, the less likely it is that intensive sampling is needed.

The process flow and control factors from the process relationship table influence the selection of a sampling plan. The process flow diagram defines a number of sources of variability at each operation. The sampling method must be adapted to the process. For example, Chapter 4 discussed the problem with using a $n = 5$ sampling plan for a three-spindle machine. The control factors help to identify when samples should be collected. If tool setup or die change are controlling factors, samples should be collected at the tool or die change.

In Chapter 6 the sensitivity of control charts to shifts in location was discussed. It was noted that many typical sampling frequencies provide inadequate protection against process shifts. The consequence of a process operating in an out-of-control state must be considered when specifying the sampling intensity. In some cases, use of the Gillette lot control system (Chap. 4) can minimize the impact of an out-of-control situation. Past performance of a process is often a good indicator of expected future performance. A process that frequently goes out of control should obviously be sampled more intensively than one that historically remains stable. Also, the past incidence of outliers may require increased sampling intensity until the cause of the problem can be eliminated.

Selection of the sampling intensity must be balanced with the selected control method. For example, in some cases it may be preferable to sample more intensively using a measurement scale check sheet rather than less intensively using a control chart. Even after considering all the preceding factors, it is sometimes useful to try several sampling plans to select the plan that provides the desired control.

14.7.4 Capability

Columns 11, 12, and 13 of the table report the capability performance of the operation using the standard capability indices discussed in Chapter 7. The C_p and C_{pk} indices conveniently summarize process capability by indicating both process potential C_p and process performance C_{pk}. When a one-sided specification limit is used, it should be understood that the C_{pk} results refer to either CPL or CPU. With each reported capability result, the date on which the data was collected should be given. Also, listing the last three capability evaluations enables assessment of any process changes. Recall that the process must be stable for the capability indices to be meaningful. When any out-of-control condition exists, the table should indicate

NS – AP X

that is,

Not stable – action plan X

The follow-up section discusses the need for this practice.

14.7.5 Gage Repeatability

The ability of a gage to repeat measurements changes over time. In some cases it is appropriate to prepare a control chart for gage repeatability, as discussed in Appendix I. Often it is sufficient to periodically conduct a gage repeatability study. This simple procedure not only indicates the condition of a gage but enables assessment of whether a low process capability could, in part, be due to poor gage repeatability. Again, reporting the dates for the last three evaluations enables assessment of changes over time. A lack of adequate repeatability may be related to inadequate measurement precision, which is indicated by the number of gaging divisions reported in column 8 of the table.

14.7.6 Follow-up System

The collection of operation control plan tables for each process operation identifies how the process will be controlled on an ongoing basis. This part of the DPS would be incomplete without a follow-up system to ensure proper implementation. It is not unusual to have a well-conceived plan fail because of a lack of follow-up. Too often there is a tendency to overlook the details required for effective implementation of any plan. Three approaches may be used to ensure effective implementation and control of the process: action instructions, alarm system, and a timing chart.

Approach 1. Action Instructions. Each of the control methods identified in the operation control plan table utilizes data to monitor some aspect of the process. If, for example, an $\bar{X}$ and R chart is utilized, the common practice is simply to start a chart at the operation. This approach is inadequate for establishing a DPS. What actions are to be performed if the process becomes unstable? In all likelihood, it can be anticipated that a chart will exhibit an out-of-control signal at some time, or why would we monitor the process? Not considering what actions should be taken when out-of-control signals occur will result in a "fire fighting" response or worse—no response. Thus it is important to develop action instructions for each of the control methods. These instructions should simply state the actions that should be taken if a specified data signal is obtained. These instructions should be part of the data collection form or posted at the operation.

Several approaches can be used to identify appropriate actions. The best method is to use knowledge of the process to identify actions that will define what process element has changed. A variation and special causes diagram (Chap. 4) can be used to associate process changes with control chart signals. These tools seek to identify special causes of variation, which can then be examined as part of the action instructions. In many cases it is possible to identify the most likely special causes. It is not necessary to wait for a problem to occur! When the special cause cannot be determined, a problem-solving team (step 4) should be used.

Another approach to identify appropriate action instructions is to have a team prepare a process flow C & E diagram. The effect being considered is "abnormal process variability." The potential causes of variability can then be addressed in the action instructions. If execution of the action instructions does not uncover the special cause problem, the problem-solving team has a head start on investigating the problem. Many potential causes have already been eliminated. Also, the information collected at the time that the special cause results in unusual process performance is extremely valuable. Most action instructions attempt to quantify the process conditions (the 5M elements). Later investigations will find this information invaluable.

14.7 Develop and Implement a Plan

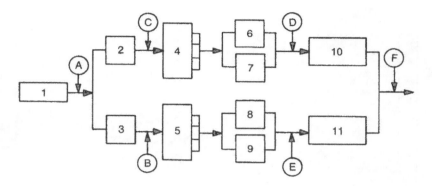

Figure 14.5 Process flow diagram with control locations.

One useful approach to help identify a problem is to collect additional data. If a control chart is being used, additional subgroups can be plotted. Similarly, other control methods can also utilize additional data. Different characteristics can be evaluated including both part and process measurements. The additional data serve three purposes. First, there is clear verification that a problem exists. Second, the duration of the problem can be quantified. Third, the values of process parameters can be used to establish a cause-and-effect relationship.

Approach 2. Alarm System. A comprehensive approach to identify the source of a process problem is to develop an alarm system. It may be necessary to increase the sampling frequency of all control methods to help isolate a problem. It may also be required to implement special control or evaluation methods, such as comparison plots, to supplement the monitoring methods. The intent of the alarm system is to seek to identify quickly what changed in the process.

Consider the example in which control chart F in Figure 14.5 went out of control and the cause could not be identified quickly. All control methods upstream of F could increase their sampling intensity. Comparison plots could be prepared for machines 4 and 5 or any of the other multiple stream operations [(2, 3), (6, 7, 8, 9), or (10, 11)]. This approach should be able to quickly determine what changed in the process. The typical approach is to evaluate various potential special causes in series. The alarm system evaluates many potential causes in parallel, which reduces the time for problem resolution.

Approach 3. Timing Chart. Recall from Chapter 13 that an action plan was used to establish the timing and responsibility for problem-solving actions. The process control timing chart is used to implement the operational control plan. Table 14.5 is an example of the chart in which who will do what by when is identified. An "x" indicates the characteristic is planned for sampling on that day. The circled "x" indicates the data were collected and reviewed. This simple chart provides a complete follow-up system for a department. Both timing and responsibility are clear.

A system sometimes used in manufacturing operations is to rotate which characteristics are charted on control charts. The timing chart makes it easy to use rotating control methods.

608   14 Building a Defect Prevention System

Table 14.5 Department Process Control Timing Chart

Operation	Part characteristic, process parameter	Control method	Sampling	Responsibility	5/1 M	T	W	T	F	M	5/7 T	W	T	F	M	5/14 T	W	T	F	M	5/21 T	W	T	F
IM																								
10																								
20																								
30	Length	Stratified $\bar{X}$, R	Four every 2 hours	Jones	⊗	⊗	⊗	⊗	⊗	⊗		x		x	x	x	x	x	x	x	x	x	x	x
		Comparison plot	One per week	Smith		⊗										x					x			
		Gage R & R	One per month	Jones				x																
	Squareness	V check sheet	One per hour	Jones	⊗	⊗	⊗	⊗	⊗	⊗	x	x	x	x	x	x	x	x	x	x	x	x	x	x
		Gage R & R	One per month	Jones									x											
	Parallelism	Run chart	One per hour	Jones	⊗	⊗	⊗	⊗	⊗	⊗	x	x	x	x	x	x	x	x	x	x	x	x	x	x
		Gage R & R	One per month	Jones									x											
	Appearance	A check sheet	Five per hour	Jones	x	⊗	x	⊗	x	x	x	x	x	x	x	x	x	x	x	x	x	x	x	x
	Tool size	Run chart	At TC	Jones		⊗		⊗		⊗	x													
		Gage R & R	One every 2 months	Jones					x															
	Spindle alignment	Run chart	One per week	Lamb			⊗				x					x					x			
		Target chart	One per month	Lamb	x															x				
		Gage R & R	One per quarter	Lamb																				
	Coolant oil	$\bar{X}$, R	Three per week	Wilson	⊗			⊗		⊗		x			x				x	x			x	
		Gage R & R	One per quarter	Wilson																				
40																								

14.8 Develop an Incoming Material DPS

Characteristic	M	T	W	T	F	M	T	W	T	F
1										
Control chart $\bar{X}$, R						x	x	x	x	x
Measurement scale check sheet	x	x	x	x	x					
2										
Control chart $\bar{X}$, R	x	x	x	x	x					
Measurement scale check sheet						x	x	x	x	x

This system provides control of more characteristics with a fixed amount of labor.

It should be emphasized that the process control plan developed in this section 7 is a dynamic plan. The part characteristics and process parameters controlled will probably change over time. Similarly, the control methods utilized will change. The timing chart provides a useful follow-up system for monitoring the operation control plans to ensure proper implementation and to provide a historical record of what controls were in place.

14.8 Develop an Incoming Material Defect Prevention System

Recall from Figure 2.1 that a work process starts with various inputs that are processed through a series of added value tasks to obtain an output. The process relationship table considered critical process inputs, but a true prevention approach must have a DPS established at the supplier of each input. These inputs should not be limited to unfinished or semifinished parts but should include all materials and tools used to operate the process. Each supplier of these inputs should implement the 10 DPS steps; only then can a manufacturer have a complete prevention system.

A significant effort is required to interact with all the suppliers of inputs to a process. Thus, it is desirable to minimize the number of suppliers. Deming (1982, pp. 27–30) discusses the benefits of a single-source, long-term relationship with a supplier. With this type of relationship, it is possible to develop an effective DPS for incoming material.

Too often, suppliers are unfamiliar with the critical requirements of a manufacturer. All characteristics on a blueprint are considered equally important. In fact, there are often only a few critical characteristics. Consider the gear blanking machining example in Table 14.4, where the first operation clamps and locates on the outer diameter (OD) and machines the inner diameter (ID). Improving the capability of the gear blank ID will have no impact on the final quality of the part. However, improving the gear blank OD would probably produce a better final part. Thus, communication between supplier and manufacturer is an important part of establishing a DPS. A good way to facilitate communication is to have joint teams prepare the process relationship table and the process control plan.

Once the table is developed and the control plan implemented, the control plan tables and timing chart tables serve as convenient review documents. When problems are encountered, the four elements of the problem analysis system can be used to guide the analysis process. Because of the detailed analysis performed before any problems arose, the teams addressing any problems will start with a good understanding of the process.

Another practice that encourages high quality of incoming material is the use of just-in-time inventory management. With minimal production float, it is essential that parts be of high quality. When poor quality causes an interruption of a production line, the problem becomes highly visible throughout the organization. The supplier or manufacturer systems are forced to resolve problems and establish a prevention system. The procedures described in Sections 14.6 and 14.7 provide a system to develop the required prevention-oriented control system.

14.9   Emphasize Management Evaluation of Systems

Implementation of the DPS steps is not performed over a short time period. Management must direct and become involved in the implementation process. Action plans should be developed to manage the transition along with periodic reviews to monitor progress. The most difficult challenge for management is establishing a prevention mindset in the organization. Continually asking how a problem could have been prevented is difficult when one is consumed by many daily problems. Again, managers are seemingly forced into doing the wrong job.

Recall Deming's assertion that the system is responsible for 85% of the problems and that only management can change the system. Many systems are not prevention oriented, and existing effective systems are often not used. Focusing on evaluating systems rather than problems or administration should be a primary role of management. In practice, the reverse is true. Utilizing the 10 DPS steps helps to focus activity on developing effective prevention systems. It is not possible for management to delegate responsibility to lead this transformation process.

14.10   Develop a Continual Improvement Mindset

The benefits of continuing improvement are not realized by most organizations. Performance indicators oscillate. Good performance is associated with the individual initiative of a few top managers. Poor performance is similarly associated with the lack of ability of a few individuals. However, it is the lack of prevention-oriented systems that makes continuing improvement difficult to sustain. The mindset is conditioned to accept that problems will always exist, and so they do! Only management can change the systems that will in turn change the mindset of the organization. Each problem is evidence that a system failed and that an opportunity for improvement exists.

Traditional management systems utilize targets for all performance indicators to evaluate a manager's performance. This approach encourages focusing on short-term improvement, often at the expense of longer term improvement. It is much easier to develop a few actions that will produce a temporary improvement than to change a non–prevention-oriented system. Changing a system can require great effort, particularly when superiors do not appreciate the rationale for the change. The mindset of middle managers can only be changed if top management changes their methods for evaluating a manager's performance.

The lack of continual improvement in most cases is not due to the lack of management's appreciation of its value. It is rather the lack of prevention-oriented systems that impedes progress. A continual improvement mindset can only be attained by institutionalizing preventive systems throughout the organization. A car that has had poor preventive maintenance may provide adequate transportation across town, but it is a poor risk for driving across the country.

Selected Bibliography

American Society for Quality Control, Automotive Division (1986). *Statistical Process Control Manual*, ASQC, Milwaukee, Wisconsin.

Box, G.E.P., Hunter, W.G. and Hunter, J.S. (1978). *Statistics for Experimenters*, John Wiley and Sons, New York.

Burr, I.W. (1976). *Statistical Quality Control Methods*, Marcel Dekker, Inc., New York.

Charbonneau, H.C. and Webster, G.L. (1978) *Industrial Quality Control*, Prentice-Hall, New York.

Clifford, P.C. (1959) "Control Charts Without Calculations: Some Modifications and Some Extensions," *Industrial Quality Control*, 15, 40–44.

Conover, W.J., Johnson, M.E., Johnson, M.M. (1981). "A Comparative Study of Tests for Homogeneity of Variances, with Applications to the Outer Continental Shelf Bidding Data," *Technometrics*, 23, 351–361.

Deming, W.E. (1986). *Out of Crisis*, MIT Press, Cambridge, Massachusetts.

Deming, W.E. (1982). *Quality, Productivity, and Competitive Position*, Massachusetts Institute of Technology, Center for Advanced Engineering Study, Cambridge, Massachusetts.

Dixon, W.J. and Massey, F.J. (1983). *Introduction to Statistical Analysis, 4th Edition*, McGraw-Hill, New York.

Draper, N. and Smith, H. (1981). *Applied Regression Analysis*, John Wiley and Sons, New York.

Ferrell, E.B. (1953). "Control Charts Using Midranges and Medians," *Industrial Quality Control*, 9, 30–32.

Ford Motor Company (1984). *Continuing Process Control and Process Capability Improvement*, Ford Motor Company, Dearborn, Michigan.

Gartside, P.S. (1972). "A Study of Methods for Comparing Several Variances," *Journal of the American Statistical Assn.*, 67, 342–346. Correction 68, p 251.

Grant, E.L. and Leavenworth, R.S. (1980). *Statistical Quality Control*, McGraw-Hill, New York.

Grubbs, F.E. (1983). "An Optimum Procedure for Setting Machines or Adjusting Processes," *Journal of Quality Technology*, 15, 186–189.

Hartley, H.O. (1950). "The Maximum F-Ratio as a Short Cut Test for Heterogeneity of Variance," *Biometrika*, 37, 308–312.

Ishikawa, K. (1982). *Guide to Quality Control*, Asian Productivity Organization, Tokyo, Japan.

Jaech, J.L. (1985). *Statistical Analysis of Measurement Errors*, John Wiley and Sons, New York.

Juran, J.M., Gryna, F.M., and Bingham, R.S. (1979). *Quality Control Handbook*, McGraw-Hill, New York.

Juran, J.M. and Gryna, F.M. (1980). *Quality Planning and Analysis*, McGraw-Hill, New York.

Kackar, R.N. (1985). "Off-Line Quality Control, Parameter Design, and the Taguchi Method," *Journal of Quality Technology*, 17, 176–188.

Kane, V.E. (1986). "Process Capability Indices," *Journal of Quality Technology*, 18, 41–52. Correction 18, p 265.

Nelson, L.S. (1982a). "Testing Variation Before and After Treatment of a Single Sample," *Journal of Quality Technology*, 14, 44–45.

Nelson, L.S. (1982b). "Control Chart for Medians," *Journals of Quality Technology*, 14, 226–227.

Nelson, L.S. (1983a). "Exact Critical Values for Use with the Analysis of Means," *Journal of Quality Technology*, 15, 40–44.

Nelson, L.S. (1983b). "A Sign Test for Correlation," *Journal of Quality Technology*, 15, 199–200.

Nelson, L.S. (1984). "The Shewhart Control Chart—Tests for Special Causes," *Journal of Quality Technology*, 16, 237–239.

Nelson, L.S. (1985). "Interpreting Shewhart $\bar{X}$ Control Charts," *Journal of Quality Technology*, 17, 114–116.

Nelson, L.S. (1986). "Control Chart for Multiple Stream Processes," *Journal of Quality Technology*, 18, 255–256.

Nelson, L.S. (1987a). "A Gap Test for Variances," *Journal of Quality Technology*, 19, 107–109.

Nelson, L.S. (1987b). "Upper 10%, 5% and 1% Points of the Maximum F-Ratio," *Journal of Quality Technology*, 19, 165–167.

Odeh, R.E. Owen, D.B., Birnbaum, Z.W., and Fisher, L. (1977). *Pocket Book of Statistical Tables*, Marcel Dekker, Inc., New York.

Ott, E.R. (1975). *Process Quality Control*, McGraw-Hill, New York.

Ott, E.R. and Snee, R.D. (1973). "Identifying Useful Differences in a Multiple-Head Machine," *Journal of Quality Technology*, 5, 47–57.

Scherkenbach, W.W. (1986). *The Deming Route to Quality and Productivity*, Mercury Press, Rockville, Maryland.

Schilling, E.G. (1973). "A Systematic Approach to the Analysis of Means," *Journal of Quality Technology*, 5, 93–108, 147–159.

Snedecor, G.W. and Cochran, W.G. (1980). *Statistical Methods*, 7th Edition, Iowa State Univ. Press, Ames, Iowa.

Tsai, P. (1988). "Variable gage repeatability and reproducibility study using the analysis of variance method," *Quality Engineering*, 1, 107–115.

Wadsworth, H.M., Stephens, K.S., and Godfrey, A.B. (1986). *Modern Methods for Quality Control and Improvement*, John Wiley and Sons, New York.

Wheeler, D.J. (1983). "Detecting a Shift in Process Average: Tables of the Power Function for $\bar{X}$ charts," *Journal of Quality Technology*, 15, 155–170.

Wheeler, D.J. and Chambers, D.S. (1986). *Understanding Statistical Process Control*, Keith Press, Knoxville, Tennessee.

Appendix I
Gage Evaluation

I.1 Measurement Process

The importance of gaging has been emphasized throughout this book. The ability of a gage to obtain meaningful measurements directly influences the operation control plan in Table 14.4, which is an integral part of a defect prevention system (DPS). Since the statistical techniques discussed in the text rely on measurements, the ability of a gage to obtain repeatable, reproducible, and accurate measurements should logically be assessed prior to data collection. Unfortunately, many of the concepts developed in this book are needed to evaluate the measurement process. Thus, this discussion appears in an appendix.

Prior to evaluating a measurement process, it is necessary to address three items. First, the intent of making the measurement must be determined. In many cases, measurements are made to assess how well a manufacturing process is meeting a part feature specification identified on an engineering drawing. Understanding the intent of assigning a tolerance to a part feature will greatly assist in evaluating the measurement process. For example, knowing whether a characteristic is a free clearance dimension or a dimension that mates with another part will provide an understanding of the engineering intent. Second, the operational definition of the measurement intent must be specified. In Chapter 4 an example was considered in which the size of a part feature may refer to the minimum, maximum, or average size. Size is the most common, seemingly straightforward part feature, yet it can have several operational definitions. To develop a good operational definition (Chap. 2), it is necessary to have a clear understanding of the measurement intent. Third, the gaging method must be specified. How does the gage arrive at the measurement? For example, measurements from a mechanical gage are greatly influenced by how a part is positioned in the gage and the location at which gage probes contact the part. Clearly specifying the entire gaging process enables evaluation of whether the process is consistent with the operational definition of the measurement intent:

$$\text{Identify measurement intent} \rightarrow \text{specify operational definition} \rightarrow \text{specify gaging process}$$

Recall from Chapter 2 that one of the major elements in the definition of a process is measurement. However, it is often not appreciated that obtaining a measurement is itself a process. A generic process flow diagram appears in Figure I.1 along with several sources of variation. Once measurement is understood as a process, quantifying the sources of

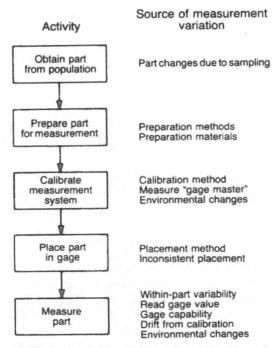

Figure I.1   Generic process flow for measurement of a part.

variation (e.g., repeatability, reproducibility, and accuracy) and evaluating the stability of the sources of variation become important (see Deming, 1986, p. 332). These topics are discussed in the following sections.

I.2   Sources of Measurement Error

The measurement of a characteristic results in a number that differs to some degree from the "true" value of the characteristic. This difference is the result of measurement error due to the sources of variability associated with the measurement process. There are five major sources of measurement error:

1. Repeatability
2. Reproducibility
3. Accuracy
4. Linearity
5. Stability

Gage characteristics are often discussed in terms of accuracy and precision. The familiar target charts in Figure I.2 illustrate these independent concepts. In the terms discussed here, accuracy and linearity relate to accuracy and repeatability and reproducibility relate to precision. A variety of methods exist to quantify the magnitude of these sources of gaging variability (e.g., Charbonneau and Webster, 1978, or American Society for Quality Control, 1986). The procedures presented here have proven easy to apply in manufacturing applications.

I.2 Sources of Measurement Error

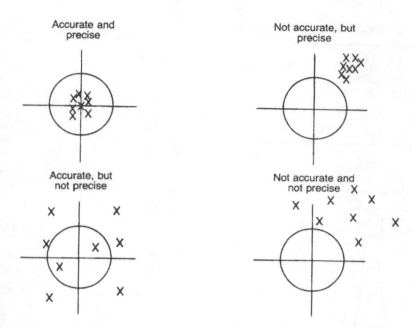

Figure I.2 Accuracy and precision concepts.

I.2.1 Repeatability

Repeated measurements of a single part by one operator result in a distribution of values. For a perfect gage, all repeated measurements would be identical. However, if sufficient gage sensitivity exists, there is generally a bell-shaped, normal distribution of remeasurement values, as shown in Figure I.3. The standard deviation of the remeasurement values is

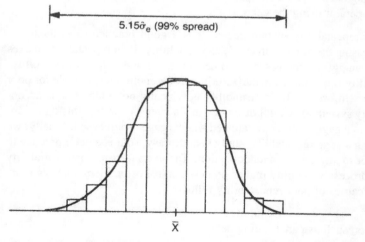

Figure I.3 Spread of repeated measurements of an individual part.

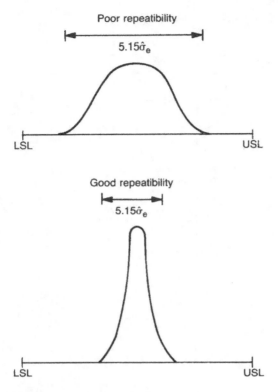

Figure I.4 Gage repeatability differences.

an indicator of the repeatability of the gage. The repeatability error standard deviation is denoted by $\hat{\sigma}_e$.

Repeatability of a gage is the variation $\hat{\sigma}_e$ associated with repeated measurements of a single part by one operator using the gage.

Figure I.4 depicts the repeatability of two gages. It is a good practice to evaluate repeatability prior to assessing the other sources of gage variability. Poor repeatability makes it difficult to obtain meaningful estimates of the other sources of measurement variability.

Poor gage repeatability is a common manufacturing problem often responsible for poor or inconsistent quality. Unfortunately, repeatability is typically not addressed as an ongoing part of the quality system. The DPS discussed in Chapter 14 corrects this omission. Also, the repeatability of a gage should be evaluated when a gage is purchased. Finally, we see here that performing a gage repeatability study is quite easy, so it is a good practice to conduct this study prior to any special data collection. The extra measurement variability added to any normal process variability makes process control or problem analysis more difficult. Some of the causes of poor repeatability follow:

Gage not clean
Worn gage components, such as part-locating details
Within-measured-part variability
Poor gage design

I.2 Sources of Measurement Error

Malfunction of gage
Improper gaging method
Environmental conditions

The sources of gaging variability are quantified here using two measurements on a group of N parts. The method for obtaining the two groups of measurements determines which sources of variability can be quantified. Remeasurement of N parts is preferable to repeated measurements of a single random part for two reasons. First, it is important to select parts spanning the specification range to effectively evaluate a gage's performance. Second, within-part variability often influences the ability of a part to be effectively measured. Selecting N parts enables sampling different within-part variability conditions. The data resulting from the gage studies can be displayed as follows:

Part	First reading	Second reading	Difference
1	X_1	Y_1	$D_1 = X_1 - Y_1$
2	X_2	Y_2	$D_2 = X_2 - Y_2$
.	.	.	.
.	.	.	.
.	.	.	.
N	X_N	Y_N	$D_N = X_N - Y_N$
			Mean: $\bar{D}$
			SD: s_D

The standard deviation s_D of the differences can then be used to estimate repeatability variability:

$$\hat{\sigma}_e = \frac{s_D}{\sqrt{2}}$$

It is convenient to relate repeatability to the tolerance range:

$$\% \text{ repeatability} = 100 \times \frac{\text{spread due to gaging error}}{\text{tolerance}}$$

$$= 100 \times \frac{5.15\hat{\sigma}_e}{\text{tolerance}}$$

$$= 100 \times \frac{3.64 s_D}{\text{USL} - \text{LSL}}$$

By convention, a 99% spread, corresponding to $5.15\hat{\sigma}_e$, is used rather than a 99.73% spread ($6\hat{\sigma}_e$) used in typical capability assessments. Procedure I.1 gives a general gage evaluation procedure that can be used to assess repeatability. Jaech (1985, Sec. 2.2) gives the statistical foundation of this procedure.

An alternative method of obtaining an estimate of repeatability is to use the range between the first and second measurements. The average range $\bar{R}$ is computed over the N parts, so that

$$\hat{\sigma}_e = \frac{\bar{R}}{d_2^*}$$

Procedure I.1 Gage Evaluation Procedure

A simple data collection format can be used to evaluate gage repeatability, reproducibility, or accuracy. In all cases, N parts are measured twice, resulting in the following data:

Part	First reading	Second reading	Difference
1	X_1	Y_1	$D_1 = X_1 - Y_1$
2	X_2	Y_2	$D_2 = X_2 - Y_2$
.	.	.	.
.	.	.	.
.	.	.	.
N	X_N	Y_N	$D_N = X_N - Y_N$
			Mean: $\bar{D}$
			SD: s_D

The sample size N should be at least 10 for a monitoring type of evaluation. More stable estimates of gaging error are obtained with $N = 30$ or more. A gage evaluation worksheet accompanies this display. The work sheet calculations are appropriate for any N. In all cases, a scatter plot of the first versus the second reading is recommended to provide a visual evaluation of gaging variability. An ongoing gage repeatability control chart is the best method of establishing gage stability and the magnitude of the gaging variability.

Study Procedure

The following steps should be used to collect the data:

Collect N parts that span the range of the specification.
Randomly label parts from 1 to N. Ensure that the labeling does not alter the characteristic to be measured.
Calibrate the gage, and verify that the gage is functioning normally.
Measure and record the first reading for all N parts.
Randomly remeasure and record the second reading for all N parts. The second gage reading should be made according to what source of gage error is being quantified:
  Same gage and operator: repeatability
  Same gage, different operator: repeatability and reproducibility
  Reference standard gage: accuracy
The individual making the second reading should not have knowledge of the first reading.
Compute the difference $D = X - Y$ between the first and second reading. This difference may be positive (+) or negative (−). Compute the mean $\bar{D}$ and standard deviation s_D of the differences.

Gage Repeatability

The same gage must be used for the first and second readings to evaluate repeatability. The repeatability standard deviation

$$\hat{\sigma}_e = \frac{s_D}{\sqrt{2}}$$

is referred to as "gage error standard deviation" on the work sheet. Also,

$$\% \text{ repeatability} = 100 \times \frac{5.15\hat{\sigma}_e}{\text{tolerance}}$$

Gage Reproducibility

Using the same gage with different operators making the first and second reading enables estimates of repeatability and reproducibility (R & R). The reproducibility standard deviation

$$\hat{\sigma}_o = \sqrt{\frac{\bar{D}^2}{2} - \frac{\hat{\sigma}_e^2}{N}}$$

is referred to as "operator measurement standard deviation" on the work sheet. If the quantity under the radical is negative, $\hat{\sigma}_o$ is assumed 0. Also,

$$\% \text{ reproducibility} = 100 \times \frac{5.15\hat{\sigma}_o}{\text{tolerance}}$$

Gage Repeatability and Reproducibility

After calculating both repeatability and reproducibility (R & R) standard deviations, a combined estimate

$$\hat{\sigma}_{R\&R} = \sqrt{\hat{\sigma}_e^2 + \hat{\sigma}_o^2}$$

is referred to as "measurement error standard deviation" on the work sheet. Also,

$$\% \text{ R \& R} = 100 \times \frac{5.15\hat{\sigma}_{R\&R}}{\text{tolerance}}$$

Gage Accuracy (Bias)

The accuracy of a gage can be evaluated if one of the readings is from a reference standard gage. A significant difference between the means (Procedure 12.5) of the first and second readings indicates a likely gaging bias of magnitude $\bar{D}$. Linearity can be assessed using a scatter plot. If a significant constant bias exists,

$$\% \text{ accuracy} = 100 \times \frac{\text{bias}}{\text{tolerance}}$$

A standard criterion for evaluating the measurement process for the percentage error is 0–10% (excellent), 10–20% (good), 20–30% (marginal), and greater than 30% (unacceptable).

Since each range is based on $n = 2$ measurements, $d_2 = 1.13$ is normally used. However, small values of N require that we use the constant d_2^* from Table I.1, which provides a better estimate of $\hat{\sigma}_e$. For $N = 10$, $d_2^* = 1.16$, which gives

$$\% \text{ repeatability} = 100 \times \frac{5.15\bar{R}/d_2^*}{\text{USL} - \text{LSL}} = 100 \times \frac{4.44\bar{R}}{\text{USL} - \text{LSL}}$$

For N greater than 10 the difference between d_2 and d_2^* is small, so $d_2 = 1.13$ can be used. It is always useful to display the results of a repeatability study using a scatter plot with a benchmark line, as shown in Figure 12.26.

Table I.1 Values of d_2^*

N	d_2^*
5	1.19
6	1.18
7	1.17
8	1.17
9	1.16
10	1.16
11	1.16
12	1.15
13	1.15
14	1.15
15	1.15
Maximum	1.13

Example I.1 Consider measuring 10 parts twice on a dimensional gage on which the specification range is 5–15. The resulting data are shown below:

Part	First reading	Second reading	Difference
1	10	12	−2
2	8	9	−1
3	5	4	1
4	11	9	2
5	12	13	−1
6	10	7	3
7	6	6	0
8	10	14	−4
9	12	11	1
10	14	14	0

$\bar{D} = -0.1 \quad s_D = 2.02$

The standard deviation of the differences is then used to compute

$$\% \text{ repeatability} = 100 \times \frac{3.64 \times 2.02}{15 - 5}$$

$$= 74\%$$

I.2 Sources of Measurement Error

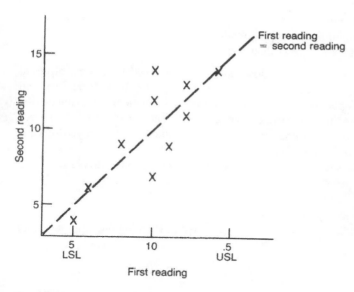

Figure I.5 Scatter plot for Example I.1.

From Procedure I.1, repeatability above 30% is unacceptable. The average range can be computed from the column of differences ignoring the sign, so that $\bar{R} = 1.5$. Using the range calculation for repeatability yields $[100(4.44 \times 1.5)]/10 = 67\%$. As discussed previously, estimates of variability require larger sample sizes (e.g., N greater than 30) to obtain more stable estimates. Thus, the two methods of computing $\hat{\sigma}_e$ can be expected to differ, with s_D providing the statistically superior method of estimation. The scatter plot in Figure I.5 provides a useful display of the data. Notice that the first reading had 3 values at 10, but the second reading's value for these parts ranged between 7 and 14. A range of 7 due to gage repeatability error is clearly unacceptable when the specification range is 10.

I.2.2 Reproducibility

Repeated measurements of a part by two or more operators may produce different mean values (we are assuming common repeatability). The standard deviation associated with the operator mean differences is denoted by $\hat{\sigma}_o$. In the ideal case, if operators had the same mean values the reproducibility variation would be zero.

> *Reproducibility* of a gage is the between-operator variation $\hat{\sigma}_o$ associated with repeated measurements of a single part by different operators using the gage.

Figure I.6 depicts the reproducibility variability due to two operators.

In an industrial setting a number of individuals may use a gage, so it is important that a gage be designed so that consistent results are obtained by different operators. Just as other measurement errors can cause a serious problem, a significant reproducibility error can also be a major concern. Fortunately, reproducibility error, once detected, can usually be minimized. In most cases, improper use of a gage can be easily corrected. It is always a good practice to use picture displays to illustrate proper gaging procedures. Some of the causes of poor reproducibility follow:

Operators using different gaging methods
Insufficient operator training in proper gaging method
Poor gage design, allowing subjective evaluations

Evaluation of reproducibility error is most easily accomplished by having different operators measure N parts. The calculations to obtain estimates of repeatability and reproducibility for the two-operator case appear in Procedure I.1, where it is assumed that different operators make the first and second gage readings. Jaech (1985, Sec. 2.2) gives the statistical justification for the estimation procedure. A more general analysis of variance approach is considered in a later section.

The standard deviation associated with operator reproducibility variability is

$$\hat{\sigma}_o = \sqrt{\frac{\bar{D}^2}{2} - \frac{\hat{\sigma}_e^2}{N}}$$

If the quantity inside the radical is negative, $\hat{\sigma}_o$ is assumed 0. Relating the reproducibility error to the tolerance,

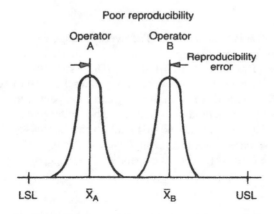

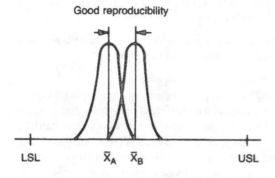

Figure I.6 Gage reproducibility differences.

I.2 Sources of Measurement Error

$$\% \text{ reproducibility} = 100 \times \frac{\text{spread due to operator measurement variability}}{\text{tolerance}}$$

$$= 100 \times \frac{5.15 \hat{\sigma}_o}{\text{tolerance}}$$

The gage measurement variability can be quantified by the combined repeatability and reproducibility (R & R) standard deviation:

$$\hat{\sigma}_{R\&R} = \sqrt{\hat{\sigma}_e^2 + \hat{\sigma}_o^2}$$

Relating this quantity to the tolerance:

$$\% \text{ Repeatability and reproducibility} = 100 \times \frac{\text{spread due to gage measurement variability}}{\text{tolerance}}$$

$$= 100 \times \frac{5.15 \hat{\sigma}_{R\&R}}{\text{tolerance}}$$

Example I.2 Considering having two operators measure a group of 10 parts for which the specification range is 0–25:

Part	Operator A	Operator B	Difference
1	16	18	−2
2	10	8	2
3	5	8	−3
4	7	7	0
5	19	23	−4
6	15	13	2
7	8	8	0
8	10	12	−2
9	12	11	1
10	17	19	−2

$\bar{D} = -0.8 \quad s_D = 2.1$

The standard deviation of the differences is used to compute

$$\% \text{ repeatability} = 100 \times \frac{3.64 \times 2.1}{25}$$

$$= 31\%$$

with $\hat{\sigma}_e = 1.48$. The reproducibility standard deviation is

$$\hat{\sigma}_o = \sqrt{\frac{(-.8)^2}{2} - \frac{(1.48)^2}{10}} = .32$$

Thus,

$$\% \text{ reproducibility} = 100 \, \frac{5.15 \times 0.32}{25}$$

$$= 7\%$$

The combined measurement variability is

$$\hat{\sigma}_{R\&R} = \sqrt{(1.48)^2 + (.32)^2} = 1.51$$

Thus,

$$\%R \& R = 100 \times \frac{5.15 \times 1.51}{25}$$

$$= 31\%$$

Repeatability is obviously the major problem with the gage.

1.2.3 Accuracy and Linearity

Gage accuracy quantifies the relationship between the "true" value of a characteristic and its measured value.

Accuracy of a gage is the difference between the gage measurement value and the "true" value of the measured characteristic.

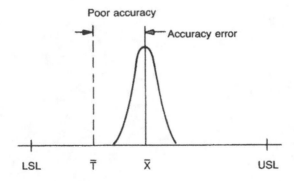

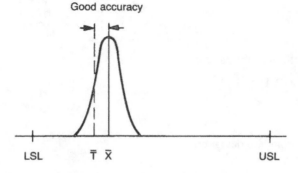

Figure I.7 Gage accuracy differences.

I.2 Sources of Measurement Error

Repeated measurements of a single part are not necessarily the same, so the mean remeasurement $\bar{X}$ is used as the gage value. The true value of a characteristic is an abstract term that will always be unknown. However, the true value is represented by the mean $\bar{T}$ of repeated measurements of a part using a gage of known high accuracy and sensitivity (reference standard, Deming, 1986, p. 292). The difference $\bar{X} - \bar{T}$ is also referred to as bias. Figure I.7 depicts the accuracy concept.

Gage linearity is directly related to accuracy.

Linearity of a gage is the change in accuracy through the operating range of the gage.

Figure I.8 shows a gage that exhibits a lack of linearity. At the midpoint of the specification limits the gage is accurate. At the lower specification limit (LSL), the gage reading is biased high. At the upper specification limit (USL), the gage is biased low.

Any lack of gage accuracy or linearity has a direct impact on product quality since machines are set up and processes monitored using a shift (bias) in the measurement system. Accuracy and linearity should be evaluated when a gage is purchased and periodically thereafter. It is good practice to use high and low masters for calibration so that gages can be mastered at the upper and lower end of the gaging range. This helps to eliminate any problems with a lack of linearity. Some of the causes for any lack of accuracy or linearity follow:

Improper calibration or setup of gage
Too long a period of time between calibration
Worn or incorrect masters

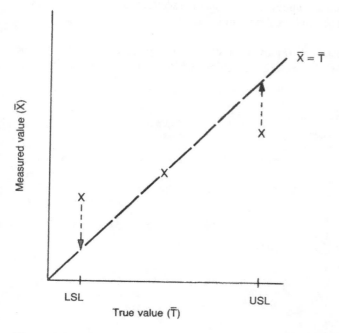

Figure I.8 Example of gage linearity. Changes in accuracy (small dashed lines) are due to lack of linearity.

Use of a single master rather than high and low master
Within-measured-part variability
Malfunction of gage
Improper gaging method
Environmental conditions (e.g., heat)

Accuracy can be evaluated using the standard approach of measuring N parts that span the operating range of the gage. The first reading is obtained from the gage being evaluated and the second from a reference standard gage. The difference in the means is due to a lack of gage accuracy. The actual computed difference is typically not exactly zero. Generally, it is of interest to evaluate whether the difference is 0 so that the gage can be assumed accurate. Procedure 12.5 gives a method to test whether the difference can be assumed to not differ significantly from zero. If the difference is significant, it is useful to relate the lack of accuracy to the tolerance range of the dimension being measured. This practice provides a perspective by which the errors can be evaluated, as shown in Figure I.7. The percentage accuracy is defined simply as

$$\% \text{ accuracy} = 100 \times \frac{\text{accuracy}}{\text{tolerance}}$$
$$= 100 \times \frac{\bar{X} - \bar{T}}{\text{USL} - \text{LSL}}$$

The positive difference is used for the calculation. Procedure I.1 gives a general gage evaluation procedure that can be used to assess accuracy. Linearity can be evaluated using this procedure and selecting parts from different points in the gaging range. Generally, accuracy and linearity are evaluated simultaneously using a scatter plot, which provides an effective display of gaging problems, as shown in Figure 12.26.

Example I.3 Consider measuring 10 parts on a dimensional gage and then having the parts remeasured on a coordinate measuring machine (CMM). The resulting data are as follows:

Part	Gage	CMM	Difference
1	14	12	2
2	15	15	0
3	19	19	0
4	11	13	−2
5	24	25	−1
6	13	16	−3
7	23	21	2
8	11	9	2
9	17	16	1
10	19	24	−5

$\bar{D} = -.4 \quad s_D = 2.37$

I.2 Sources of Measurement Error

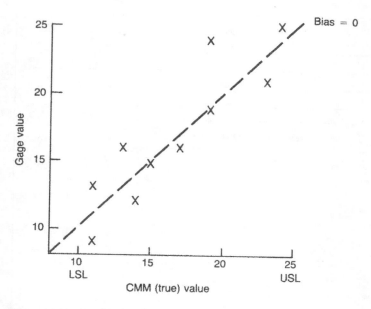

Figure I.9 Accuracy and linearity evaluation for Example I.3.

Following steps 6–8 in Procedure 12.5 results in the following accuracy evaluation for the gage:

$$T = \frac{\sqrt{10} \times 0.4}{2.37}$$
$$= .53$$

Since this difference is less than 2.26 from step 8, we conclude there is no mean difference between the two gages. Although concluding a 0 bias exists, it is important to examine the scatter plot to evaluate whether any lack of linearity exists. Figure I.9 indicates a reasonably balanced scatter around the bias = 0 benchmark line. Thus, there seems to be no evidence for a lack of linearity.

I.2.4 Gage Evaluation Procedure

The evaluation method in Procedure I.1 can be used to assess repeatability, reproducibility, and accuracy. The associate gage evaluation work sheet provides an easy-to-use data collection and analysis form. If the total percentage of gage error is above 30%, corrective actions should be attempted. Experience has shown that most cases exceeding 30% error can be corrected. Simple actions, such as cleaning or replacing worn parts, often significantly reduce gaging variability.

I.2.5 ANOVA Approach*

The method discussed in Procedure I.1 is a special case of a more general analysis of variance (ANOVA) approach for estimating repeatability and reproducibility variability. A common situation is to have two operators perform a gage repeatability study using

Gage Evaluation Work Sheet

Date: _____
Part no./name: _____ Department: _____
Characteristic: _____ Operation: _____
Specification: _____ Gage type: _____
Gage operator: _____ Gage no.: _____

Part	First reading X	Second reading Y	Difference $D = X - Y$
1			
2			
3			
4			
5			
6			
7			
8			
9			
10			

Mean of differences ($\bar{D}$) $A =$ _____
Standard deviation of differences (s_D) $B =$ _____

Study procedure

Collect N parts that span the range of the specification.
Randomly label parts from 1 to 10. Ensure that labeling does not alter the characteristics to be measured.
Calibrate the gage, and verify that the gage is functioning normally.
Measure and record the first reading for all 10 parts.
Randomly remeasure and record the second reading for all N parts. The individual making the second reading should *not* have knowledge of the first reading.
Compute the difference $D = X - Y$ between the first and second readings. This difference may be positive (+) or negative (−). Compute the mean A and standard deviation B of the differences. Perform the following calculations:

Tolerance = $\underbrace{}_{\text{Upper specification}} - \underbrace{}_{\text{Lower specification}} = \underbrace{}_{\text{TOL}}$

Gage variability—repeatability

Same gage used for first and second measurements:

Gage error standard deviation $\hat{\sigma}_e = 0.71 \times \underbrace{}_{B} = \underbrace{}_{C}$

% repeatability $= 515 \times \left(\underbrace{}_{C} \div \underbrace{}_{\text{TOL}} \right) = \underbrace{}$

Operator gaging variability—reproducibility

Same gage, different operator used for first and second measurement:

Operator measurement standard deviation $\hat{\sigma}_o = \sqrt{\left(\underline{} \div 2\right) - \left(\underline{} \div \underline{}\right)} = \underline{}$
$ A^2 C^2 N D$

If the quantity under the square root is negative, operator variability $\hat{\sigma}_o$ is assumed 0.

% reproducibility = $515 \times \left(\underline{} \div \underline{}\right) = \underline{}$
$ D \text{TOL}$

Measurement variability—repeatability and reproducibility (R & R)

Same gage, different operator used for first and second measurements:

Measurement error standard deviation $\hat{\sigma}_{R\&R} = \sqrt{\underline{} + \underline{}} = \underline{}$
$ C^2 D^2 E$

% R & R = $515 \times \left(\underline{} \div \underline{}\right) = \underline{}$
$ E \text{TOL}$

Gage accuracy

Reference standard gage used for second reading:

Accuracy (bias) = $\dfrac{\underline{}}{A}$

Test value for bias: $\dfrac{\underline{}}{\sqrt{N}} \times \left(\dfrac{\underline{}}{A} \div \dfrac{\underline{}}{B}\right) = \dfrac{\underline{}}{T}$

N	t
5	2.78
10	2.26
15	2.14
20	2.09
25	2.06
30	2.05
Maximum	1.96

If T is larger than t, a significant bias probably exists and the percentage accuracy should be assessed for its importance:

% accuracy = $100 \times \left(\underline{} \div \underline{}\right) = \underline{}$
$ A \text{TOL}$

A standard criterion for evaluating the measurement process for the percentage error is 0–10% (excellent), 10–20% (good), 20–30% (marginal), and greater than 30% (unacceptable).

Procedure I.2 Gage Repeatability and Reproducibility

Let two gage repeatability studies be performed on a single gage by two operators. The same N parts are analyzed separately by each operator. This gage repeatability and reproducibility (R & R) analysis combines the information from the two studies to obtain estimates of repeatability error $\hat{\sigma}_e$, reproducibility of operators $\hat{\sigma}_o$, and a combined repeatability and reproducibility $\hat{\sigma}_{R\&R}$. The data can be arranged as follows:

	Operator A			Operator B			
	First reading	Second reading	Average	First reading	Second reading	Average	
Part	X	Y	$\tfrac{1}{2}(X+Y)$	X'	Y'	$\tfrac{1}{2}(X'+Y')$	Grand average
1	X_1	Y_1	$\tfrac{1}{2}(X_1+Y_1)$	X'_1	Y'_1	$\tfrac{1}{2}(X'_1+Y'_1)$	$\tfrac{1}{4}(X_1+Y_1+X'_1+Y'_1)$
2	X_2	Y_2	$\tfrac{1}{2}(X_2+Y_2)$	X'_2	Y'_2	$\tfrac{1}{2}(X'_2+Y'_2)$	$\tfrac{1}{4}(X_2+Y_2+X'_2+Y'_2)$
.	.	.	.	.	.	.	.
.	.	.	.	.	.	.	.
.	.	.	.	.	.	.	.
N	X_N	Y_N	$\tfrac{1}{2}(X_N+Y_N)$	X'_N	Y'_N	$\tfrac{1}{2}(X'_N+Y'_N)$	$\tfrac{1}{4}(X_N+Y_N+X'_N+Y'_N)$
Mean			$\bar{X}_A$			$\bar{X}_B$	$\bar{X}_P$
Standard deviation			s_A			s_B	s_P

s_T = standard deviation of all X, Y, X', and Y' values

The number of parts N should be a minimum of 10 and preferably 30 or more to obtain stable estimates of variability. The parts should ideally span the range of the specification. Also, the second readings should not be biased by knowledge of the first reading. Operators should be unaware of each other's measurements.

The following steps should be used to estimate the two sources of measurement variability:

1. Compute the average of the first and second reading for operator A, $\tfrac{1}{2}(X_i + Y_i)$. Compute the mean $\bar{X}_A$ and standard deviation s_A for these N values. Make the same calculations for operator B. Compute the grand average of the first and second readings for both operators, $\tfrac{1}{4}(X_i + Y_i + X'_i + Y'_i)$. Compute the mean $\bar{X}_P$ and standard deviation s_P for these N part values. Compute the standard deviation s_T of all X, Y, X', and Y' values (4N readings).
2. Compute the estimated standard deviations for repeatability and reproducibility.

Repeatability (error):
$$\hat{\sigma}_e = \sqrt{\frac{(4N-1)s_T - 4(N-1)s_P^2 - N(\bar{X}_A - \bar{X}_B)^2}{3N-1}}$$

Reproducibility (operator):
$$\hat{\sigma}_o = \sqrt{\frac{N(\bar{X}_A - \bar{X}_B)^2 - \hat{\sigma}_e^2}{2N}}$$

Repeatability and reproducibility:
$$\hat{\sigma}_{R\&R} = \sqrt{\hat{\sigma}_e^2 + \hat{\sigma}_o^2}$$

The standard 99% spread is obtained by multiplying these standard deviations by 5.15. If a negative value is obtained under the radical, the estimated standard deviation is assumed zero.

3. Compute the percentage error due to the three sources:

I.2 Sources of Measurement Error

$$\% \text{ repeatability} = 100 \times \frac{5.15\hat{\sigma}_e}{\text{USL} - \text{LSL}}$$

$$\% \text{ reproducibility} = 100 \times \frac{5.15\hat{\sigma}_o}{\text{USL} - \text{LSL}}$$

$$\% \text{ R \& R} = 100 \times \frac{5.15\hat{\sigma}_{R\&R}}{\text{USL} - \text{LSL}}$$

where the upper and lower specification limits are denoted by USL and LSL, respectively. A standard criterion for evaluating the measurement process for the percentage error is 0–10% (excellent), 10–20% (good), 20–30% (marginal), and greater than 30% (unacceptable).

Procedure I.1. The ANOVA procedure combining the analysis information is given in Procedure I.2. A comparable procedure using ranges is given in the *Statistical Process Control Manual* (American Society for Quality Control, 1986).

Example I.4 Consider the following data obtained when operators measured the same 10 parts; the specification range is 0–25.

Part	Operator A First reading	Operator A Second reading	Operator A Average	Operator B First reading	Operator B Second reading	Operator B Average	Grand average
1	16	14	15.0	18	15	16.5	15.75
2	10	9	9.5	8	9	8.5	9.00
3	5	7	6.0	8	8	8.0	7.00
4	7	7	7.0	7	8	7.5	7.25
5	19	21	20.0	23	22	22.5	21.25
6	15	14	14.5	13	15	14.0	14.25
7	8	5	6.5	8	10	9.0	7.75
8	10	11	10.5	12	12	12.0	11.25
9	12	11	11.5	11	10	10.5	11.00
10	17	19	18.0	19	18	18.5	18.25
Mean			$\bar{X}_A = 11.85$			$\bar{X}_B = 12.70$	$\bar{X}_P = 12.275$
Standard deviation			$s_A = 4.888$			$s_B = 5.073$	$s_P = 4.935$

1. The means and standard deviations are shown. The standard deviation of all 40 values is $s_T = 4.873$.
2. The estimated standard deviations are as follows:

Repeatability

$$\hat{\sigma}_e = \sqrt{\frac{(4\cdot 10 - 1)(4.873)^2 - (4)(10 - 1)(4.935)^2 - (10)(11.85 - 12.70)^2}{(3\cdot 10 - 1)}}$$

$$= \sqrt{\frac{926.10 - 876.75 - 7.225}{29}}$$

$$= 1.21$$

Reproducibility

$$\hat{\sigma}_o = \sqrt{\frac{(10)(11.85 - 12.70)^2 - (1.21)^2}{20}}$$

$$= 0.55$$

Repeatability and reproducibility

$$\hat{\sigma}_{R\&R} = \sqrt{(1.21)^2 + (0.55)^2}$$

$$= 1.33$$

3. Since USL $-$ LSL $= 25 - 0 = 25$, the percentage error due to each source of variation is

$$\% \text{ repeatability} = 100 \times \frac{5.15 \times 1.21}{25} = 25\%$$

$$\% \text{ reproducibility} = 100 \times \frac{5.15 \times 0.55}{25} = 11\%$$

$$\% \text{ R \& R} = 100 \times \frac{5.15 \times 1.33}{25} = 27\%$$

This gaging process would be judged marginal.

This estimated gage repeatability utilizes the combined repeatability for each operator. Computed separately using the Procedure I.1 work sheet, the estimated repeatability is

% repeatability for operator $A = 26\%$

% repeatability for operator $B = 23\%$

The combined estimate of 25% is a better estimate of the true gage repeatability since it is based on more data (i.e., the error sums of squares and operator, part interaction sums of squares).

The more general case with N parts, M operators, and n repeated measurements of the N parts by each operator can be evaluated using a standard two-factor (operator and part) ANOVA approach (Tsai, 1988). The ANOVA table appears as follows:

Source	Degrees of freedom	Sum of squares	Mean square
Operator	$M - 1$	SS_O	$MS_O = SS_O/(M - 1)$
Part	$N - 1$	SS_P	$MS_P = SS_P/(N - 1)$
Error	$NMn - N - M + 1$	$SS_{\text{pool}} = SS_{O\times E} + SS_E$	$MS_{\text{pool}} = SS_{\text{pool}}/(NMn - N - M + 1)$
Total	$NMn - 1$	SS_T	

I.3 Gage Sensitivity

This analysis assumes no interaction exists between operators and parts so that the operator, part interaction sum of squares ($SS_{O \times E}$) is pooled with the error sum of squares (SS_E). The table can be obtained from many computer programs. The sources of measurement variability then becomes

$$\text{Repeatability} \quad \hat{\sigma}_e = \sqrt{MS_{\text{pool}}}$$

$$\text{Reproducibility} \quad \hat{\sigma}_o = \sqrt{\frac{MS_O - MS_{\text{pool}}}{Nn}}$$

$$\text{Repeatability and reproducibility} \quad \hat{\sigma}_{R\&R} = \sqrt{\hat{\sigma}_e^2 + \hat{\sigma}_o^2}$$

Again, if any quantity under the radical is negative, the estimated standard deviation is assumed 0.

I.3 Gage Sensitivity

The sensitivity of a gage refers to its ability to discriminate between parts. For example, a gage that reads in units of .0001 inch is more sensitive than one that reads to .001 inch. In Chapter 4 we noted that a lack of gage sensitivity can cause false out-of-control signals due to a range estimate that is too low. In most manufacturing situations, it is desirable to have at least 10 units within the specification limits (Chap. 2). This practice enables assessment of whether a process is capable ($C_p = 1.0$ or greater). However, for effective process control, 10 units should be present within the process spread ($6\hat{\sigma}$).

Wheeler and Chambers (1986) give a useful criterion for evaluating the adequacy of a gage's sensitivity for use with control charts. If there are four or fewer possible values within the range control limits, inadequate sensitivity exists. For example, if $UCL_R = 3.2$ and the data are recorded as integers, then the range values below UCL_R are 0, 1, 2, and 3 so the gage has inadequate sensitivity. The consequence of any insensitivity is that $\bar{R}$ will be too low, causing control chart limits to be too narrow. These narrow limits result in false out-of-control signals. Also, process capability estimates will be overstated.

Lack of gage sensitivity is often a problem with a process that has high capability (e.g., $C_p = 2.0$ or higher). This is illustrated by considering a simple example. Suppose a process has a tolerance of $\pm .0005$ inch and the gaging is designed so that the increments of measurement are .0001 inch. For a capable process with $C_p = 1.0$, the process spread is .001, which contains the desirable 10 measurement scale divisions. In this case,

$$6\hat{\sigma} = .001 \quad \text{or} \quad \hat{\sigma} = .000167$$

The upper control limit for a subgroup size of $n = 5$ is

$$UCL_R = D_4 \bar{R} \quad \hat{\sigma} = \frac{\bar{R}}{d_2}$$
$$= D_4 d_2 \hat{\sigma}$$
$$= 2.12 \times 2.33 \times .000167$$
$$= .00082$$

Thus, there are nine possible values below UCL_R, which indicates adequate gaging sensitivity exists. Suppose the process is improved so that $C_p = 2.0$; only 50% of the tolerance is now being used. In this case,

$$6\hat{\sigma} = .0005$$
$$\hat{\sigma} = .000083$$

and

$$\text{UCL}_R = .00041$$

which implies that the gage sensitivity is now marginal since there are five possible values below UCL_R. Table I.2 considers $C_p = 1.33$ (75% of tolerance) and different subgroup sizes using coded data ($\times 10{,}000$). As the capability increases, fewer values are within the range control limit, which indicates an increasing sensitivity problem. The smaller subgroup sizes make the problem worse. Fortunately, as the process capability increases the need for intensive use of process control decreases and it is possible to use alternative control methods. The discussion associated with Table 14.4 presents some possibilities. Also, Table 14.4 lists the gaging divisions and process capability so that gage sensitivity problems can be anticipated.

The possibility that out-of-control signals may result from poor gage sensitivity is not known to many users of control charts. Appreciation of this can be gained by considering an example that compares different measurement sensitivities. Consider the situation in which the diameter of a part has a specification of $1.345 \pm .001$ inch. The measurements are coded so that

$$\text{Actual value} = 1.345 + .0001 \times \text{gage value}$$

where $\text{LSL} = -10$ and $\text{USL} = 10$. Consider four different measurement scales with decreasing sensitivity:

Gaging units	Number of gaging intervals	Gage scale
.0001	20	−10 ─── 0 ─── 10
.0002	10	−10 ─── 0 ─── 10
.0005	4	−10 ─── 0 ─── 10
.001	2	−10 ─── 0 ─── 10

Table I.2 Range Upper Control Limit for Ten Measurement Scale Divisions Within the Specification Limits

Capability C_p	Subgroup size			
	2	3	4	5
1.0 ($\hat{\sigma} = 1.67$)	6.2	7.3	7.8	8.2
1.3 ($\hat{\sigma} = 1.25$)	4.6	5.4	5.9	6.1
2.0 ($\hat{\sigma} = .83$)	3.1	3.6	3.9	4.1

I.3 Gage Sensitivity

The impact of the different measurement sensitivities can be illustrated by generating 20 subgroups of size $n = 4$ from a normal distribution with mean 0 and $\sigma = 3.33$. The data for the 20- and 4-interval measurement scales appear in Table I.3. The same data were used to create the results for each of the remaining two measurement scales by rounding to the appropriate scale division. Thus, the measurements in the third subgroup represent the following values:

Actual value	Coded actual value	Gage value for different intervals			
		20	10	4	2
1.34508	.8	1	0	0	0
1.34443	−5.7	−6	−6	−5	−10
1.34510	1.0	1	2	0	0
1.34488	−1.2	−1	−2	0	0

Table I.3 Artificial Data Showing Influence of Measurement Sensitivity

Part	20 Measurement intervals				$\bar{X}$	R	4 Measurement intervals				$\bar{X}$	R
	Measurements						Measurements					
1	−4	−1	−1	−2	−2.00	3	−5	0	0	0	−1.25	5
2	6	−2	0	2	1.50	8	5	0	0	0	1.25	5
3	1	−6	1	−1	−1.25	7	0	−5	0	0	−1.25	5
4	0	0	−2	0	−.50	2	0	0	0	0	.00	0
5	4	0	1	−1	1.00	5	5	0	0	0	1.25	5
6	−2	2	0	0	.00	4	0	0	0	0	.00	0
7	−5	0	0	1	−1.00	6	−5	0	0	0	−1.25	5
8	−1	0	−2	3	.00	5	0	0	0	5	1.25	5
9	5	5	−2	6	3.50	8	5	5	0	5	3.75	5
10	1	−1	−1	−2	−.75	3	0	0	0	0	.00	0
11	1	−3	−7	3	−1.50	10	0	−5	−5	5	−1.25	10
12	−4	3	−4	−1	−1.50	7	−5	5	−5	0	−1.25	10
13	−8	2	3	−4	−1.75	11	−10	0	5	−5	−2.50	15
14	−1	2	0	5	1.50	6	0	0	0	5	1.25	5
15	−1	0	0	2	.25	3	0	0	0	0	.00	0
16	6	0	2	1	2.25	6	5	0	0	0	1.25	5
17	−2	−2	2	0	−.50	4	0	0	0	0	.00	0
18	3	−1	0	1	.75	4	5	0	0	0	1.25	5
19	1	2	3	1	1.75	2	0	0	5	0	1.25	5
20	−3	1	−1	−3	−1.50	4	−5	0	0	−5	−2.50	5
					$\bar{\bar{X}} = .013$	$\bar{R} = 5.4$					$\bar{\bar{X}} = .062$	$\bar{R} = 4.75$

Table I.4 Control Chart Values for Different Gage Sensitivities

Units	$\bar{X}$	$LCL_{\bar{X}}$	$UCL_{\bar{X}}$	$\bar{R}$	UCL_R	No. Possible R Values	$\hat{\sigma} = \bar{R}/d_2$	Individuals SD
.0001	.01	−3.9	3.9	5.4	12.3	13	2.62	2.78
.0002	.05	−4.0	4.0	5.5	12.5	7	2.67	2.77
.0005	.06	−3.4	3.5	4.75	10.8	3	2.31	2.92
.001	.00	−2.2	2.2	3.0	6.8	1	1.46	2.76

The control chart values appear in Table I.4 and the charts in Figure I.10. The decreased measurement sensitivity causes $\bar{R}$ to be too low, which makes out-of-control signals more likely owing to the reduced spread of the control limits. Out-of-control signals are due to the insensitivity of the measurement scale, not special causes. If we were to look for special causes, none would exist, creating a lack of confidence in the charts. Moderate insensitivity of the measurement system may or may not produce a meaningful chart—the user cannot trust the chart.

Three alternatives should be considered. First, the standard deviation s of the combined individual measurements is not as heavily influenced as $\bar{R}$ by a lack of gage sensitivity, as seen in Table I.4. Using $\bar{R} = d_2 s$, it is possible to get a better estimate of $\bar{R}$ that can then be used to compute control chart limits. Second, alternative control methods, such as a measurement scale check sheet, may be appropriate. Finally, a more sensitive gage can be purchased. If the process capability is high, this action may not be required.

I.4 Gage Stability

Chapter 3 emphasized the only method of assessing the stability of a process was to utilize control charts and demonstrate that no special causes of variation are present (i.e., no out-of-control signals). The stability of the gaging process should also be an area of concern. As with a production process, the stability of the gaging process can only be evaluated using control charts. To assess gage stability it is necessary to remeasure the same parts on an ongoing basis or conduct an ongoing gage repeatability study. Since remeasurement of the same parts is involved, there is a natural human tendency for operators to be influenced by prior measurements. The data collection plan for the control chart should seek to minimize this influence. Three types of control charts are considered here.

Gage Target Control Chart. The simplest type of gage control chart involves ongoing repeated measurements of the same n (say, $n = 5$) parts. First, n parts spanning the specification limits should be selected. These parts should then be measured by a sensitive and accurate gage that is typically considered a reference standard for the measurement. Let the assumed "true" values of the n parts be denoted $T_1, T_2, \ldots, T_n$. The average of all the n parts $\bar{T}$ is the target value for the average of the repeated measurements. The second step involves a baseline gage repeatability study using a reasonably large number of parts (say, $N = 30$) and the best gage operator (Procedure I.1). Care should be taken to ensure

I.4 Gage Stability

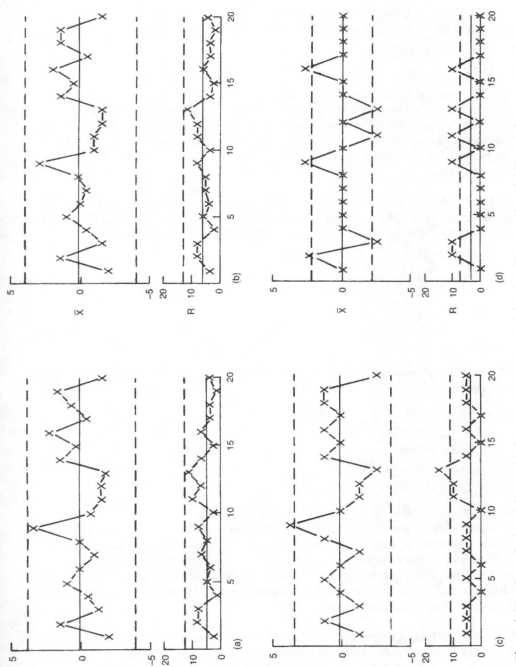

Figure I.10 Control charts for evaluating measurement sensitivity: (a) 20 intervals; (b) 10 intervals; (c) 4 intervals; (d) 2 intervals.

that the gage is in good operating condition. The intent of the study is to obtain a good estimate of the repeatability error $\hat{\sigma}_e$. The control limits can now be simply calculated as

$$\text{UCL}_{\bar{X}} = \bar{T} + A\hat{\sigma}_e$$
$$\text{LCL}_{\bar{X}} = \bar{T} - A\hat{\sigma}_e$$

where values of A appear in Appendix III.

Maintaining the chart involves remeasurement of the n parts and plotting the mean $\bar{X}$. Trends on one side of $\bar{T}$ indicate an accuracy (drift) problem. Extreme variability around $\bar{T}$ indicates a change in repeatability. This control chart approach is most suitable for automatic or digital output gages since recorded measurements are not influenced by operator recall.

Gage Difference Control Chart. A slight modification to the target control chart can provide a more explicit monitoring of gage repeatability. The true values $T_1, T_2, \ldots, T_n$ are subtracted from the individual gage measurements to form the differences

$$D_1 = X_1 - T_1$$
$$D_2 = X_2 - T_2$$
$$\vdots$$
$$D_n = X_n - T_n$$

The mean $\bar{D}$ and standard deviation of these differences can be plotted on a $\bar{X}$ and s chart with control limits:

$$\text{UCL}_{\bar{D}} = A\hat{\sigma}_e$$
$$\text{LCL}_{\bar{D}} = -A\hat{\sigma}_e$$
$$\text{UCL}_R = B_6\hat{\sigma}_e$$
$$\text{Centerline} = c_4\hat{\sigma}_e$$
$$\text{LCL}_R = B_5\hat{\sigma}_e$$

where the control chart constants appear in Appendix III.

Maintaining the control chart involves remeasurement of the n parts, subtracting the true values to obtain the differences and plotting the mean and standard deviation of the differences as with the target chart method. Since $\hat{\sigma}_e$ is used as a baseline, any deterioration in the accuracy or repeatability is reflected on the mean and standard deviation control chart, respectively. Of course, a standard control chart for the D values can also be used, which would assume $\hat{\sigma}_e$ is not known.

Gage Repeatability Control Chart. A disadvantage of the previous methods is that it is necessary to use the same n parts or evaluate T_i for every new subgroup. An alternative approach is to conduct an ongoing gage repeatability evaluation. Using the gage evaluation work sheet, the difference $\bar{D}$ and the percentage repeatability can be obtained on an ongoing basis. These values can be plotted on a standard $\bar{X}$ and s control chart. The control limits are

$$\text{UCL}_{\bar{D}} = \bar{\bar{D}} + A\sqrt{2}\hat{\sigma}_e$$
$$\text{LCL}_{\bar{D}} = \bar{\bar{D}} - A\sqrt{2}\hat{\sigma}_e$$

I.4 Gage Stability

$$\text{Baseline \% repeatability} = \frac{515\hat{\sigma}_e}{\text{tolerance}}$$

$$\text{UCL}_\% = B_6 \text{ (baseline \% repeatability)}$$
$$\text{Centerline} = c_4 \text{ (baseline \% repeatability)}$$
$$\text{LCL}_\% = B_5 \text{ (baseline \% repeatability)}$$

where the control chart constants can be obtained from Appendix III and $\bar{\bar{D}}$ is the overall mean of subgroup $\bar{D}$ values, which is normally assumed to be 0. Again, $\hat{\sigma}_e$ is from the baseline study when the gage was known to be functioning normally.

If the same operator and gage are used to make the repeated measurements, the $\bar{D}$ should be close to zero. However, the more interesting application of this control chart is to have different operators perform the two groups of measurements, in which case $\bar{D}$ may not be close to zero. This approach is then an ongoing evaluation of the production measurement system, but accuracy is not assessed.

Example I.5 A baseline study was conducted with $N = 30$ parts, and it was determined that $\hat{\sigma}_e = 0.58$ and tolerance = 20.

$$\text{Baseline \% repeatability} = \frac{515\hat{\sigma}_e}{\text{tolerance}}$$
$$= \frac{515 \times 0.58}{20}$$
$$= 15\%$$

The study was conducted after the gage had been reconditioned and was considered to be operating satisfactorily. An experienced operator performed the study. A gage evaluation control chart was established to ensure stability of the measurement process. The sampling plan consisted of an operator measuring five parts at the beginning of a shift. At the end of the shift, a second operator remeasured the parts. The differences from the gage evaluation work sheet were plotted on the control chart in Figure I.11. The control limits are as follows:

$$\text{UCL}_{\bar{D}} = 1.34 \times \sqrt{2} \times 0.58 = 1.10$$
$$\bar{\bar{D}} = 0$$
$$\text{LCL}_{\bar{D}} = -1.34 \times \sqrt{2} \times 0.58 = -1.10$$
$$\text{UCL}_\% = 1.96 \times 15 = 29.4\%$$
$$\text{Centerline} = .94 \times 15 = 14.1\%$$

The gaging process appears to have remained stable in both reproducibility and repeatability over time and a number of operators. The $n = 5$ ongoing gage repeatability study results in observed percentage repeatabilities ranging between 25.5 and 7.3%. This variability emphasizes that making a decision about a gage based on only a few repeated measurements is not good practice.

A supplier was continually having shipments rejected by a manufacturing plant. A gage evaluation control chart was established using a simple procedure. Included with each shipment were $n = 5$ tagged parts that were labeled with the supplier's measured value. The plant then measured the parts and recorded the bias $\bar{D}$ and percentage repeatability

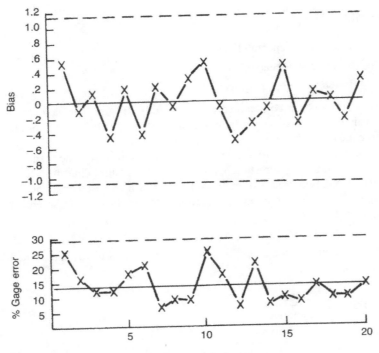

Figure I.11 Gage evaluation control chart.

on a gage evaluation control chart. It was quickly determined that the two gaging systems did not agree. The problem was corrected, and the control chart verified the required gaging system stability.

I.5 Gage Purchase

The initial purchase of a gage provides a unique opportunity to design and test a gage so that all engineering and production requirements are met. Correcting gaging problems once a gage is placed in production can be a time-consuming, expensive process. More importantly, the quality of the products produced will suffer if gages are not designed and qualified properly. Control of the measurement process that monitors product quality starts with the initial gage purchase.

Procedure I.3 gives a method for purchasing a gage that addresses the topics discussed in this appendix. Central to this evaluation procedure is the repeatability and accuracy study presented in Procedure I.1. The common practice of evaluating a gage's quality by repeatedly remeasuring a gage master provides minimal information on how the gage will perform on the production floor. In general, it is the failure of gage vendors to adequately address normal within-part variability that increases repeatability error. The presence of minor burrs, flash, or other surface irregularities can cause significant problems, as Example I.6 illustrates. In most cases a minor change in the gage can significantly reduce

Procedure I.3 Gage-Purchasing Procedure

The purchase of a gage starts with defining the initial requirement for a measurement and ends with the gage functioning properly on the production floor. The gage buyer and vendor have a responsibilty to work together to address the following topics:

1. Define the measurement requirements.

    Identify the functional requirement for the characteristics being measured. Does the measured characteristic meet the intent of the customer's functional requirement? Why is control of the characteristic required?

    Is the measurement requirement operationally defined so that no possibility of misinterpretation exists?

2. Develop the gaging concept.

    How is the characteristic to be manufactured? If possible, locate the part in the gage corresponding to the method of manufacturing.

    The gage design should address the expected within-part variability that could influence the positioning of the part (e.g., machining irregularities, such as burrs).

    Gage correlation problems should be addressed if the gaging concept differs from other gages measuring the same characteristic in other machining or assembly departments or at a part supplier.

    The gage should have at least 10 scale divisions within the specification limits.

3. Evaluate the gage masters.

    A gage master should be available for the upper and lower specification limits.

    Gage masters should be measured by an accepted reference standard.

    A gage mastering procedure should be developed.

    The gage should be allowed to run an extended period of time to ensure there is no drift from an initial mastering of the gage.

    The time between required mastering of the gage should be determined so that no drift is experienced during production operations.

4. Perform the gage evaluation test.

    Select $N = 30$ parts that span the specification range and exhibit normal within-part variability. Use the gage evaluation Procedure I.1 to evaluate repeatability, reproducibility, and accuracy.

    The combined gaging error should be less than 20% of the tolerance, with a target of less than 10%.

5. Identify preventive maintenance requirements.

    A checklist of PM requirements for the gage should be developed.

    Appropriate spare parts should be determined.

6. Develop training materials.

    Prepare training materials for gage operation.

    Prepare operator certification for gage operation, including reproducibility studies.

7. Provide visual cue boards.

    Develop a display board to be located adjacent to the gage that specifies the following:

    Part characteristic (with tolerances) being measured indicated on a drawing

    Mastering instructions and required intervals

    Coding of the measured values

    Operating instructions

    Boards should use picture cues when possible.

8. Conduct a production tryout.

    The gage should be tested in the production setting to ensure correct operation.

    Gage evaluation should be conducted to ensure that no changes have occurred.

    A plan should be developed to conduct repeatability and accuracy studies on an ongoing basis.

gaging variability. Only by performing repeatability studies on production parts will these problems be identified. In a small percentage of cases, major gaging flaws are uncovered. In either case, the gage buyer is well served to follow Procedure I.3.

Example I.6 A gage to measure a height dimension was evaluated and found to have 76% repeatability error. Further investigation determined that some parts had a burr on the locating surface that caused the inconsistency. This burr was removed in later processing and did not affect other machining operations. It was decided that a simple modification to the gaging head would eliminate the problem, as shown in Figure I.12. The repeatability after modification was 13%. The data are as follows:

	Before modification			After modification		
	1	2	Difference	1	2	Difference
	−1.5	−1.0	−.5	0	0	0
	0	0	0	1.0	1.0	0
	2.0	−1.0	3.0	1.0	1.0	0
	−.5	0	−.5	0	0	0
	2.0	2.5	−.5	1.0	1.0	0
	1.0	1.0	0	1.0	1.0	0
	0	0	0	.5	0	.5
	1.0	3.0	−2.0	0	0	0
	−2.0	−2.5	.5	0	0	0
	−.5	−1.0	.5	1.5	1.0	.5
Mean			$\bar{D} = .05$			$\bar{D} = .1$
SD			$s_D = 1.26$			$s_D = .21$

The tolerance was ±3, so the second study did not select parts that adequately spanned the specification range. However, there was obvious improvement in the gaging error.

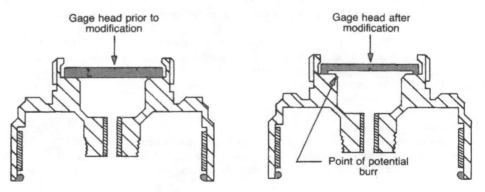

Figure I.12 Gage design for Example I.6.

Appendix II
Geometric Dimensioning and Tolerancing

II.1 Introduction

Geometric dimensioning and tolerancing (GDT) is a technical data base through which product engineering and manufacturing organizations can communicate via product drawings, whether on paper or on computer graphics terminals. It is the engineering product methodology to geometrically *describe design intent* and provide the documentation base for the design of the quality and production systems. It is a technique of communication between design, manufacturing, and inspection via symbology that guarantees a uniform interpretation of the requirements necessary for producing the component.

GDT to a recognized standard provides the dimensions of a component and its tolerances in a language that eliminates confusing and inconsistent notes, implied datums, local interpretations, and incomplete specifications. Engineering requirements can be readily defined with the specific setup required to verify acceptance. Many of the current product drawings released to manufacturing locations fail to identify an adequate "datum reference frame" from which repeatable measurements can be made. Until recently, the full potential of GDT was not explored. There has been only minimal training in this discipline, and most of these efforts were directed toward engineering drafting personnel.

The ANSI Y14.5M-1982 standard, properly applied, provides a method of dimensioning and tolerancing part features with reference to the actual function or relationship of these features. When effectively utilized, it can enhance economics, productivity, quality, and repeatability for process control.

The advantages of using GDT symbology to a recognized national standard are numerous. Several noteworthy examples are as follows:

Symbols are more quickly drawn and, because of their compactness, can be applied at or near the place on the drawing to which they apply.
With combinations of symbols, requirements may be clearly stated that would otherwise require long notes.

Based upon ANSI Y14.5M-1982. This appendix is intended to serve as a reference only in understanding geometric dimensioning and tolerancing and is a condensed version of some of the material. For additional information, refer to American National Standard ANSI Y14.5M-1982, *Dimensioning and Tolerancing,* published by the American Society of Mechanical Engineers, United Engineering Center, 345 East 47th Street, New York, NY 10017. Ford GDT Group, March 20, 1987.

Symbols are clearly in view and uniformly interpretable at the place of application on the drawing.

The symbol approach tends to provide more order and precision to the general drawing appearance.

The use of symbolic methods is in keeping with the objectives of international standardization, in which barriers of language, as well as technical understanding, must be overcome.

II.2 Common Terms and Definitions

Basic Dimension. This is a numerical value used to describe the theoretically exact size, profile, orientation, or location of a feature or datum target. It is the basis from which permissible variations are established by tolerances on other dimensions, in notes, or in the feature control frame.

Maximum Material Condition (MMC). Denoted Ⓜ this is the condition in which a feature of size contains the *maximum* amount of material within the stated limits of size (e.g., minimum hole diameter and maximum shaft diameter).

Least Material Condition (LMC). Denoted Ⓛ this is the condition in which a feature of size contains the *least* amount of material within the stated limits of size (e.g., maximum hole diameter and minimum shaft diameter).

Regardless of Feature Size (RFS). Denoted Ⓢ this term is used to indicate that a geometric tolerance or datum reference applies at any increment of size of the feature within its size tolerance. (No additional tolerance of position or form is available, no matter what the size.)

Full Indicator Movement (FIM). This is the total movement of an indicator when appropriately applied to a surface to measure its variations [formerly called total indicator reading (TIR) or full indicator reading (FIR)].

Virtual Condition. This is the boundary generated by the collective effects of the specified MMC limit of size of a feature and any applicable geometric tolerances.

External feature: MMC size *plus* any applicable geometric tolerances
Internal feature: MMC size *minus* any applicable geometric tolerances

Feature Control Frame. The feature control frame shown in Figures II.1 and II.2 is where the specific geometric control for an individual feature is identified. It consists of (1) type of control (geometric characteristic), (2) tolerance zone (shape and tolerance), (3) tolerance zone modifiers (i.e., MMC, LMC, or RFS) if applicable, and (4) datum references (if applicable) and any datum modifiers (if applicable).

II.3 Datums

A datum is a theoretically exact point, axis, or plane derived from the true geometric counterpart of a specified datum feature. A datum is the origin from which the location of

II.4 Datum Reference Frame

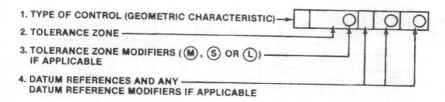

Figure II.1 Feature control frame.

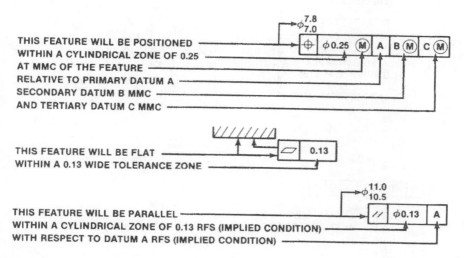

Figure II.2 Feature control frame breakdown.

geometric characteristics of features of a part is established. A *datum feature* is an actual feature of a part used to established a datum. A *simulated datum* is the manufacturing or inspection equipment used to simulate a true datum. These datum concepts are illustrated in Figure II.3.

The datum reference features illustrated in Figure II.4 are *specified*, and their order of precedence is dictated in the feature control frame.

Targeted datums are those specified reference features that have specific point(s), line(s), or area(s) dimensioned on the part to be used to establish the datum, as shown in Figures II.5 through II.7.

II.4 Datum Reference Frame

The part features selected as datums are identified as such on the drawing. These datum features, used as a set, fix the part in relation to three mutually perpendicular planes, jointly called a datum reference frame. Figures II.4 and II.8 illustrate this practice.

The selection of datum features identifies the relationships of features important to the part *function* and design intent. If the drawing does not convey the feature relationships

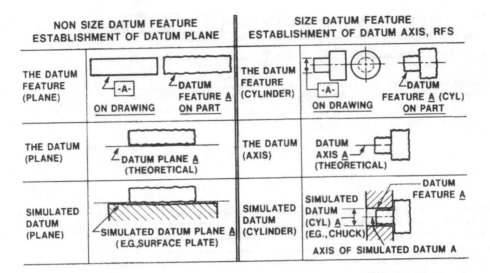

Figure II.3 Datums.

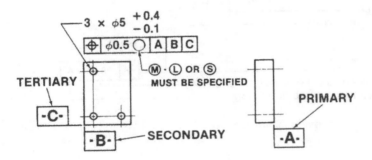

Figure II.4 Example of specified datums.

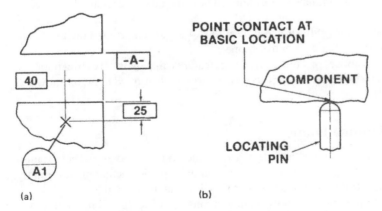

Figure II.5 A datum target point is indicated by the symbol ×, which is dimensionally located on a direct view of the surface. (a) Drawing callout; (b) interpretation.

II.4 Datum Reference Frame

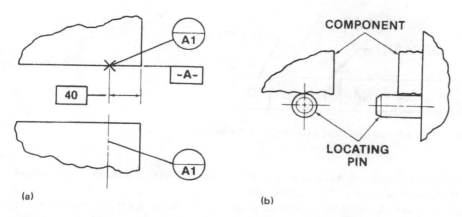

Figure II.6 A datum target line is indicated by the symbol × on an edge view of the surface, a phantom line on the direct view, or both. (a) Drawing callout; (b) interpretation.

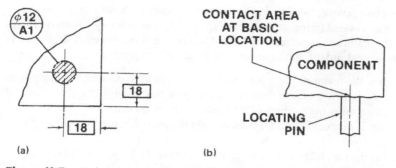

Figure II.7 A datum target area is indicated by section lines inside a phantom outline of the desired shape, with controlling dimensions added. The diameter of circular areas is given in the upper half of the datum target symbol. (a) Drawing callout; (b) interpretation.

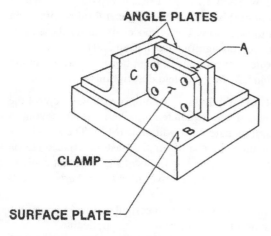

Figure II.8 Setup using datums.

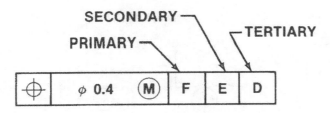

Figure II.9 Order of preference for datum references.

essential to part function, there is no assurance that manufacturing and inspection personnel will use identical procedures for locating parts.

Machine tables and movements, surface plates, and so on, are not true planes but are of such high quality that they simulate datum planes. Measurements are made from planes, axes, and points in the processing equipment.

A problem arises however, when the processing equipment has no particular sequence or datums specified (primary, secondary, and tertiary). To ensure uniform manufacturing and processing of a component, the drawing specifies the sequence or rank of datums by its position in the feature control frame as shown in Figure II.9. The primary datum is always the *first* datum in the feature control frame. It should be noted that the sequence of datums is *not* based on alphabetical order but on the basis of functional requirements.

The *primary datum* is the feature(s) of a part on which a minimum of three contact points must be obtained. It is a physical feature on the part that establishes the origin of dimensional relationships from which other features are evaluated.

The *secondary datum* is the part feature(s) used to align the part. A minimum of two points of contact are required.

The *tertiary datum* is the part feature used to position the part. Only one point of contact is required.

II.5 Rules

These general rules have been established by ANSI Y14.5M-1982, to provide users with a better understanding and proper application of this standard. Five general rules have been established that all users of this standard must follow. For purposes of clarity these rules have been numbered; however, the number has no significance in itself and does not appear in the standard. A brief listing of these rules are referenced here; however, users should review Sections 2.7–2.11 in the ANSI standard for complete descriptions.

1. Unless otherwise specified, the limits of an individual feature of size will control the form as well as the size. No element of the actual feature shall extend beyond a boundary of perfect form at the maximum material condition (MMC). The actual size of the feature measured at any cross section shall be within the least material condition (LMC) limit of size. The control of form prescribed by limits of size does not apply to commercial stock sizes established by industry or government standards or parts subject to free-state variation.
2. For tolerance of position, MMC, LMC, or RFS must be specified on the drawing with respect to the individual tolerance, datum reference, or both, as applicable.
3. For all other geometric controls, when no modifier is specified, RFS applies with

II.6 Main Geometric Characteristics

respect to the individual tolerance datum reference, or both. When MMC is required, it must be specified on the drawing.

4. When tolerances of form or position are expressed by symbols or notes, each such tolerance applicable to a screw thread shall be understood to apply to the axis of the thread derived from the pitch cylinder. If the pitch cylinder is not desired, the applicable feature must be stated below the control frame (e.g., major ϕ or minor ϕ). For gears and splines, a qualifying notation *must* be added to the symbol or note (e.g., major ϕ, minor ϕ, or PD). Nothing is implied.
5. When a feature is designated to be used as a datum, these datum features apply at their virtual condition when used to verify a feature relationship even though the feature control frame specifies MMC for the datum(s).

II.6 Main Geometric Characteristics

II.6.1 Form Tolerances

Form tolerances are used to state how far an actual surface or feature is permitted to vary from the desired form implied on a drawing. Form tolerances are applicable to single features or elements of single features only; therefore, the feature control frame for form tolerances does not specify a datum reference frame. The four types of form control are (1) straightness, (2) flatness, (3) circularity, and (4) cylindricity.

1. Straightness. A condition in which an element of a surface or an axis is a straight line. The considered element or axis must lie within the tolerance zone specified in the feature control frame. There are three variations of straightness; each is illustrated in Figures II.10 through II.12.
2. Flatness. A condition on a surface with all elements in one plane (Fig. II.13). The entire surface must lie between two parallel planes. The distance between the two planes is dictated by the tolerance specified in the feature control frame. Note that the orientation of the planes to the surface is established by the surface itself and has no relationship to any other feature on the part.

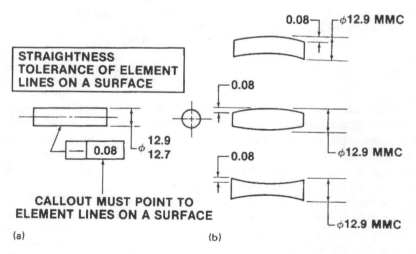

Figure II.10 Straightness case 1. (a) Drawing callout; (b) interpretation.

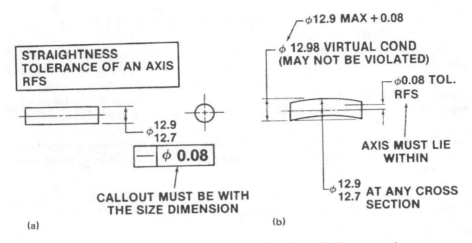

Figure II.11 Straightness case 2. (a) Drawing callout; (b) interpretation.

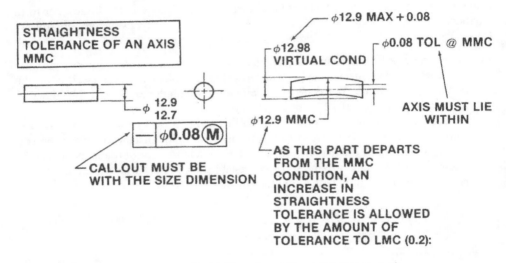

Figure II.12 Straightness case 3. (a) Drawing callout; (b) interpretation.

3. Circularity. A condition on a surface of revolution (cylinder, cone, or sphere) where all points of the surface intersected by any plane perpendicular to a common axis (cylinder or cone) or passing through a common center (sphere) are equidistant from the axis of that center (Fig. II.14). Each circular element of the surface must lie between two concentric circles. The width (radial separation) between the two circles is the tolerance zone specified in the feature control frame.

4. Cylindricity. A condition on a surface of revolution in which all points of the surface are equidistant from a common axis (Fig. II.15). Each circular element of the surface must lie between two concentric cylinders. The width between the two cylinders is the tolerance zone specified in the feature control frame. Cylindricity may be applied only to cylindrical features.

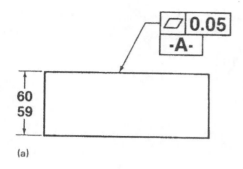
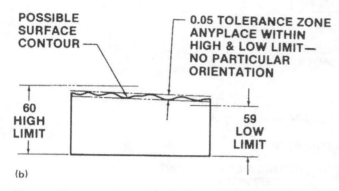

Figure II.13 Flatness. (a) Drawing callout; (b) interpretation.

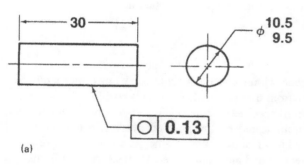
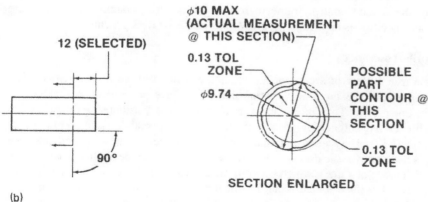

Figure II.14 Circularity. (a) Drawing callout; (b) interpretation.

651

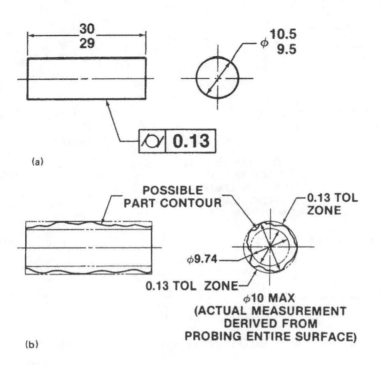

Figure II.15 Cylindricity. (a) Drawing callout; (b) interpretation.

II.6.2 Position Tolerances

A positional tolerance defines a zone within which the center, axis, or center plane of a feature of size is permitted to vary from true (theoretically exact) position. Basic dimensions establish the true position from specified datum features and between interrelated features. A positional tolerance is indicated by the position symbol, a tolerance, and appropriate datum references placed in a feature control frame, as shown in Figure II.16.

The material condition of the feature and any datums that are features of size must be identified in the feature control frame to reflect the material condition at which the tolerance zone specified is applicable (e.g., MMC, LMC, or RFS per rule 2).

II.6.3 Profile Tolerances

Profile is the true outline of a component or feature defined by basic dimensions. The profile tolerance specifies the total variation that the surface or element may exhibit in relation to its true profile. Tolerance boundaries for profile are distributed in relation to the true profile bilaterally, unilaterally, or in some cases with a special distribution of the total tolerance (Fig. II.17). They are not affected by how the feature is dimensioned since the true profile is defined by basic dimensions. Profile tolerances may be applied to individual element lines (profile of a line), which is two-dimensional, or to the entire surface (profile of a surface), which is three-dimensional. Profile tolerances are unique in that they may be used to control size, form, position, orientation, and runout within the specified tolerance. Profile tolerances may be defined as an individual feature (perfect counterpart of itself with no datum reference frame) or in relation to other features (with a datum reference frame).

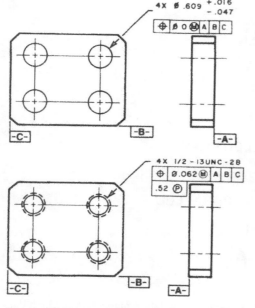

Figure II.16 Example of position tolerances.

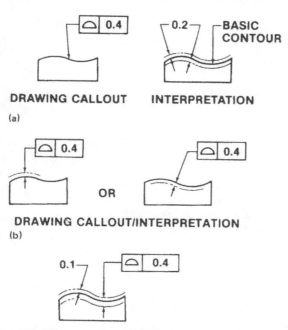

Figure II.17 Profile tolerancing. (a) Bilateral tolerance: where a bilateral tolerance is intended, it is only necessary to show the feature control symbol with an arrow directed to the surface. (b) Unilateral tolerance: for a unilateral tolerance, a phantom line is drawn parallel to the basic profile to clearly indicate whether the tolerance zone is inside or outside the basic profile. (c) Special case: for an unequally disposed profile tolerance, phantom lines are also required to indicate the distribution of the tolerance.

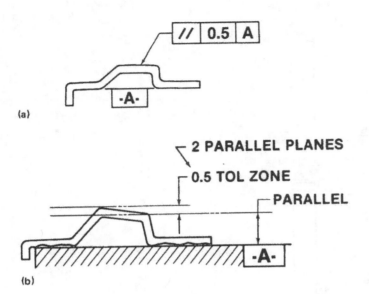

Figure II.18 Parallelism: two parallel planes, parallel to a datum plane within which the feature must lie. (a) Drawing callout; (b) interpretation.

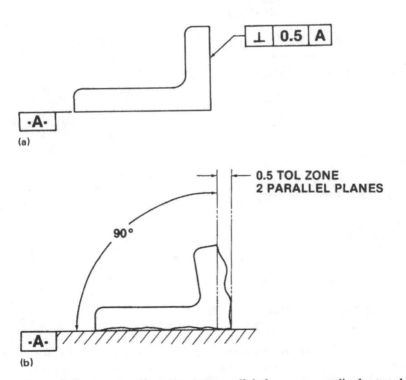

Figure II.19 Perpendicularity: two parallel planes, perpendicular to a datum plane within which the surface of a feature must lie. (a) Drawing callout; (b) interpretation.

II.6 Main Geometric Characteristics

II.6.4 Orientation Tolerances

The orientation tolerances are parallelism, perpendicularity, and angularity. These tolerances control the orientation of an individual feature to another. The feature must be related to at least one datum in the feature control frame. In some instances more than one datum reference is required to stabilize the tolerance zone.

Parallelism: the condition of a surface or axis equidistant at all points from a datum of reference (Fig. II.18). The tolerance zone may be (1) between two parallel planes, (2) between two parallel lines, or (3) a cylindrical shape.

Perpendicularity: the condition of a surface, axis, or median plane exactly 90° with respect to a datum of reference (Fig. II.19). The tolerance zone may be (1) between two parallel planes, (2) between two parallel lines, or (3) a cylindrical shape.

Angularity: the condition of a surface or axis at a specified angle (other than 90°) from a datum of reference (Fig. II.20). The tolerance zone is the distance between two parallel planes, inclined at a specified basic angle in which the surface or axis of the feature must lie.

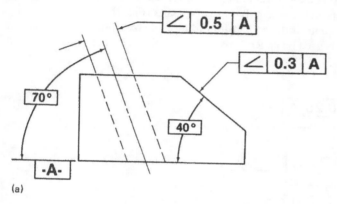

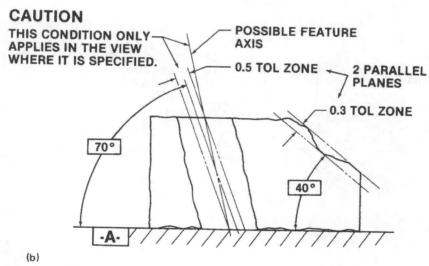

Figure II.20 Angularity. (a) Drawing callout; (b) interpretation.

II.6.5 Coaxial Tolerances

Coaxial tolerances identify permissible variation in the relationship of two or more features of revolution with coincident axes. There are four types of tolerances available to specify coaxiality: (1) runout, either circular or total runout; (2) position; (3) profile of a surface; and (4) concentricity.

1. Runout controls the composite errors of form, orientation, and position of one or more features of a part in relation to a datum axis. The datum axis must be established by a feature of sufficient length, two features having sufficient axial separation, or a feature and a face perpendicular to its axis. It is used on an RFS basis and is a relatively inexpensive way to control the relationship of two coaxial features. Size is always verified independently from a runout tolerance. In *circular runout*, tolerance applies to *each* circular element of the feature, individually, as a part is rotated 360° about the datum axis (Fig. II.21). In *total runout*, tolerance applies to *all* circular elements, simultaneously, as a part is rotated 360° about the datum axis (Fig. II.22).
2. Position tolerance controls axial variation on an MMC basis. Typically it is used to ensure interchangeability in the assembly of mating parts (Fig. II.23).
3. The profile of a surface controls the composite variation of size, form, orientation and position (Fig. II.24). It is used on an RFS basis and is a relatively inexpensive method

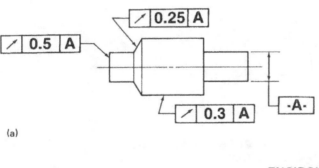

(a)

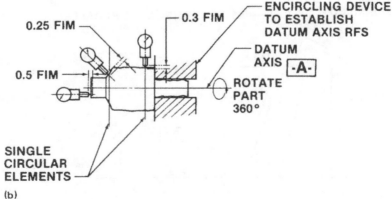

(b)

Figure II.21 Circular runout: each circular element of the feature must be within the runout tolerance. (a) Drawing callout; (b) interpretation.

II.6 Main Geometric Characteristics

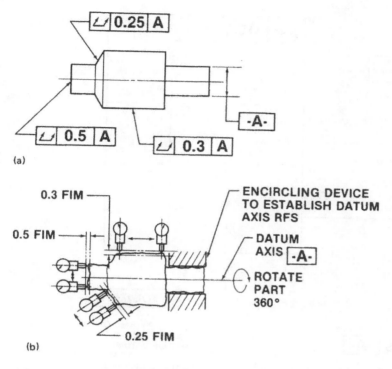

Figure II.22 Total runout: all surface elements across the entire surface must be within the runout tolerance. (a) Drawing callout; (b) interpretation.

of controlling coaxiality. Since verification of profile is compared to the feature's ideal geometry, variables data can be readily obtained and analyzed to determine what caused the variation (e.g., size, form, position, or orientation).

4. Concentricity is the condition in which the axes of all cross-sectional elements of a surface of revolution are common to the axis of a datum feature (Fig. II.25). The concentricity tolerance zone is a cylindrical zone whose axis coincides with the datum axis and within which all cross-sectional axes of the feature being controlled must lie. The selection for the control of coaxial features should be based on function and cost. Concentricity is expensive and position, runout, or profile should be considered first.

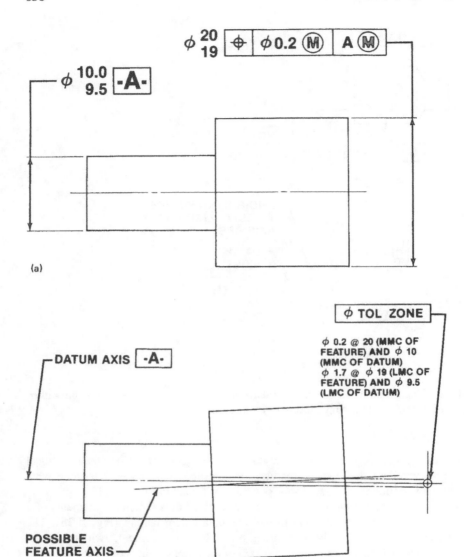

Figure II.23 Coaxial position. (a) Drawing callout; (b) interpretation.

II.6 Main Geometric Characteristics

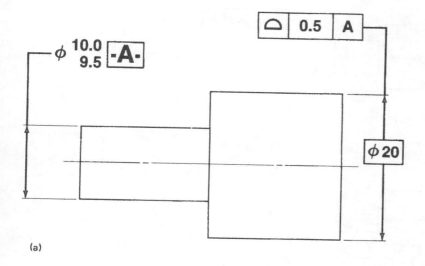

(a)

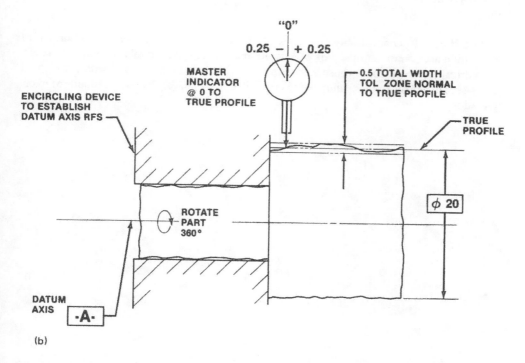

(b)

Figure II.24 Coaxial profile. (a) Drawing callout; (b) interpretation.

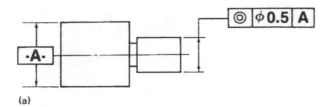

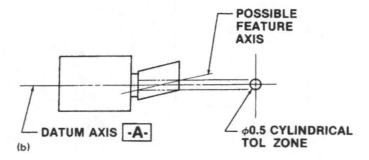

Figure II.25 In concentricity, the axes of all cross-sectional elements of a surface of revolution are common to the axis of a datum feature. A concentricity tolerance zone is a cylindrical tolerance zone whose axis coincides with the datum axis and within which all cross-sectional axes of the feature being controlled must lie. (a) Drawing callout; (b) interpretation.

Appendix III
Tables

Table III.1 Control Chart Constants

Subgroup size n	Standard $\bar{X}$, R chart				Standard $\bar{X}$, s chart					Standard deviation (σ), known $\bar{X}$, s chart				Median A_6	Modified control limits V_1
	A_2	D_3	D_4	d_2	A_3	B_3	B_4	c_4		A	B_5	B_6			
2	1.880	0	3.267	1.128	2.659	0	3.267	.798		2.121	0	2.606		—	1.406
3	1.023	0	2.575	1.693	1.954	0	2.568	.886		1.732	0	2.276		1.187	1.090
4	.729	0	2.282	2.059	1.628	0	2.266	.921		1.500	0	2.088		—	.971
5	.577	0	2.115	2.326	1.427	0	2.089	.940		1.342	0	1.964		.691	.905
6	.483	0	2.004	2.534	1.287	.030	1.970	.952		1.225	.029	1.874		—	.862
7	.419	.076	1.924	2.704	1.182	.118	1.882	.959		1.134	.113	1.806		.509	.830
8	.373	.136	1.864	2.847	1.099	.185	1.815	.965		1.061	.179	1.751		—	.805
9	.337	.184	1.816	2.970	1.032	.239	1.761	.969		1.000	.232	1.707		.412	.786
10	.308	.223	1.777	3.078	.975	.284	1.716	.973		.949	.276	1.669		—	.769
11	.285	.256	1.744	3.173	.927	.321	1.679	.975		.905	.313	1.637		.350	.755
12	.266	.284	1.716	3.258	.886	.354	1.646	.978		.866	.346	1.610		—	.744
13	.249	.308	1.692	3.336	.850	.382	1.618	.979		.832	.374	1.585		.306	.733
14	.235	.329	1.671	3.407	.817	.406	1.594	.981		.802	.399	1.563		—	.724
15	.223	.348	1.652	3.472	.789	.428	1.572	.982		.775	.421	1.544		.274	.715

Table III.1 (Continued)

Standard $\bar{X}$ and R Chart (Procedure 3.2)

	Location	Variability
UCL	$\bar{\bar{X}} + A_2\bar{R}$	$D_4\bar{R}$
Centerline	$\bar{\bar{X}}$	$\bar{R}$
LCL	$\bar{\bar{X}} - A_2\bar{R}$	$D_3\bar{R}$

Estimated population standard deviation $\hat{\sigma} = \bar{R}/d_2$.

Standard $\bar{X}$ and s Chart (Procedure 3.3)

	Location	Variability
UCL	$\bar{\bar{X}} + A_3\bar{s}$	$B_4\bar{s}$
Centerline	$\bar{\bar{X}}$	$\bar{s}$
LCL	$\bar{\bar{X}} - A_3\bar{s}$	$B_3\bar{s}$

Estimated population standard deviation $\hat{\sigma} = \bar{s}/c_4$.

$\bar{X}$ and s Chart Assuming Population Standard Deviation σ Is Known

	Location	Variability
UCL	$\bar{\bar{X}} + A\sigma$	$B_6\sigma$
Centerline	$\bar{\bar{X}}$	$c_4\sigma$
LCL	$\bar{\bar{X}} - A\sigma$	$B_5\sigma$

If population mean is assumed known, replace $\bar{\bar{X}}$ by known value.

Median $\tilde{X}$ and R Chart (Procedure 3.4)

	Location	Variability
UCL	$\bar{\tilde{X}} + A_6\bar{R}$	$D_4\bar{R}$
Centerline	$\bar{\tilde{X}}$	$\bar{R}$
LCL	$\bar{\tilde{X}} - A_6\bar{R}$	$D_3\bar{R}$

Estimated population standard deviation $\hat{\sigma} = \bar{R}/d_2$.

Modified Control Limits $\bar{X}$ and R Chart (Procedure 4.2)

	Location	Variability
URL	$USL - V_1\bar{R}$	$D_4\bar{R}$
Centerline	$\bar{\bar{X}}$	$\bar{R}$
LRL	$LSL + V_1\bar{R}$	$D_3\bar{R}$

Estimated population standard deviation $\hat{\sigma} = \bar{R}/d_2$.

Let n denote the number of measurements made on each of k subgroups. For each subgroup calculate

$$\bar{X} = \frac{X_1 + X_2 + \cdots + X_n}{n}$$

$$R = \text{maximum}(X_1, X_2, \ldots, X_n) - \text{minimum}(X_1, X_2, \ldots, X_n)$$

or

$$s = \sqrt{\frac{(X_1 - \bar{X})^2 + (X_2 - \bar{X})^2 + \cdots + (X_n - \bar{X})^2}{n - 1}}$$

Assuming $(\bar{X}_i, R_i)$ or $(\bar{X}_i, s_i)$ is available for each subgroup,

$$\bar{\bar{X}} = \frac{\bar{X}_1 + \bar{X}_2 + \cdots + \bar{X}_k}{k}$$

$$\bar{R} = \frac{R_1 + R_2 + \cdots + R_k}{k}$$

$$\bar{s} = \frac{s_1 + s_2 + \cdots + s_k}{k}$$

Table III.2 Normal Distribution Probability Values

p	x	p	x
.50	.00000	.925	1.43953
.51	.02507	.930	1.47579
.52	.05015	.935	1.51410
.53	.07527	.940	1.55477
.54	.10043	.945	1.59819
.55	.12566	.950	1.64485
.56	.15097	.955	1.69540
.57	.17637	.960	1.75069
.58	.20189	.965	1.81191
.59	.22754	.970	1.88079
.60	.25335	.975	1.95996
.61	.27932	.980	2.05375
.62	.30548	.985	2.17009
.63	.33185	.990	2.32635
.64	.35846		
.65	.38532	.991	2.36562
.66	.41246	.992	2.40892
.67	.43991	.993	2.45726
.68	.46770	.994	2.51214
.69	.49585	.995	2.57583
.70	.52440	.996	2.65207
.71	.55338	.997	2.74778
.72	.58284	.998	2.87816
.73	.61281	.999	3.09023
.74	.64335		
.75	.67449	.9991	3.12139
.76	.70630	.9992	3.15591
.77	.73885	.9993	3.19465
.78	.77219	.9994	3.23888
.79	.80642	.9995	3.29053
.80	.84162	.9996	3.35279
.81	.87790	.9997	3.43161
.82	.91537	.9998	3.54008
.83	.95417	.9999	3.71902
.84	.99446		
.85	1.03643	.9999 5	3.89059
.86	1.08032	.9999 9	4.26489
.87	1.12639	.9999 95	4.41717
.88	1.17499	.9999 99	4.75342
.89	1.22653	.9999 995	4.89164
.900	1.28155	.9999 999	5.19934
.905	1.31058	.9999 9995	5.32672
.910	1.34076	.9999 9999	5.61200
.915	1.37220	.9999 9999 5	5.73073
.920	1.40507	.9999 9999 9	5.99781

Source: From Odeh et al. (1977), with permission. Probability p observed value is less than or equal to x.

Table III.3 *t*-Distribution Critical Values

Degrees of freedom *f*	Two-tail significance level (%)			
	10	5	2	1
1	6.3138	12.7062	31.8205	63.6567
2	2.9200	4.3027	6.9646	9.9248
3	2.3534	3.1824	4.5407	5.8409
4	2.1319	2.7766	3.7470	4.6041
5	2.0151	2.5706	3.3651	4.0322
6	1.9432	2.4469	3.1427	3.7075
7	1.8946	2.3646	2.9980	3.4995
8	1.8595	2.3060	2.8965	3.3554
9	1.8331	2.2622	2.8214	3.2498
10	1.8125	2.2281	2.7638	3.1693
11	1.7959	2.2010	2.7181	3.1058
12	1.7823	2.1788	2.6810	3.0545
13	1.7709	2.1604	2.6503	3.0123
14	1.7613	2.1448	2.6245	2.9768
15	1.7531	2.1314	2.6025	2.9467
16	1.7459	2.1199	2.5835	2.9208
17	1.7396	2.1098	2.5669	2.8982
18	1.7341	2.1009	2.5524	2.8784
19	1.7291	2.0930	2.5395	2.8609
20	1.7247	2.0860	2.5280	2.8453
21	1.7207	2.0796	2.5176	2.8314
22	1.7171	2.0739	2.5083	2.8188
23	1.7139	2.0687	2.4999	2.8073
24	1.7109	2.0639	2.4922	2.7969
25	1.7081	2.0595	2.4851	2.7874
26	1.7056	2.0555	2.4786	2.7787
27	1.7033	2.0518	2.4727	2.7707
28	1.7011	2.0484	2.4671	2.7633
29	1.6991	2.0452	2.4620	2.7564
30	1.6973	2.0423	2.4573	2.7500
31	1.6955	2.0395	2.4528	2.7440
32	1.6939	2.0369	2.4487	2.7385
33	1.6924	2.0345	2.4448	2.7333
34	1.6909	2.0322	2.4411	2.7284
35	1.6896	2.0301	2.4377	2.7238
36	1.6883	2.0281	2.4345	2.7195
37	1.6871	2.0262	2.4314	2.7154
38	1.6860	2.0244	2.4286	2.7116
39	1.6849	2.0227	2.4258	2.7079
40	1.6839	2.0211	2.4233	2.7045
42	1.6820	2.0181	2.4185	2.6981
44	1.6802	2.0154	2.4141	2.6923
46	1.6787	2.0129	2.4102	2.6870
48	1.6772	2.0106	2.4066	2.6822
50	1.6759	2.0086	2.4033	2.6778
55	1.6730	2.0040	2.3961	2.6682
60	1.6706	2.0003	2.3901	2.6603
65	1.6686	1.9971	2.3851	2.6536
70	1.6669	1.9944	2.3808	2.6479
75	1.6654	1.9921	2.3771	2.6430
80	1.6641	1.9901	2.3739	2.6387
85	1.6630	1.9883	2.3710	2.6349
90	1.6620	1.9867	2.3685	2.6316
95	1.6611	1.9853	2.3662	2.6286
100	1.6602	1.9840	2.3642	2.6259
Maximum	1.6449	1.9600	2.3263	2.5758

Source: From Odeh et al. (1977), with permission.

Table III.4 F-Distribution Critical Values[a]

5% Level

f_2	1	2	3	4	5	6	7	8	9	10	12	15	20	24	30	40	60	120
2	18.51	19.00	19.16	19.25	19.30	19.33	19.35	19.37	19.38	19.40	19.41	19.43	19.45	19.45	19.46	19.47	19.48	19.49
3	10.13	9.55	9.28	9.12	9.01	8.94	8.89	8.85	8.81	8.79	8.74	8.70	8.66	8.64	8.62	8.59	8.57	8.55
4	7.71	6.94	6.59	6.39	6.26	6.16	6.09	6.04	6.00	5.96	5.91	5.86	5.80	5.77	5.75	5.72	5.69	5.66
5	6.61	5.79	5.41	5.19	5.05	4.95	4.88	4.82	4.77	4.74	4.68	4.62	4.56	4.53	4.50	4.46	4.43	4.40
6	5.99	5.14	4.76	4.53	4.39	4.28	4.21	4.15	4.10	4.06	4.00	3.94	3.87	3.84	3.81	3.77	3.74	3.70
7	5.59	4.74	4.35	4.12	3.97	3.87	3.79	3.73	3.68	3.64	3.57	3.51	3.44	3.41	3.38	3.34	3.30	3.27
8	5.32	4.46	4.07	3.84	3.69	3.58	3.50	3.44	3.39	3.35	3.28	3.22	3.15	3.12	3.08	3.04	3.01	2.97
9	5.12	4.26	3.86	3.63	3.48	3.37	3.29	3.23	3.18	3.14	3.07	3.01	2.94	2.90	2.86	2.83	2.79	2.75
10	4.96	4.10	3.71	3.48	3.33	3.22	3.14	3.07	3.02	2.98	2.91	2.85	2.77	2.74	2.70	2.66	2.62	2.58
11	4.84	3.98	3.59	3.36	3.20	3.09	3.01	2.95	2.90	2.85	2.79	2.72	2.65	2.61	2.57	2.53	2.49	2.45
12	4.75	3.89	3.49	3.26	3.11	3.00	2.91	2.85	2.80	2.75	2.69	2.62	2.54	2.51	2.47	2.43	2.38	2.34
13	4.67	3.81	3.41	3.18	3.03	2.92	2.83	2.77	2.71	2.67	2.60	2.53	2.46	2.42	2.38	2.34	2.30	2.25
14	4.60	3.74	3.34	3.11	2.96	2.85	2.76	2.70	2.65	2.60	2.53	2.46	2.39	2.35	2.31	2.27	2.22	2.18
15	4.54	3.68	3.29	3.06	2.90	2.79	2.71	2.64	2.59	2.54	2.48	2.40	2.33	2.29	2.25	2.20	2.16	2.11
16	4.49	3.63	3.24	3.01	2.85	2.74	2.66	2.59	2.54	2.49	2.42	2.35	2.28	2.24	2.19	2.15	2.11	2.06
17	4.45	3.59	3.20	2.96	2.81	2.70	2.61	2.55	2.49	2.45	2.38	2.31	2.23	2.19	2.15	2.10	2.06	2.01
18	4.41	3.55	3.16	2.93	2.77	2.66	2.58	2.51	2.46	2.41	2.34	2.27	2.19	2.15	2.11	2.06	2.02	1.97
19	4.38	3.52	3.13	2.90	2.74	2.63	2.54	2.48	2.42	2.38	2.31	2.23	2.16	2.11	2.07	2.03	1.98	1.93
20	4.35	3.49	3.10	2.87	2.71	2.60	2.51	2.45	2.39	2.35	2.28	2.20	2.12	2.08	2.04	1.99	1.95	1.90
21	4.32	3.47	3.07	2.84	2.68	2.57	2.49	2.42	2.37	2.32	2.25	2.18	2.10	2.05	2.01	1.96	1.92	1.87
22	4.30	3.44	3.05	2.82	2.66	2.55	2.46	2.40	2.34	2.30	2.23	2.15	2.07	2.03	1.98	1.94	1.89	1.84
23	4.28	3.42	3.03	2.80	2.64	2.53	2.44	2.37	2.32	2.27	2.20	2.13	2.05	2.01	1.96	1.91	1.86	1.81
24	4.26	3.40	3.01	2.78	2.62	2.51	2.42	2.36	2.30	2.25	2.18	2.11	2.03	1.98	1.94	1.89	1.84	1.79
25	4.24	3.39	2.99	2.76	2.60	2.49	2.40	2.34	2.28	2.24	2.16	2.09	2.01	1.96	1.92	1.87	1.82	1.77
26	4.23	3.37	2.98	2.74	2.59	2.47	2.39	2.32	2.27	2.22	2.15	2.07	1.99	1.95	1.90	1.85	1.80	1.75
27	4.21	3.35	2.96	2.73	2.57	2.46	2.37	2.31	2.25	2.20	2.13	2.06	1.97	1.93	1.88	1.84	1.79	1.73
28	4.20	3.34	2.95	2.71	2.56	2.45	2.36	2.29	2.24	2.19	2.12	2.04	1.96	1.91	1.87	1.82	1.77	1.71
29	4.18	3.33	2.93	2.70	2.55	2.43	2.35	2.28	2.22	2.18	2.10	2.03	1.94	1.90	1.85	1.81	1.75	1.70
30	4.17	3.32	2.92	2.69	2.53	2.42	2.33	2.27	2.21	2.16	2.09	2.01	1.93	1.89	1.84	1.79	1.74	1.68
40	4.08	3.23	2.84	2.61	2.45	2.34	2.25	2.18	2.12	2.08	2.00	1.92	1.84	1.79	1.74	1.69	1.64	1.58
48	4.04	3.19	2.80	2.57	2.41	2.29	2.21	2.14	2.08	2.03	1.96	1.88	1.79	1.75	1.70	1.64	1.59	1.52
60	4.00	3.15	2.76	2.53	2.37	2.25	2.17	2.10	2.04	1.99	1.92	1.84	1.75	1.70	1.65	1.59	1.53	1.47
90	3.95	3.10	2.71	2.47	2.32	2.20	2.11	2.04	1.99	1.94	1.86	1.78	1.69	1.64	1.59	1.53	1.46	1.39
120	3.92	3.07	2.68	2.45	2.29	2.18	2.09	2.02	1.96	1.91	1.83	1.75	1.66	1.61	1.55	1.50	1.43	1.35
Maximum	3.84	3.00	2.60	2.37	2.21	2.10	2.01	1.94	1.88	1.83	1.75	1.67	1.57	1.52	1.46	1.39	1.32	1.22

III.4 F-Distribution

1% Level

f_2																		
2	98.50	99.00	99.17	99.25	99.30	99.33	99.36	99.37	99.39	99.40	99.42	99.43	99.45	99.46	99.47	99.47	99.48	99.49
3	34.12	30.92	29.46	28.71	28.24	27.91	27.67	27.49	27.35	27.23	27.05	26.87	26.69	26.60	26.50	26.41	26.32	26.22
4	21.20	18.00	16.69	15.98	15.52	15.21	14.98	14.80	14.66	14.55	14.37	14.20	14.02	13.93	13.84	13.75	13.65	13.56
5	16.26	13.27	12.06	11.39	10.97	10.67	10.46	10.29	10.16	10.05	9.89	9.72	9.55	9.47	9.38	9.29	9.20	9.11
6	13.75	10.92	9.78	9.15	8.75	8.47	8.26	8.10	7.98	7.87	7.72	7.56	7.40	7.31	7.23	7.14	7.06	6.97
7	12.25	9.55	8.45	7.85	7.46	7.19	6.99	6.84	6.72	6.62	6.47	6.31	6.16	6.07	5.99	5.91	5.82	5.74
8	11.26	8.65	7.59	7.01	6.63	6.37	6.18	6.03	5.91	5.81	5.67	5.52	5.36	5.28	5.20	5.12	5.03	4.95
9	10.56	8.02	6.99	6.42	6.06	5.80	5.61	5.47	5.35	5.26	5.11	4.96	4.81	4.73	4.65	4.57	4.48	4.40
10	10.04	7.56	6.55	5.99	5.64	5.39	5.20	5.06	4.94	4.85	4.71	4.56	4.41	4.33	4.25	4.17	4.08	4.00
11	9.65	7.21	6.22	5.67	5.32	5.07	4.89	4.74	4.63	4.54	4.40	4.25	4.10	4.02	3.94	3.86	3.78	3.69
12	9.33	6.93	5.95	5.41	5.06	4.82	4.64	4.50	4.39	4.30	4.16	4.01	3.86	3.78	3.70	3.62	3.54	3.45
13	9.07	6.70	5.74	5.21	4.86	4.62	4.44	4.30	4.19	4.10	3.96	3.82	3.66	3.59	3.51	3.43	3.34	3.25
14	8.86	6.51	5.56	5.04	4.69	4.46	4.28	4.14	4.03	3.94	3.80	3.66	3.51	3.43	3.35	3.27	3.18	3.09
15	8.68	6.36	5.42	4.89	4.56	4.32	4.14	4.00	3.89	3.80	3.67	3.52	3.37	3.29	3.21	3.13	3.05	2.96
16	8.53	6.23	5.29	4.77	4.44	4.20	4.03	3.89	3.78	3.69	3.55	3.41	3.26	3.18	3.10	3.02	2.93	2.84
17	8.40	6.11	5.18	4.67	4.34	4.10	3.93	3.79	3.68	3.59	3.46	3.31	3.16	3.08	3.00	2.92	2.83	2.75
18	8.29	6.01	5.09	4.58	4.25	4.01	3.84	3.71	3.60	3.51	3.37	3.23	3.08	3.00	2.92	2.84	2.75	2.66
19	8.18	5.93	5.01	4.50	4.17	3.94	3.77	3.63	3.52	3.43	3.30	3.15	3.00	2.92	2.84	2.76	2.67	2.58
20	8.10	5.85	4.94	4.43	4.10	3.87	3.70	3.56	3.46	3.37	3.23	3.09	2.94	2.86	2.78	2.69	2.61	2.52
21	8.02	5.78	4.87	4.37	4.04	3.81	3.64	3.51	3.40	3.31	3.17	3.03	2.88	2.80	2.72	2.64	2.55	2.46
22	7.95	5.72	4.82	4.31	3.99	3.76	3.59	3.45	3.35	3.26	3.12	2.98	2.83	2.75	2.67	2.58	2.50	2.40
23	7.88	5.66	4.76	4.26	3.94	3.71	3.54	3.41	3.30	3.21	3.07	2.93	2.78	2.70	2.62	2.54	2.45	2.35
24	7.82	5.61	4.72	4.22	3.90	3.67	3.50	3.36	3.26	3.17	3.03	2.89	2.74	2.66	2.58	2.49	2.40	2.31
25	7.77	5.57	4.68	4.18	3.85	3.63	3.46	3.32	3.22	3.13	2.99	2.85	2.70	2.62	2.54	2.45	2.36	2.27
26	7.72	5.53	4.64	4.14	3.82	3.59	3.42	3.29	3.18	3.09	2.96	2.81	2.66	2.58	2.50	2.42	2.33	2.23
27	7.68	5.49	4.60	4.11	3.78	3.56	3.39	3.26	3.15	3.06	2.93	2.78	2.63	2.55	2.47	2.38	2.29	2.20
28	7.64	5.45	4.57	4.07	3.75	3.53	3.36	3.23	3.12	3.03	2.90	2.75	2.60	2.52	2.44	2.35	2.26	2.17
29	7.60	5.42	4.54	4.04	3.73	3.50	3.33	3.20	3.09	3.00	2.87	2.73	2.57	2.49	2.41	2.33	2.23	2.14
30	7.56	5.39	4.51	4.02	3.70	3.47	3.30	3.17	3.07	2.98	2.84	2.70	2.55	2.47	2.39	2.30	2.21	2.11
40	7.31	5.18	4.31	3.83	3.51	3.29	3.12	2.99	2.89	2.80	2.66	2.52	2.37	2.29	2.20	2.11	2.02	1.92
48	7.19	5.08	4.22	3.74	3.43	3.20	3.04	2.91	2.80	2.71	2.58	2.44	2.28	2.20	2.12	2.02	1.93	1.82
60	7.08	4.98	4.13	3.65	3.34	3.12	2.95	2.82	2.72	2.63	2.50	2.35	2.20	2.12	2.03	1.94	1.84	1.73
90	6.93	4.85	4.01	3.53	3.23	3.01	2.84	2.72	2.61	2.52	2.39	2.24	2.09	2.00	1.92	1.82	1.72	1.60
120	6.85	4.79	3.95	3.48	3.17	2.96	2.79	2.66	2.56	2.47	2.34	2.19	2.03	1.95	1.86	1.76	1.66	1.53
Maximum	6.63	4.61	3.78	3.32	3.02	2.80	2.64	2.51	2.41	2.32	2.18	2.04	1.88	1.79	1.70	1.59	1.47	1.32

[a] f_1 = numerator degrees of freedom, f_2 = denominator degrees of freedom.

Source: From Odeh et al. (1977), with permission.

Table III.5 Analysis of Means Constants h

5% Level

f	2	3	4	5	6	7	8	9	10	11	12	13	14	15	16	17	18	19	20	f
3	3.18	4.18																		3
4	2.78	3.56	3.89																	4
5	2.57	3.25	3.53	3.72																5
6	2.45	3.07	3.31	3.49	3.62															6
7	2.37	2.94	3.17	3.33	3.45	3.56														7
8	2.31	2.86	3.07	3.21	3.33	3.43	3.51													8
9	2.26	2.79	2.99	3.13	3.24	3.33	3.41	3.48												9
10	2.23	2.74	2.93	3.07	3.17	3.26	3.33	3.40	3.45											10
11	2.20	2.70	2.88	3.01	3.12	3.20	3.27	3.33	3.39	3.44										11
12	2.18	2.67	2.85	2.97	3.07	3.15	3.22	3.28	3.33	3.38	3.42									12
13	2.16	2.64	2.81	2.94	3.03	3.11	3.18	3.24	3.29	3.34	3.38	3.42								13
14	2.15	2.62	2.79	2.91	3.00	3.08	3.14	3.20	3.25	3.30	3.34	3.37	3.41							14
15	2.13	2.60	2.76	2.88	2.97	3.05	3.11	3.17	3.22	3.26	3.30	3.34	3.37	3.40						15
16	2.12	2.58	2.74	2.86	2.95	3.02	3.09	3.14	3.19	3.23	3.27	3.31	3.34	3.37	3.40					16
17	2.11	2.57	2.73	2.84	2.93	3.00	3.06	3.12	3.16	3.21	3.25	3.28	3.31	3.34	3.37	3.40				17
18	2.10	2.55	2.71	2.82	2.91	2.98	3.04	3.10	3.14	3.18	3.22	3.26	3.29	3.32	3.35	3.37	3.40			18
19	2.09	2.54	2.70	2.81	2.89	2.96	3.02	3.08	3.12	3.16	3.20	3.24	3.27	3.30	3.32	3.35	3.37	3.40		19
20	2.09	2.53	2.68	2.79	2.88	2.95	3.01	3.06	3.11	3.15	3.18	3.22	3.25	3.28	3.30	3.33	3.35	3.37	3.40	20
24	2.06	2.50	2.65	2.75	2.83	2.90	2.96	3.01	3.05	3.09	3.13	3.16	3.19	3.22	3.24	3.27	3.29	3.31	3.33	24
30	2.04	2.47	2.61	2.71	2.79	2.85	2.91	2.96	3.00	3.04	3.07	3.10	3.13	3.16	3.18	3.20	3.22	3.25	3.27	30
40	2.02	2.43	2.57	2.67	2.75	2.81	2.86	2.91	2.95	2.98	3.01	3.04	3.07	3.10	3.12	3.14	3.16	3.18	3.20	40
60	2.00	2.40	2.54	2.63	2.70	2.76	2.81	2.86	2.90	2.93	2.96	2.99	3.02	3.04	3.06	3.08	3.10	3.12	3.14	60
120	1.98	2.37	2.50	2.59	2.66	2.72	2.77	2.81	2.84	2.88	2.91	2.93	2.96	2.98	3.00	3.02	3.04	3.06	3.08	120
Maximum	1.96	2.34	2.47	2.56	2.62	2.68	2.72	2.76	2.80	2.83	2.86	2.88	2.90	2.93	2.95	2.97	2.98	3.00	3.02	Maximum

III.5 Analysis of Means Constants h

1% Level

df \ k	3	4	5	6	7	8	9	10	11	12	13	14	15	16	17	18	19	20	Maximum
3	5.84	7.51																	
4	4.60	5.74	6.21																
5	4.03	4.93	5.29	5.55															
6	3.71	4.48	4.77	4.98	5.16														
7	3.50	4.18	4.44	4.63	4.78	4.90													
8	3.36	3.98	4.21	4.38	4.52	4.63	4.72												
9	3.25	3.84	4.05	4.20	4.33	4.43	4.51	4.59											
10	3.17	3.73	3.92	4.07	4.18	4.28	4.36	4.43	4.49										
11	3.11	3.64	3.82	3.96	4.07	4.16	4.23	4.30	4.36	4.41									
12	3.06	3.57	3.74	3.87	3.98	4.06	4.13	4.20	4.25	4.31	4.35								
13	3.01	3.51	3.68	3.80	3.90	3.98	4.05	4.11	4.17	4.22	4.26	4.30							
14	2.98	3.46	3.63	3.74	3.84	3.92	3.98	4.04	4.09	4.14	4.18	4.22	4.26						
15	2.95	3.42	3.58	3.69	3.79	3.86	3.92	3.98	4.03	4.08	4.12	4.16	4.19	4.22					
16	2.92	3.38	3.54	3.65	3.74	3.81	3.87	3.93	3.98	4.02	4.06	4.10	4.14	4.17	4.20				
17	2.90	3.35	3.50	3.61	3.70	3.77	3.83	3.89	3.93	3.98	4.02	4.05	4.09	4.12	4.14	4.17			
18	2.88	3.33	3.47	3.58	3.66	3.73	3.79	3.85	3.89	3.94	3.97	4.01	4.04	4.07	4.10	4.12	4.15		
19	2.86	3.30	3.45	3.55	3.63	3.70	3.76	3.81	3.86	3.90	3.94	3.97	4.00	4.03	4.06	4.08	4.11	4.13	
20	2.85	3.28	3.42	3.53	3.61	3.67	3.73	3.78	3.83	3.87	3.90	3.94	3.97	4.00	4.02	4.05	4.07	4.09	4.12
24	2.80	3.21	3.35	3.45	3.52	3.58	3.64	3.69	3.73	3.77	3.80	3.83	3.86	3.89	3.91	3.94	3.96	3.98	4.00
30	2.75	3.15	3.28	3.37	3.44	3.50	3.55	3.59	3.63	3.67	3.70	3.73	3.76	3.78	3.81	3.83	3.85	3.87	3.89
40	2.70	3.09	3.21	3.29	3.36	3.42	3.46	3.50	3.54	3.58	3.60	3.63	3.66	3.68	3.70	3.72	3.74	3.76	3.78
60	2.66	3.03	3.14	3.22	3.29	3.34	3.38	3.42	3.46	3.49	3.51	3.54	3.56	3.59	3.61	3.63	3.64	3.66	3.68
120	2.62	2.97	3.07	3.15	3.21	3.26	3.30	3.34	3.37	3.40	3.42	3.45	3.47	3.49	3.51	3.53	3.55	3.56	3.58
Maximum	2.58	2.91	3.01	3.08	3.14	3.18	3.22	3.26	3.29	3.32	3.34	3.36	3.38	3.40	3.42	3.44	3.45	3.47	3.48

Source: From Nelson (1983a), with permission. Values for $k = 2$ are standard t-distribution values.

Table III.6 Variability Comparison Critical Values[a]

n	5% Level		1% Level	
	C_1	C_2	C_1	C_2
5	.60	2.87	.52	4.40
10	.69	1.83	.62	2.28
15	.73	1.58	.67	1.85
20	.76	1.46	.70	1.67
25	.78	1.39	.72	1.56
30	.80	1.34	.74	1.49
35	.81	1.31	.76	1.44
40	.82	1.28	.77	1.40
45	.83	1.26	.78	1.37
50	.83	1.25	.79	1.34
60	.85	1.22	.81	1.30
70	.86	1.20	.82	1.28
80	.86	1.18	.82	1.25
90	.87	1.17	.83	1.24
100	.88	1.16	.84	1.22

[a] Let s denote the standard deviation of n values. The population standard deviation likely (at the 5% or 1% levels) lies in the following interval. Upper variation limit: UVL = $C_2 s$. Lower variation limit: LVL = $C_1 s$. See Procedure 11.1.

III.7 F_{max} Ratio

Table III.7 F_{max} Ratio Critical Values[a]

n	\\					k						
	2	3	4	5	6	7	8	9	10	11	12
						5% Level						
3	39.0	87.5	142	202	266	333	403	475	550	626	704
4	15.4	27.8	39.5	50.9	62.0	72.8	83.5	93.9	104	114	124
5	9.60	15.5	20.6	25.2	29.5	33.6	37.5	41.2	44.8	48.3	51.6
6	7.15	10.8	13.7	16.3	18.7	20.9	22.9	24.8	26.7	28.4	30.0
7	5.82	8.36	10.4	12.1	13.6	15.0	16.3	17.5	18.6	19.7	20.7
8	4.99	6.94	8.44	9.70	10.8	11.8	12.7	13.5	14.3	15.1	15.7
9	4.43	6.00	7.19	8.17	9.02	9.77	10.5	11.1	11.7	12.2	12.7
10	4.03	5.34	6.31	7.11	7.79	8.40	8.94	9.44	9.90	10.3	10.7
11	3.72	4.85	5.67	6.34	6.91	7.41	7.86	8.27	8.64	8.99	9.32
13	3.28	4.16	4.79	5.30	5.72	6.09	6.42	6.72	6.99	7.24	7.48
16	2.86	3.53	4.00	4.37	4.67	4.94	5.17	5.38	5.57	5.75	5.91
21	2.46	2.95	3.28	3.53	3.74	3.92	4.08	4.22	4.35	4.46	4.57
31	2.07	2.40	2.61	2.77	2.90	3.01	3.11	3.19	3.27	3.34	3.40
61	1.67	1.84	1.96	2.04	2.11	2.17	2.21	2.25	2.29	2.32	2.35
Maximum	1.00	1.00	1.00	1.00	1.00	1.00	1.00	1.00	1.00	1.00	1.00

Table III.7 (Continued)

n	\multicolumn{11}{c}{k}
	2	3	4	5	6	7	8	9	10	11	12
	\multicolumn{11}{c}{1% Level}
3	199	449	729	1036	1362	1705	2063	2432	2813	3204	3604
4	47.5	84.6	120	154	187	219	251	282	313	343	373
5	23.2	36.7	48.4	59.1	69.0	78.3	87.2	95.7	104	112	119
6	14.9	22.1	27.9	33.0	37.6	41.9	45.8	49.5	53.1	56.4	59.6
7	11.1	15.6	19.2	22.2	24.9	27.3	29.6	31.7	33.6	35.5	37.2
8	8.89	12.1	14.6	16.6	18.4	20.0	21.5	22.8	24.1	25.3	26.4
9	7.50	9.94	11.8	13.3	14.6	15.7	16.8	17.7	18.6	19.5	20.2
10	6.54	8.49	9.93	11.1	12.1	13.0	13.8	14.5	15.2	15.8	16.4
11	5.85	7.46	8.64	9.59	10.4	11.1	11.7	12.3	12.8	13.3	13.8
13	4.91	6.10	6.95	7.63	8.20	8.69	9.13	9.53	9.89	10.2	10.5
16	4.07	4.93	5.52	5.99	6.37	6.71	7.00	7.27	7.51	7.73	7.93
21	3.32	3.90	4.29	4.60	4.85	5.06	5.25	5.42	5.57	5.70	5.83
31	2.63	2.99	3.23	3.41	3.56	3.68	3.79	3.88	3.97	4.04	4.12
61	1.96	2.15	2.26	2.35	2.42	2.47	2.52	2.57	2.61	2.64	2.67
Maximum	1.00	1.00	1.00	1.00	1.00	1.00	1.00	1.00	1.00	1.00	1.00

[a] Let $s_1, s_2, \ldots, s_k$ denote the standard deviations of k groups where n values were obtained from each group. The F_{max} ratio is defined as

$$F_{max} \text{ ratio} = \frac{\text{maximum}(s_1, s_2, \ldots, s_k)}{\text{minimum}(s_1, s_2, \ldots, s_k)}$$

If the F_{max} ratio is greater than the value in this table, a statistically significant difference exists between the standard deviations of the groups (at the 5 or 1% level).

Source: From Nelson, L. S. (1987b). Upper 10%, 5% and 1% Points of the Maximum F-Ratio. *Journal of Quality Technology* **19**, 165–167.

III.8 Sign Test for Correlation

Table III.8 Sign Test for Correlation Critical Values[a]

SS	c	SS	c	SS	c
\multicolumn{6}{c}{5% Level}						
20–22	5	54–55	19	86–87	33
23–24	6	56–57	20	88–89	34
25–27	7	58–60	21	90–91	35
28–29	8	61–62	22	92–93	36
30–32	9	63–64	23	94–96	37
33–34	10	65–66	24	97–98	38
35–36	11	67–69	25	99–101	39
37–39	12	70–71	26	110	43
40–41	13	72–73	27	120	48
42–43	14	74–76	28	130	52
44–46	15	77–78	29	140	57
47–48	16	79–80	30	150	61
49–50	17	81–82	31	200	84
51–53	18	83–85	32		
\multicolumn{6}{c}{1% Level}						
18–20	3	54–56	17	87–89	31
21–23	4	57–58	18	90–91	32
24–25	5	59–60	19	92–93	33
26–28	6	61–63	20	94–96	34
29–31	7	64–65	21	97–98	35
32–33	8	66–68	22	99–100	36
34–36	9	69–70	23	110	40
37–38	10	71–72	24	120	44
39–41	11	73–75	25	130	48
42–43	12	76–77	26	140	53
44–46	13	78–79	27	150	57
47–48	14	80–82	28	200	79
59–51	15	83–84	29		
52–53	16	85–86	30		

[a] Critical values are derived from standard two-tail binomial percentage points with $p = 0.5$. For sample size above 100 use an asymptotic approximation. See Procedure 12.2 for the evaluation procedure.

Table III.9 Correlation Coefficient r Critical Values[a]

N	5% Level minimum value	1% Level minimum value	N	5% Level minimum value	1% Level minimum value
5	.878	.959	25	.396	.505
6	.811	.917	30	.361	.463
7	.755	.875	35	.334	.430
8	.707	.834	40	.312	.403
9	.666	.798	45	.294	.380
10	.632	.765	50	.279	.361
11	.602	.735	60	.254	.330
12	.576	.708	70	.235	.306
13	.553	.684	80	.220	.286
14	.532	.661	90	.207	.270
15	.514	.641	100	.197	.257
16	.497	.623	150	.160	.210
17	.482	.606	200	.139	.182
18	.468	.590	300	.113	.149
19	.456	.575	400	.098	.129
20	.444	.561	500	.088	.115

[a] If the computed value of the correlation coefficient (ignoring sign) is greater than the tabled value, the correlation between the two characteristics is probably not zero (Procedure 12.3).
Source: From Odeh et al. (1977), with permission.

Table III.10 Random Numbers

```
10 09 73 25 33    76 52 01 35 86    34 67 35 48 76    80 95 90 91 17    39 29 27 49 45
37 54 20 48 05    64 89 47 42 96    24 80 52 40 37    20 63 61 04 02    00 82 29 16 65
08 42 26 89 53    19 64 50 93 03    23 20 90 25 60    15 95 33 47 64    35 08 03 36 06
99 01 90 25 29    09 37 67 07 15    38 31 13 11 65    88 67 67 43 97    04 43 62 76 59
12 80 79 99 70    80 15 73 61 47    64 03 23 66 53    98 95 11 68 77    12 17 17 68 33

66 06 57 47 17    34 07 27 68 50    36 69 73 61 70    65 81 33 98 85    11 19 92 91 70
31 06 01 08 05    45 57 18 24 06    35 30 34 26 14    86 79 90 74 39    23 40 30 97 32
85 26 97 76 02    02 05 16 56 92    68 66 57 48 18    73 05 38 52 47    18 62 38 85 79
63 57 33 21 35    05 32 54 70 48    90 55 35 75 48    28 46 82 87 09    83 49 12 56 24
73 79 64 57 53    03 52 96 47 78    35 80 83 42 82    60 93 52 03 44    35 27 38 84 35

98 52 01 77 67    14 90 56 86 07    22 10 94 05 58    60 97 09 34 33    50 50 07 39 98
11 80 50 54 31    39 80 82 77 32    50 72 56 82 48    29 40 52 42 01    52 77 56 78 51
83 45 29 96 34    06 28 89 80 83    13 74 67 00 78    18 47 54 06 10    68 71 17 78 17
88 68 54 02 00    86 50 75 84 01    36 76 66 79 51    90 36 47 64 93    29 60 91 10 62
99 59 46 73 48    87 51 76 49 69    91 82 60 89 28    93 78 56 13 68    23 47 83 41 13

65 48 11 76 74    17 46 85 09 50    58 04 77 69 74    73 03 95 71 86    40 21 81 65 44
80 12 43 56 35    17 72 70 80 15    45 31 82 23 74    21 11 57 82 53    14 38 55 37 63
74 35 09 98 17    77 40 27 72 14    43 23 60 02 10    45 52 16 42 37    96 28 60 26 55
69 91 62 68 03    66 25 22 91 48    36 93 68 72 03    76 62 11 39 90    94 40 05 64 18
09 89 32 05 05    14 22 56 85 14    46 42 75 67 88    96 29 77 88 22    54 38 21 45 98

91 49 91 45 23    68 47 92 76 86    46 16 28 35 54    94 75 08 99 23    37 08 92 00 48
80 33 69 45 98    26 94 03 68 58    70 29 73 41 35    53 14 03 33 40    42 05 08 23 41
44 10 48 19 49    85 15 74 79 54    32 97 92 65 75    57 60 04 08 81    22 22 20 64 13
12 55 07 37 42    11 10 00 20 40    12 86 07 46 97    96 64 48 94 39    28 70 72 58 15
63 60 64 93 29    16 50 53 44 84    40 21 95 25 63    43 65 17 70 82    07 20 73 17 90

61 19 69 04 46    26 45 74 77 74    51 92 43 37 29    65 39 45 95 93    42 58 26 05 27
15 47 44 52 66    95 27 07 99 53    59 36 78 38 48    82 39 61 01 18    33 21 15 94 66
94 55 72 85 73    67 89 75 43 87    54 62 24 44 31    91 19 04 25 92    92 92 74 59 73
42 48 11 62 13    97 34 40 87 21    16 86 84 87 67    03 07 11 20 59    25 70 14 66 70
23 52 37 83 17    73 20 88 98 37    68 93 59 14 16    26 25 22 96 63    05 52 28 25 62

04 49 35 24 94    75 24 63 38 24    45 86 25 10 25    61 96 27 93 35    65 33 71 24 72
00 54 99 76 54    64 05 18 81 59    96 11 96 38 96    54 69 28 23 91    23 28 72 95 29
35 96 31 53 07    26 89 80 93 54    33 35 13 54 62    77 97 45 00 24    90 10 33 93 33
59 80 80 83 91    45 42 72 68 42    83 60 94 97 00    13 02 12 48 92    78 56 52 01 06
46 05 88 52 36    01 39 09 22 86    77 28 14 40 77    93 91 08 36 47    70 61 74 29 41

32 17 90 05 97    87 37 92 52 41    05 56 70 70 07    86 74 31 71 57    85 39 41 18 38
69 23 46 14 06    20 11 74 52 04    15 95 66 00 00    18 74 39 24 23    97 11 89 63 38
19 56 54 14 30    01 75 87 53 79    40 41 92 15 85    66 67 43 68 06    84 96 28 52 07
45 15 51 49 38    19 47 60 72 46    43 66 79 45 43    59 04 79 00 33    20 82 66 95 41
94 86 43 19 94    36 16 81 08 51    34 88 88 15 53    01 54 03 54 56    05 01 45 11 76

98 08 62 48 26    45 24 02 84 04    44 99 90 88 96    39 09 47 34 07    35 44 13 18 80
33 18 51 62 32    41 94 15 09 49    89 43 54 85 81    88 69 54 19 94    37 54 87 30 43
80 95 10 04 06    96 38 27 07 74    20 15 12 33 87    25 01 62 52 98    94 62 46 11 71
79 75 24 91 40    71 96 12 82 96    69 86 10 25 91    74 85 22 05 39    00 38 75 95 79
18 63 33 25 37    98 14 50 65 71    31 01 02 46 74    05 45 56 14 27    77 93 89 19 36

74 02 94 39 02    77 55 73 22 70    97 79 01 71 19    52 52 75 80 21    80 81 45 17 48
54 17 84 56 11    80 99 33 71 43    05 33 51 29 69    56 12 71 92 55    36 04 09 03 24
11 66 44 98 83    52 07 98 48 27    59 38 17 15 39    09 97 33 34 40    88 46 12 33 56
48 32 47 79 28    31 24 96 47 10    02 29 53 68 70    32 30 75 75 46    15 02 00 99 94
69 07 49 41 38    87 63 79 19 76    35 58 40 44 01    10 51 82 16 15    01 84 87 69 38
```

Reproduced with permission from *A Million Random Digits with 100,000 Normal Deviates* by the RAND Corporation (New York: The Free Press, 1955). Copyright 1955 and 1983 by the RAND Corporation.

RANDOM NUMBERS (CONTINUED)

```
09 18 82 00 97   32 82 53 95 27   04 22 08 63 04   83 38 98 73 74   64 27 85 80 44
90 04 58 54 97   51 98 15 06 54   94 93 88 19 97   91 87 07 61 50   68 47 66 46 59
73 18 95 02 07   47 67 72 62 69   62 29 06 44 64   27 12 46 70 18   41 36 18 27 60
75 76 87 64 90   20 97 18 17 49   90 42 91 22 72   95 37 50 58 71   93 82 34 31 78
54 01 64 40 56   66 28 13 10 03   00 68 22 73 98   20 71 45 32 95   07 70 61 78 13

08 35 86 99 10   78 54 24 27 85   13 66 15 88 73   04 61 89 75 53   31 22 30 84 20
28 30 60 32 64   81 33 31 05 91   40 51 00 78 93   32 60 46 04 75   94 11 90 18 40
53 84 08 62 33   81 59 41 36 28   51 21 59 02 90   28 46 66 87 95   77 76 22 07 91
91 75 75 37 41   61 61 36 22 69   50 26 39 02 12   55 78 17 65 14   83 48 34 70 55
89 41 59 26 94   00 39 75 83 91   12 60 71 76 46   48 94 97 23 06   94 54 13 74 08

77 51 30 38 20   86 83 42 99 01   68 41 48 27 74   51 90 81 39 80   72 89 35 55 07
19 50 23 71 74   69 97 92 02 88   55 21 02 97 73   74 28 77 52 51   65 34 46 74 15
21 81 85 93 13   93 27 88 17 57   05 68 67 31 56   07 08 28 50 46   31 85 33 84 52
51 47 46 64 99   68 10 72 36 21   94 04 99 13 45   42 83 60 91 91   08 00 74 54 49
99 55 96 83 31   62 53 52 41 70   69 77 71 28 30   74 81 97 81 42   43 86 07 28 34

33 71 34 80 07   93 58 47 28 69   51 92 66 47 21   58 30 32 98 22   93 17 49 39 72
85 27 48 68 93   11 30 32 92 70   28 83 43 41 37   73 51 59 04 00   71 14 84 36 43
84 13 38 96 40   44 03 55 21 66   73 85 27 00 91   61 22 26 05 61   62 32 71 84 23
56 73 21 62 34   17 39 59 61 31   10 12 39 16 22   85 49 65 75 60   81 60 41 88 80
65 13 85 68 06   87 64 88 52 61   34 31 36 58 61   45 87 52 10 69   85 64 44 72 77

38 00 10 21 76   81 71 91 17 11   71 60 29 29 37   74 21 96 40 49   65 58 44 96 98
37 40 29 63 97   01 30 47 75 86   56 27 11 00 86   47 32 46 26 05   40 03 03 74 38
97 12 54 03 48   87 08 33 14 17   21 81 53 92 50   75 23 76 20 47   15 50 12 95 78
21 82 64 11 34   47 14 33 40 72   64 63 88 59 02   49 13 90 64 41   03 85 65 45 52
73 13 54 27 42   95 71 90 90 35   85 79 47 42 96   08 78 98 81 56   64 69 11 92 02

07 63 87 79 29   03 06 11 80 72   96 20 74 41 56   23 82 19 95 38   04 71 36 69 94
60 52 88 34 41   07 95 41 98 14   59 17 52 06 95   05 53 35 21 39   61 21 20 64 55
83 59 63 56 55   06 95 89 29 83   05 12 80 97 19   77 43 35 37 83   92 30 15 04 98
10 85 06 27 46   99 59 91 05 07   13 49 90 63 19   53 07 57 18 39   06 41 01 93 62
39 82 09 89 52   43 62 26 31 47   64 42 18 08 14   43 80 00 93 51   31 02 47 31 67

59 58 00 64 78   75 56 97 88 00   88 83 55 44 86   23 76 80 61 56   04 11 10 84 08
38 50 80 73 41   23 79 34 87 63   90 82 29 70 22   17 71 90 42 07   95 95 44 99 53
30 69 27 06 68   94 68 81 61 27   56 19 68 00 91   82 06 76 34 00   05 46 26 92 00
65 44 39 56 59   18 28 82 74 37   49 63 22 40 41   08 33 76 56 76   96 29 99 08 36
27 26 75 02 64   13 19 27 22 94   07 47 74 46 06   17 98 54 89 11   97 34 13 03 58

91 30 70 69 91   19 07 22 42 10   36 69 95 37 28   28 82 53 57 93   28 97 66 62 52
68 43 49 46 88   84 47 31 36 22   62 12 69 84 08   12 84 38 25 90   09 81 59 31 46
48 90 81 58 77   54 74 52 45 91   35 70 00 47 54   83 82 45 26 92   54 13 05 51 60
06 91 34 51 97   42 67 27 86 01   11 88 30 95 28   63 01 19 89 01   14 97 44 03 44
10 45 51 60 19   14 21 03 37 12   91 34 23 78 21   88 32 58 08 51   43 66 77 08 83

12 88 39 73 43   65 02 76 11 84   04 28 50 13 92   17 97 41 50 77   90 71 22 67 69
21 77 83 09 76   38 80 73 69 61   31 64 94 20 96   63 28 10 20 23   08 81 64 74 49
19 52 35 95 15   65 12 25 96 59   86 28 36 82 58   69 57 21 37 98   16 43 59 15 29
67 24 55 26 70   35 58 31 65 63   79 24 68 66 86   76 46 33 42 22   26 65 59 08 02
60 58 44 73 77   07 50 03 79 92   45 13 42 65 29   26 76 08 36 37   41 32 64 43 44

53 85 34 13 77   36 06 69 48 50   58 83 87 38 59   49 36 47 33 31   96 24 04 36 42
24 63 73 87 36   74 38 48 93 42   52 62 30 79 92   12 36 91 86 01   03 74 28 38 73
83 08 01 24 51   38 99 22 28 15   07 75 95 17 77   97 37 72 75 85   51 97 23 78 67
16 44 42 43 34   36 15 19 90 73   27 49 37 09 39   85 13 03 25 52   54 84 65 47 59
60 79 01 81 57   57 17 86 57 62   11 16 17 85 76   45 81 95 29 79   65 13 00 48 60
```

RANDOM NUMBERS (CONTINUED)

03	99	11	04	61	93	71	61	68	94	66	08	32	46	53	84	60	95	82	32	88	61	81	91	61
38	55	59	55	54	32	88	65	97	80	08	35	56	08	60	29	73	54	77	62	71	29	92	38	53
17	54	67	37	04	92	05	24	62	15	55	12	12	92	81	59	07	60	79	36	27	95	45	89	09
32	64	35	28	61	95	81	90	68	31	00	91	19	89	36	76	35	59	37	79	80	86	30	05	14
69	57	26	87	77	39	51	03	59	05	14	06	04	06	19	29	54	96	96	16	33	56	46	07	80
24	12	26	65	91	27	69	90	64	94	14	84	54	66	72	61	95	87	71	00	90	89	97	57	54
61	19	63	02	31	92	96	26	17	73	41	83	95	53	82	17	26	77	09	43	78	03	87	02	67
30	53	22	17	04	10	27	41	22	02	39	68	52	33	09	10	06	16	88	29	55	98	66	64	85
03	78	89	75	99	75	86	72	07	17	74	21	65	31	66	35	20	83	33	74	87	53	90	88	23
48	22	86	33	79	85	78	34	76	19	53	15	26	74	33	35	66	35	29	72	16	81	86	03	11
60	36	59	46	53	35	07	53	39	49	42	61	42	92	97	01	91	82	83	16	98	95	37	32	31
83	79	94	24	02	56	62	33	44	42	34	99	44	13	74	70	07	11	47	36	09	95	81	80	65
32	96	00	74	05	36	40	98	32	32	99	38	54	16	00	11	13	30	75	86	15	91	70	62	53
19	32	25	38	45	57	62	05	26	06	66	49	76	86	46	78	13	86	65	59	19	64	09	94	13
11	22	09	47	47	07	39	93	74	08	48	50	92	39	29	27	48	24	54	76	85	24	43	51	59
31	75	15	72	60	68	98	00	53	39	15	47	04	83	55	88	65	12	25	96	03	15	21	91	21
88	49	29	93	82	14	45	40	45	04	20	09	49	89	77	74	84	39	34	13	22	10	97	85	08
30	93	44	77	44	07	48	18	38	28	73	78	80	65	33	28	59	72	04	05	94	20	52	03	80
22	88	84	88	93	27	49	99	87	48	60	53	04	51	28	74	02	28	46	17	82	03	71	02	68
78	21	21	69	93	35	90	29	13	86	44	37	21	54	86	65	74	11	40	14	87	48	13	72	20
41	84	98	45	47	46	85	05	23	26	34	67	75	83	00	74	91	06	43	45	19	32	58	15	49
46	35	23	30	49	69	24	89	34	60	45	30	50	75	21	61	31	83	18	55	14	41	37	09	51
11	08	79	62	94	14	01	33	17	92	59	74	76	72	77	76	50	33	45	13	39	66	37	75	44
52	70	10	83	37	56	30	38	73	15	16	52	06	96	76	11	65	49	98	93	02	18	16	81	61
57	27	53	68	98	81	30	44	85	85	68	65	22	73	76	92	85	25	58	66	88	44	80	35	84
20	85	77	31	56	70	28	42	43	26	79	37	59	52	20	01	15	96	32	67	10	62	24	83	91
15	63	38	49	24	90	41	59	36	14	33	52	12	66	65	55	82	34	76	41	86	22	53	17	04
92	69	44	82	97	39	90	40	21	15	59	58	94	90	67	66	82	14	15	75	49	76	70	40	37
77	61	31	90	19	88	15	20	00	80	20	55	49	14	09	96	27	74	82	57	50	81	69	76	16
38	68	83	24	86	45	13	46	35	45	59	40	47	20	59	43	94	75	16	80	43	85	25	96	93
25	16	30	18	89	70	01	41	50	21	41	29	06	73	12	71	85	71	59	57	68	97	11	14	03
65	25	10	76	29	37	23	93	32	95	05	87	00	11	19	92	78	42	63	40	18	47	76	56	22
36	81	54	36	25	18	63	73	75	09	82	44	49	90	05	04	92	17	37	01	14	70	79	39	97
64	39	71	16	92	05	32	78	21	62	20	24	78	17	59	45	19	72	53	32	83	74	52	25	67
04	51	52	56	24	95	09	66	79	46	48	46	08	55	58	15	19	11	87	82	16	93	03	33	61
83	76	16	08	73	43	25	38	41	45	60	83	32	59	83	01	29	14	13	49	20	36	80	71	26
14	38	70	63	45	80	85	40	92	79	43	52	90	63	18	38	38	47	47	61	41	19	63	74	80
51	32	19	22	46	80	08	87	70	74	88	72	25	67	36	66	16	44	94	31	66	91	93	16	78
72	47	20	00	08	80	89	01	80	02	94	81	33	19	00	54	15	58	34	36	35	35	25	41	31
05	46	65	53	06	93	12	81	84	64	74	45	79	05	61	72	84	81	18	34	79	98	26	84	16
39	52	87	24	84	82	47	42	55	93	48	54	53	52	47	18	61	91	36	74	18	61	11	92	41
81	61	61	87	11	53	34	24	42	76	75	12	21	17	24	74	62	77	37	07	58	31	91	59	97
07	58	61	61	20	82	64	12	28	20	92	90	41	31	41	32	39	21	97	63	61	19	96	79	40
90	76	70	42	35	13	57	41	72	00	69	90	26	37	42	78	46	42	25	01	18	62	79	08	72
40	18	82	81	93	29	59	38	86	27	94	97	21	15	98	62	09	53	67	87	00	44	15	89	97
34	41	48	21	57	86	88	75	50	87	19	15	20	00	23	12	30	28	07	83	32	62	46	86	91
63	43	97	53	63	44	98	91	68	22	36	02	40	08	67	76	37	84	16	05	65	96	17	34	88
67	04	90	90	70	93	39	94	55	47	94	45	87	42	84	05	04	14	98	07	20	28	83	40	60
79	49	50	41	46	52	16	29	02	86	54	15	83	42	43	46	97	83	54	82	59	36	29	59	38
91	70	43	05	52	04	73	72	10	31	75	05	19	30	29	47	66	56	43	82	99	78	29	34	78

RANDOM NUMBERS (CONTINUED)

```
94 01 54 68 74   32 44 44 82 77   59 82 09 61 63   64 65 42 58 43   41 14 54 28 20
74 10 88 82 22   88 57 07 40 15   25 70 49 10 35   01 75 51 47 50   48 96 83 86 03
62 88 08 78 73   95 16 05 92 21   22 30 49 03 14   72 87 71 73 34   39 28 30 41 49
11 74 81 21 02   80 58 04 18 67   17 71 05 96 21   06 55 40 78 50   73 95 07 95 52
17 94 40 56 00   60 47 80 33 43   25 85 25 89 05   57 21 63 96 18   49 85 69 93 26

66 06 74 27 92   95 04 35 26 80   46 78 05 64 87   09 97 15 94 81   37 00 62 21 86
54 24 49 10 30   45 54 77 08 18   59 84 99 61 69   61 45 92 16 47   87 41 71 71 98
30 94 55 75 89   31 73 25 72 60   47 67 00 76 54   46 37 62 53 66   94 74 64 95 80
69 17 03 74 03   86 99 59 03 07   94 30 47 18 03   26 82 50 55 11   12 45 99 13 14
08 34 58 89 75   35 84 18 57 71   08 10 55 99 87   87 11 22 14 76   14 71 37 11 81

27 76 74 35 84   85 30 18 89 77   29 49 06 97 14   73 03 54 12 07   74 69 90 93 10
13 02 51 43 38   54 06 61 52 43   47 72 46 67 33   47 43 14 39 05   31 04 85 66 99
80 21 73 62 92   98 52 52 43 35   24 43 22 48 96   43 27 75 88 74   11 46 61 60 82
10 87 56 20 04   90 39 16 11 05   57 41 10 63 68   53 85 63 07 43   08 67 08 47 41
54 12 75 73 26   26 62 91 90 87   24 47 28 87 79   30 54 02 78 86   61 73 27 54 54

60 31 14 28 24   37 30 14 26 78   45 99 04 32 42   17 37 45 20 03   70 70 77 02 14
49 73 97 14 84   92 00 39 80 86   76 66 87 32 09   59 20 21 19 73   02 90 23 32 50
78 62 65 15 94   16 45 39 46 14   39 01 49 70 66   83 01 20 98 32   25 57 17 76 28
66 69 21 39 86   99 83 70 05 82   81 23 24 49 87   09 50 49 64 12   90 19 37 95 68
44 07 12 80 91   07 36 29 77 03   76 44 74 25 37   98 52 49 78 31   65 70 40 95 14

41 46 88 51 49   49 55 41 79 94   14 92 43 96 50   95 29 40 05 56   70 48 10 69 05
94 55 93 75 59   49 67 85 31 19   70 31 20 56 82   66 98 63 40 99   74 47 42 07 40
41 61 57 03 60   64 11 45 86 60   90 85 06 46 18   80 62 05 17 90   11 43 63 80 72
50 27 39 31 13   41 79 48 68 61   24 78 18 96 83   55 41 18 56 67   77 53 59 98 92
41 39 68 05 04   90 67 00 82 89   40 90 20 50 69   95 08 30 67 83   28 10 25 78 16

25 80 72 42 60   71 52 97 89 20   72 68 20 73 85   90 72 65 71 66   98 88 40 85 83
06 17 09 79 65   88 30 29 80 41   21 44 34 18 08   68 98 48 36 20   89 74 79 88 82
60 80 85 44 44   74 41 28 11 05   01 17 62 88 38   36 42 11 64 89   18 05 95 10 61
80 94 04 48 93   10 40 83 62 22   80 58 27 19 44   92 63 84 03 33   67 05 41 60 67
19 51 69 01 20   46 75 97 16 43   13 17 75 52 92   21 03 68 28 08   77 50 19 74 27

49 38 65 44 80   26 60 42 35 54   21 78 54 11 01   91 17 81 01 74   29 42 09 04 38
06 31 28 89 40   15 99 56 93 21   47 45 86 48 09   98 18 98 18 51   29 65 18 42 15
60 94 20 03 07   11 89 79 26 74   40 40 56 80 32   96 71 75 42 44   10 70 14 13 93
92 32 99 89 32   78 28 44 63 47   71 20 99 20 61   39 44 89 31 36   25 72 20 85 64
77 93 66 35 74   31 38 45 19 24   85 56 12 96 71   58 13 71 78 20   22 75 13 65 18

38 10 17 77 56   11 65 71 38 97   95 88 95 70 67   47 64 81 38 85   70 66 99 34 06
39 64 16 94 57   91 33 92 25 02   92 61 38 97 19   11 94 75 62 03   19 32 42 05 04
84 05 44 04 55   99 39 66 36 80   67 66 76 06 31   69 18 19 68 45   38 52 51 16 00
47 46 80 35 77   57 64 96 32 66   24 70 07 15 94   14 00 42 31 53   69 24 90 57 47
43 32 13 13 70   28 97 72 38 96   76 47 96 85 62   62 34 20 75 89   08 89 90 59 85

64 28 16 18 26   18 55 56 49 37   13 17 33 33 65   78 85 11 64 99   87 06 41 30 75
66 84 77 04 95   32 35 00 29 85   86 71 63 87 46   26 31 37 74 63   55 38 77 26 81
72 46 13 32 30   21 52 95 34 24   92 58 10 22 62   78 43 86 62 76   18 39 67 35 38
21 03 29 10 50   13 05 81 62 18   12 47 05 65 00   15 29 27 61 39   59 52 65 21 13
95 36 26 70 11   06 65 11 61 36   01 01 60 08 57   55 01 85 63 74   35 82 47 17 08

40 71 29 73 80   10 40 45 54 52   34 03 06 07 26   75 21 11 02 71   36 63 36 84 24
58 27 56 17 64   97 58 65 47 16   50 25 94 63 45   87 19 54 60 92   26 78 76 09 39
89 51 41 17 88   68 22 42 34 17   73 95 97 61 45   30 34 24 02 77   11 04 97 20 49
15 47 25 06 69   48 13 93 67 32   46 87 43 70 88   73 46 50 98 19   58 86 93 52 20
12 12 08 61 24   51 24 74 43 02   60 88 35 21 09   21 43 73 67 86   49 22 67 78 37
```

Table III.11 Equivalent Hardness Values for Steel

Identification diameter (mm)	Brinell hardness[a]	Rockwell hardness[b]			
		A scale	B scale	C scale	D scale
2.25	—	84.1	—	65.3	74.8
2.35	—	82.2	—	61.7	72.0
2.40	—	81.2	—	60.0	70.7
2.45	—	80.5	—	58.7	69.7
2.50	—	80.7	—	59.1	70.0
2.55	—	79.8	—	57.3	68.7
2.60	—	78.8	—	55.6	67.4
2.65	—	78.0	—	54.0	66.1
2.70	—	77.1	—	52.5	65.0
2.75	495	76.7	—	51.6	64.3
2.80	477	75.9	—	50.3	63.2
2.85	461	75.1	—	48.8	61.9
2.90	444	74.3	—	47.2	61.0
2.95	429	73.4	—	45.7	59.7
3.00	415	72.8	—	44.5	58.8
3.05	401	72.0	—	43.1	57.8
3.10	388	71.4	—	41.8	56.8
3.15	375	70.6	—	40.4	55.7
3.20	363	70.0	—	39.1	54.6
3.25	352	69.3	(110.0)[c]	37.9	53.8
3.30	341	68.7	(109.0)	36.6	52.8
3.35	331	68.1	(108.5)	35.5	51.9
3.40	321	67.5	(108.0)	34.3	51.0
3.45	311	66.9	(107.5)	33.1	50.0
3.50	302	66.3	(107.0)	32.1	49.3
3.55	293	65.7	(106.0)	30.9	48.3
3.60	285	65.3	(105.5)	29.9	47.6
3.65	277	64.6	(104.5)	28.8	46.7
3.70	269	64.1	(104.0)	27.6	45.9
3.75	262	63.6	(103.0)	26.6	45.0
3.80	255	63.0	(102.0)	25.4	44.2
3.85	248	62.5	(101.0)	24.2	43.2
3.90	241	61.8	100.0	22.8	42.0
3.95	235	61.4	99.0	21.7	41.4
4.00	229	60.8	98.2	20.5	40.5
4.05	223	—	97.3	(18.8)	—
4.10	217	—	96.4	(17.5)	—
4.20	207	—	94.6	(15.2)	—
4.30	197	—	92.8	(12.7)	—
4.40	187	—	90.7	(10.0)	—
4.50	179	—	89.0	(8.0)	—
4.60	170	—	86.8	(5.4)	—
4.70	163	—	85.0	(3.3)	—
4.80	156	—	82.9	(0.9)	—
4.90	149	—	80.8	—	—
5.00	143	—	78.7	—	—
5.10	137	—	76.4	—	—
5.20	131	—	74.0	—	—
5.30	126	—	72.0	—	—
5.40	121	—	69.8	—	—
5.50	116	—	67.6	—	—
5.60	111	—	65.7	—	—

[a] Brinell hardness is based on the diameter of an impressed indentation of a standard ball (10 mm ball, 3000 kg load).
[b] Rockwell hardness scales are based on the following loads and penetrators; A, 60 kg load, brale penetrator; B, 100 kg load, 1/16 inch diameter ball; C, 150 kg load, brale penetrator; D, 100 kg load, brale penetrator.
[c] Numbers in parentheses are beyond the normal range of the measurement scale.

Table III.12 Metric to U.S. Conversion Factors

U.S. units × constant = metric units	Metric units × constant = U.S. units

Length

inch × 25.40 = millimeters (mm)	mm × .03937 = inches (in.)
× 2.540 = centimeters (cm)	cm × .3937 = inches
microinch × .0254 = micrometers (μm)	μm × 39.37 = microinches (μ in.)
foot × .3048 = meters (m)	m × 3.281 = feet (ft)
yard × .9144 = meters (m)	m × 1.0936 = yards (yd)
mile × 1.6093 = kilometers (km)	km × .62137 = miles (mi)

Area

square inch × 645.16 = square millimeters (mm²)	mm² × .001550 = square inches (in.²)
× 6.4516 = square centimeters (cm²)	cm² × .1550 = square inches (in.²)
square foot × .0929 = square meters (m²)	m² × 10.764 = square feet (ft²)
square yard × .8361 = square meters (m²)	m² × 1.196 = square yards (yd²)
acre × .4047 = hectares (10⁴ m²) (ha)	ha × 2.471 = acres
square miles × 2.590 = square kilometers (km²)	km² × .3861 = square miles

Volume

cubic inch × 16387 = cubic millimeters (mm³)	mm³ × .00006102 = cubic inches (in.³)
× 16.387 = cubic centimeters (cm³)	cm³ × .06102 = cubic inches (in.³)
× .016387 = liters	liter × 61.024 = cubic inches (in.³)
quart × .94635 = liters	liter × 1.0567 = quarts (qt)
gallon × 3.7854 = liters	liter × .2642 = gallons (gal)
cubic foot × 28.317 = liters	liter × .03531 = cubic feet (ft³)
× .028317 = cubic meters (m³)	m³ × 35.315 = cubic feet (ft³)
fluid ounce × 29.573 = milliliters (ml)	ml × .03381 = fluid ounces (fl oz)
cubic yard × .7646 = cubic meters (m³)	m³ × 1.3080 = cubic yards (yd³)

Mass and Weight

ounce, avoirdupois × 28.350 = grams (g)	g × .35274 = ounces avoirdupois (oz avdp)
pound, avoirdupois × .45359 = kilograms (kg)	kg × 2.2046 = pounds avoirdupois (lb avdp)
short ton (2000 pounds) × 907.18 = kilograms (kg)	kg × .0011023 = short tons
× .90718 = metric tons (t)	t × .11023 = short tons
ounce, apothecary's or ounce, troy × 31.103 = grams (g)	g × .032151 = troy ounces (oz t)
pound, apothecary's × .37324 = kilograms (kg)	kg × 2.6792 = apothecary's pound (lb ap)

III.12 Metric to U.S. Conversion

Force or load (newton = kg m/s²)

ounce × .27801 = newtons (N) N × 3.5969 = ounces
pound × 4.4482 = newtons (N) N × .22481 = pounds

Energy or work (joule = Ws)

foot-pounds × 1.3558 = absolute joules (J) J × .73756 = foot pounds
calorie × 4.184 = joules (J) J × .23901 = calories
British thermal unit (BTU) × 1054.4 = joules (J), Systeme International J × .00094845 = BTU
watt-hour × 3600 = joules (J) J × .00027778 = watt-hours

Torque

pound-inch × .11298 = newton-meters (N-m) N-m × 8.8507 = pound-inches
pound-feet × 1.3558 = Newton-meter (N-m) N-m × .73756 = pound-feet

Velocity

miles/hour (mph) × 1.6093 = kilometers/hour (km/h) km/h × .62137 = mph
feet/second (fps) × .3048 = meters/second (m/s) m/s × 3.2808 = fps
miles/hour (mph) × .44704 = meters/second (m/s) m/s × 2.2369 = mph

Acceleration

feet per second squared × .3048 = meters per second squared (m/s²) m/s² × 3.2808 = feet per second squared
inches per second squared × .0254 = meters per second squared (m/s²) m/s² × 39.370 = inches per second squared

Power

horsepower (hp) × .7457 = kilowatts (kW) (1 hp = 550 foot-pounds per second = 745.70 W) kW × 1.3410 = hp

Light

foot-candle (fc) × 10.764 = lumens per square meter (lm/m²) lm/m² × .092903 = fc

Pressure or stress (pascal = newtons per square meter)

pounds per square inch (psi) × 70.307 = grams per square centimeter (g/cm²) g/cm² × .01422 = psi
× 6.895 = kilopascals (kPa) kPa × .1450 = psi

Temperature

Degrees fahrenheit (°F) = 1.8(degree centigrade) + 32 Degree centigrade (°C) = .556(°F − 32)

INDEX

A

Accuracy, 29, 614, 623–626
Action instructions, 606–607
Action plan, 542, 553–555, 558, 566
Adjustable characteristic, 124
Administrative applications:
   examples, 77–83, 374–377
   tools, 9–10, 19
Alarm system, 607
Analysis of means, 424–427
Angularity, 655
Attribute data:
   control charts (*see* Control charts, count, percentage)
   defined, 26
Automatic compensating systems, 138

B

Before/after comparison, 189, 358, 428–429, 479–480, 498–499, 511–515
Bias (*see* Accuracy)
Brainstorming, 543–544

C

Capability
   CPU/CPL, 271–272
   definitions, 266–272, 286–287
   C_{pk}, 271–275
   C_p, 268–270
   k, 272
   improvements, 279–280
   specification limits, 72
   stability, 277–280
   subgroup selection, 278–280
   target chart, 508–509
Cause-and-effect diagram:
   alternative solutions, 569
   benefits, 555
   definition, 542–543
   effect, 544
   5M, 548–549
   potential causes, 545–546, 552–555, 558, 564–565
   preparation, 556
   problem solving, 540–542
   process flow, 546–547
   stratification, 549–552
   testing, 483–486, 518–523
   types, 546–552
Characteristic matrix, 597–598
Characteristic tree, 596–597
Check sheet:
   checklist, 329–330
   classification, 320–323
   frequency, 325–326
   location, 323–325
   measurements scale, 326–329
   types, 320
Circularity, 650

Coded data, 36–38, 181–184
Common-cause (*see* Variation)
Comparison methods:
  location, 419–428
  two groups, 428–432, 495–498
  variability, 410–418
Comparison principle, 186–188, 382, 405–406
Computers, 10, 384, 590
Concentricity, 657
Containment actions, 564
Control chart:
  histograms, 191–200
  patterns, 220–231
  placement, 35, 122–123
  reaction, 73–75
  sampling plan, 99–104, 227, 283–285
  scatter plot, 506–507
  sensitivity, 230–231
  simulation, 238–242
  stratified, 102–105, 148–149, 231–238
  start-up, 103, 105
  types:
    count, 63–65
    dot, 288–292
    mean and range, 49–53
    mean and standard deviation, 54–55
    median and range, 58–60
    percentage, 60–63
  versus lot sampling, 98
  zones, 71, 105
Control limits (*see* Appendix III, Table 1)
  width, 71–72, 119, 279
Controlled variation:
  definition, 66
  defects, 36
  within a process, 35
Consecutive sampling:
  control limits, 100–101
  definition, 98–99
Correlation (*see also* Scatter plot):
  coefficient (r), 471–475
  patterns, 457–460
  sign test, 465–471
Customer, 561–562, 570–571, 587, 589–590
Cylindricity, 650

D

Data summaries, 542, 578–579
Datum, 644–645
  reference frame, 645–648
Defect Prevention:
  ingredients, 6–10
  organizational environment, 4–5
  system, 585–610
  vital systems, 10–11
Designed experiments, 406

E

Employee participation, 7–8, 540

F

F max, 411–413
Feature control frame, 644
Fishbone diagram (*see* Cause-and-effect diagram)
5M, (*see also* Cause-and-effect diagram)
  control levels, 599–601
  definition, 12–13
  variation, 110–111
5W2H, 331, 561–562
Flatness, 649
Full indicator movement (FIM), 644

G

Gage (*see also* Measurement)
  control, 589
  difference control chart, 638
  purchase, 640–642
  repeatability control chart, 638–640
  target control chart, 636–637
Geometric dimensioning and tolerance, 130, 643–660

Gillette system, 139–140
Graphs:
  bar, 387–395
  line, 395–398
  pie, 398–399
  types, 383–384

H

Histogram:
  capability calculation, 285, 287–288
  control charts, 191–200
  definition, 170–171
  instable process, 200–202
  mean and standard deviation, 175
  mixtures, 226–230
Housekeeping, 10, 588

I

Improvement:
  comparisons, 405–406
  continual, 5
  Pareto analysis, 349–355
  process capability, 278
Incoming material, 609–610
Indicators, 384–386, 590–592
Ishikawa diagram (*see* Cause-and-effect diagram)

J

Job Shop, 140–143

L

Latent cause, 568
LC chart, 420–424

Least material condition, (LMC), 644
Linearity, 614, 624
Location:
  capability, 272, 280
  comparison, 419–428
  definition, 38
  estimates, 38–44
  instability, 216–218
  shift, 220–221, 223
  targeting, 128–129
  trends, 223–224

M

Machine tryout, 293–295, 445–448
Management:
  evaluation, 558–559, 576–577, 610
  obligations, 73–75, 592
  participation, 6–7, 540
Maximum material condition (MMC), 644
Mean, 38–39, 41
  shifting (*see* Modified control limits)
Measurement, 26–33, 401
  accuracy and precision, 614–615
  ANOVA analysis, 630–633
  errors, 614–629
  process, 613–614
  recording errors, 226
  recording variation, 190–191
  scale, 31
  scatter plot, 486–488, 523–524
  sensitivity, 71, 130, 603, 633–636
  stability, 636–640
Median, 38–39, 41
Modified control limits, 130–135
Multiple populations (*see* Stratification)

N

Nonadjustable characteristic, 124
Normal distribution, 178–180

O

Operation-control plan table, 601–609
Operational definition, 27–28
Out-of-control (*see also* Variation, Special cause)
   bunching, 229–230
   delay, 230–231
   hugging the centerline, 105, 109, 228
   inconsistent instability, 229
   intervention, 225–226
   location shift, 220–221
   mixtures, 226–230
   signals, 67–72, 220
   zones (*see* Zones, Control chart)
Outlier, 188–189
Overcontrol, 125–127

P

Parallelism, 655
Pareto Diagram:
   comparison, 355–358
   construction, 345–347, 350
   cost, 358–360
   cumulative percentages, 347
   exploding, 354
   percentages, 347–349
Pareto Principle, 344–345
Performance indicators, 360, 372–373, 384–388
Periodic sampling, 99
Perpendicularity, 655
Position (true) tolerance, 652
Prediction line, 490–495
Preventive maintenance, 10, 588, 591, 603
Problem analysis:
   C & E diagram, 483–486
   process, 557–559
   reporting, 576–579
   roadblocks, 581–582
   steps, 541, 574–575
   system, 540–542, 557–575
   tools, 9, 576

Problem definition, 560–564
Problem reaction wheel, 3–4
Process:
   audits, 295–297
   definition, 12–14
   flow chart (*see* Process flow diagram), 13
   key elements, 14–16
   relationships, 593–601
   robustness, 480–483
   sampling, 97–98
   variation, 15, 35, 44
Process control:
   definition (*see* stable process)
   methods, 603–604
Process flow diagram, 14–15
   alternate flows, 14
   sources of variation, 15, 21
   targeting (*see* Target flows)
Process limits, 42–43, 178, 326
Process relationship table, 593–601
Process spread, 41–43, 73, 179–180, 198–199, 267, 269, 326
Profile tolerance, 652

Q

Quality circles (*see* Teams)

R

Randon sampling, 99
Range:
   calculation, 38–39, 41
   chart (*see* Control chart), 53
   within variation, 111
Regardless of feature size (RFS), 644
Regression, 488
   linear (*see* Prediction line), 490–495
Rejection limits (*see* Modified control limits)
Repeatability, 29, 606, 614–618
   control chart, 638–640
Reproducibility, 29, 614, 620–623

Index

Root cause, 568
Run chart:
  multiple point, 46–47
  standard, 44–46

S

Sampling, 97–98
  balanced, 406–408
  comparison, 119, 143–148
  intensity, 604–605
  plan, 113, 123–124, 400–401
  types, 98–101
  variation, 35
Scatter plot (*see also* Correlation)
  benchmark line, 475–477
  construction, 460–465
  control chart, 506–507
  patterns, 457–460, 474
  potential problems, 488–489
  stratified plot, 477–479
Short runs, 138–139
Sign test, 465–471
Special cause (*see* Variation)
Specification limits:
  control charts, 72–73
  defects, 36
  histograms, 185
  in-process, 483, 515–518
Stable process (*see also* Controlled variation)
  control limits, 72
  definition, 47, 53, 213–215
Standard deviation:
  calculation, 38–39, 41, 179
  chart (*see* Control chart), 54–55
  coded data, 175
  of a mean, 194–195
  pooled, 418
  within variation, 111
Statistical control (*see* Stable)
Statistical problem solving, 8–10
Statistical significance, 408–410
Stem and leaf, 173, 176–177
Stratification:
  definition, 381–382
  histograms, 186–188
  Pareto diagrams, 355–357, 360–365
  problem definition, 561–562
  scatter plot, 477–479
Stratified sampling, 99–100
Straightness, 649
Subgroup:
  rational, 101–102
  size, 55, 57, 105–110
  types (*see also* Capability), 98–101

T

Taguchi methods, 406
Target charts, 507–511
Target flows, 23–26, 123
Targeting (*see also* Adjustable characteristics), 128–129, 511–515
Teams, 7–8, 543, 559–560, 592–593
Tool:
  differences, 130–131
  wear, 136–138
Training, 10
Transformations, 180–184
Troubleshooting (*see also* Problem analysis), 99–100, 120–122, 382, 405–406, 480–481, 518–523

V

VC chart, 415, 417
Variability:
  comparison, 410–418
  definition, 38
  estimates, 38–44
  instability, 218–220
  shift, 222–225
Variable data:
  control chart (*see* Control chart, dot, mean, median)
  defined, 26
Variation (*see also* Controlled variation)
  between and within (*see also* Variation diagram), 110–114

Variation (*continued*)
   common-cause, 65–67, 72–75, 363–364, 382
   definition, 38
   diagram, 114–120
   estimates, 38–44
   limits, 412–413, 415
   special-cause, 65–67, 72–75, 110, 279–280, 363–364, 382
     diagram, 114–120
   within part, 129–130
Virtual condition, 644

Z

Zones:
   control chart, 71, 105, 107
   histogram, 178–180, 195–199